T0350296

THE THREE-DIMENSIONAL NAVIER–STOKES EQUATIONS

A rigorous but accessible introduction to the mathematical theory of the three-dimensional Navier–Stokes equations, this book provides self-contained proofs of some of the most significant results in the area, many of which can only be found in research papers. Highlights include the existence of global-in-time Leray–Hopf weak solutions and the local existence of strong solutions; the conditional local regularity results of Serrin and others; and the partial regularity results of Caffarelli, Kohn, and Nirenberg.

Appendices provide background material and proofs of some 'standard results' that are hard to find in the literature. A substantial number of exercises are included, with full solutions given at the end of the book. As the only introductory text on the topic to treat all of the mainstream results in detail, this book is an ideal text for a graduate course of one or two semesters. It is also a useful resource for anyone working in mathematical fluid dynamics.

James C. Robinson is a Professor of Mathematics at the University of Warwick.

José L. Rodrigo is a Professor of Mathematics at the University of Warwick.

Witold Sadowski is an Assistant Professor in the Institute of Applied Mathematics at the University of Warsaw.

The Three-Dimensional Navier–Stokes Equations

Classical Theory

JAMES C. ROBINSON
University of Warwick

JOSÉ L. RODRIGO
University of Warwick

WITOLD SADOWSKI
University of Warsaw

CAMBRIDGE
UNIVERSITY PRESS

University Printing House, Cambridge CB2 8BS, United Kingdom

Cambridge University Press is part of the University of Cambridge.

It furthers the University's mission by disseminating knowledge in the pursuit of education, learning and research at the highest international levels of excellence.

www.cambridge.org
Information on this title: www.cambridge.org/9781107019669

© James C. Robinson, José L. Rodrigo, and Witold Sadowski 2016

First published 2016

A catalogue record for this publication is available from the British Library

ISBN 978-1-107-01966-9 Hardback

To Moomin – JCR

To Sam & Sofia – JLR

To Dorota – WS

Contents

Preface

The purpose of this book is to provide a rigorous but accessible introduction
to the mathematical theory of the three-dimensional Navier–Stokes equations,
suitable for graduate students, by giving self-contained proofs of what we see
as three of the most significant results in the area:

- the existence of global-in-time Leray–Hopf weak solutions, i.e. weak solutions that satisfy the strong energy inequality (Leray, 1934; Hopf, 1951); and the local-in-time existence of strong solutions;
- local regularity results due to Serrin (1962) and others: if

$$u \in L^r((a, b); L^s(U)), \qquad \frac{2}{r} + \frac{3}{s} \leq 1,$$

  then u is spatially smooth within U; and
- the partial regularity result of Caffarelli, Kohn, & Nirenberg (1982) that guarantees that the one-dimensional parabolic Hausdorff measure of the set of space–time singularities of any suitable weak solution is zero.

We end with a result that makes use of many of the properties of solutions
that we prove throughout the book, the almost-everywhere uniqueness of the
Lagrangian particle trajectories for suitable weak solutions.

We also treat the following topics that often fall into the category of 'standard
results' but turn out to be hard to find in the required form:

- the Helmholtz–Weyl decomposition on $\mathbb{T}^3$;
- the L^p boundedness of the Leray projector on $\mathbb{T}^3$;
- estimates on the pressure in the periodic case;
- weak solutions as distributional solutions;
- smoothness and uniqueness for $u \in L^r(0, T; L^s)$, $\frac{2}{r} + \frac{3}{s} = 1, 3 < s \leq \infty$;
- the local existence of solutions in $H^{1/2}$ and L^3 via energy estimates;

– local estimates for the velocity given the vorticity (via Biot–Savart);
– the validity of the local energy inequality; and
– local and maximal regularity results for solutions of the heat equation.

We assume knowledge of the basic language and results common in the rigorous study of PDEs such as provided by Evans (1998), Renardy & Rogers (2004), or Robinson (2001), among many.

Particularly in the earlier parts of the book, where much of the material is well known and has been presented many times, we have chosen not to clutter the exposition with frequent and exhaustive references. Instead historical discussion and suggestions for further reading are delayed until the Notes at the end of each chapter. After the Notes there is generally a selection of exercises, which either expand on material in the main text or contain steps that would interrupt the flow of an argument. Full solutions are given at the end of the book. This makes the first part of the book, which treats the classical existence, uniqueness, and regularity theory, particularly suitable for a first course on the Navier–Stokes equations.

There are many other books that cover some of the material here: our presentation has been particularly influenced by Chemin, Desjardins, Gallagher, & Grenier (2006); Constantin & Foias (1988); Doering & Gibbon (1995); Galdi (2000, 2011); and Temam (1977, 1983). We touch only briefly on the analysis of the equation in critical spaces, which has been the focus of much research in the last two decades; this topic is covered in detail in the books by Cannone (1995, see also his 2003 review article) and Lemarié-Rieusset (2002).

We would like to thank all our mentors, colleagues, and students, who over the years have fostered our interest in this subject. We would particularly like to acknowledge the academic support and friendship of John Gibbon, Charles Fefferman, and Grzegorz Łukaszewicz. We are very grateful to David McCormick, Wojciech Ożański, Benjamin Pooley, and Mikołaj Sierżęga, who read various drafts of the book and gave us many helpful comments.

During the writing of this book JCR was supported by an EPSRC Leadership Fellowship EP/G007470/1, which also funded collaborative visits for JCR and JLR to work with WS (and vice versa). JLR is currently supported by the European Research Council, grant no. 616797. We would all like to thank the Warsaw Center of Mathematics and Computer Science for their financial support towards our meetings in Warsaw.

Introduction

The three-dimensional incompressible Navier–Stokes equations form the fundamental mathematical model of fluid dynamics. Derived from basic physical principles under the assumption of a linear relationship between the stress and the rate-of-strain in the fluid, their applicability to real-life problems is undisputed. However, a rigorous mathematical theory for these equations is still far from complete: in particular, there is no guarantee of the global existence of unique solutions.

The aim of this introductory chapter is to give an idea of the derivation of the model and the physical significance of the terms in the equations along with an informal overview of what is currently known and what the problems are in advancing the theory further. Because we are considering the modelling behind the equation the style of presentation in this chapter is somewhat looser than in the rest of the book, which concentrates on what can be proved rigorously about this canonical model.

The Navier–Stokes equations

The Navier–Stokes equations govern the time evolution of the three-component velocity field u and the scalar pressure p in a fluid of uniform density lying within some region $\Omega \subseteq \mathbb{R}^3$: the conservation of linear momentum leads to

$$\partial_t u - \nu \Delta u + (u \cdot \nabla) u + \nabla p = 0, \tag{1}$$

while the conservation of mass yields the divergence-free (incompressibility) condition

$$\nabla \cdot u = 0. \tag{2}$$

1

The parameter $v > 0$ is known as the 'kinematic viscosity'; we will discuss this further later, but for now we note that in the main part of the book we will take $v = 1$. The case $v = 0$ yields the Euler equations for inviscid ('ideal') fluid flow, a model whose mathematical treatment requires very different tools and which will hence play very little part in what follows.

Equation (1) is a system of three equations for the time evolution of the three components of u. However, it should be noticed that in equation (1) there are four unknowns: u_1, u_2, u_3, and p. There is no explicit equation for the time evolution of p, however; rather p has to be chosen to ensure that the incompressibility condition (2) holds. Thus (2) is an indispensable part of the model, since without it the whole system would be underdetermined.

We will consider the problem on three types of domains:

(i) the physical case of a bounded domain $\Omega \subset \mathbb{R}^3$ that has a smooth boundary (in this case we refer to Ω as 'a smooth bounded domain') equipped with Dirichlet (no-slip) boundary conditions $u|_{\partial\Omega} = 0$;
(ii) the whole space $\mathbb{R}^3$ with suitable decay at infinity (e.g. $u \in L^p(\mathbb{R}^3)$ for some $1 \le p < \infty$); and
(iii) the torus $\mathbb{T}^3 = (\mathbb{R}/2\pi\mathbb{Z})^3$, sometimes described as the case of 'periodic boundary conditions'; for mathematical convenience we will also impose the zero-average condition $\int_{\mathbb{T}^3} u = 0$.

The latter two cases have the advantage that they do not involve any boundaries, which greatly simplifies the analysis in many instances: when we carry out an analysis restricted to these two cases we will refer to the equations 'in the absence of boundaries'. The domains in cases (i) and (iii) have the advantage of being bounded and of allowing the use of the Poincaré inequality ($\|u\|_{L^2} \le c\|\nabla u\|_{L^2}$); again, there are situations in which a common analysis can be performed, such as the Galerkin-based existence argument we employ in Chapter 4.

Generally our approach in this book is to present the arguments in the situation in which they are simplest (often on the torus) and then to discuss the necessary changes required to deal with the other cases.

When we seek a solution of the Navier–Stokes equations (1)–(2) with an initial condition u_0 we want a divergence-free velocity u (satisfying the boundary conditions) and a pressure p that together[1] satisfy equation (1) for all $t > 0$ and at every point in the domain Ω. Moreover, $u(t)$ must tend to u_0 as t tends to zero.

---

[1] However, if we take the divergence of (1) then $-\Delta p = \partial_i\partial_j(u_iu_j)$, and so p is determined by u (see Chapter 5). We often consider the equation in a form in which it suffices to solve for u alone.

The main question concerning the model of fluid flow embodied in the 3D Navier–Stokes equations is whether for each sensible initial condition there is exactly one solution defined for arbitrarily large times. Unfortunately, the answer to this question is not known and we are currently unable to rule out the possibility that a perfectly smooth initial condition might evolve according to the 3D Navier–Stokes equations and yet blow up in a finite time.[2] It is not surprising, therefore, that one of the seven Millennium Problems announced by the Clay Institute addresses the question of the global regularity of solutions to the 3D Navier–Stokes equations (Fefferman, 2000): a prize of one million dollars is offered for either a proof of the existence of regular solutions that exist for all $t > 0$ or a counterexample to such an existence theorem (in the absence of boundaries).

Physical derivation of the equations

We now give a very quick idea of the derivation of the equations, and of the meaning of the various terms. More details can be found in Batchelor (1999), Chorin & Marsden (1993), Doering & Gibbon (1995), or Majda & Bertozzi (2002), for example. We assume that the fluid occupies a region $\Omega \subseteq \mathbb{R}^3$.

We begin with the incompressibility condition. Assuming that the density of the fluid ρ is constant, the net flux of mass across the boundary of any spatial region must be zero: in this case, for any sufficiently regular region $U \subset \Omega$ we must have

$$\rho \int_{\partial U} u(x, t) \cdot n \, \mathrm{d}S = 0,$$

where n is the unit outward normal to U. Using the Divergence Theorem we obtain[3]

$$\rho \int_U \nabla \cdot u(x, t) \, \mathrm{d}x = 0.$$

Since this must hold for any sufficiently regular region $U \subset \Omega$, it follows that $\nabla \cdot u = 0$ at every point in Ω.

To derive the equations for the conservation of momentum, we consider a volume of fluid $V(t)$ which we now allow to move with the flow, i.e. we take some initial volume $V(0)$ and consider

$$V(t) = \{X(t; x_0) : x_0 \in V(0)\}, \tag{3}$$

---

[2] In two dimensions such behaviour is not possible, see Section 6.5.

[3] Throughout this book we use $\mathrm{d}x$ to denote the volume element in spatial integrals, rather than $\mathrm{d}V$ or the more cumbersome $\mathrm{d}x^3$.

where $X(t; x_0)$ denotes the solution of the equations for fluid particle trajectories

$$\dot{X}(t) = u(X, t), \qquad X(0) = x_0. \tag{4}$$

(We consider the solutions of these equations in more detail in Chapter 17.)

Note that on a fluid trajectory, for any function $f(x, t)$ we have

$$\frac{\mathrm{d}}{\mathrm{d}t} f(X(t), t) = \frac{\partial f}{\partial t}(X(t), t) + \frac{\partial f}{\partial x_j}(X(t), t)\dot{X}_j(t)$$

$$= \frac{\partial f}{\partial t}(X(t), t) + \frac{\partial f}{\partial x_j}(X(t), t)u_j(X(t), t),$$

where we have used the Einstein summation convention (summing over repeated indices). We write the second term in the notationally convenient form $(u \cdot \nabla)f$, and then we have

$$\frac{\mathrm{d}}{\mathrm{d}t} f(X(t), t) = \partial_t f + (u \cdot \nabla)f,$$

where we write ∂_t for $\partial/\partial t$ and the terms on the right-hand side are evaluated at $(X(t), t)$. Therefore the rate of change of momentum for the fluid in $V(t)$ is

$$\frac{\mathrm{d}}{\mathrm{d}t} \left\{ \rho \int_{V(0)} u(X(t; \alpha), t) \, \mathrm{d}\alpha \right\} = \rho \int_{V(0)} \frac{\mathrm{d}}{\mathrm{d}t} u(X(t; \alpha), t) \, \mathrm{d}\alpha$$

$$= \rho \int_{V(t)} \{\partial_t u + (u \cdot \nabla)u\} \, \mathrm{d}x, \tag{5}$$

noting that due to the incompressibility constraint the factor $|\det(\nabla X)|$ introduced by changing variables in the integral is 1 (this is not immediately obvious and is discussed in more detail in Section 17.1).

We now consider the force on a volume V of the fluid, which is given in terms of the stress tensor σ by the vector F whose ith component is

$$F_i = \int_{\partial V} \sigma_{ij} n_j \, \mathrm{d}S. \tag{6}$$

The particular form of the stress tensor in the Navier–Stokes equations is based on the modelling assumption that σ is a homogeneous, isotropic, and linear function of the 'rate-of-strain tensor' whose components are

$$E_{ij} = \frac{1}{2} \left(\frac{\partial u_i}{\partial x_j} + \frac{\partial u_j}{\partial x_i} \right). \tag{7}$$

(A fluid for which this is true is called a 'Newtonian fluid'.) In three dimensions this means that σ must be of the form

$$-qI + 2\mu E + \gamma \operatorname{tr}(E)I,$$

where μ and γ are fixed scalar constants (the factor of 2 is for later convenience to compensate for the factor of $1/2$ in the definition of E), tr denotes the trace, and q is a scalar that can vary in space and time. Since u is divergence free,

$$\mathrm{tr}(E) = E_{ii} = \nabla \cdot u = 0,$$

and so the stress tensor must take the form

$$\sigma = -qI + 2\mu E. \tag{8}$$

The pressure term $-qI$ provides a uniform force in all directions, while the second term corresponds to forces resulting from differences in the local velocity in the flow, i.e. shear forces, due to the viscous nature of the fluid. To see how this term corresponds to these shear forces we consider the following simple situation, as described by Feynman, Leighton, & Sands (1970). Suppose that we have a fluid contained between two solid plane surfaces and we keep the lower surface stationary while moving the other at a constant speed v (slow enough to prevent any turbulent effects) as in Figure 1.

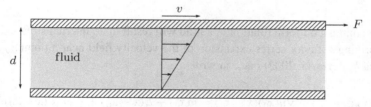

Figure 1. Experimental setup to measure the viscosity of a fluid (after Feynman, Leighton, & Sands, 1970). The fluid is contained between two parallel plates a distance d apart, and the top plate is moved with a constant velocity v.

It is an empirical fact that if the distance between the bottom plane and the upper plane is d then the force F per unit area of the upper plate required to maintain the motion is proportional to the speed v and inversely proportional to the distance d:

$$\frac{F}{A} = \mu \frac{v}{d}.$$

The constant of proportionality μ is known as the dynamic viscosity and is an intrinsic property of the fluid.

Now consider the situation locally, imagining a rectangular cell in the fluid whose faces are parallel with the flow as in Figure 2. In this case the shear force across this cell will be

$$\frac{\delta F}{\delta A} = \mu \frac{\delta v}{\delta y},$$

leading to a local shear force $\mu \, \partial v / \partial y$, in line with the form of the stress tensor in (8).

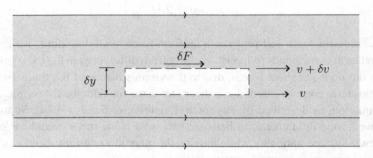

Figure 2. Local stresses caused by local velocity differences. (Figure after Feynman, Leighton, & Sands, 1970.)

By contrast, uniform rotation of a fluid will result in no internal stresses. By performing a Taylor series expansion of the velocity field near a point x_0 (cf. Majda & Bertozzi, 2002) one can write

$$u(x) \simeq u(x_0) + \nabla u(x_0)(x - x_0) = u(x_0) + E(x - x_0) + \frac{1}{2} \omega \times (x - x_0),$$

where E is the rate-of-strain tensor from (7) and $\omega = \operatorname{curl} u$ is the vorticity. This offers another indication that the combination of velocity gradients in E is the relevant quantity for determining the local stress.

Given the form of the stress tensor in (8), it follows from (6) that the ith component of the force on a volume V is given (using the Divergence Theorem) by

$$\int_{\partial V} \sigma_{ij} n_j \, dS = \int_V \partial_j \sigma_{ij} \, dx$$

$$= \int_V \partial_j \left\{ -q\delta_{ij} + \mu(\partial_i u_j + \partial_j u_i) \right\} dx$$

$$= \int_V -\partial_i q + \mu \Delta u_i \, dx,$$

writing ∂_j for $\frac{\partial}{\partial x_j}$ and using the fact that $\partial_j u_j = \nabla \cdot u = 0$. So the total force on the volume V is

$$\int_V -\nabla q + \mu \Delta u \, dx. \tag{9}$$

Therefore if we consider the change in momentum of the fluid in the volume $V(t)$ we have, combining (5) and (9),

$$\rho \int_{V(t)} \{\partial_t u + (u \cdot \nabla)u\} \, dx = \int_{V(t)} -\nabla q + \mu \Delta u \, dx.$$

Since $V(t)$ is arbitrary[4] we can deduce that

$$\rho \, \partial_t u - \mu \Delta u + \rho(u \cdot \nabla)u + \nabla q = 0,$$

with

$$\nabla \cdot u = 0. \tag{10}$$

Dividing the first of these equations through by ρ we obtain

$$\partial_t u - \nu \Delta u + (u \cdot \nabla)u + \nabla p = 0, \tag{11}$$

where $\nu = \mu/\rho$ is termed the 'kinematic viscosity' and we have set $p = q/\rho$.

Throughout what follows we will consider the equations with $\nu = 1$. We can do this since we are primarily interested in the question of the existence and uniqueness of solutions for all positive times. Note that if the pair $(u(x, t), p(x, t))$ solves (10)–(11) on the time interval $[0, T)$ then (u_ν, p_ν) with

$$u_\nu(x, t) = \nu^{-1}u(x, t/\nu) \qquad \text{and} \qquad p_\nu(x, t) = \nu^{-2}p(x, t/\nu)$$

solves the equations with $\nu = 1$ on the time interval $[0, \nu T)$ but still on the same spatial domain. Therefore if one could prove global existence of smooth solutions ('$T = \infty$') for the choice $\nu = 1$ the same would follow for any other choice of $\nu > 0$.

Rescaling of solutions

There is another 'rescaling transformation' that is particularly significant for the Navier–Stokes equations when they are posed on the whole space, but is also important in the local regularity theory we develop in Part II.

Suppose first that $u(x, t)$ is a solution of the linear heat equation

$$\partial_t u - \Delta u = 0, \qquad x \in \mathbb{R}^3. \tag{12}$$

---

[4] Given any choice of volume V, we can find an initial volume $V(0)$ such that the recipe in (3) yields $V(t) = V$, by solving (4) backwards from time t to time zero.

It is simple to check that for any $\lambda > 0$ and any $\alpha \in \mathbb{R}$ the function

$$u_{\lambda,\alpha} = \lambda^{\alpha} u(\lambda x, \lambda^2 t) \tag{13}$$

is again a solution of (12).

When we consider the Navier–Stokes equations we need the nonlinear term $(u \cdot \nabla)u$ to transform in the same way as $\partial_t u$ and Δu; it is easy to check that this requires the choice $\alpha = 1$ in (13). Therefore if $u(x, 0) = u_0(x)$ gives rise to a solution $u(x, t)$ of the Navier–Stokes equations on $\mathbb{R}^3$, with corresponding pressure $p(x, t)$, then the rescaled functions

$$u_{\lambda}(x, t) := \lambda u(\lambda x, \lambda^2 t) \qquad \text{and} \qquad p_{\lambda}(x, t) := \lambda^2 p(\lambda x, \lambda^2 t)$$

still solve the equations, but with rescaled initial data $u_{0,\lambda}(x) = \lambda u_0(\lambda x)$. For $\lambda > 1$ this corresponds to shrinking (spatial) distances by a factor of λ^{-1} and increasing the speed by a factor of λ; it is clear that the time (=distance/speed) should therefore shrink by a factor of λ^{-2}.

Note that if an initial condition u_0 gives rise to the solution $u(x, t)$ and the rescaled initial condition $u_{0,\lambda}$ gives rise to the solution $u_{\lambda}(x, t)$ then

$$u_{\lambda} \text{ is regular on } [0, T_{\lambda}] \quad \Leftrightarrow \quad u \text{ is regular on } [0, \lambda^2 T_{\lambda}]. \tag{14}$$

Spaces of functions in which the norm is unchanged by the rescaling

$$u(x) \mapsto \lambda u(\lambda x) \tag{15}$$

are termed 'critical spaces'. These are the natural spaces in which to try to prove 'small data' results, i.e. the global existence of smooth solutions when the norm of the initial condition is small, since the norm of the data is unaffected by the rescaling transformation (15). We give two examples of such results, in the Sobolev space $\dot{H}^{1/2}$ (Chapter 10) and the Lebesgue space L^3 (Chapter 11).

Critical spaces are important since in some cases local existence in a critical space can be used to deduce global existence. Suppose that X is a critical space and for initial data in X regular solutions exist on the time interval $(0, \tau)$, where $\tau = \tau(\|u_0\|_X)$, i.e. the local existence time depends only on the norm in X. Given $u_0 \in X$ and any $T > 0$, choose λ such that

$$\lambda^2 \tau(\|u_0\|_X) > T;$$

then $\|u_{0,\lambda}\|_X = \|u_0\|_X$ since X is a critical space and the corresponding rescaled solution u_{λ} is regular on $(0, \tau(\|u_{0,\lambda}\|_X))$. It follows from (14) that the solution u is regular on $(0, \lambda^2 \tau(\|u_0\|_X)) \supset (0, T]$.

None of the local existence results for the 3D Navier–Stokes equations in critical spaces obtain such an existence time depending only on the norm; but this programme has proved successful in other contexts, for example semilinear wave equations (Shatah & Struwe, 1994) and isentropic compressible fluids (Danchin, 2000).

Diffusion, advection, and pressure

The regularity of solutions of the 3D Navier–Stokes equations is determined in a competition between the three terms corresponding to diffusion, advection, and pressure. The first of these terms $(-\Delta u)$ contributes a smoothing effect that we expect to help prevent the solutions from blowing up; the second (the nonlinear term $(u \cdot \nabla)u$) corresponds to advection by the fluid and seems to be a source of danger; we will often eliminate the third term (∇p) in our formulation, but its influence remains.

Diffusion: the Laplacian

Let us consider only the first two terms in the equation, i.e. the time derivative and the Laplacian term. It is well known that the solutions of the heat equation

$$\partial_t u - \Delta u = 0$$

are smooth and global in time even if the initial condition is only an element of L^2 (i.e. $\int |u|^2 < \infty$, so that u has finite kinetic energy); see Appendix D. Let us recall the formal reasoning that leads to this result since it will be worthwhile to compare it later with corresponding attempts to prove a similar theorem for the Navier–Stokes equations.

We will consider the equation posed on the torus $\mathbb{T}^3$. Multiplying the heat equation by u and integrating over $\mathbb{T}^3$ it follows that

$$\frac{1}{2} \frac{\mathrm{d}}{\mathrm{d}t} \int_{\mathbb{T}^3} |u|^2 + \int_{\mathbb{T}^3} |\nabla u|^2 = 0,$$

and so after an integration in time we obtain the energy equality

$$\frac{1}{2} \|u(t)\|_{L^2}^2 + \int_0^t \|\nabla u(s)\|_{L^2}^2 \, \mathrm{d}s = \frac{1}{2} \|u(0)\|_{L^2}^2.$$

Thus the kinetic energy

$$\frac{1}{2} \|u(t)\|_{L^2}^2 = \frac{1}{2} \int_{\mathbb{T}^3} |u(t, x)|^2 \, \mathrm{d}x$$

is dissipated over time. Moreover, the integral

$$\int_0^T \|\nabla u(s)\|_{L^2}^2 \, ds$$

(which represents the total energy dissipation) is bounded by the initial energy, so it is easy to deduce that ∇u belongs to L^2 for almost every time.

This allows us to improve the regularity of u by a bootstrapping argument: if we restart the time evolution from some time $\tau > 0$ at which $\nabla u(\tau) \in L^2$ and then multiply the heat equation by $-\Delta u$ we obtain

$$\frac{1}{2}\frac{d}{dt}\int_{\mathbb{T}^3} |\nabla u|^2 + \int_{\mathbb{T}^3} |\Delta u|^2 = 0$$

after an integration by parts. From this we can conclude that the L^2 norm of ∇u also decays in time (for $t > \tau$), and that Δu is in L^2 for almost every time $t > \tau$. Continuing in this way we can show that u is smooth for any $t > 0$.

Moreover, it is worth noticing that the energy in the heat equation is dissipated faster in the smaller length scales (i.e. higher frequencies). The easiest way to see this is to solve the heat equation using Fourier series; we can write the solution in the form

$$u(x, t) = \sum_{k \in \mathbb{Z}^3} \hat{u}_k(0) e^{-|k|^2 t} e^{ik \cdot x},$$

when the initial condition u_0 is given by

$$u_0(x) = \sum_{k \in \mathbb{Z}^3} \hat{u}_k(0) e^{ik \cdot x}.$$

The larger $|k|$, the smaller the factor $e^{-|k|^2 t}$, and hence the more rapid the damping effect of the diffusion. Thus the Laplacian tames wild oscillations in the small scales (large $|k|$) and has a smoothing effect on the solution.[5]

Advection: the nonlinear term

The character of the nonlinear term is very much opposite to that of the Laplacian. The simplest model containing a time derivative and a nonlinear term similar to $(u \cdot \nabla)u$ is the scalar Burgers equation

$$u_t + uu_x = 0.$$

---

[5] It should be noted that there are ODE models $\dot{x} = f(x)$, $x \in \mathbb{R}^2$, that have global solutions, but in which the introduction of diffusion leads to the unexpected blowup of solutions of the PDE $\partial_t u - \Delta u = f(u)$, see Weinberger (1999), for example.

Of course, observations drawn from such an extreme simplification are not directly applicable to the Navier–Stokes equations, but nevertheless it is noteworthy that it is easy, using the method of characteristics, to find examples of smooth initial conditions for which solutions of the Burgers equation develop shocks (jump discontinuities in u) in a finite time (see Section 3.3 in Renardy & Rogers (2004) for example).

There is also another argument for seeing the nonlinear term as the source of potential danger. Counteracting the effect of the Laplacian, which dissipates energy in the smaller scales, the nonlinear term transfers energy from the larger scales to the smaller scales. Indeed, let us again consider the Navier–Stokes equations on the three-dimensional torus and assume (see Doering & Gibbon, 1995) that at a certain time u can be expanded as a finite sum of Fourier modes

$$u(x) = \sum_{|k| \leq N} \hat{u}_k e^{ik \cdot x}$$

(with the understanding that $\hat{u}_k = 0$ for $|k| > N$). Straightforward computations show that

$$(u \cdot \nabla)u = \sum_{|m| \leq 2N} \hat{v}_m e^{im \cdot x},$$

where $\hat{v}_m$ is given by

$$\hat{v}_m = \sum_{|l| \leq N} (\hat{u}_{m-l} \cdot l) i \hat{u}_l.$$

Therefore smaller length scales are immediately activated by the nonlinear term. The energy pumped to such small scales could potentially induce fine oscillations that might perhaps lead to the blowup of the gradient of u.

Seen from this perspective the main role of the viscosity coefficient ν in equation (11) is to reflect the strength of the Laplacian in its struggle with the nonlinear term; this idea is in accord with the experimental observation that for large values of viscosity the flow is laminar, easily predictable and very regular, while for small viscosities the flow becomes turbulent, see Batchelor (1999), for example.

The pressure

The effect of the pressure is more subtle, but significant. If we ignore the divergence-free condition and drop the pressure term from the Navier–Stokes equations, we obtain the viscous Burgers equation

$$\partial_t u - \Delta u + (u \cdot \nabla)u = 0.$$

Classical solutions of this equation obey a maximum principle (Kiselev & Ladyzhenskaya, 1957) which can be used as the basis of a global existence proof (see Pooley & Robinson, 2016, for example).

To see this, assume that u is not identically zero, fix $\alpha > 0$, and consider the equation satisfied by $|v|$, where $v(x, t) = e^{-\alpha t}u(x, t)$, which is

$$\frac{\partial}{\partial t}|v|^2 + 2\alpha|v|^2 + (u \cdot \nabla)|v|^2 - 2v \cdot \Delta v = 0. \tag{16}$$

Noting that $2v \cdot \Delta v = \Delta|v|^2 - 2|\nabla v|^2$ it follows that if $|v|^2$ has a local maximum at $(x, t) \in \mathbb{T}^3 \times (0, T]$ then the left-hand side of (16) is strictly positive unless $|v(x, t)| = 0$. Since this is not possible, it must be the case that $\|v(\cdot, t)\|_{L^\infty} \leq \|v(\cdot, 0)\|_{L^\infty}$, which means that

$$\|u(t)\|_{L^\infty} \leq e^{\alpha t}\|u(0)\|_{L^\infty}.$$

Since $\alpha > 0$ was arbitrary it follows that $\|u(t)\|_{L^\infty} \leq \|u(0)\|_{L^\infty}$ for all $t > 0$ (while u remains a classical solution).

The pressure therefore plays an extremely important role in the regularity problem for the Navier–Stokes equations; but in the standard formulation of the Navier–Stokes problem the pressure is eliminated using the observation that the gradient of any function ϕ is orthogonal to any divergence-free function,

$$\nabla \cdot u = 0 \quad \Rightarrow \quad \int u \cdot \nabla\phi \, dx = 0.$$

By projecting the Navier–Stokes equations onto the space of divergence-free functions (where the solution must belong) we therefore obtain an equation in which the pressure plays no obvious role, and we will do this throughout most of Part I of this book. But even though the pressure is an 'invisible' term in the standard formulation of the problem, we cannot forget that it plays a crucial role in the problem of regularity of solutions, and we will see in Part II that it is inescapable when we come to study the equations locally.

Overview of the contents of the book

Part I of this book covers the classical existence results for weak and strong solutions, which date back to Leray (1934) and Hopf (1951), estimates for the pressure, and local existence of solutions in the critical spaces $\dot{H}^{1/2}$ and L^3. Part II treats conditional local regularity results due to Serrin (1962) and others and the partial regularity results for suitable weak solutions due primarily to Caffarelli, Kohn, & Nirenberg (1982).

The appendices provide some useful background material from functional analysis (Appendix A), harmonic analysis (Appendix B), the theory of elliptic and parabolic PDEs (Appendices C and D), and a measurable selection theorem required in the final chapter of the book (Appendix E).

Part I

Weak and Strong Solutions

Overview of Part I

In the first part of the book we treat classical results on the existence of weak and strong solutions and investigate some of the properties of these solutions. With the exception of Chapters 5 (on the pressure) and 11 (local existence in L^3), and parts of Chapter 2 required for these two chapters, none of the results in Part I require any techniques from harmonic analysis.

We recall some basic definitions and results on relevant function spaces in Chapter 1. Chapter 2 introduces the Helmholtz–Weyl decomposition whereby any function $v \in L^2$ can be written as the sum of a divergence-free vector field and a gradient; this decomposition allows us to eliminate the pressure from the governing equations throughout most of Part I. By working in the periodic setting we are able to give a particularly simple explicit proof of the existence of such a decomposition (Theorem 2.6).

In Chapter 3 we give our weak formulation of the Navier–Stokes equations (see Definition 3.3): u is a weak solution if

(i) $u \in L^\infty(0, T; L^2) \cap L^2(0, T; H^1)$ for every $T > 0$ and
(ii) u satisfies

$$-\int_0^s \langle u, \partial_t \varphi \rangle + \int_0^s \langle \nabla u, \nabla \varphi \rangle + \int_0^s \langle (u \cdot \nabla) u, \varphi \rangle = \langle u_0, \varphi(0) \rangle - \langle u(s), \varphi(s) \rangle$$

for a.e. $s > 0$, for a suitable class of divergence-free test functions φ.

We use the Galerkin method to prove the existence of at least one weak solution in Theorem 4.4; we then show in Theorem 4.6 that the solution we have constructed satisfies the strong energy inequality

$$\frac{1}{2} \|u(t)\|^2 + \int_s^t \|\nabla u(r)\|^2 \, dr \le \frac{1}{2} \|u(s)\|^2 \qquad \text{for all} \qquad t \ge s$$

for almost every $s \in [0, \infty)$, including $s = 0$. Weak solutions that satisfy the strong energy inequality, termed 'Leray–Hopf weak solutions', turn out to be extremely important in what follows.

17

The use of divergence-free test functions in (ii), above, eliminates the pressure from the equations. In Chapter 5 we show that in the absence of boundaries the pressure can be obtained, given a weak solution u, by solving the equation $-\Delta p = \partial_i \partial_j (u_i u_j)$, and that the resulting pair (u, p) satisfies the governing equations in the sense of distributions. Using results from harmonic analysis we also show how in the absence of boundaries (on $\mathbb{T}^3$ and $\mathbb{R}^3$) one can estimate p given u, in particular proving that

$$\|p\|_{L^r} \leq C_r \|u\|_{L^{2r}}^2, \qquad 1 < r < \infty.$$

In Chapter 6 we show that if $u_0 \in H^1$ then there is a more regular 'strong' solution that satisfies $u \in L^\infty(0, T; H^1) \cap L^2(0, T; H^2)$ for some $T > 0$ (Theorem 6.8). While they retain their regularity these strong solutions are unique in the class of Leray–Hopf weak solutions (Theorem 6.10). In Chapter 7 we use bootstrapping arguments to show that any strong solution is smooth on the time interval (ε, T) (Theorem 7.5). By combining results from these two chapters we show in Theorem 8.14 that Leray–Hopf weak solutions are in fact smooth on a collection of open time intervals of full measure in $(0, \infty)$, and we give an upper bound on the dimension of the set of singular times (if it is not empty). We also show that if u satisfies the 'Serrin condition'

$$u \in L^r(0, T; L^s), \qquad \frac{2}{r} + \frac{3}{s} \leq 1, \qquad 2 \leq r < \infty, \quad 3 < s \leq \infty, \quad (17)$$

then u is smooth on $(0, T]$ (Theorem 8.17); this condition also ensures uniqueness in the class of Leray–Hopf weak solutions (Theorem 8.19).

Chapter 9 shows that for any $T > 0$ the initial data that give rise to a strong solution on $(0, T)$ form an open subset of H^1 (Theorem 9.1), and that the Galerkin approximations used in Chapter 5 to prove existence of a solution must converge strongly in $L^\infty(0, T; L^2) \cap L^2(0, T; H^1)$ when the limiting solution is strong (Theorem 9.3).

Chapter 10 discusses scaling properties of solutions and defines a critical space as one in which the norm is unaffected by the rescaling $u(x) \mapsto \lambda u(\lambda x)$: one such space is $\dot{H}^{1/2}(\mathbb{R}^3)$, and we show in Corollary 10.2 that initial conditions in $\dot{H}^{1/2}(\mathbb{T}^3)$ lead to regular solutions on $(0, T)$ for some $T > 0$, and global-in-time regular solutions when $\|u_0\|_{\dot{H}^{1/2}}$ is sufficiently small. We prove a similar result in the larger critical space $L^3(\mathbb{T}^3)$ in Theorem 11.4, using the pressure estimates from Chapter 5.

1

Function spaces

In this chapter we introduce our (mostly standard) notation and recall some basic facts about the function spaces we use most often in this book. Less standard material is covered in Section 1.7.1 (Sobolev spaces on $\mathbb{T}^3$) and Section 1.9 (Bochner spaces).

1.1 Domain of the flow

We consider almost exclusively flows in a three-dimensional domain. We focus mainly on two non-physical domains without boundaries:

- the whole space $\mathbb{R}^3$ and
- the three-dimensional torus $\mathbb{T}^3$.

As discussed in the Introduction, we will supplement the problem with appropriate additional conditions: on the whole space we will require some decay at infinity (for example, $u \in L^2(\mathbb{R}^3)$, which corresponds to finite kinetic energy), while on the torus it is convenient to take solutions with zero average, $\int_{\mathbb{T}^3} u = 0$. We concentrate in this chapter on defining function spaces on these domains, taking these constraints into account.

Throughout the book we also state the corresponding results for the physical case

- $\Omega \subset \mathbb{R}^3$ is a simply-connected bounded open set with a smooth boundary, which we call a 'smooth bounded domain' for short; on this kind of domain we always impose[1] the Dirichlet boundary condition $u|_{\partial\Omega} = 0$.

---

[1] "It turns out – although it is not at all self-evident – that in all circumstances where it has been experimentally checked, *the velocity of a fluid is exactly zero at the surface of a solid.*" (Feynman, Leighton, & Sands, 1970, their italics)

We highlight when results in this chapter require bounded domains or the absence of boundaries; an analysis on the torus can take advantage of both of these simplifications, which is why many of the results in the book (particularly in Part I) are proved in this case: the exposition is simplified but the essential difficulties remain.

Functions defined on the torus $\mathbb{T}^3 = (\mathbb{R}/2\pi\mathbb{Z})^3$ can be realised as periodic functions defined on all of $\mathbb{R}^3$, i.e.

$$u(x + 2\pi k) = u(x) \qquad \text{for all} \quad k \in \mathbb{Z}^3; \tag{1.1}$$

it is also convenient at times to identify u with its restriction to the fundamental domain $[0, 2\pi)^3$.

We will usually not distinguish in notation between spaces of scalar and vector-valued functions; when some ambiguity could arise we will use a notation like $[L^p]^n$ (this example denotes n-component vector-valued functions with each component in L^p) and in Chapter 2, in which the distinction between scalar and vector functions is significant, we will define $\mathbb{L}^p := [L^p]^3$.

1.2 Derivatives

We will use the notation ∂_j for the partial derivative corresponding to the jth coordinate. Two combinations of first derivatives will be particularly significant in what follows: if $u: \mathbb{R}^3 \to \mathbb{R}^3$ is a vector field then we can define the divergence of u as

$$\operatorname{div} u = \nabla \cdot u = \partial_i u_i$$

(summing, as ever, over repeated indices) and the curl of u as

$$\operatorname{curl} u = \nabla \wedge u = \begin{vmatrix} \underline{e}_1 & \underline{e}_2 & \underline{e}_3 \\ \partial_1 & \partial_2 & \partial_3 \\ u_1 & u_2 & u_3 \end{vmatrix},$$

where $\underline{e}_1$, $\underline{e}_2$, and $\underline{e}_3$ are the unit vectors along the coordinate axes. For calculations it is sometimes convenient to express the curl in the form

$$(\operatorname{curl} u)_i = \epsilon_{ijk}\partial_j u_k,$$

where ϵ_{ijk} is the Levi–Civita tensor defined by

$$\epsilon_{ijk} = \begin{cases} 1 & ijk \text{ is an even permutation of } 123 \\ -1 & ijk \text{ is an odd permutation of } 123 \\ 0 & \text{otherwise.} \end{cases} \tag{1.2}$$

We also define

$$\delta_{ij} = \begin{cases} 1 & i = j \\ 0 & i \neq j. \end{cases} \tag{1.3}$$

The equality

$$\epsilon_{ijk}\epsilon_{ilm} = \delta_{jl}\delta_{km} - \delta_{jm}\delta_{kl}, \tag{1.4}$$

which is useful for proving vector identities such as

$$\operatorname{curl}\operatorname{curl} u = \nabla(\nabla \cdot u) - \Delta u$$

(see Exercise 2.2), can be easily (if painfully) checked by hand.

For higher-order derivatives we will employ multi-index notation. We write

$$\alpha = (\alpha_1, \alpha_2, \ldots, \alpha_n) \quad \text{and} \quad |\alpha| = \alpha_1 + \alpha_2 + \cdots + \alpha_n,$$

where the α_i are non-negative integers. For a vector $x = (x_1, \ldots, x_n)$ we define $x^\alpha = x_1^{\alpha_1} \cdots x_n^{\alpha_n}$, and similarly we set

$$\partial^\alpha = \partial_1^{\alpha_1} \cdots \partial_n^{\alpha_n}.$$

We write $\alpha \leq \beta$ if $\alpha_i \leq \beta_i$ for all $i = 1, \ldots, n$, and define the factorial $\alpha! = \alpha_1! \cdots \alpha_n!$. With this notation the Leibniz formula for the differentiation of a product can be written as

$$\partial^\alpha(fg) = \sum_{0 \leq \beta \leq \alpha} \binom{\alpha}{\beta} (\partial^\beta f)(\partial^{\alpha-\beta} g),$$

where

$$\binom{\alpha}{\beta} = \frac{\alpha!}{\beta!(\alpha - \beta)!}$$

and in a mild abuse of notation we have written 0 for $(0, \ldots, 0)$.

Finally we write

$$|\nabla u| := \left(\sum_{i,j=1}^{3} |\partial_i u_j|^2 \right)^{1/2}.$$

1.3 Spaces of continuous and differentiable functions

Let Ω be a bounded open subset of $\mathbb{R}^n$. We use the following standard notation.

- $C^0(\Omega)$ is the space of continuous functions on Ω. The space $C^0(\overline{\Omega})$ of continuous functions on $\overline{\Omega}$ with norm

$$\|u\|_{C^0(\overline{\Omega})} := \sup_{x \in \overline{\Omega}} |u(x)|$$

is a Banach space.

- $C^{0,\gamma}(\overline{\Omega}), 0 < \gamma \leq 1$, is the space of all uniformly γ-Hölder continuous functions on $\overline{\Omega}$, i.e. functions $f : \overline{\Omega} \to \mathbb{R}$ for which there exists $C > 0$ such that

$$|f(x) - f(y)| \leq C|x - y|^\gamma \qquad \text{for every} \quad x, y \in \overline{\Omega}.$$

The space $C^{0,\gamma}(\overline{\Omega})$ of γ-Hölder continuous functions on $\overline{\Omega}$ with norm

$$\|u\|_{C^0(\overline{\Omega})} + \sup_{x,y \in \overline{\Omega}} \frac{|u(x) - u(y)|}{|x - y|^\gamma}$$

is a Banach space.

- $C^k(\Omega)$ is the space of all k-times continuously differentiable functions on Ω. The space $C^k(\overline{\Omega})$, of all k-times continuously differentiable functions with derivatives up to order k continuous on $\overline{\Omega}$ is a Banach space when equipped with the norm

$$\|u\|_{C^k(\overline{\Omega})} := \sum_{|\alpha| \leq k} \|\partial^\alpha u\|_{C^0(\overline{\Omega})}.$$

- $C^{k,\gamma}(\overline{\Omega}), 0 < \gamma \leq 1$, consists of all functions $u \in C^k(\Omega)$ for which all the kth derivatives are Hölder continuous with exponent γ, i.e. $\partial^\alpha u \in C^{0,\gamma}(\overline{\Omega})$ for every multi-index α with $|\alpha| = k$.

We will also use the following spaces of smooth (infinitely differentiable) functions.

- $C_c^\infty(\Omega)$ is the space of all smooth functions with compact support in Ω,

$$C_c^\infty(\Omega) := \{\varphi : \varphi \in C^\infty(\Omega), \quad \text{supp}\,\varphi \subset\subset \Omega\},$$

where $A \subset\subset B$ is used to denote the fact that A is a compact subset of B. At times, to maintain a unified notation across all choices of domains, we will also write $C_c^\infty(\mathbb{T}^3)$, but in this case the subscript c is redundant since all functions defined on $\mathbb{T}^3$ have compact support.

- $\mathscr{S}(\mathbb{R}^n)$ is the space of Schwartz functions on $\mathbb{R}^n$, that is the space of all smooth functions such that

$$p_{k,\alpha}(u) := \sup_{x \in \mathbb{R}^n} |x|^k |\partial^\alpha u(x)|$$

is finite for every choice of $k = 0, 1, 2, \ldots$ and multi-index $\alpha \geq 0$. We denote by $\mathscr{S}'(\mathbb{R}^n)$ the space of all tempered distributions on $\mathbb{R}^n$, that is the collection of all bounded linear functionals f on $\mathscr{S}(\mathbb{R}^n)$ that are continuous in the sense that $f(u_n) \to 0$ as $n \to \infty$ if[2] $(u_n) \in \mathscr{S}(\mathbb{R}^n)$ with $p_{k,\alpha}(u_n) \to 0$ as $n \to \infty$ (for all k and α as above). For more details see Friedlander & Joshi (1999), for example.

When X is a Banach space we denote the space of all continuous functions from $[0, T]$ into X by $C([0, T]; X)$; equipped with the norm

$$\|u\|_{C([0,T];X)} := \sup_{t \in [0,T]} \|u(t)\|_X$$

this is again a Banach space. For functions in $C([0, T]; X)$ we have the following version of the Arzelà–Ascoli Theorem, which will be one of the key ingredients in the proof of a simple version of the Aubin–Lions Lemma (Theorem 4.11) that we will use to show the existence of weak solutions of the Navier–Stokes equations in Chapter 4.

Theorem 1.1 (Arzelà–Ascoli Theorem) *Let X be a Banach space and (u_n) a sequence of functions in $C([0, T]; X)$ such that*

(i) *for each $t \in [0, T]$ there exists a compact set $K(t) \subset X$ such that for every $n \in \mathbb{N}$ we have $u_n(t) \in K(t)$;*
(ii) *the functions u_n are equicontinuous: for every $\varepsilon > 0$ there exists $\delta > 0$ such that for every $n \in \mathbb{N}$*

$$|s - t| \leq \delta \quad \Rightarrow \quad \|u_n(s) - u_n(t)\|_X \leq \varepsilon.$$

Then there exists a subsequence (u_{n_k}) and a function $u \in C([0, T]; X)$ such that

$$u_{n_k} \to u \quad in \quad C([0, T]; X).$$

1.4 Lebesgue spaces

Given a measurable subset Ω of $\mathbb{R}^n$, a function $f \colon \Omega \to \mathbb{R}$ is (Lebesgue) measurable if $f^{-1}(U)$ is (Lebesgue) measurable for every Borel subset U of $\mathbb{R}$. It is equivalent in this case to require that $f^{-1}(U)$ is measurable for every open set $U \subset \mathbb{R}$, or for every closed set $U \subset \mathbb{R}$.

---

[2] Here and in what follows we use shorthand notation $(u_n) \in X$ to denote a sequence $(u_n)_{n=1}^{\infty}$ such that $u_n \in X$ for every $n \in \mathbb{N}$.

By $L^p(\Omega)$, where $1 \le p < \infty$, we denote the standard Lebesgue space of measurable p-integrable (scalar or vector-valued) functions with the norm

$$\|u\|_{L^p} := \left(\int_\Omega |u(x)|^p \, \mathrm{d}x \right)^{1/p}.$$

The space $L^2(\Omega)$ is a Hilbert space when equipped with the inner product

$$\langle u, v \rangle := \int_\Omega u(x) \cdot v(x) \, \mathrm{d}x;$$

since we use this space so frequently we reserve the notation $\| \cdot \|$ for the L^2 norm, and usually omit the L^2 subscript. We use the same notation $\langle f, g \rangle$ to denote $\int_\Omega f(x) \cdot g(x) \, \mathrm{d}x$ whenever $f \cdot g \in L^1$ (see also Section 1.8).

The space $L^\infty(\Omega)$ of essentially bounded functions on Ω is equipped with the standard norm

$$\|u\|_{L^\infty} := \operatorname{ess\,sup} |u| = \inf\{a \in \mathbb{R} : \mu(\{x \in \Omega : |u(x)| \ge a\}) = 0\},$$

where μ denotes the Lebesgue measure.

For $1 \le p \le \infty$ the space $L^p_{\mathrm{loc}}(\Omega)$ consists of those functions that are contained in $L^p(K)$ for every compact subset K of Ω.

We now recall some elementary facts about Lebesgue spaces, and in particular mention some inequalities that will be used frequently in what follows.

Theorem 1.2 (Hölder's inequality) *Let Ω be a measurable set in $\mathbb{R}^n$, either bounded or unbounded. If $u \in L^p(\Omega)$ and $v \in L^q(\Omega)$, where*

$$\frac{1}{p} + \frac{1}{q} = 1, \qquad 1 \le p, q \le \infty,$$

then $uv \in L^1(\Omega)$ and

$$\int_\Omega |u(x)v(x)| \, \mathrm{d}x \le \|u\|_{L^p(\Omega)} \|v\|_{L^q(\Omega)}.$$

The Hölder inequality can be used in any domain since its proof is just a simple application of Young's inequality

$$ab \le \frac{a^p}{p} + \frac{b^q}{q} \qquad \text{for} \qquad \frac{1}{p} + \frac{1}{q} = 1, \quad 1 \le p, q \le \infty \qquad (1.5)$$

(see Exercise 1.1 for a proof of (1.5)). We often use Young's inequality 'with ε': with (p, q) as in (1.5), for any $\varepsilon > 0$ we have

$$ab \le \varepsilon a^p + C(\varepsilon)b^q;$$

in fact $C(\varepsilon) = q^{-1}(p\varepsilon)^{-q/p}$ but this explicit form is rarely required.

Applying Hölder's inequality with $v \equiv 1$ on Ω shows that Lebesgue spaces are nested on domains of finite measure.

Corollary 1.3 *If $\Omega \subset \mathbb{R}^n$ is a set of finite measure then*

$$L^q(\Omega) \subset L^p(\Omega) \qquad if \qquad 1 \leq p \leq q \leq \infty.$$

It is important to observe that this is not the case when Ω does not have finite measure (see Exercise 1.2).

In the context of the Navier–Stokes equations we often need to estimate integrals of products of three functions (usually the nonlinear term $(u \cdot \nabla)u$ multiplied by some 'test' function). Therefore in addition to the standard version of Hölder's inequality we will also use the following variant (for the proof see Exercise 1.3).

Theorem 1.4 *Let Ω be a measurable set in $\mathbb{R}^n$. If $u \in L^p(\Omega)$, $v \in L^q(\Omega)$, and $w \in L^r(\Omega)$ with*

$$\frac{1}{p} + \frac{1}{q} + \frac{1}{r} = 1, \qquad 1 \leq p, q, r \leq \infty,$$

then $uvw \in L^1(\Omega)$ and

$$\int_\Omega |u(x)v(x)w(x)| \, \mathrm{d}x \leq \|u\|_{L^p} \|v\|_{L^q} \|w\|_{L^r}.$$

The two most frequent uses of this in what follows[3] will be with exponents $(6, 2, 3)$,

$$\|uvw\|_{L^1} \leq \|u\|_{L^6} \|v\|_{L^2} \|w\|_{L^3}, \tag{1.6}$$

and $(4, 2, 4)$,

$$\|uvw\|_{L^1} \leq \|u\|_{L^4} \|v\|_{L^2} \|w\|_{L^4}.$$

We will often estimate the L^6 norm by the norm in H^1 using the 3D Sobolev embedding $H^1(\Omega) \subset L^6(\Omega)$ (see Theorem 1.7 or Theorem 1.18), and the L^3 and L^4 norms using the very useful technique of Lebesgue interpolation from the following theorem, whose proof is another application of Hölder's inequality (see Exercise 1.4).

---

[3] Since we will often be estimating expressions of the form $\int |u||\nabla v||w|$ (see Exercise 1.5) we will usually want to put the L^2 norm on the gradient term; this is the reason for the ordering of the norms in the two examples here.

Theorem 1.5 (Lebesgue Interpolation) *Let Ω be a measurable set in $\mathbb{R}^n$. If $1 \le p \le r \le q \le \infty$ and $u \in L^p(\Omega) \cap L^q(\Omega)$ then $u \in L^r(\Omega)$ with*

$$\|u\|_{L^r} \le \|u\|_{L^p}^{\alpha} \|u\|_{L^q}^{1-\alpha}, \quad where \quad \frac{1}{r} = \frac{\alpha}{p} + \frac{1-\alpha}{q}. \tag{1.7}$$

The case that we will use most frequently will be the interpolation of L^3 between L^2 and L^6,

$$\|u\|_{L^3} \le \|u\|_{L^2}^{1/2} \|u\|_{L^6}^{1/2}. \tag{1.8}$$

It is worth remembering that the Hölder and interpolation inequalities are valid in any measurable subset of $\mathbb{R}^n$, bounded or unbounded, and are dimension independent.

1.5 Fourier expansions

For real-valued functions defined on $\mathbb{T}^3$ it is often useful to consider their Fourier expansion, i.e. to write a function u defined on $\mathbb{T}^3$ in the form

$$u(x) = \sum_{k \in \mathbb{Z}^3} \hat{u}_k e^{ik \cdot x}, \tag{1.9}$$

where the coefficients $\hat{u}_k$ are complex numbers satisfying the condition

$$\hat{u}_k = \overline{\hat{u}_{-k}} \quad \text{for all} \quad k \in \mathbb{Z}^3, \tag{1.10}$$

imposed to ensure that u is real.

The main issue is the convergence of the series expansion in (1.9). In some cases this is relatively simple: we can write any $u \in L^2(\mathbb{T}^3)$ in the form (1.9), with the understanding that the sum converges in $L^2(\mathbb{T}^3)$, since the collection $\{e^{ik \cdot x}\}$ forms an orthogonal basis for (complex-valued functions in) $L^2(\mathbb{T}^3)$; if needed, the Fourier coefficients $\hat{u}_k$ can be computed explicitly via the integral

$$\hat{u}_k = \frac{1}{(2\pi)^3} \int_{\mathbb{T}^3} e^{-ik \cdot x} u(x) \, dx, \tag{1.11}$$

using the orthogonality of the basis elements. When u is an m-component vector-valued function the coefficients $\hat{u}_k$ are elements of $\mathbb{C}^m$ and condition (1.10) holds for all components of $\hat{u}_k$.

The characterisation of functions in $L^2(\mathbb{T}^3)$ in terms of their Fourier coefficients is straightforward: $u \in L^2(\mathbb{T}^3)$ if and only if $\sum |\hat{u}_k|^2 < \infty$. However, in other L^p spaces the situation is somewhat more complicated.

Let (C_N) be an increasing sequence of subsets of $\mathbb{Z}^3$ such that

$$\bigcup_{N=1}^{\infty} C_N = \mathbb{Z}^3.$$

Then

$$\left\| u - \sum_{k \in C_N} \hat{u}_k e^{ik \cdot x} \right\|_{L^p} \to 0 \qquad \text{as} \qquad N \to \infty$$

for every $u \in L^p(\mathbb{T}^3)$, $1 \le p < \infty$, if and only if the maps

$$u \mapsto \sum_{k \in C_N} \hat{u}_k e^{ik \cdot x}$$

are uniformly bounded (in N) from $L^p(\mathbb{T}^3)$ into $L^p(\mathbb{T}^3)$. Boundedness of the 'truncated sum operators' implies convergence since trigonometric polynomials are dense in $L^p(\mathbb{T}^3)$, while the converse follows using the Principle of Uniform Boundedness (see Proposition 1.9 in Muscalu & Schlag, 2013, for example).

In the one-dimensional case (functions on $\mathbb{T}$) the Fourier expansions converge in every L^p, $1 < p < \infty$, see Theorem 3.20 and Exercise 3.7 in Muscalu & Schlag (2013), for example. In higher dimensions this can only be guaranteed when $p = 2$ (see Corollary 3.6.10 in Grafakos (2008); the proof relies on a result of Fefferman (1971) for the equivalent phenomenon in the context of the Fourier transform). However, one can obtain convergence in $L^p(\mathbb{T}^3)$, $p \ne 2$, by considering 'square sums' of Fourier components, i.e. using the one-dimensional result repeatedly.

Theorem 1.6 *Let* $Q_N = [-N, N]^3 \cap \mathbb{Z}^3$. *For every* $u \in L^1(\mathbb{T}^3)$ *and every* $N \in \mathbb{N}$ *define*

$$S_N(u) := \sum_{k \in Q_N} \hat{u}_k e^{ik \cdot x},$$

where the Fourier coefficients $\hat{u}_k$ *are given by (1.11). Then for every* $1 < p < \infty$ *there is a constant* $C_p > 0$, *independent of* N, *such that*

$$\|S_N(u)\|_{L^p} \le C_p \|u\|_{L^p} \quad \text{for all} \quad u \in L^p(\mathbb{T}^3)$$

and $S_N(u) \to u$ *in* $L^p(\mathbb{T}^3)$.

1.6 Sobolev spaces $W^{k,p}$

We begin with some well-known results about the Sobolev spaces $W^{k,p}$ for integer values of k. If not stated otherwise we assume below that $\Omega \subseteq \mathbb{R}^n$ is either the whole space or a bounded domain with at least Lipschitz boundary.

We recall that a function $g \in L^1_{\text{loc}}(\Omega)$ is the weak derivative $\partial_i u$ of a function $u \in L^1_{\text{loc}}(\Omega)$ if

$$\int_\Omega g(x)\varphi(x)\,\mathrm{d}x = -\int_\Omega u(x)\partial_i\varphi(x)\,\mathrm{d}x \qquad \text{for all} \quad \varphi \in C_c^\infty(\Omega).$$

Let $1 \le p < \infty$. The space $W^{1,p}(\Omega)$ consists of all those functions $u \in L^p(\Omega)$ all of whose first weak derivatives $\partial_i u$ exist and are in $L^p(\Omega)$ (we often write $\nabla u \in L^p(\Omega)$ for short). The norm in $W^{1,p}(\Omega)$ is given by

$$\|u\|_{W^{1,p}} := \left(\int_\Omega |u(x)|^p + \sum_{i=1}^n \int_\Omega |\partial_i u(x)|^p \,\mathrm{d}x \right)^{1/p}. \tag{1.12}$$

The space $C^\infty(\Omega) \cap W^{1,p}(\Omega)$ is dense in $W^{1,p}(\Omega)$.

The space $W_0^{1,p}(\Omega)$, with $1 \le p < \infty$ is defined as the closure of $C_c^\infty(\Omega)$ in the norm (1.12). The space $W_0^{1,p}(\Omega)$ is the subset of $W^{1,p}(\Omega)$ that consists of functions vanishing on the boundary (in the sense of trace). Usually we have $W_0^{1,p}(\Omega) \neq W^{1,p}(\Omega)$, but on the whole space these coincide:

$$W_0^{1,p}(\mathbb{R}^n) = W^{1,p}(\mathbb{R}^n).$$

Suppose now that $k \ge 2$ and $1 \le p < \infty$. The space $W^{k,p}(\Omega)$ is the space of functions $u \in W^{k-1,p}(\Omega)$ all of whose first weak derivatives also belong to $W^{k-1,p}(\Omega)$. The standard norm in $W^{k,p}(\Omega)$ is given by

$$\|u\|_{W^{k,p}}^p := \sum_{|\alpha| \le k} \|\partial^\alpha u\|_{L^p}^p. \tag{1.13}$$

The space $W_0^{k,p}(\Omega)$ is the closure of $C_c^\infty(\Omega)$ in the $W^{k,p}$ norm.

The following theorem gives the fundamental embedding results for the spaces $W^{1,p}$, valid on any sufficiently regular domain.

Theorem 1.7 *Suppose that $u \in W^{1,p}(\Omega)$, where $\Omega = \mathbb{T}^n$ or $\Omega \subseteq \mathbb{R}^n$ (bounded or unbounded).*

(i) *Sobolev embedding: if $1 \le p < n$ then we have a continuous embedding*

$$W^{1,p}(\Omega) \subset L^{p^*}(\Omega), \qquad \text{where} \quad p^* = np/(n-p).$$

Furthermore there exists a constant $c > 0$ such that if $u \in W_0^{1,p}$ then

$$\|u\|_{L^{p^*}} \leq c\|\nabla u\|_{L^p}.$$

(ii) *Morrey's Theorem: if $p > n$ then all functions in $W^{1,p}(\Omega)$ are bounded and continuous on $\overline{\Omega}$ with*

$$W^{1,p}(\Omega) \subset C^{0,1-n/p}(\overline{\Omega}).$$

On a bounded domain we also have the following compact embedding $W^{1,p}(\Omega) \subset\subset L^q(\Omega)$.

Theorem 1.8 (Rellich–Kondrachov Theorem) *Take $p \geq 1$ and let $q \in [1, p^*)$, where $p^* = np/(n - p)$. If Ω is a bounded domain with Lipschitz boundary then from any sequence that is bounded in $W^{1,p}(\Omega)$ we can choose a subsequence that converges strongly in $L^q(\Omega)$.*

On a bounded domain Ω or on $\mathbb{T}^3$ we also have the Poincaré inequality, provided that we include some additional condition (zero boundary data or zero average).

Theorem 1.9 (Poincaré inequality) *Let $1 \leq p < \infty$ and let Ω be a smooth bounded domain. Then the Poincaré inequality*

$$\|u\|_{L^p} \leq C\|\nabla u\|_{L^p}$$

holds

  (i) *for all $u \in W_0^{1,p}(\Omega)$;*
  (ii) *for all $u \in W^{1,p}(\Omega)$ such that $\int_\Omega u = 0$; and*
  (iii) *for all $u \in W^{1,p}(\mathbb{T}^3)$ such that $\int_{\mathbb{T}^3} u = 0$.*

We do not have such an inequality on $\mathbb{R}^3$: the simple example of the constant function $u \equiv 1$ shows that this inequality does not hold in general, but constant functions are not included in $W^{1,p}(\mathbb{R}^3)$. Taking a non-zero function $u \in C_c^\infty(\mathbb{R}^3)$ and then considering the family $u_k(x) = u(kx)$ shows that the inequality also fails to hold uniformly on $W^{1,p}(\mathbb{R}^3)$.

1.7 Sobolev spaces H^s with $s \geq 0$

Throughout the book we will most often use the L^2-based Sobolev spaces $W^{k,2}$; we adopt the notation

$$H^k(\Omega) := W^{k,2}(\Omega) \qquad \text{and} \qquad H_0^k(\Omega) := W_0^{k,2}(\Omega)$$

for all $k = 1, 2, 3, \ldots$ The advantage of these spaces is that they are Hilbert spaces with the inner product

$$\langle u, v \rangle_{H^k} := \sum_{0 \leq |\alpha| \leq k} \langle \partial^\alpha u, \partial^\alpha v \rangle.$$

The corresponding norm

$$\|u\|_{H^k}^2 = \langle u, u \rangle_{H^k} = \sum_{0 \leq |\alpha| \leq k} \|\partial^\alpha u\|^2 \qquad (1.14)$$

clearly coincides with the $W^{k,2}$ norm as defined in (1.13). We refer to this norm as 'the standard H^k norm', but in many situations (as we will now see) it is convenient to use an alternative, but equivalent, norm.

We now recall some basic properties of these spaces when defined on the whole space $\mathbb{R}^3$ and on the torus $\mathbb{T}^3$; we thereby motivate the generalisation of the definition of H^k to non-integer values of k.

1.7.1 Sobolev spaces on $\mathbb{T}^3$

Let us first consider the case of a torus. Since we concentrate on L^2-based Sobolev spaces, it is convenient to work with the Fourier expansion of u,

$$u(x) = \sum_{k \in \mathbb{Z}^3} \hat{u}_k e^{ik \cdot x}, \qquad \hat{u}_k = \overline{\hat{u}_{-k}}, \qquad (1.15)$$

with convergence of the expansion in $L^2(\mathbb{T}^3)$ assured if $\sum_{k \in \mathbb{Z}^3} |\hat{u}_k|^2 < \infty$ (see Section 1.5).

Let u be given by (1.15) and assume for simplicity that u is a scalar function. Then formally

$$\|u\|^2 = \int_{\mathbb{T}^3} |u(x)|^2 \, dx = \int_{\mathbb{T}^3} \sum_{k,l} \hat{u}_k \hat{u}_l e^{i(k+l) \cdot x} \, dx$$

$$= \sum_k \int_{\mathbb{T}^3} |\hat{u}_k|^2 \, dx = (2\pi)^3 \sum_k |\hat{u}_k|^2.$$

Moreover, using the fact that $\partial_j e^{ik \cdot x} = ik_j e^{ik \cdot x}$ we can (again formally) compute the gradient of u to obtain

$$\|\nabla u\|^2 = \sum_{j=1}^3 \|\partial_j u\|^2 = (2\pi)^3 \sum_{k \in \mathbb{Z}^3} |k|^2 |\hat{u}_k|^2.$$

It follows that

$$\|u\|_{H^1(\mathbb{T}^3)}^2 = \|u\|_{L^2(\mathbb{T}^3)}^2 + \|\nabla u\|_{L^2(\mathbb{T}^3)}^2 = (2\pi)^3 \sum_{k \in \mathbb{Z}^3} (1 + |k|^2) |\hat{u}_k|^2.$$

Generalising this simple observation, we can define the space $H^s(\mathbb{T}^3)$ for any $s \geq 0$ in the following way. The norm defined in (1.16) is equivalent to the standard definition in (1.14) when $s \in \mathbb{N}$, but the expression makes sense for any $s \geq 0$.

Definition 1.10 For $s \geq 0$ the Sobolev space $H^s(\mathbb{T}^3)$ consists of all functions given by (1.15) for which the norm $\| \cdot \|_{H^s}$ defined by

$$\|u\|_{H^s}^2 := (2\pi)^3 \sum_{k \in \mathbb{Z}^3} (1 + |k|^{2s}) |\hat{u}_k|^2 \qquad (1.16)$$

is finite.

It is immediate from the definition that $H^{s_2}(\mathbb{T}^3) \subset H^{s_1}(\mathbb{T}^3)$ if $s_2 > s_1$.

It is also useful to consider homogeneous Sobolev spaces, in which we drop the L^2 part of the norm. For any $u \in H^s(\mathbb{T}^3)$ the expression

$$\|u\|_{\dot{H}^s}^2 := (2\pi)^3 \sum_{k \in \dot{\mathbb{Z}}^3} |k|^{2s} |\hat{u}_k|^2, \qquad (1.17)$$

where we exclude $k = (0, 0, 0)$ from the summation by setting

$$\dot{\mathbb{Z}}^3 := \{k \in \mathbb{Z}^3 : |k| \neq 0\},$$

is clearly well defined. However, $\| \cdot \|_{\dot{H}_s}$ does not define a norm on $H^s(\mathbb{T}^3)$ since it is equal to zero for every constant function. On the other hand, (1.17) does define a norm on the subset of $H^s(\mathbb{T}^3)$ consisting of zero-mean functions. We therefore make the following definition.

Definition 1.11 The space $\dot{L}^2(\mathbb{T}^3)$ consists of all functions $u \in L^2(\mathbb{T}^3)$ that have zero mean,

$$\fint_{\mathbb{T}^3} u(x)\, dx = 0.$$

It is easy to see that $\dot{L}^2(\mathbb{T}^3)$ is given by the collection of all (real) functions with Fourier expansion in which $\hat{u}_0 = 0$, i.e. all real functions of the form

$$\sum_{k \in \dot{\mathbb{Z}}^3} \hat{u}_k e^{ik \cdot x} \qquad \text{with} \qquad \sum_{k \in \dot{\mathbb{Z}}^3} |\hat{u}_k|^2 < \infty.$$

Definition 1.12 The homogeneous Sobolev space $\dot{H}^s(\mathbb{T}^3)$ is defined for $s \geq 0$ as

$$\dot{H}^s(\mathbb{T}^3) := H^s(\mathbb{T}^3) \cap \dot{L}^2(\mathbb{T}^3).$$

The nesting of homogeneous spaces, $\dot{H}^{s_2}(\mathbb{T}^3) \subset \dot{H}^{s_1}(\mathbb{T}^3)$ if $s_2 > s_1$, holds in the periodic case, but we will see that the same is not true on the whole space.

It is useful to notice that

$$\|u\|_{\dot{H}^1}^2 = (2\pi)^3 \sum_{k \in \dot{\mathbb{Z}}^3} |k|^2 |\hat{u}_k|^2 = (2\pi)^3 \sum_{k \in \dot{\mathbb{Z}}^3} |k\hat{u}_k|^2 = \|\nabla u\|^2$$

and

$$\|u\|_{\dot{H}^2}^2 = (2\pi)^3 \sum_{k \in \dot{\mathbb{Z}}^3} |k|^4 |\hat{u}_k|^2 = (2\pi)^3 \sum_{k \in \dot{\mathbb{Z}}^3} \left||k|^2 \hat{u}_k\right|^2 = \|\Delta u\|^2.$$

For later use (e.g. in Chapter 10) it is also convenient to express the $\dot{H}^s$ norm in another way, by writing it as $\|\Lambda^s u\|$, where Λ^s is defined by

$$\Lambda^s u := \sum_{k \in \dot{\mathbb{Z}}^3} |k|^s \hat{u}_k \mathrm{e}^{\mathrm{i}k \cdot x} \quad \text{when} \quad u = \sum_{k \in \dot{\mathbb{Z}}^3} \hat{u}_k \mathrm{e}^{\mathrm{i}k \cdot x}. \tag{1.18}$$

Notice that $\Lambda^2 u = -\Delta u$ and so $\Lambda = (-\Delta)^{1/2}$; loosely speaking, one can think of Λ^s as a 'derivative of order s'. In this case the inner product in $\dot{H}^s$ can be written as

$$\langle u, v \rangle_{\dot{H}^s} = \langle \Lambda^s u, \Lambda^s v \rangle.$$

1.7.2 Sobolev spaces on $\mathbb{R}^3$

In order to formulate the corresponding definitions of H^s and $\dot{H}^s$ on the whole space $\mathbb{R}^3$ we replace Fourier series by the Fourier transform. We recall that one can define[4] the Fourier transform of a function $u \in L^1(\mathbb{R}^3)$ by

$$\hat{u}(\xi) = \int_{\mathbb{R}^3} \mathrm{e}^{-2\pi \mathrm{i}\xi \cdot x} u(x) \, \mathrm{d}x,$$

and then for all $u \in L^1(\mathbb{R}^3) \cap L^2(\mathbb{R}^3)$ we have the Plancherel identity

$$\|u\|^2 = \int_{\mathbb{R}^3} |\hat{u}(\xi)|^2 \, \mathrm{d}\xi.$$

Using this identity we can extend the definition of the Fourier transform to all functions in $L^2(\mathbb{R}^3)$ by a density argument.

When $f, \hat{f} \in L^1(\mathbb{R}^3)$ we can write f as the inverse Fourier transform of $\hat{f}$,

$$f(x) = (\hat{f})^{\vee}(x) := \int_{\mathbb{R}^3} \mathrm{e}^{2\pi \mathrm{i}\xi \cdot x} \hat{f}(\xi) \, \mathrm{d}\xi; \tag{1.19}$$

arguing as above, we can extend the operation $^{\vee}$ to $\hat{f} \in L^2(\mathbb{R}^3)$ by a density argument; in this case we cannot write the integral term in (1.19), but we still have the identity $f = (\hat{f})^{\vee}$.

[4] There are a variety of definitions, depending on where one chooses to place the factor of 2π. One advantage of this definition is that it makes the Fourier transform an isometry from L^2 into itself, i.e. the Plancherel identity holds as stated here with no multiplicative constant.

Since $\widehat{\partial_i u}(\xi) = 2\pi i\xi_i \hat{u}(\xi)$ we have

$$\|\nabla u\|^2 = \int_{\mathbb{R}^3} |\widehat{\nabla u}(\xi)|^2 \, d\xi = (2\pi)^2 \int_{\mathbb{R}^3} |\xi|^2 |\hat{u}(\xi)|^2 \, d\xi;$$

we can define the space $H^1(\mathbb{R}^3)$ as the subspace of $L^2(\mathbb{R}^3)$ consisting of all functions for which the norm

$$\|u\|^2_{H^1(\mathbb{R}^3)} = \int_{\mathbb{R}^3} (1 + |\xi|^2)|\hat{u}(\xi)|^2 \, d\xi$$

is finite. This leads to the following definition of the space $H^s(\mathbb{R}^3)$ for non-integer values of $s \geq 0$.

Definition 1.13 Let $s \geq 0$. The Sobolev space $H^s(\mathbb{R}^3)$ is defined by

$$H^s(\mathbb{R}^3) := \left\{ u \in L^2(\mathbb{R}^3) : \|u\|^2_{H^s} < \infty \right\},$$

where

$$\|u\|^2_{H^s} := \int_{\mathbb{R}^3} (1 + |\xi|^{2s})|\hat{u}(\xi)|^2 \, d\xi. \tag{1.20}$$

As in the periodic case, the norm defined in (1.20) is equivalent to the standard H^k norm when $s = k \in \mathbb{N}$, but s need not be an integer for the definition to make sense. Clearly $H^{s_2}(\mathbb{R}^3) \subset H^{s_1}(\mathbb{R}^3)$ if $s_2 > s_1$.

For any $u \in H^s(\mathbb{R}^3)$ we define the homogeneous H^s norm of u by setting

$$\|u\|^2_{\dot{H}^s} := \int_{\mathbb{R}^3} |\xi|^{2s} |\hat{u}(\xi)|^2 \, d\xi. \tag{1.21}$$

There are some subtleties in defining the homogeneous space $\dot{H}^s(\mathbb{R}^3)$, since only knowing that $\|u\|_{\dot{H}^s}$ is finite still allows bad behaviour of $\hat{u}$ at the origin in Fourier space.[5] One way to circumvent this is to add an extra condition on $\hat{u}$, as in the following definition.

Definition 1.14 The homogeneous Sobolev space $\dot{H}^s(\mathbb{R}^3)$ is defined as

$$\left\{ u \in \mathscr{S}' : \hat{u} \in L^1_{\text{loc}}(\mathbb{R}^3) \quad \text{and} \quad \int_{\mathbb{R}^3} |\xi|^{2s} |\hat{u}(\xi)|^2 \, d\xi < \infty \right\},$$

where $\mathscr{S}'$ is the collection of all tempered distributions.

The homogeneous spaces on $\mathbb{R}^3$ are not nested, but we nevertheless have a very useful interpolation property which we now recall. The result follows

---

[5] We use the definition favoured by Bahouri, Chemin, & Danchin (2011), see also Chemin et al. (2006). This avoids the complexities that arise from problems with attaching meaning to $|\xi|^s \hat{u}$ when one only knows that $\hat{u} \in \mathscr{S}'$, see the discussion in Chapter 6 of Grafakos (2009). The space $\dot{H}^s(\mathbb{R}^n)$ is complete if $s < n/2$ and in this case it can also be defined as the completion of $\mathscr{S}$ in the $\dot{H}^s(\mathbb{R}^n)$ norm; it is not complete when $s > n/2$.

from a straightforward use of Hölder's inequality, see Exercise 1.6. Note that the correct exponents on the right-hand side can be found simply by 'counting derivatives' (as expressed by the requirement that $s = \theta s_1 + (1 - \theta)s_2$).

Lemma 1.15 (Sobolev interpolation inequality for $\mathbb{T}^3$ or $\mathbb{R}^3$) *Suppose that* $0 \le s_1 \le s \le s_2$, $0 \le \theta \le 1$ *and* $s = \theta s_1 + (1 - \theta)s_2$. *If* $u \in \dot{H}^{s_1} \cap \dot{H}^{s_2}$ *then*

$$\|u\|_{\dot{H}^s} \le \|u\|_{\dot{H}^{s_1}}^{\theta} \|u\|_{\dot{H}^{s_2}}^{1-\theta} \tag{1.22}$$

and if $u \in H^{s_2}$ *then*

$$\|u\|_{H^s} \le c\|u\|_{H^{s_1}}^{\theta} \|u\|_{H^{s_2}}^{1-\theta}. \tag{1.23}$$

1.7.3 Sobolev spaces on bounded domains

The key to proving many results for Sobolev spaces on bounded domains is the following extension theorem. The uniformity statement in the theorem is not standard, so we provide a proof; we will use this uniformity in our construction of weak solutions on the whole of $\mathbb{R}^3$ in Chapter 4.

Theorem 1.16 *Let* $\Omega \subset \mathbb{R}^3$ *be a smooth bounded domain. For each* $k \in \mathbb{N}$ *there exists a linear operator* $E_k \colon H^k(\Omega) \to H^k(\mathbb{R}^3)$ *and a constant* C_k *such that*

(i) $E_k u|_\Omega = u$;
(ii) $\|E_k u\|_{H^j(\mathbb{R}^3)} \le C_k \|u\|_{H^j(\Omega)}$, $j = 0, \ldots, k$.

The constant C_k *depends on* Ω *but can be taken uniformly for all domains of the form* $R\Omega := \{Rx : x \in \Omega\}$ *with* $R \ge 1$.

Proof We prove only the remark about the uniformity for domains $R\Omega$, where $R \ge 1$; for the construction of an extension operator see Adams & Fournier (2003), Constantin & Foias (1988), Evans (1998), Robinson (2001), or Stein (1970), among others. Suppose that we have an extension operator for $H^k(\Omega)$ as in the statement of the theorem; we will show that as a consequence there exist extension operators

$$E_k^R \colon H^k(R\Omega) \to H^k(\mathbb{R}^3)$$

satisfying (i) and (ii) with a different (but uniform) constant C_k'.

Given $u \colon R\Omega \to \mathbb{R}$ let $v(x) = u(Rx)$ so that $v \colon \Omega \to \mathbb{R}$ and

$$\|v\|_{\dot{H}^j(\Omega)} = R^{j-3/2} \|u\|_{\dot{H}^j(R\Omega)}, \qquad j = 0, \ldots, k,$$

and hence in particular

$$\|v\|_{H^j(\Omega)} \le R^{j-3/2} \|u\|_{H^j(R\Omega)}, \qquad j = 0, \ldots, k$$

(note that we now have inhomogeneous norms in the inequality). To extend a function $u \in H^k(R\Omega)$ we set $E_k^R u(x) := [E_k v](x/R)$; then for any $j = 0, \ldots, k$ we have

$$\|[E_k v](x/R)\|_{\dot{H}^j(\mathbb{R}^3)} = R^{-j+3/2} \|E_k v\|_{\dot{H}^j(\mathbb{R}^3)} \leq C_k R^{-j+3/2} \|v\|_{H^j(\Omega)}$$
$$\leq C_k \|u\|_{H^j(R\Omega)}.$$

It follows that for any $j = 0, \ldots, k$ we have

$$\|[E_k v](x/R)\|_{H^j(\mathbb{R}^3)} \leq \sum_{l=0}^{j} \|[E_k v](x/R)\|_{\dot{H}^l(R\Omega)}$$
$$\leq \sum_{l=0}^{j} C_k \|u\|_{H^l(R\Omega)} \leq (k+1) C_k \|u\|_{H^j(R\Omega)},$$

so we can obtain the required uniformity by choosing $C_k' = (k+1) C_k$. $\qquad \square$

It is an immediate consequence of this extension theorem that the interpolation inequality (1.23) extends to integer-valued Sobolev spaces on bounded domains.

Lemma 1.17 (Sobolev interpolation inequality on Ω) *Let $s, s_1, s_2 \in \mathbb{N}$ with $0 \leq s_1 \leq s \leq s_2$, and choose $\theta \in [0,1]$ such that $s = \theta s_1 + (1-\theta)s_2$. Then for any $u \in H^{s_2}$*

$$\|u\|_{H^s} \leq c \|u\|_{H^{s_1}}^{\theta} \|u\|_{H^{s_2}}^{1-\theta}. \tag{1.24}$$

Proof Use Theorem 1.16 to extend u to a function $E_{s_2}[u] \in H^{s_2}(\mathbb{R}^n)$ such that

$$\|E_{s_2}[u]\|_{H^{s_2}(\mathbb{R}^n)} \leq C_{s_2} \|u\|_{H^{s_2}(\Omega)} \quad \text{and} \quad \|E_{s_2}[u]\|_{H^{s_1}(\mathbb{R}^n)} \leq C_{s_2} \|u\|_{H^{s_1}(\Omega)},$$

from which it follows using (1.23) that

$$\|u\|_{H^s(\Omega)} \leq \|E_{s_2}[u]\|_{H^s(\mathbb{R}^n)}$$
$$\leq c \|E_{s_2}[u]\|_{H^{s_1}(\mathbb{R}^n)}^{\theta} \|E_{s_2}[u]\|_{H^{s_2}(\mathbb{R}^n)}^{1-\theta}$$
$$\leq c \|u\|_{H^{s_1}(\Omega)}^{\theta} \|u\|_{H^{s_2}(\Omega)}^{1-\theta}. \qquad \square$$

1.7.4 Sobolev embedding results

As s increases the space H^s consists of more and more regular functions. In some sense the critical exponent in three dimensions is $s = 3/2$, which is related to the fact that

$$\sum_{k\in\mathbb{Z}^3}(1+|k|^{2s})^{-1} \qquad \text{and} \qquad \int_{\mathbb{R}^3}(1+|x|^{2s})^{-1}\,\mathrm{d}x$$

are finite for $s > 3/2$ and infinite for $s < 3/2$ (see Exercise 1.8).

We now give some important Sobolev embeddings for the spaces H^s.

Theorem 1.18 (Sobolev embeddings) *Let $\Omega \subset \mathbb{R}^3$ be any smooth bounded domain, $\mathbb{R}^3$, or $\mathbb{T}^3$.*

(i) *If $0 \le s < 3/2$ then $H^s(\Omega) \subset L^{6/(3-2s)}(\Omega)$ and there exists $c_s > 0$ such that*

$$\|u\|_{L^{6/(3-2s)}} \le c_s \|u\|_{H^s} \qquad \text{for all} \qquad u \in H^s(\Omega);$$

hence when Ω has finite measure $H^s \subset L^p$ for $1 \le p \le 6/(3-2s)$.

(ii) *If $s > 3/2$ then $H^s(\Omega) \subset C^0(\overline{\Omega})$ and there exists a constant $c_s > 0$ such that*

$$\|u\|_{C^0} \le c_s \|u\|_{H^s} \qquad \text{for all} \qquad u \in H^s(\Omega).$$

Furthermore H^s is a Banach algebra: if $u, v \in H^s$ then $uv \in H^s$ with

$$\|uv\|_{H^s} \le c_s \|u\|_{H^s} \|v\|_{H^s}.$$

There is a useful variant of (i) on the whole space, and on domains when the functions are zero on the boundary or have zero average. Note that the norm on the right-hand side only needs to be in the homogeneous space $\dot{H}^s$.

Theorem 1.19 *There exists a constant c (depending on Ω in the case of a smooth bounded domain) such that for $0 \le s < 3/2$*

$$\|u\|_{L^{6/(3-2s)}} \le c\|u\|_{\dot{H}^s} \qquad \text{for all} \qquad \begin{cases} u \in \dot{H}^s(\mathbb{R}^3) \text{ or } \dot{H}^s(\mathbb{T}^3) \\ u \in H_0^s(\Omega) \\ u \in H^s(\Omega) \text{ with } \int_\Omega u = 0. \end{cases}$$

The 'missing' case in Theorem 1.18 is $s = 3/2$, for which the embedding $H^s \subset L^\infty$ fails. The following inequality, which would follow by Sobolev interpolation (Lemma 1.15) if this embedding did hold, sometimes provides a useful replacement. (A more general version of this result is also true, see Exercise 1.10.)

Theorem 1.20 (Agmon's inequality) *If $u \in H^2(\mathbb{R}^3)$, $H^2(\mathbb{T}^3)$, or $H^2(\Omega)$, then u is (almost everywhere equal to) a continuous function and*

$$\|u\|_{L^\infty} \le c\|u\|_{H^1}^{1/2}\|u\|_{H^2}^{1/2}.$$

As a consequence of Theorem 1.18 functions that belong to Sobolev spaces of arbitrary order are smooth if the boundary of the domain is sufficiently regular.

Theorem 1.21 *If $u \in H^k(\mathbb{R}^n)$ for all $k = 1, 2, 3, \ldots$ then u is smooth. If Ω is a bounded open set with a smooth boundary or $\Omega = \mathbb{T}^3$ and $u \in H^k(\Omega)$ for all $k = 1, 2, 3, \ldots$ then $u \in C^\infty(\overline{\Omega})$ (i.e. $u \in C^k(\overline{\Omega})$ for every k).*

1.8 Dual spaces

If X is a Banach space then we denote the dual space of X (the space of all bounded linear functionals on X) by X^* and use the notation $\langle f, u \rangle_{X^* \times X}$ to denote the dual pairing between elements $f \in X^*$ and $u \in X$; we drop the $X^* \times X$ subscript when this is clear from the context. The norm in X^* is given by

$$\|f\|_{X^*} := \sup_{\|u\|_X = 1} |\langle f, u \rangle_{X^* \times X}|. \qquad (1.25)$$

We recall that the dual space of $L^p(\Omega)$, $1 \le p < \infty$, is (isometrically isomorphic to) the space $L^q(\Omega)$, where $1/p + 1/q = 1$. For $1 < p < \infty$ these spaces are reflexive; and for $1 < q \le \infty$ we have

$$\|u\|_{L^q} = \sup_{\|f\|_{L^p} = 1} \int_\Omega f(x) u(x) \, dx,$$

which sometimes provides a useful way to bound $\|u\|_{L^q}$. As already remarked in Section 1.4, we usually omit the subscript $L^q \times L^p$ in the duality pairing in this case, writing $\langle f, g \rangle$ for $\int fg$.

The dual space of $H_0^k(\Omega)$, where Ω is a bounded open set, is denoted by $H^{-k}(\Omega)$. To be consistent with this notation we denote the dual of $\dot{H}^s(\mathbb{T}^3)$ by $H^{-s}(\mathbb{T}^3)$, where now s can be any positive number; the norm in the space $H^s(\mathbb{T}^3)$ for negative s is still given by the formula (1.17). Indeed, suppose that $s > 0$, $u \in \dot{H}^s$, and v is given by a formal expression

$$v(x) = \sum_{k \in \dot{\mathbb{Z}}^3} \hat{v}_k e^{ikx} \quad \text{with} \quad \sum_{k \in \dot{\mathbb{Z}}^3} |k|^{-2s} |\hat{v}_k|^2 < \infty.$$

Then we can attach meaning to the expression

$$\langle v, u \rangle_{H^{-s} \times \dot{H}^s} := \sum_{k \in \dot{\mathbb{Z}}^3} \hat{v}_k \hat{u}_{-k},$$

because, using (1.25),

$$\left|\langle u, v\rangle_{H^{-s}\times\dot{H}^s}\right| \leq \sum_{k\in\mathbb{Z}^3}|\hat{v}_k\hat{u}_{-k}| \leq \sum_{k\in\mathbb{Z}^3}\left(|k|^{-s}|\hat{v}_k|\right)\left(|k|^s|\hat{u}_{-k}|\right)$$

$$\leq \left(\sum_{k\in\mathbb{Z}^3}|k|^{-2s}|\hat{v}_k|^2\right)^{1/2}\left(\sum_{k\in\mathbb{Z}^3}|k|^{2s}|\hat{u}_k|^2\right)^{1/2},$$

where in the last inequality we used the Cauchy–Schwarz inequality for sums.

Note that if X and Y are linear spaces and $X \subset Y$ then $Y^* \subset X^*$. It follows from the three-dimensional Sobolev embedding $H^s \subset L^{6/(3-2s)}$ (part (i) of Theorem 1.18) that $L^{6/(3+2s)} \subset H^{-s}$ and in particular

$$L^{6/5} \subset H^{-1}, \tag{1.26}$$

a result that will be useful in Chapter 5.

1.9 Bochner spaces

The solutions of the Navier–Stokes equations (and other evolutionary partial differential equations) are usually defined as elements of Bochner spaces, such as $L^\infty(0, T; L^2)$ or $L^2(0, T; H^1)$. Therefore we will briefly recall the definition and basic properties of such spaces. Concise proofs of the results in this section are not so easy to find; Roubíček (2013) is a good source.

1.9.1 Banach-space-valued integration

Throughout this section we let X be a separable Banach space with norm $\|\cdot\|_X$. Assuming separability of X greatly simplifies the presentation, but we will also mention results for the non-separable case.

Definition 1.22 A function $u\colon [0, T] \to X$ is said to be *strongly measurable* if $u(t) = \lim_{n\to\infty} s_n(t)$ for almost all $t \in [0, T]$, where $s_n\colon [0, T] \to X$ are simple functions of the form $\sum_{i=1}^m \chi_{E_i}(t)u_i$, with E_i Lebesgue measurable and $u_i \in X$. (Here χ_A denotes the characteristic function of the set A.)

A simpler condition to check is the following.

Definition 1.23 A function $u\colon [0, T] \to X$ is said to be *weakly measurable* if $t \mapsto \langle f, u(t)\rangle$ is measurable for all $f \in X^*$.

If X is separable then these two properties are equivalent, a result due to Pettis (1938) (see also Chapter V, Section 4, in Yosida, 1980).

Theorem 1.24 *If X is separable then u:* $[0, T] \to X$ *is strongly measurable if and only if it is weakly measurable.*

Definition 1.25 Let X be a Banach space. For $1 \leq p < \infty$ the *Bochner space* $L^p(0, T; X)$ consists of all strongly measurable functions for which the norm

$$\|u\|_{L^p(0,T;X)} := \left(\int_0^T \|u(s)\|_X^p \, ds \right)^{1/p}$$

is finite. The space $L^\infty(0, T; X)$ is the space of all strongly measurable functions for which the norm

$$\|u\|_{L^\infty(0,T;X)} := \operatorname*{ess\,sup}_{t \in [0,T]} \|u(t)\|_X$$

is finite.

In Part II we will often treat functions with some local integrability properties, i.e. $u \in L^r(t_1, t_2; L^s(U))$, for some $t_2 > t_1$ and some $U \subset \mathbb{R}^3$. As a more concise notation we will sometimes write[6]

$$u \in L_t^r L_x^s(G), \qquad \text{where} \quad G = U \times (t_1, t_2).$$

While we can identify $L^p(0, T; L^p(\Omega))$ with $L^p(\Omega \times (0, T))$ for $1 \leq p < \infty$, this is not the case when $p = \infty$, see Exercise 1.11.

The completeness of $L^p(0, T; X)$ is a consequence of the construction of the space in a way that parallels the construction of the standard Lebesgue spaces of integrable functions.

Theorem 1.26 *If X is a Banach space and* $1 \leq p \leq \infty$ *then the space* $L^p(0, T; X)$ *is complete.*

If $u \in L^1(0, T; X)$ then we can define the Bochner integral of u, which is an element of X; one way to do this when X is reflexive[7] is given in the following lemma.

Lemma 1.27 *If X is a reflexive Banach space and* $u \in L^1(0, T; X)$ *then there exists a unique* $g \in X$ *such that*

$$\langle f, g \rangle = \int_0^T \langle f, u(s) \rangle \, ds \tag{1.27}$$

---

[6] Some care needs to be taken, since some authors choose to reverse this ordering, e.g. use $L^{3,\infty}$ to denote functions in $L_t^\infty L_x^3$.

[7] If $g = \int_0^T u(s) \, ds$ is the Bochner integral of u then the equality in (1.27) still holds, even when X is not reflexive. However, (1.27) does not define g in a unique way, and in this case the Bochner integral must be constructed as the limit of the integrals of X-valued simple functions, as in the standard development of the Lebesgue integral, see Bourbaki (2004) or Yosida (1980).

for any $f \in X^*$. In this case we define

$$\int_0^T u(s)\, ds := g.$$

Proof   Define a map $\Gamma: X^* \to \mathbb{R}$ by setting

$$\Gamma(f) = \int_0^T \langle f, u(s) \rangle\, ds$$

for each $f \in X^*$. This mapping Γ is clearly linear, and it is bounded since

$$|\Gamma(f)| \leq \int_0^T |\langle f, u(s) \rangle|\, ds \leq \int_0^T \|f\|_{X*} \|u(s)\|_X\, ds$$
$$\leq \left(\int_0^T \|u(s)\|_X\, ds \right) \|f\|_{X*},$$

and so $\Gamma \in X^{**}$. Since X is reflexive, it follows that there exists an element $g \in X$ such that for every $f \in X^*$

$$\langle f, g \rangle = \Gamma(f) = \int_0^T \langle f, u(s) \rangle\, ds,$$

as required.   □

Bochner-integrable functions can be approximated by various classes of smoother functions. The following lemma will be useful later.

Lemma 1.28   *If B is a countable orthonormal basis of a Hilbert space H then the set*

$$X = \left\{ u : \ u = \sum_{j=1}^N c_j(t) v_j, N \in \mathbb{N}, \ c_j \in C^\infty([0, T]), \ v_j \in B \right\}$$

is dense in $L^2(0, T; H)$.

Proof   Take any $u \in L^2(0, T; H)$ and choose $\delta > 0$. Define

$$c_k(t) = \langle u(t), v_k \rangle_H \quad \text{and} \quad u_N = \sum_{j=1}^N c_j(t) v_j.$$

Then

$$\|u\|_{L^2(0,T;H)}^2 = \int_0^T \sum_{j=1}^\infty c_j^2(t) < \infty.$$

Clearly

$$\|u_N\|_{L^2(0,T;H)}^2 \le \|u\|_{L^2(0,T;H)}^2 \quad \text{and} \quad \|u - u_N\|_{L^2(0,T;H)} \to 0.$$

First, take N so large that $\|u - u_N\|_{L^2(0,T;H)}^2 \le \delta/2$. Now, for each $j = 1, 2, \dots, N$, extend c_j by zero outside $(0, T)$ to yield functions $\tilde{c}_j$ defined on $\mathbb{R}$; choose $\varepsilon > 0$ so small that for the standard mollifications $c_j^\varepsilon = \rho_\varepsilon * \tilde{c}_j$ we have

$$\|c_j - c_j^\varepsilon\|_{L^2(0,T)}^2 \le \delta/(2N) \qquad \text{and} \qquad c_j^\varepsilon \in C^\infty[0, T]$$

for all $j = 1, \dots, N$ (see Theorem A.9). Let $u_N^\varepsilon = \sum_{j=1}^N c_j^\varepsilon(t)v_j \in X$. Then

$$\|u_N^\varepsilon - u_N\|_{L^2(0,T;H)}^2 = \int_0^T \sum_{j=1}^N |c_j(t) - c_j^\varepsilon(t)|^2 \, dt = \sum_{j=1}^N \|c_j^\varepsilon - c_j\|_{L^2(0,T)}^2 \le \delta/2.$$

The result now follows from the triangle inequality. □

When X^* is separable it is easy to identify the dual of $L^p(0, T; X)$.

Theorem 1.29 *Let X be a Banach space with X^* separable, take $p \in [1, \infty)$, and let (p, p') be conjugate indices. Then the dual space of $L^p(0, T; X)$ is (isometrically isomorphic to) $L^{p'}(0, T; X^*)$, with the natural pairing*

$$\langle f, u \rangle_{L^{p'}(0,T;X^*) \times L^p(0,T;X)} := \int_0^T \langle f(t), u(t) \rangle_{X^* \times X} \, dt.$$

Note that we still have $L^{p'}(0, T; X^*) \subset [L^p(0, T; X)]^*$ even when X^* is not separable.

Finally we recall the definition of the weak (time) derivative of a Bochner function and some consequences of the definition.

Definition 1.30 *If X is a Banach space we say that $u \in L^1(0, T; X)$ has* weak time derivative *$\partial_t u \in L^1(0, T; X)$ if*

$$\int_0^T u(s)\partial_t \varphi(s) \, ds = -\int_0^T \partial_t u(s)\varphi(s) \, ds \tag{1.28}$$

for all $\varphi \in C_c^\infty(0, T)$.

In this definition note that the function φ is a scalar function, and that (1.28) is understood as an equality between two Bochner integrals, i.e. as an equality between two elements of X.

The following lemma (Lemma 1.1 in Chapter 3 of Temam, 1977; see also Lemma 1.3.1 in Chapter IV of Sohr, 2001) provides equivalent ways of interpreting the weak derivative of Definition 1.30.

Lemma 1.31 *Suppose that $u, g \in L^1(0, T; X)$. Then the following are equivalent*

(i) *$\partial_t u = g$ in the sense of Definition 1.30;*
(ii) *there exists $\xi \in X$ such that*

$$u(t) = \xi + \int_0^t g(s)\, \mathrm{d}s \qquad \text{for almost every } t \in [0, T];$$

(iii) *for every $f \in X^*$*

$$\frac{\mathrm{d}}{\mathrm{d}t} \langle f, u \rangle = \langle f, g \rangle,$$

where the time derivative is taken in the weak sense.

In particular, if (i)–(iii) hold then u is almost everywhere equal to a continuous function from $[0, T]$ into X.

The following corollary will be particularly useful in what follows.

Corollary 1.32 *Suppose that $u, g \in L^1(0, T; X)$ and that for every $f \in X^*$*

$$\langle f, u(t_2) \rangle - \langle f, u(t_1) \rangle = \int_{t_1}^{t_2} \langle f, g(s) \rangle\, \mathrm{d}s \qquad (1.29)$$

for almost every pair (t_1, t_2) with $0 \le t_1 \le t_2 \le T$. Then $\partial_t u = g$ in the sense of Definition 1.30.

Proof It follows from (1.29) that $\langle f, u(t) \rangle$ is absolutely continuous and almost everywhere differentiable with derivative $\langle f, g(t) \rangle$. By assumption $\langle f, u(t) \rangle$ and $\langle f, g(t) \rangle$ are both elements of $L^1(0, T)$. It follows that $\langle f, g \rangle$ is the weak time derivative of $\langle f, u \rangle$, and hence (by (iii) implies (i) of Lemma 1.31) that $\partial_t u = g$. □

Note that if X is reflexive one could similarly take $u, g \in L^1(0, T; X^*)$ and require

$$\langle u(t_2), \varphi \rangle - \langle u(t_1), \varphi \rangle = \int_{t_1}^{t_2} \langle g(s), \varphi \rangle\, \mathrm{d}s \qquad (1.30)$$

for every $\varphi \in X$ and almost every (t_1, t_2); clearly it is also sufficient to require (1.30) only for φ in some dense subset of X.

1.9.2 Sobolev-type theorems for Bochner spaces

Lemma 1.31 shows that if $u, \partial_t u \in L^1(0, T; X)$ then $u \in C^0([0, T]; X)$. The following results are in the same spirit, but allow for u and its derivative to belong to different spaces, as in the above definition. They also include the useful integration-by-parts formula (1.31).

Theorem 1.33 *If $u \in L^2(0, T; H_0^1)$ and $\partial_t u \in L^2(0, T; H^{-1})$ then we have $u \in C^0([0, T]; L^2)$ with*

$$\sup_{t \in [0,T]} \|u(t)\|_{L^2} \leq C(\|u\|_{L^2(0,T;H_0^1)} + \|\partial_t u\|_{L^2(0,T;H^{-1})}).$$

Moreover $t \mapsto \|u(t)\|^2$ is absolutely continuous and

$$\frac{1}{2} \frac{d}{dt} \|u\|^2 = \langle \partial_t u, u \rangle_{H^{-1} \times H_0^1}.$$

For the proof of the result in this form see Lemma 10.4 in Renardy & Rogers (2004), for example. A more general version can be found in Chapter 7 of Roubíček (2013).

Corollary 1.34 *If $u, v \in L^2(0, T; H_0^1)$ and $\partial_t u, \partial_t v \in L^2(0, T; H^{-1})$ then the mapping $t \mapsto \langle u, v \rangle$ is absolutely continuous and*

$$\langle u(t), v(t) \rangle = \langle u(s), v(s) \rangle + \int_s^t \langle \partial_t u, v \rangle + \int_s^t \langle \partial_t v, u \rangle. \tag{1.31}$$

Proof We have $(u + v) \in L^2(0, T; H_0^1)$ and $\partial_t (u + v) \in L^2(0, T; H^{-1})$, and so $t \mapsto \|u + v\|^2$ is absolutely continuous. Hence

$$2\langle u, v \rangle = \|u + v\|^2 - \|v\|^2 - \|u\|^2$$

is absolutely continuous and

$$\frac{1}{2} \frac{d}{dt} \left(\|u\|^2 + 2\langle u, v \rangle + \|v\|^2 \right) = \langle \partial_t (u + v), u + v \rangle$$

for almost all t and the result follows on integrating between s and t. $\qquad \square$

The following result allows a generalisation of a portion of Theorem 1.33 to higher order Sobolev spaces. The result is not a direct corollary, and requires some additional argumentation, see Theorem 4 in Section 5.9.2 of Evans (1998) or Theorem 7.2 in Robinson (2001), for example.

Proposition 1.35 *If $\partial_t u \in L^2(0, T; H^{k-1})$ and $u \in L^2(0, T; H^{k+1})$ then we have $u \in C^0([0, T]; H^k)$ with*

$$\sup_{t \in [0,T]} \|u(t)\|_{H^k} \le C \left(\|u\|_{L^2(0,T;H^{k+1})} + \|\partial_t u\|_{L^2(0,T;H^{k-1})} \right).$$

Notes

The standard theory of Sobolev spaces summarised in this chapter can be found in many places, including Adams & Fournier (2003), Evans (1998), and Evans & Gariepy (1992). Fractional Sobolev spaces on bounded domains are naturally defined using the theory of interpolation spaces, see Bergh & Löfström (1976) or Lunardi (2009), for example. Some discussion of Bochner integration can be found in Rudin (1991) and Roubíček (2013), with full details given in Bourbaki (2004).

Exercises

1.1 By writing $ab = \exp(\log(ab))$ and using the fact that the exponential is a convex function, prove Young's inequality

$$ab \le \frac{a^p}{p} + \frac{b^q}{q} \quad \text{for} \quad \frac{1}{p} + \frac{1}{q} = 1, \quad 1 < p, q < \infty.$$

1.2 Given any p, q with $1 \le q < p \le \infty$, provide an example of a function $u \in L^p(\mathbb{R}^3)$ such that $u \notin L^q(\mathbb{R}^3)$.

1.3 Prove the three-term Hölder inequality

$$\int_\Omega |u(x)v(x)w(x)| \, dx \le \|u\|_{L^p} \|v\|_{L^q} \|w\|_{L^r}$$

for $u \in L^p$, $v \in L^q$, $w \in L^r$ with $p^{-1} + q^{-1} + r^{-1} = 1$.

1.4 Use the Hölder inequality to show that if $1 \le p \le r \le q \le \infty$ and $u \in L^p(\Omega) \cap L^q(\Omega)$ then $u \in L^r(\Omega)$ with

$$\|u\|_{L^r} \le \|u\|_{L^p}^\alpha \|u\|_{L^q}^{1-\alpha}, \quad \text{where} \quad \frac{1}{r} = \frac{\alpha}{p} + \frac{1-\alpha}{q}.$$

1.5 Let $u = (u_1, u_2, u_3)$, $v = (v_1, v_2, v_3)$, and $w = (w_1, w_2, w_3)$. Show that

$$|[(u \cdot \nabla)v] \cdot w| \le |u||\nabla v||w|,$$

where $|\nabla v| := \left(\sum_{i,j} |\partial_i v_j|^2 \right)^{1/2}$.

1.6 Let $0 \leq s_1 \leq s \leq s_2$, and take $0 \leq \theta \leq 1$ such that $s = \theta s_1 + (1 - \theta)s_2$. Use Hölder's inequality and the Fourier expansion of u to prove that if $u \in \dot{H}^{s_2}(\mathbb{T}^3)$ then

$$\|u\|_{\dot{H}^s} \leq \|u\|_{\dot{H}^{s_1}}^{\theta} \|u\|_{\dot{H}^{s_2}}^{1-\theta}.$$

1.7 Show that if $f : (0, \infty) \to \mathbb{R}$ is a non-increasing function then there are constants C_1 and C_2 such that

$$C_1 \int_1^{\infty} r^2 f(r) \, dr \leq \sum_{k \in \mathbb{Z}^3} f(|k|) \leq C_2 \left(f(1) + \int_1^{\infty} r^2 f(r) \, dr \right).$$

1.8 Show that

$$\int_{\mathbb{R}^d} \frac{1}{1 + |x|^{2s}} \, dx < \infty$$

if and only if $s > d/2$, and, using Exercise 1.7, show that

$$\sum_{k \in \mathbb{Z}^3} \frac{1}{1 + |k|^{2s}} < \infty$$

if and only if $s > 3/2$.

1.9 Prove that for each $s > 3/2$ there exists a constant C_s such that whenever $u \in H^s(\mathbb{T}^3)$ then $u \in C^0(\mathbb{T}^3)$ with

$$\|u\|_{L^\infty} \leq C_s \|u\|_{H^s}. \tag{1.32}$$

1.10 By splitting the Fourier expansion of u into low and high modes, prove that for any $s > 3/2$ and $0 \leq r < 3/2$ there is a constant $C_{s,r}$ such that if $u \in H^s(\mathbb{T}^3)$ then

$$\|u\|_{L^\infty} \leq C_{s,r} \|u\|_{H^r}^{\alpha} \|u\|_{H^s}^{1-\alpha}, \qquad \text{where} \quad \alpha = (s - \tfrac{3}{2})/(s - r). \tag{1.33}$$

1.11 Show that the function defined on $(0, 1) \times (0, 1)$ by

$$f(x, t) = \begin{cases} 0 & x < t \\ 1 & x \geq t, \end{cases}$$

which is clearly an element of $L^\infty((0, 1) \times (0, 1))$, is not an element of $L_t^\infty((0, 1); L_x^\infty(0, 1))$. [Hint: define $F : (0, 1) \to L_x^\infty(0, 1)$ by setting $F(t)(x) = f(x, t)$, and show that $F^{-1}(B_{1/2}(F(t))) = \{t\}$ for any $t \in (0, 1)$. Hence find an open subset U of $L_x^\infty(0, 1)$ such that $F^{-1}(U)$ is not measurable.] (This example is due to Juan Arias de Reyna.)

1.12 Suppose that $0 \leq l_1 \leq k_1 \leq k_2 \leq l_2$. Show that

$$\|u\|_{H^{k_1}} \|u\|_{H^{k_2}} \leq \|u\|_{H^{l_1}} \|u\|_{H^{l_2}} \quad \text{for} \quad k_1 + k_2 = l_1 + l_2. \quad (1.34)$$

1.13 Prove that the norm

$$\|u\|_{H^s}^2 = \|u\|^2 + \|u\|_{\dot{H}^s}^2$$

is equivalent to the standard norm $\|u\|_{W^{s,2}}$ on both $\mathbb{T}^3$ and $\mathbb{R}^3$ for $s \in \mathbb{N}$.

2

The Helmholtz–Weyl decomposition

Since solutions of the Navier–Stokes equations are required to be divergence free, it is very useful to define and characterise subspaces of the standard Lebesgue and Sobolev spaces that consist of divergence-free functions. Although we usually will not distinguish in notation between spaces of scalar and vector-valued functions, we make an exception in this chapter since now the distinction between scalars and vectors is crucial (e.g. in (2.1) below, g is a scalar and h a vector). Instead of using the more clumsy notation $[L^2]^3$ and $[H^s]^3$ we introduce

$$\mathbb{L}^2 := [L^2]^3 \qquad \text{and} \qquad \mathbb{H}^s := [H^s]^3.$$

Our goal is to show that any $u \in \mathbb{L}^2$ can be written in a unique way as the sum

$$u = h + \nabla g, \tag{2.1}$$

where the vector function h is divergence free (in a weak sense) and g belongs to H^1. In other words

$$\mathbb{L}^2 = H \oplus G,$$

where H is a space of divergence-free functions (satisfying an appropriate boundary condition) and G is the space of gradients of functions in H^1.

This theorem is a variant of a well-known result due to Helmholtz (1858) that a C^2 vector field f on $\mathbb{R}^3$ that decays sufficiently fast at infinity can be written uniquely as

$$f = \operatorname{curl} h + \nabla g.$$

Since $\operatorname{div}(\operatorname{curl} h) = 0$ the decomposition of a sufficiently regular function into the curl and the gradient is an example of the decomposition in which we are interested in this chapter.

Since we will usually consider the Navier–Stokes equations 'in the absence of boundaries', i.e. on $\mathbb{R}^3$ or on $\mathbb{T}^3$, we will prove the validity of the Helmholtz–Weyl decomposition in the case of a torus. The corresponding results for the whole space can be proved similarly (Exercise 2.4); we will discuss the case of a bounded domain more briefly. We will also introduce two concepts related to such a decomposition: the Leray projector $\mathbb{P}$ onto divergence-free functions, defined by setting $\mathbb{P}(u) = h$ when $u = h + \nabla g$, and the Stokes operator $A = -\mathbb{P}\Delta$.

2.1 The Helmholtz–Weyl decomposition on the torus

In the case of a torus we will decompose only the homogeneous space $\dot{\mathbb{L}}^2$ (rather than all of $\mathbb{L}^2$) since we will always include the zero-average condition when considering the Navier–Stokes equations in this setting.

In order to prove the decomposition $\dot{\mathbb{L}}^2 = H \oplus G$ we need to define the appropriate spaces H and G in this context. We begin with $H(\mathbb{T}^3)$. If u is given by $u(x) = \hat{u}_k \mathrm{e}^{\mathrm{i}k \cdot x}$ then a straightforward computation shows that

$$\mathrm{div}\, u(x) = \mathrm{i}(k \cdot \hat{u}_k)\mathrm{e}^{\mathrm{i}k \cdot x}.$$

So this function u is divergence free if and only if $\hat{u}_k$ is orthogonal to k. This leads us to the following definition.

Definition 2.1 We define the space $H = H(\mathbb{T}^3)$ as

$$\left\{ u \in \dot{\mathbb{L}}^2 : u = \sum_{k \in \dot{\mathbb{Z}}^3} \hat{u}_k \mathrm{e}^{\mathrm{i}k \cdot x}, \ \hat{u}_{-k} = \overline{\hat{u}_k} \text{ and } k \cdot \hat{u}_k = 0 \text{ for all } k \in \dot{\mathbb{Z}}^3 \right\}$$

and equip H with the $\mathbb{L}^2$-norm.

Since in this definition of the space H we implicitly consider functions given as the limits of smooth functions, we need to clarify in what sense these limits are 'divergence free'.

Definition 2.2 A function $u \in L^1(\Omega)$ is weakly divergence free if

$$\langle u, \nabla \varphi \rangle = 0 \quad \text{for every} \quad \varphi \in C_c^\infty(\Omega).$$

Lemma 2.3 *Each $u \in H(\mathbb{T}^3)$ is weakly divergence free and moreover*

$$\langle u, \nabla \varphi \rangle = 0 \quad \text{for all} \quad \varphi \in H^1(\mathbb{T}^3).$$

Proof If $u \in H$ is given by

$$u(x) = \sum_{l \in \mathbb{Z}^3} \hat{u}_l e^{il \cdot x}$$

and $\varphi_k(x) = e^{-ik \cdot x}$ then

$$\int_{\mathbb{T}^3} u(x) \cdot \nabla \varphi_k(x) \, dx = \int_{\mathbb{T}^3} \hat{u}_k e^{ik \cdot x} \cdot \nabla e^{-ik \cdot x} = -i \int_{\mathbb{T}^3} \hat{u}_k \cdot k = 0,$$

since $\hat{u}_k \cdot k = 0$ for all $k \in \dot{\mathbb{Z}}^3$. The result follows since any $\varphi \in H^1(\mathbb{T}^3)$ can be approximated arbitrarily closely in H^1 by real-valued finite linear combinations of the functions φ_k. $\qquad\square$

Definition 2.4 The space G on the torus $\mathbb{T}^3$ is defined as

$$G(\mathbb{T}^3) = \left\{ u \in \dot{\mathbb{L}}^2 : u = \nabla g, \text{ where } g \in \dot{H}^1(\mathbb{T}^3) \right\}.$$

We assume here that g belongs to $\dot{H}^1$ (rather than to H^1) in order to obtain the uniqueness of g in the resulting Helmholtz–Weyl decomposition.

Corollary 2.5 *The spaces H and G are orthogonal, i.e.*

$$\langle h, \nabla g \rangle = 0$$

for every $h \in H$ and $\nabla g \in G$.

We can now state the first result of this chapter, which is the existence of the Helmholtz–Weyl decomposition on the torus. In this simple case we are able to prove the existence of such a decomposition with very explicit calculations.

Theorem 2.6 (Helmholtz–Weyl decomposition on $\mathbb{T}^3$) *The space $\dot{\mathbb{L}}^2$ can be written as*

$$\dot{\mathbb{L}}^2 = H \oplus G,$$

i.e. every function $u \in \dot{\mathbb{L}}^2$ can be written in a unique way as

$$u = h + \nabla g, \tag{2.2}$$

where the vector-valued function h belongs to H and the scalar function g belongs to $\dot{H}^1$. Moreover,

(i) *the functions h and ∇g are orthogonal in $\mathbb{L}^2$,*

$$\langle h, \nabla g \rangle = 0,$$

  and

(ii) *if in addition u belongs to $\dot{\mathbb{H}}^s$, $s > 0$, then $h \in \dot{\mathbb{H}}^s$ and $g \in \dot{H}^{s+1}$.*

Proof Take $u \in \mathbb{\dot{L}}^2$ and write it in the form

$$u(x) = \sum_{k \in \mathbb{Z}^3} \hat{u}_k e^{ik \cdot x}.$$

Write the vector coefficients $\hat{u}_k$ as the linear combination of k and a vector w_k perpendicular to k in $\mathbb{C}^3$,

$$\hat{u}_k = \alpha_k k + w_k.$$

Here $\alpha_k = \hat{u}_k \cdot k / |k|^2 \in \mathbb{C}$, $w_k \in \mathbb{C}^3$, and $w_k \cdot k = 0$. Notice that

$$|\hat{u}_k|^2 = |\alpha_k|^2 |k|^2 + |w_k|^2. \tag{2.3}$$

We therefore have

$$\begin{aligned}
u(x) = \sum_{k \in \mathbb{Z}^3} \hat{u}_k e^{ik \cdot x} &= \sum_{k \in \mathbb{Z}^3} (\alpha_k k + w_k) e^{ik \cdot x} \\
&= \sum_{k \in \mathbb{Z}^3} (-i\alpha_k) \nabla e^{ik \cdot x} + \sum_{k \in \mathbb{Z}^3} w_k e^{ik \cdot x} \\
&= \nabla g(x) + h(x),
\end{aligned}$$

where

$$g(x) := \sum_{k \in \mathbb{Z}^3} (-i\alpha_k) e^{ik \cdot x} \quad \text{and} \quad h(x) := \sum_{k \in \mathbb{Z}^3} w_k e^{ik \cdot x}.$$

Since $w_k \cdot k = 0$ for every k we conclude that h is (weakly) divergence free. Thus to show that we have obtained the required decomposition as in (2.2) we only need to show that $g \in \dot{H}^1$ and $h \in \mathbb{\dot{L}}^2$. To this end we notice that

$$\|g\|_{\dot{H}^1}^2 = \sum_{k \in \mathbb{Z}^3} |\alpha_k|^2 |k|^2, \quad \text{and} \quad \|h\|^2 = \sum_{k \in \mathbb{Z}^3} |w_k|^2.$$

Taking into account (2.3) we get

$$\|h\|^2 + \|\nabla g\|^2 = \|u\|^2.$$

From this equality (and also directly from Corollary 2.5) it follows that g belongs to $\dot{H}^1$, and h belongs to H.

Multiplying (2.3) by $|k|^{2s}$ it follows that if $u \in \mathbb{\dot{H}}^s$ then $h \in \mathbb{\dot{H}}^s$ and $g \in \dot{H}^{s+1}$.

Finally, the uniqueness of this representation follows easily, since if

$$u = h_1 + \nabla g_1 = h_2 + \nabla g_2$$

then, using Corollary 2.5, we have

$$(h_1 - h_2) + (\nabla g_1 - \nabla g_2) = 0 \quad \Rightarrow \quad \|h_1 - h_2 + \nabla g_1 - \nabla g_2\|^2 = 0$$
$$\Rightarrow \quad \|h_1 - h_2\|^2 + \|\nabla g_1 - \nabla g_2\|^2 = 0,$$

so $h_1 = h_2$ and $\nabla g_1 = \nabla g_2$. Since both g_1 and g_2 have mean zero it follows that $g_1 = g_2$. $\qquad\square$

We notice that the decomposition of $\mathbb{L}^2(\mathbb{T}^3)$ is now straightforward, since we obviously have

$$\mathbb{L}^2(\mathbb{T}^3) = \mathbb{R}^3 \oplus H(\mathbb{T}^3) \oplus G(\mathbb{T}^3)$$

(such a decomposition is also unique; see Exercise 2.5).

Theorem 2.7 *The space $\mathbb{L}^2(\mathbb{R}^3)$ can be decomposed as*

$$\mathbb{L}^2(\mathbb{R}^3) = H(\mathbb{R}^3) \oplus G(\mathbb{R}^3),$$

where

$$H(\mathbb{R}^3) := \{u : \ u \in \mathbb{L}^2(\mathbb{R}^3), \ \operatorname{div} u = 0\},$$
$$G(\mathbb{R}^3) := \{w \in \mathbb{L}^2(\mathbb{R}^3) : \ w = \nabla g \ \text{and} \ g \in \dot{H}^1(\mathbb{R}^3)\}.$$

Proof See Exercise 2.4. $\qquad\square$

The Helmholtz–Weyl decomposition allows us to define the 'Leray projector', the orthogonal projector onto the space of divergence-free functions.

Definition 2.8 On the torus and on the whole space the Leray projector $\mathbb{P}$ is given by

$$\mathbb{P}u = v \quad \Leftrightarrow \quad u = v + \nabla w,$$

where $v \in H$ and $\nabla w \in G$.

On the whole space we have

$$\mathbb{P} \colon \mathbb{L}^2(\mathbb{R}^3) \to H(\mathbb{R}^3),$$

while on the torus we prefer to have a restricted domain of $\mathbb{P}$:

$$\mathbb{P} \colon \dot{\mathbb{L}}^2(\mathbb{T}^3) \to H(\mathbb{T}^3),$$

because on $\mathbb{T}^3$ we will always consider solutions of the Navier–Stokes equations with zero mean.

The Leray projector can be computed in a very straightforward way when we consider functions in the absence of boundaries. For example, if u belongs to $\dot{\mathbb{L}}^2(\mathbb{T}^3)$ and is given by $u(x) = \sum_{k \in \mathbb{Z}^3} \hat{u}_k e^{ik \cdot x}$ then the Leray projection of u is given by the formula

$$\mathbb{P}u(x) = \sum_{k \in \mathbb{Z}^3} \left(\hat{u}_k - \frac{\hat{u}_k \cdot k}{|k|^2} k \right) e^{ik \cdot x}.$$

On the torus (and on the whole space) the Leray projector commutes with any derivative. This follows from the linearity of $\mathbb{P}$ and the fact that differentiation of a function u given by $u(x) = \hat{u}_k e^{ik \cdot x}$ reduces to multiplication by a constant.

Lemma 2.9 *The Leray projector $\mathbb{P}$ on the torus and on the whole space commutes with any derivative:*

$$\mathbb{P}(\partial_j u) = \partial_j(\mathbb{P}u) \qquad j = 1, 2, 3$$

for all $u \in \dot{\mathbb{H}}^1$.

Proof We consider the case of a torus; for the proof in case of the whole space see Exercise 2.6. Let u be given by $u(x) = \hat{u}_k e^{ik \cdot x}$. For any such u we have[1]

$$\mathbb{P}\partial_j(u) = \mathbb{P}\left(ik_j u\right) = ik_j \mathbb{P}u = \partial_j \mathbb{P}u.$$

By the linearity of $\mathbb{P}$ the result holds for any finite combination of such functions. The general case now follows since $\mathbb{P} \colon \dot{H}^s \to \dot{H}^s$, $s = 0, 1$, and $\partial_j \colon \dot{\mathbb{H}}^1 \to \dot{\mathbb{L}}^2$ are all continuous. $\qquad\square$

2.2 The Helmholtz–Weyl decomposition in $\Omega \subset \mathbb{R}^3$

We will now show how the results of the previous section can be adapted to the case when Ω is an open bounded set in $\mathbb{R}^3$, only sketching the proofs. In order to define the space $H = H(\Omega)$ we first introduce the space $C_{c,\sigma}^\infty$ of divergence-free smooth functions with compact support in Ω,

$$C_{c,\sigma}^\infty(\Omega) := \{\varphi \in [C_c^\infty(\Omega)]^3 : \ \operatorname{div} \varphi = 0\}.$$

Definition 2.10 Let Ω be any open subset of $\mathbb{R}^3$ (bounded or unbounded). The space $H = H(\Omega)$ is given by

$$H := \text{the completion of } C_{c,\sigma}^\infty(\Omega) \text{ in the norm of } \mathbb{L}^2(\Omega).$$

We equip the space H with the $\mathbb{L}^2$-norm.

---

[1] We are ignoring here the fact that u of this form is in fact complex valued. To make the proof fully rigorous we could consider $u(x) = \hat{u}_k e^{ik \cdot x} + \hat{u}_{-k} e^{-ik \cdot x}$.

(We note that when $\Omega = \mathbb{R}^3$ this gives an equivalent definition of the space $H(\mathbb{R}^3)$ defined already in Theorem 2.7.)

Since the divergence-free condition is satisfied by every function in $C_{c,\sigma}^\infty(\Omega)$ it is easy to see that the limit functions (i.e. elements of H) must also be (weakly) divergence free. In fact a little more is true, as we now show.

Lemma 2.11 *The functions in $H(\Omega)$ are weakly divergence free. Moreover if $u \in H$ then*

$$\int_\Omega u(x) \cdot \nabla \psi(x) \, dx = 0 \qquad for\ all \qquad \psi \in H^1(\Omega). \qquad (2.4)$$

Proof   Let $u \in H$, $\psi \in H^1(\Omega)$, $\varphi_n \in C_{c,\sigma}^\infty(\Omega)$ with $\varphi_n \to u$ in H. Then

$$\int_\Omega \varphi_n(x) \cdot \nabla \psi(x) \, dx = -\int_\Omega (\text{div}\, \varphi_n)(x)\psi(x) \, dx = 0$$

for every n, and to finish the proof it is enough to pass to the limit as $n \to \infty$.   □

Another key property of functions in $H(\Omega)$ is that their normal component is well defined at the boundary (in some weak sense). We will formulate a slightly stronger result concerning the larger space

$$E(\Omega) := \{u \in \mathbb{L}^2(\Omega) : \ \text{div}\, u \in L^2(\Omega)\},$$

which we equip with the norm $\|u\|_E := \|u\| + \|\text{div}\, u\|$. On a smooth bounded domain the space E can also be defined as the completion of $C^1(\overline{\Omega})$ in the norm $\|\cdot\|_E$.

Lemma 2.12 *Let Ω be a bounded open set with smooth boundary.[2] If $u \in E(\Omega)$ then the normal component of u is well defined on the boundary as a bounded linear functional on the space of traces of functions in $H^1(\Omega)$ and the Gauss formula holds*

$$\langle \text{div}\, u, v \rangle + \langle u, \nabla v \rangle = \int_{\partial\Omega} (u \cdot n)v, \qquad u \in E(\Omega), \ v \in H^1(\Omega).$$

Proof   We sketch the argument; for more details see Proposition 1.4 in Constantin & Foias (1988) or Theorem 1.2 in Temam (1977).

Suppose that $\varphi_n \in C^1(\overline{\Omega})$ with

$$\varphi_n \to u \ \text{in} \ \mathbb{L}^2 \ \text{and} \ \text{div}\, \varphi_n \to \text{div}\, u \ \text{in} \ L^2.$$

---

[2] The smoothness of boundary can be easily relaxed to, for example, C^2 regularity.

Then from the Divergence Theorem it follows that

$$\int_\Omega \text{div}\,(\tilde{v}\varphi_n) = \langle \text{div}\,\varphi_n, \tilde{v} \rangle + \langle \varphi_n, \nabla \tilde{v} \rangle = \int_{\partial\Omega} [\varphi_n(s) \cdot n(s)]v(s)\,ds$$

for any $\tilde{v} \in H^1(\Omega)$ and $v = \tilde{v}|_{\partial\Omega}$ in the sense of trace. The limit (as $n \to \infty$) on the left-hand side is well defined, so we can introduce a bounded linear functional $u \cdot n$ acting on the space of traces of functions in H^1 (that is on the space $H^{1/2}(\partial\Omega)$) by

$$F_u(v) = \langle u \cdot n, v \rangle_{\partial\Omega} := \langle \text{div}\,u, \tilde{v} \rangle + \langle u, \nabla \tilde{v} \rangle, \qquad (2.5)$$

where $\tilde{v} \in H^1(\Omega)$ and $\tilde{v}|_{\partial\Omega} = v$. This definition[3] does not depend on the choice of $\tilde{v}$ because if $\tilde{v}_1, \tilde{v}_2 \in H^1$ and $\tilde{v}_1|_{\partial\Omega} = \tilde{v}_2|_{\partial\Omega} = v$ then $\tilde{v}_1 - \tilde{v}_2 \in H_0^1$ and

$$\langle \text{div}\,u, \tilde{v}_1 - \tilde{v}_2 \rangle + \langle u, \nabla(\tilde{v}_1 - \tilde{v}_2) \rangle = 0$$

so

$$F_u(v) = \langle \text{div}\,u, \tilde{v}_1 \rangle + \langle u, \nabla \tilde{v}_1 \rangle = \langle \text{div}\,u, \tilde{v}_2 \rangle + \langle u, \nabla \tilde{v}_2 \rangle.$$

Hence it is easy to see that $F_u(v)$ is well defined for every $u \in E(\Omega)$ and every $v \in H^{1/2}(\partial\Omega)$ and that if $u \in C^1(\overline{\Omega})$ then the functional F_u is given by a normal component of u at the boundary. $\qquad\qquad\square$

Corollary 2.13 *Let Ω be a smooth bounded domain. If $u \in H(\Omega)$ then we have $u \cdot n = 0$ on $\partial\Omega$ in the sense of Lemma 2.12.*

Proof Since $\text{div}\,u = 0$ for $u \in H$, the corollary follows immediately from (2.4) and (2.5). $\qquad\qquad\square$

We have therefore shown that

$$u \in H \quad \Rightarrow \quad u \in \mathbb{L}^2, \; \text{div}\,u = 0 \; \text{in} \; \Omega \; \text{and} \; u \cdot n = 0 \; \text{on} \; \partial\Omega.$$

The converse is also true for a large class of domains as stated in the following lemma (see Theorem III.2.3 in Galdi, 2011).

Lemma 2.14 *If Ω is a bounded open set in $\mathbb{R}^3$ with Lipschitz boundary then H coincides with the space of divergence-free functions in $\mathbb{L}^2$ such that $u \cdot n = 0$ on $\partial\Omega$, where n is the normal vector to the boundary.*

We now introduce the orthogonal complement of H in $\mathbb{L}^2(\Omega)$.

---

[3] More precisely, $\tilde{v} = l(v)$, where l is a lifting operator from $H^{1/2}(\partial\Omega)$ into $H^1(\Omega)$.

Definition 2.15 Let Ω be an open bounded set in $\mathbb{R}^3$ with Lipschitz boundary. The space $G = G(\Omega)$ is defined as

$$G := \{w \in \mathbb{L}^2(\Omega) : \ w = \nabla g \text{ and } g \in H^1(\Omega)\}.$$

We now give a sketch of the proof of the Helmholtz–Weyl decomposition of $\mathbb{L}^2(\Omega)$. Details and further discussion can be found, for example, in Temam (1977, Theorems 1.4 and 1.5).

Theorem 2.16 (Helmholtz–Weyl decomposition on $\Omega \subset \mathbb{R}^3$) *Let Ω be an open bounded set in* $\mathbb{R}^3$ *with Lipschitz boundary. Then*

$$\mathbb{L}^2(\Omega) = H(\Omega) \oplus G(\Omega),$$

where H and G are given by Definitions 2.10 and 2.15.

Proof Given $u \in \mathbb{L}^2(\Omega)$, let $\phi_1 \in H^1$ be the weak solution of

$$\Delta \phi_1 = \operatorname{div} u \quad \text{in } \Omega, \qquad \phi_1|_{\partial\Omega} = 0.$$

(Notice that the problem is well posed since $\operatorname{div} u \in H^{-1}$ when $u \in \mathbb{L}^2$.) Then

$$u = v + \nabla \phi_1,$$

where v is divergence free since

$$\operatorname{div} v = \operatorname{div} u - \Delta \phi_1 = 0.$$

However, we do not yet have $v \in H(\Omega)$, since it may have non-zero normal component on $\partial\Omega$. Therefore we need to find the Helmholtz–Weyl decomposition of v

$$v = h + \nabla \phi_2,$$

where $h \in H$. The key ingredient is the fact that when $v \in \mathbb{L}^2(\Omega)$ is divergence free the boundary condition $v \cdot n$ can be understood in the sense of Lemma 2.12. Since h must have zero normal component on $\partial\Omega$ we require

$$v \cdot n = \nabla \phi_2 \cdot n \quad \text{on } \partial\Omega,$$

see Figure 2.1.

Furthermore, v and h are divergence free so $\nabla \phi_2$ must also be divergence free, and it follows that ϕ_2 is harmonic. Thus we look for a (weak) solution $\phi_2 \in H^1(\Omega)$ of the Neumann problem

$$\Delta \phi_2 = 0 \quad \text{in } \Omega \qquad \text{with} \qquad \nabla \phi_2 \cdot n = v \cdot n \quad \text{on } \partial\Omega, \qquad (2.6)$$

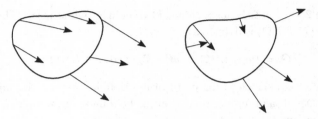

Figure 2.1. If a vector field v (see the left picture) is divergence free and we have $v = h + \nabla\phi_2$, where $h \in H$, then ϕ_2 is harmonic and the normal derivative of ϕ_2 is equal on the boundary to the normal component of v (right picture).

and define $h = u - \nabla(\phi_1 + \phi_2)$. It is easy to see that $\nabla \cdot h = 0$ and $h \cdot n = 0$. Furthermore

$$\|u\|^2 = \langle h + \nabla(\phi_1 + \phi_2), h + \nabla(\phi_1 + \phi_2) \rangle = \|h\|^2 + \|\nabla(\phi_1 + \phi_2)\|^2$$

since h and $\nabla(\phi_1 + \phi_2)$ are orthogonal in $\mathbb{L}^2$. $\qquad\square$

The definition of the Leray projector is not changed in the case of $\Omega \subseteq \mathbb{R}^3$.

Definition 2.17 Let Ω be an open set in $\mathbb{R}^3$. Then the Leray projector

$$\mathbb{P} \colon \mathbb{L}^2 \to H$$

is defined by

$$\mathbb{P}(u) = h \quad \Leftrightarrow \quad u = h + \nabla g, \quad u \in H, \ \nabla g \in G.$$

Similarly, as in the case of a torus, if Ω is a smooth bounded domain and $u \in \mathbb{H}^m(\Omega)$ then

$$\|\mathbb{P}u\|_{\mathbb{H}^m} \leq c\|u\|_{\mathbb{H}^m}$$

(see Remark 1.6. in Temam, 1977). It is a trivial observation that the Leray projector is symmetric.

Theorem 2.18 *On every domain*

$$\langle \mathbb{P}u, v \rangle = \langle u, \mathbb{P}v \rangle \quad \text{for all} \ u, v \in \mathbb{L}^2.$$

Proof If $u = h_1 + \nabla g_1$ and $v = h_2 + \nabla g_2$ then

$$\langle \mathbb{P}u, v \rangle = \langle h_1, h_2 + \nabla g_2 \rangle = \langle h_1, h_2 \rangle = \langle h_1 + \nabla g_1, h_2 \rangle = \langle u, \mathbb{P}v \rangle,$$

so $\mathbb{P}$ is symmetric. $\qquad\square$

However, unlike in the periodic and whole space cases, the Leray projector defined on $\Omega \subset \mathbb{R}^3$ does not necessarily commute with derivatives, as the following example shows.

Example 2.19 Consider $u(x_1, x_2, x_3) = (x_2, -x_1, 0)$ on the cylinder

$$Q = \{(x_1, x_2, x_3) \in \mathbb{R}^3 : x_1^2 + x_2^2 < 1, -1 < x_3 < 1\}.$$

Then u is divergence free and its normal component is zero on ∂Q, and so $\mathbb{P}u = u$. It follows that

$$\partial_1 \mathbb{P}u = \partial_1 u = (0, -1, 0),$$

while

$$\mathbb{P}\partial_1 u = \mathbb{P}(0, -1, 0) = (0, 0, 0),$$

since $(0, -1, 0) = \nabla(-x_2)$.

2.3 The Stokes operator

Having defined the Leray projector we can now define another key object in the theory of the Navier–Stokes equations, the Stokes operator. To this end let us first introduce the space of divergence-free functions in $\mathbb{H}^1$ that vanish on the boundary.

Definition 2.20 On the whole space and on the torus the space V is given by

$$V := H \cap \mathbb{H}^1.$$

On a smooth bounded domain in $\mathbb{R}^3$ we define $V = V(\Omega)$ by

$$V := H \cap \mathbb{H}_0^1.$$

The space V is equipped with the norm $\| \cdot \|_V = \| \cdot \|_{\mathbb{H}^1}$.

In the absence of boundaries the only difference between the spaces H and V is the 'upgraded' regularity of functions in V. When Ω is a bounded open subset of $\mathbb{R}^3$ there is also another difference that comes from the boundary conditions: every function in V vanishes on the boundary, while for functions in H only their normal component is required to be zero at $\partial\Omega$.

Definition 2.21 The Stokes operator A is defined by

$$Au = -\mathbb{P}\Delta u,$$

with its domain given by

$$D(A) = V \cap \mathbb{H}^2.$$

Let us first take a closer look at the Stokes operator defined on the torus, which is the simplest case. Since the Leray projector on $\mathbb{T}^3$ commutes with derivatives we have

$$-\mathbb{P}\Delta u = -\Delta \mathbb{P}u = -\Delta u$$

for all $u \in D(A)$ on $\mathbb{T}^3$; a similar result holds on the whole space.

Theorem 2.22 *On the whole space $\mathbb{R}^3$ and on the torus $\mathbb{T}^3$ we have*

$$-\mathbb{P}\Delta u = -\Delta \mathbb{P}u$$

for all $u \in \mathbb{H}^2$. Hence

$$Au = -\Delta u$$

for all $u \in D(A)$ on $\mathbb{T}^3$ and $\mathbb{R}^3$.

From Theorem 2.22 it follows that on the torus the Stokes operator is given by the explicit formula

$$Au(x) = \sum_{k \in \dot{\mathbb{Z}}^3} |k|^2 \hat{u}_k e^{ik \cdot x}, \quad u \in D(A),$$

so $Au = \Lambda^2 u$, where Λ is given by (1.18). However, in the case of a bounded domain, the Leray projector does not commute with derivatives, and we cannot exclude the possibility that $-\mathbb{P}\Delta u \neq -\Delta \mathbb{P}u$ (see Exercise 2.8).

Suppose now that Ω is a smooth bounded domain. It follows from the definition of A that $Au = f \in H$ if and only if there exists a function $p \in H^1$ such that u and p solve the so-called *Stokes problem*

$$\begin{aligned}
-\Delta u + \nabla p &= f \quad \text{in } \Omega, \\
\operatorname{div} u &= 0 \quad \text{in } \Omega, \\
u &= 0 \quad \text{on } \partial\Omega.
\end{aligned}$$

Multiplying the equation $-\Delta u + \nabla p = f$ by $v \in V$ and integrating over Ω we obtain the weak formulation of the Stokes problem

$$\langle \nabla u, \nabla v \rangle = \langle f, v \rangle \quad \text{for all } v \in V.$$

In a similar way as can be done for the Laplace operator one can prove the existence and uniqueness of solutions of this system of equations (see

Exercise 2.9) and the regularity estimate

$$\|u\|_{\mathbb{H}^{m+2}} + \|p\|_{H^{m+1}} \le c\|f\|_{\mathbb{H}^m},$$

which is valid under the assumption that $f \in \mathbb{H}^m$ when the domain has C^{m+2} boundary. This is much harder to prove; see, for example, Proposition 2.2 in Temam (1977) which uses Theorem 10.5 in Agmon, Douglis, & Nirenberg (1964) to provide essential elliptic estimates. Since $f = Au$ we have

$$\|u\|_{\mathbb{H}^{m+2}} \le c\|Au\|_{\mathbb{H}^m}$$

if $Au \in \mathbb{H}^m$. In particular (when $m = 0$)

$$\|u\|_{\mathbb{H}^2} \le c\|Au\|$$

for all $u \in D(A)$. This estimate is one of the most often used in the theory of the Navier–Stokes equations. It is valid also on $\mathbb{T}^3$, while on the whole space the $\mathbb{L}^2$ norm of Au bounds the $\mathbb{L}^2$ norm of all second derivatives of u.

Theorem 2.23 *On the torus $\mathbb{T}^3$ and on a smooth bounded domain we have*

$$\|u\|_{\mathbb{H}^{m+2}} \le c\|Au\|_{\mathbb{H}^m}$$

for all $u \in V$ such that $Au \in \mathbb{H}^m$ and all $m \in \mathbb{N}$. On the whole space we have

$$\|\partial_i \partial_j u\| \le c\|Au\| \quad \text{for all } u \in D(A), \ i, j = 1, 2, 3.$$

(Recall that on the torus the definition of $D(A)$ includes the condition that $\int_{\mathbb{T}^3} u = 0$.)

Proof On a smooth bounded domain the result follows, as we mentioned before, from the regularity results for the Stokes problem (for more details see Galdi, 2011).

On the torus we have

$$\|Au\|_{\mathbb{H}^m}^2 = \left\| \sum_{k \in \mathbb{Z}^3} |k|^2 \hat{u}_k e^{ik \cdot x} \right\|_{\mathbb{H}^m}^2$$
$$= (2\pi)^3 \sum_{k \in \mathbb{Z}^3} |k|^{2m} |k|^4 |\hat{u}_k|^2 = \|u\|_{\mathbb{H}^{m+2}}^2.$$

Since for every $u \in D(A)$ on the torus we have $\int_{\mathbb{T}^3} u = 0$ it follows that the homogenous norm $\| \cdot \|_{\dot{H}^m}$ on $\mathbb{T}^3$ is equivalent to the full norm $\| \cdot \|_{H^m}$ and the assertion is proved.

On the whole space

$$\|\partial_i \partial_j u\| = \|(2\pi)^2 k_i k_j \hat{u}(k)\| \le \|(2\pi)^2 |k|^2 \hat{u}(k)\| = \|Au\|$$

for $u \in D(A)$.   □

Consider now the case of the torus, and write $u \in D(A)$ as

$$u(x) = \sum_{k \in \mathbb{Z}^3} \hat{u}_k e^{ik \cdot x}.$$

This expansion can be viewed as an expansion into eigenfunctions of the Stokes operator, since

$$A\left(\hat{u}_k e^{ik \cdot x}\right) = -\mathbb{P}\Delta \hat{u}_k e^{ik \cdot x} = \mathbb{P}\hat{u}_k |k|^2 e^{ik \cdot x} = \hat{u}_k |k|^2 e^{ik \cdot x},$$

where we have used the fact that $\hat{u}_k \cdot k = 0$ for $u \in H$. So every divergence-free function u, given by $u(x) = \hat{u}_k e^{ik \cdot x}$, where $k \cdot \hat{u}_k = 0$, is an 'eigenfunction' of the Stokes operator corresponding to the eigenvalue $|k|^2$. These are not true eigenfunctions since they are complex valued, but this is easily circumvented. In fact the following theorem states that when the domain of the flow is bounded then every $u \in H$ can be approximated arbitrarily well by finite linear combinations of the eigenfunctions of the Stokes operator.

Theorem 2.24 *Let Ω be the torus $\mathbb{T}^3$, or a smooth bounded domain in $\mathbb{R}^3$. Then there exists a family of functions $\mathcal{N} = \{a_1, a_2, a_3, \ldots\}$ such that*

(i) *$\mathcal{N}$ is an orthonormal basis in $H(\Omega)$;*
(ii) *$a_j \in D(A) \cap C^\infty(\overline{\Omega})$ are eigenfunctions of the Stokes operator, that is $Aa_j = \lambda_j a_j$ for all $j \in \mathbb{N}$ with*

$$0 < \lambda_1 \le \lambda_2 \le \lambda_3 \le \cdots \le \lambda_j \le \cdots \qquad \text{and} \qquad \lambda_j \to \infty; \quad \text{and}$$

(iii) *$\mathcal{N}$ is an orthogonal basis in $V(\Omega)$.*

Proof To show (i) and (ii) when $\Omega \subset \mathbb{R}^3$ is a smooth bounded domain, one can use the Rellich–Kondrachov Theorem to show that A^{-1} is compact. Moreover, A^{-1} is also positive and self-adjoint, so the assertions (i) and (ii) of Theorem 2.24 follow from the Hilbert–Schmidt Theorem and the regularity results for A^{-1} (for more details see, for example, Section 2.6 in Temam, 1977).

In the case of the torus we can give a very direct and simple proof of (i) and (ii) by just rewriting the Fourier expansion in an appropriate form. First for each pair of vectors $k \in \mathbb{Z}^3$ and $-k \in \mathbb{Z}^3$ we choose vectors m_k and m_{-k} in such a way that they are orthogonal to each other and also perpendicular to k, see Figure 2.2.

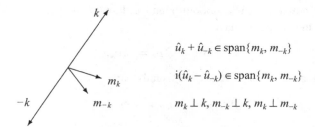

Figure 2.2. The orthogonal vectors m_k and m_{-k} span the plane perpendicular to k.

Take any $u \in H$ and let the Fourier coefficients of u be given by $\hat{u}_k$, $k \in \dot{\mathbb{Z}}^3$, with $\hat{u}_k \cdot k = 0$. Then

$$
\begin{aligned}
u(x) &= \sum_{k \in \dot{\mathbb{Z}}^3} \hat{u}_k \mathrm{e}^{\mathrm{i}k \cdot x} \\
&= \frac{1}{2} \sum_{k \in \dot{\mathbb{Z}}^3} \hat{u}_k \mathrm{e}^{\mathrm{i}k \cdot x} + \frac{1}{2} \sum_{k \in \dot{\mathbb{Z}}^3} \hat{u}_{-k} \mathrm{e}^{-\mathrm{i}k \cdot x} \\
&= \sum_{k \in \dot{\mathbb{Z}}^3} \underbrace{\frac{1}{2}(\hat{u}_k + \hat{u}_{-k})}_{\in \mathbb{R}^3} \cos(k \cdot x) + \sum_{k \in \dot{\mathbb{Z}}^3} \underbrace{\frac{1}{2}\mathrm{i}(\hat{u}_k - \hat{u}_{-k})}_{\in \mathbb{R}^3} \sin(k \cdot x) \\
&= \sum_{k \in \dot{\mathbb{Z}}^3} (a_k m_k + b_k m_{-k}) \cos(k \cdot x) + \sum_{k \in \dot{\mathbb{Z}}^3} (c_k m_k + d_k m_{-k}) \sin(k \cdot x) \\
&= \sum_{k \in \dot{\mathbb{Z}}^3} (a_k m_k + b_{-k} m_k) \cos(k \cdot x) + \sum_{k \in \dot{\mathbb{Z}}^3} (c_k m_k - d_{-k} m_k) \sin(k \cdot x) \\
&= \sum_{k \in \dot{\mathbb{Z}}^3} \alpha_k m_k \cos(k \cdot x) + \sum_{k \in \dot{\mathbb{Z}}^3} \beta_k m_k \sin(k \cdot x),
\end{aligned}
$$

where we have used the fact that cosine is even and sine is odd.

Normalising the vectors $m_k \cos(k \cdot x)$ and $m_k \sin(k \cdot x)$ in $\mathbb{L}^2$ we obtain an orthonormal basis of H consisting of smooth eigenfunctions of the Stokes operator.

Finally to prove assertion (iii) suppose that $u, v \in \mathcal{N}$ on the torus or on a smooth bounded domain. Then

$$
\langle \nabla u, \nabla v \rangle = \langle u, -\Delta v \rangle = \langle \mathbb{P}u, -\Delta v \rangle = \langle u, -\mathbb{P}\Delta v \rangle = \langle u, Av \rangle = \lambda_v \langle u, v \rangle,
$$

where λ_v is the eigenvalue corresponding to v. Since $\mathcal{N}$ is orthonormal in H it follows now that $\mathcal{N}$ must be orthogonal in $\mathbb{H}^1$, as required. $\qquad \square$

Suppose now that Ω is a smooth bounded domain and that $a_k \in \mathcal{N}$. Let us notice that in general we have

$$-\Delta a_k + \nabla p_k = \lambda_k a_k,$$

where p_k does not vanish. Hence we usually have

$$A a_k - (-\Delta a_k) = \nabla p_k \neq 0,$$

so the Leray projector and the negative Laplacian on a bounded domain Ω do not commute even on $D(A)$. (On the torus we have $p_k = 0$ for every k.)

Furthermore, we note that

$$A(A a_k) = A(\lambda_k a_k) = \lambda_k^2 a_k, \quad A a_k = \lambda_k a_k \in V,$$

so a_k are also eigenfunctions of the operator A^2 with the corresponding eigenvalues λ_k^2. Generalising this observation we can define the powers of the Stokes operator on the torus and in a smooth bounded domain Ω in a very straightforward way.

Definition 2.25 Let $u \in V$ on the torus $\mathbb{T}^3$ or in a smooth bounded domain in $\mathbb{R}^3$ be given by

$$u = \sum_{j=1}^{\infty} \tilde{u}_j a_j, \tag{2.7}$$

where $a_j \in \mathcal{N}$ and $\tilde{u}_j \in \mathbb{R}$. For $\alpha \geq 0$ define the powers of the Stokes operator by

$$A^{\alpha} u = \sum_{j=1}^{\infty} \lambda_j^{\alpha} \tilde{u}_j a_j,$$

where $\lambda_j > 0$ and $a_j \in \mathcal{N}$ are as in Theorem 2.24. The domain of the operator

$$A^{\alpha} : D(A^{\alpha}) \to H$$

is the subset of H consisting of all u given by (2.7) such that

$$\sum_{j=1}^{\infty} \lambda_j^{2\alpha} |\tilde{u}_j|^2 < \infty.$$

Lemma 2.26 *Let Ω be a smooth bounded domain or the torus. If $u, v \in D(A^s)$ then*

$$\langle A^{s_1} u, A^{s_2} v \rangle = \langle u, A^s v \rangle, \quad s_1 + s_2 = s, \quad s_1 \geq 0, \quad s_2 \geq 0.$$

Proof See Exercise 2.10. $\qquad\qquad\qquad\qquad\qquad\qquad\qquad\qquad\Box$

In the case of the torus the domain of A^α coincides with $\mathbb{H}^{2\alpha} \cap H$, due to the following result.

Theorem 2.27 *In the case of the torus* $D(A^{s/2}) = \dot{H}^s(\mathbb{T}^3)$ *and*

$$\|u\|_{\dot{H}^s} = \|A^{s/2}u\|.$$

Proof Simply note that $A^{s/2}u = \Lambda^s u$, which follows from (1.18) and the argument used in the proof of Theorem 2.24.   □

When Ω is a smooth bounded domain and $u \in D(A)$ we obtain, using Exercise 2.10,

$$
\begin{aligned}
\|\nabla u\|^2 &= \langle \nabla u, \nabla u \rangle = \langle Au, u \rangle \\
&= \langle A^{1/2}u, A^{1/2}u \rangle = \|A^{1/2}u\|^2,
\end{aligned}
$$

so the norm of u in H^1 is equivalent to the L^2 norm of $A^{1/2}u$ for $u \in D(A)$ (recall that if $u \in D(A)$ then u vanishes on the boundary). In consequence

$$D(A^{1/2}) = V.$$

Such a simple characterisation of $D(A^{1/2})$ implies that, even without referring to Theorem 2.23, we can derive a Poincaré-type inequality between $\|Au\|$ and $\|\nabla u\|$ for $u \in D(A)$ (see Exercise 2.11).

However, in general the characterisation of $D(A^\alpha)$ in the case of a smooth bounded domain is more complicated. If $m \geq 2$ is even then

$$\mathbb{H}^m(\Omega) \cap \mathbb{H}_0^{m-1}(\Omega) \cap V \subseteq D(A^{m/2}) \subseteq \mathbb{H}^m \cap V,$$

and if $m \geq 1$ is odd then

$$\mathbb{H}_0^m(\Omega) \cap V \subseteq D(A^{m/2}) \subseteq \mathbb{H}^m \cap V,$$

where $\mathbb{H}_0^m$ is the completion of $C_c^\infty(\Omega)$ in the norm of $\mathbb{H}^m$. For the proof and more details see Proposition 4.12 in Constantin & Foias (1988) and Exercises 2.12 and 2.13.

2.4 The Helmholtz–Weyl decomposition of $\mathbb{L}^q$

Finally, we will discuss (and prove in the case of $\mathbb{T}^3$) a much more difficult result, which is the Helmholtz–Weyl decomposition of the space $\mathbb{L}^q(\Omega)$, where $1 < q < \infty$. To this end let us first introduce the spaces H_q and G_q. On an open set Ω the space H_q is defined by

$$H_q := \text{the completion of } C_{c,\sigma}^\infty(\Omega) \text{ in the norm of } \mathbb{L}^q$$

and the space G_q is given by

$$G_q := \{u \in \mathbb{L}^q : u = \nabla g, \ g \in L^q_{\text{loc}}(\Omega)\}.$$

In the case of the torus we decompose the space

$$\dot{\mathbb{L}}^q(\mathbb{T}^3) = \left\{u \in \mathbb{L}^q(\mathbb{T}^3) : \int_{\mathbb{T}^3} u = 0\right\}$$

into

$$H_q(\mathbb{T}^3) = \left\{u \in \mathbb{L}^q(\mathbb{T}^3) : \text{div } u = 0\right\} \quad \text{and}$$

$$G_q(\mathbb{T}^3) = \left\{\nabla g : g \in W^{1,q}(\mathbb{T}^3), \int_{\mathbb{T}^3} g = 0\right\}.$$

The proof of such a decomposition in the periodic setting is based on the same splitting of u into two infinite sums, known already from the $\dot{\mathbb{L}}^2$ case (see Section 2.1):

$$\sum_{k \in \mathbb{Z}^3} \hat{u}_k e^{ik \cdot x} = \sum_{k \in \mathbb{Z}^3} \left(\hat{u}_k - \frac{\hat{u}_k \cdot k}{|k|^2} k\right) e^{ik \cdot x} + \nabla \sum_{k \in \mathbb{Z}^3} \left(\frac{\hat{u}_k \cdot k}{|k|^2}\right) e^{ik \cdot x}. \quad (2.8)$$

The problem we face now is to show that both sums on the right-hand side are well defined as elements of $\mathbb{L}^q$. We will show this by passing to the limit in both elements of the Helmholtz–Weyl decomposition of $S_N(u)$, the 'square sum' truncation of Fourier modes (see Theorem 1.6).

Theorem 2.28 *Let Ω be the torus $\mathbb{T}^3$, a bounded domain of class C^2, or the whole space $\mathbb{R}^3$. Then for any $q \in (1, \infty)$ we have*

$$\mathbb{L}^q(\Omega) = H_q(\Omega) \oplus G_q(\Omega),$$

that is every element $u \in \mathbb{L}^q$ can be written[4] uniquely as

$$u = h + \nabla g,$$

where $h \in H_q$ and $\nabla g \in G_q$.

Moreover, the operator $\mathbb{P}_q$ defined by

$$\mathbb{P}_q(u) = h \quad \Leftrightarrow \quad u = h + \nabla g, \ h \in H_q, \ \nabla g \in G_q$$

is bounded from $\mathbb{L}^q$ into $\mathbb{L}^q$ with

$$\|h\|_{L^q} + \|\nabla g\|_{L^q} \leq c_q \|u\|_{L^q}.$$

---

[4] In case of the torus we decompose $\dot{\mathbb{L}}^q$ rather than $\mathbb{L}^q$.

Proof   We will consider only the case of the torus. First we prove the uniqueness of the Helmholtz–Weyl decomposition of $\dot{\mathbb{L}}^q(\mathbb{T}^3)$: if

$$u = h_1 + \nabla g_1 = h_2 + \nabla g_2$$

for $h_1, h_2, \in H_q(\mathbb{T}^3)$ and $\nabla g_1, \nabla g_2 \in G_q(\mathbb{T}^3)$ then

$$h_1 - h_2 = \nabla g_2 - \nabla g_1.$$

Clearly $h_1 - h_2 \in H_q$ and $\nabla g_1 - \nabla g_2 \in G_q$. However, see Exercise 2.14,

$$H_q \cap G_q = \{0\}.$$

Therefore $h_1 - h_2 = \nabla g_1 - \nabla g_2 = 0$ and so $h_1 = h_2$ and $\nabla g_1 = \nabla g_2$. Since g_1 and g_2 have zero mean it follows that $g_1 = g_2$.

To prove existence of the Helmholtz–Weyl decomposition, using (2.8) we write $S_N(u)$, defined as in Theorem 1.6, as a sum of a divergence-free part and a gradient:

$$S_N(u) = h_N + \nabla g_N. \tag{2.9}$$

Since h_N and ∇g_N are finite sums of Fourier modes we obviously have $h_N \in H_q$ and $\nabla g_N \in G_q$ for every $q > 1$.

Although the decomposition of $S_N(u)$ is rather trivial, the difficulty remains when we try to decompose an infinite sum of Fourier modes. Hence our aim now is to show that the norms of h_N and ∇g_N are uniformly bounded in $\mathbb{L}^q$. To this end we take the divergence of equation (2.9), and using the fact that h_N is divergence free, we obtain

$$\text{div}\, S_N(u) = \Delta g_N.$$

It follows that g_N is the unique solution of the problem

$$-\Delta g_N = -\text{div}\, S_N(u)$$

$$\int_{T^3} g_N = 0.$$

Hence, using the fact that operators $(-\Delta)^{-1}$ and div commute on the torus (see Exercise 2.15),

$$\partial_i g_N = -\partial_i (-\Delta)^{-1} \text{div}\, S_N(u)$$
$$= -\partial_i \, \text{div}(-\Delta)^{-1} S_N(u).$$

From elliptic estimates in the Calderón–Zygmund (Theorem B.3) and Theorem C.5 we obtain

$$\|\nabla g_N\|_{\mathbb{L}^q} \leq c \|S_N(u)\|_{\mathbb{L}^q}.$$

From this estimate and Theorem 1.6 it follows that

$$\|\nabla g_N\|_{\mathbb{L}^q} \leq c\|u\|_{\mathbb{L}^q}.$$

Therefore the triangle inequality yields

$$\|h_N\|_{\mathbb{L}^q} = \|S_N(u) - \nabla g_N\|_{\mathbb{L}^q}$$
$$\leq \|S_N(u)\|_{\mathbb{L}^q} + \|\nabla g_n\|_{\mathbb{L}^q} \leq c\|u\|_{\mathbb{L}^q}.$$

Hence we have[5]

$$h_N \rightharpoonup h \text{ in } \mathbb{L}^q, \quad g_N \rightharpoonup g \text{ in } W^{1,q},$$

because $g_N \in W^{1,q}$ has zero mean. Since $S_N(u) \to u$ in $\mathbb{L}^q$, we can pass to the (weak) limit on both sides of (2.9) as $N \to \infty$ to get

$$u = h + \nabla g.$$

Clearly h is weakly divergence free and from the estimates on h_N and ∇g_N we have

$$\|h\|_{L^q} + \|\nabla g\|_{L^q} \leq c\|u\|_{L^q}.$$

Therefore defining

$$\mathbb{P}_q(u) := h$$

we notice that the linear operator

$$\mathbb{P}_q \colon \mathbb{L}^q \to H_q$$

is well defined and bounded.

The proof of this theorem in the case of a bounded domain can be found in Fujiwara & Morimoto (1977) or as Theorem III.1.2 in Galdi (2011). It follows similar lines to the proof of Theorem 2.16, but attaching meaning to the boundary condition $v \cdot n$ in (2.6) in this case is much more difficult. □

Notes

The Helmholtz–Weyl decomposition can be defined for open subsets of $\mathbb{R}^n$ for any $n \geq 2$. In particular, the Helmholtz–Weyl decomposition

$$\mathbb{L}^2(\Omega) = H(\Omega) \oplus G(\Omega)$$

---

[5] We immediately obtain this convergence at least for a subsequence. However, the whole sequence is in fact weakly convergent because of the uniqueness of the Helmholtz–Weyl decomposition of u.

is valid in arbitrary open domains $\Omega \subset \mathbb{R}^n$, $n \geq 2$ (see Theorem III.1.1 in Galdi, 2011). Here the space H is defined as before: it is the completion of smooth, divergence-free functions with compact support in the norm of $[L^2]^n$. The space G is given by

$$G := \left\{ u \in \mathbb{L}^2(\Omega) : u = \nabla w, \quad w \in L^2_{\text{loc}}(\Omega) \right\}.$$

Notice that the difference with a bounded domain is that we demand only local integrability of w (see Lemma 2.5.3 in Sohr, 2001).

The decomposition

$$\mathbb{L}^q(\Omega) = H_q(\Omega) \oplus G_q(\Omega)$$

for $1 < q < \infty$ holds in many types of C^1 domains including bounded and exterior domains (for other types of domains on which this decomposition holds see Farwig & Sohr, 1996, for example). However, in some unbounded domains (for example in infinite and non-convex cones with smoothed vertex) there exist values of $q \neq 2$ for which the Helmholtz–Weyl decomposition does not hold (Bogovskiĭ, 1986).

In general the validity of the Helmholtz–Weyl decomposition is equivalent to the unique solvability of the generalised version of the Neumann problem:

$$\begin{cases} \Delta p = \operatorname{div} u & \text{in } \Omega \\ \frac{\partial p}{\partial n} = u \cdot n & \text{on } \partial\Omega, \end{cases}$$

where u is given and belongs to $\mathbb{L}^q$, $1 < q < \infty$ (see Lemma III.1.2 in Galdi 2011, or Simader & Sohr, 1992). The Helmholtz–Weyl decomposition in $\mathbb{L}^q$ and $\mathbb{L}^{q'}$, (q, q') conjugate, is also equivalent to the validity (in both $\mathbb{L}^q$ and $\mathbb{L}^{q'}$) of the gradient estimate

$$\|\nabla u\|_{\mathbb{L}^q} \leq C \sup \frac{|\langle \nabla u, \nabla v \rangle|}{\|\nabla v\|_{\mathbb{L}^{q'}}},$$

where $\nabla v \in \mathbb{L}^{q'}$ and $1 < q < \infty$, see Simader, Sohr, & Varnhorn (2014).

In bounded and exterior domains, in the half space, and in the whole space $\mathbb{R}^3$ the space $V(\Omega)$ coincides with the space $\tilde{V}(\Omega)$, where

$$\tilde{V}(\Omega) := \{u \in H^1_0(\Omega), \quad \operatorname{div} u = 0\}.$$

However, there exist domains Ω such that the space of divergence-free functions in $H^1_0(\Omega)$ is larger than $V(\Omega)$. An example of such a domain is given by

$$\Omega := \{x = (x_1, x_2, x_3) \in \mathbb{R}^3 : x_1 \neq 0 \quad \text{or} \quad x_2^2 + x_3^2 < 1\}.$$

This observation is due to Heywood (1976), where further discussion, including consequences for the uniqueness of the flow, can be found (see also Galdi, 2011, Theorem III.4.4).

Exercises

2.1 Take $h \in C_c^1(\mathbb{R}^3)$ and $g = (g_1, g_2, g_3) \in [C_c^1(\mathbb{R}^3)]^3$. Prove that

$$\int h(\nabla \times g) = \int g \times \nabla h.$$

2.2 For a C^2 vector-valued function $A = (A_1, A_2, A_3)$ prove the vector identity

$$\Delta A = \nabla(\operatorname{div} A) - \nabla \times (\nabla \times A).$$

2.3 Prove that for a smooth $\mathbb{R}^3$-valued function w with compact support in $\mathbb{R}^3$ we have

$$w = -\nabla u + \operatorname{curl} v,$$

where

$$u(x) = \frac{1}{4\pi} \int \frac{\operatorname{div} w(y)}{|x - y|} \, dy$$

and

$$v(x) = \frac{1}{4\pi} \int \frac{\nabla \times w(y)}{|x - y|} \, dy.$$

Hint: Use Theorem C.4 and Exercises 2.1 and 2.2.

2.4 Denote by $\hat{u}$ the Fourier transform of u. Use the Fourier transform and its inverse to prove that the Helmholtz–Weyl decomposition of $\mathbb{L}^2$ is valid on $\mathbb{R}^3$. (Hint: split $\hat{u}(k) = \alpha(k) + \beta(k)$, where $\alpha(k)$ is parallel to k and $\beta(k)$ is orthogonal to k.)

2.5 Prove that there is no function $g \in H^1(\mathbb{T}^3)$ such that $f = \nabla g$, where $f(x) = c \in \mathbb{R}^3$ for every $x \in \mathbb{T}^3$ and $c \neq 0$.

2.6 Prove that the Leray projector $\mathbb{P}$ commutes with partial derivatives on the whole space $\mathbb{R}^3$.

2.7 Decompose $u = (x_2, 0, 2x_3)$ into a divergence-free part and the gradient of some function.

2.8 Let Ω be given by

$$\Omega = \left\{ (x_1, x_2, x_3) \in \mathbb{R}^3 : x_1^4 + x_2^4 < 1; -1 < x_3 < 1 \right\}.$$

Prove that function

$$u(x_1, x_2, x_3) = (x_2^3, -x_1^3, 0)$$

belongs to $H(\Omega)$ and

$$\Delta \mathbb{P} u \neq \mathbb{P} \Delta u.$$

2.9 Let Ω be a smooth bounded domain and take $f \in V^*(\Omega)$. Prove that the Stokes problem has a unique weak solution in V, i.e. show that there exists a unique $u \in V$ such that

$$\langle \nabla u, \nabla v \rangle = \langle f, v \rangle \quad \text{for all} \quad v \in V.$$

2.10 Let Ω be a smooth bounded domain. Prove that if $u, v \in D(A^s)$ then

$$\langle A^{s_1} u, A^{s_2} v \rangle = \langle u, A^s v \rangle, \quad s_1 + s_2 = s, \quad s_1 \geq 0, \quad s_2 \geq 0.$$

2.11 Let Ω be a smooth bounded domain. Prove, without referring to Theorem 2.23, that there is $c > 0$ such that

$$\|\nabla u\| \leq c \|Au\| \qquad \text{for all} \quad u \in D(A).$$

2.12 Let a_k be an eigenfunction of the Stokes operator on a smooth bounded domain. Prove that there is $C > 0$ (independent of k) such that

$$\|a_k\|_{H^m} \leq C \lambda_k^{m/2}.$$

2.13 Take $m \in \mathbb{N}$ and let A be the Stokes operator on a smooth bounded domain or on the torus $\mathbb{T}^3$. Prove that there is $c > 0$ such that
   (i) $\|u\|_{H^m} \leq c \|A^{m/2} u\|$ for $u \in D(A^{m/2})$;
   (ii) $A^m u = \mathbb{P}(-\Delta)^m u$, $u \in D(A^m)$; and
   (iii) $\|A^m u\| \leq c \|u\|_{H^{2m}}$, $u \in D(A^m)$.

2.14 Prove that

$$H_q(\mathbb{T}^3) \cap G_q(\mathbb{T}^3) = \{0\}.$$

2.15 Prove that on the torus we have

$$(-\Delta)^{-1} \operatorname{div} S_N(u) = \operatorname{div} (-\Delta)^{-1} S_N(u)$$

for every $u \in L^1(\mathbb{T}^3)$.

3

Weak formulation

It is a standard approach in the theory of PDEs to recast the original problem in an appropriate 'weak' form, and then to look for a solution of this reformulated equation, i.e. a 'weak solution'. Usually this reformulation allows one to exploit particular techniques that are not available for the equation in its classical form, and hence to find a weak solution. The hope is that once the existence of a weak solution is assured, it is then possible to show that a weak solution is in fact smooth, and hence a classical solution. For the Navier–Stokes equations we will be able to prove the existence of weak solutions for all positive times, but not their smoothness. Existence of smooth solutions will require more regularity of the initial data and we will be able to show existence of such 'strong solutions' only locally in time, see Chapters 6 and 7. Smooth solutions that exist for all time are only known to exist for small initial data, see Section 6.4.

The main aim of this chapter is to define an appropriate notion of a weak solution of the Navier–Stokes equations and to explain in what sense such a weak solution satisfies the initial condition.

3.1 Weak formulation

In the weak formulation of the Navier–Stokes equations we not only move some derivatives onto the test function, as is typical in the weak form of many PDEs, but we also eliminate one of the terms in the equation, namely the pressure. To do so we introduce a special class of test functions; note that the definition of $\mathcal{D}_\sigma$ is valid on all the domains considered in our book.

Definition 3.1 The space of test functions $\mathcal{D}_\sigma$ on the space–time domain $\Omega \times [0, \infty)$ is given by

$$\mathcal{D}_\sigma = \{\varphi \in C_c^\infty(\Omega \times [0, \infty)) : \operatorname{div} \varphi(t) = 0 \quad \text{for all } t \in [0, \infty)\}.$$

Notice that a test function $\varphi \in \mathcal{D}_\sigma$ must be divergence free (this is why we include the σ subscript) but does not have to vanish at time $t = 0$.

Suppose now that u is a smooth, divergence-free solution[1] of the Navier–Stokes equations

$$\partial_t u - \Delta u + (u \cdot \nabla)u + \nabla p = 0.$$

Multiplying this equation by a test function $\varphi \in \mathcal{D}_\sigma$, integrating in space and then integrating by parts[2] in the second term we obtain

$$\langle \partial_t u, \varphi \rangle + \langle \nabla u, \nabla \varphi \rangle + \langle (u \cdot \nabla)u, \varphi \rangle + \langle \nabla p, \varphi \rangle = 0.$$

The last term vanishes because φ is divergence free,

$$\langle \nabla p, \varphi \rangle = -\langle p, \operatorname{div} \varphi \rangle = 0.$$

Now we integrate in time between 0 and s. Using integration by parts in the first term,

$$\int_0^s \langle \partial_t u, \varphi \rangle = \langle u(s), \varphi(s) \rangle - \langle u(0), \varphi(0) \rangle - \int_0^s \langle u, \partial_t \varphi \rangle,$$

we obtain for any $s \geq 0$

$$\int_0^s -\langle u, \partial_t \varphi \rangle + \int_0^s \langle \nabla u, \nabla \varphi \rangle + \int_0^s \langle (u \cdot \nabla)u, \varphi \rangle = \langle u(0), \varphi(0) \rangle - \langle u(s), \varphi(s) \rangle$$

(3.1)

for all $\varphi \in \mathcal{D}_\sigma$.

Every classical solution of the Navier–Stokes equations satisfies (3.1) for all test functions. However, the converse is not true: see Exercise 3.1 for an example of a function that satisfies the weak formulation of the Navier–Stokes equations (3.1) but not its classical version.

Therefore to ensure that the class of allowable 'weak solutions' is not too broad we need to include some further conditions; we impose an energy-based constraint, consistent with certain properties of classical solutions, which we now derive. First, we prove a fundamental property of the nonlinear term.

Lemma 3.2 *Let $u \in V$ and $v, w \in H^1$. Then*

$$\langle (u \cdot \nabla)v, w \rangle = -\langle (u \cdot \nabla)w, v \rangle.$$

In particular

$$\langle (u \cdot \nabla)v, v \rangle = 0.$$

---

[1] We will call such solutions 'classical solutions'.

[2] In the case of the torus we make use of the periodicity; on an open subset of $\mathbb{R}^3$ we use the fact that φ is compactly supported.

Proof First, observe that for $u \in C_{c,\sigma}^{\infty}(\Omega)$ and $v, w \in C^1(\Omega)$ we have from an integration by parts (using Einstein's summation convention)

$$\langle (u \cdot \nabla)v, w \rangle = \int u_j(\partial_j v_i)w_i$$

$$= -\int (\partial_j u_j)v_i w_i - \int u_j v_i(\partial_j w_i)$$

$$= -\langle (u \cdot \nabla)w, v \rangle.$$

For $u \in V$ and $v, w \in H^1$ we can use a density argument since the trilinear form

$$b(u, v, w) = \langle (u \cdot \nabla)v, w \rangle$$

is continuous on $H^1 \times H^1 \times H^1$, see Exercise 3.2.

To show that $\langle (u \cdot \nabla)v, v \rangle = 0$ we notice that

$$\langle (u \cdot \nabla)v, v \rangle = -\langle (u \cdot \nabla)v, v \rangle$$

so $2\langle (u \cdot \nabla)v, v \rangle = 0$. $\qquad\square$

We are now ready to specify our additional requirements for a weak solution. Assume for a moment that u is a classical solution of the Navier–Stokes equations; then taking the L^2 inner product with u we obtain

$$\frac{1}{2}\frac{d}{dt}\|u\|^2 + \|\nabla u\|^2 + \langle (u \cdot \nabla)u, u \rangle = 0$$

(the term $\langle \nabla p, u \rangle$ vanishes as in the computations above because the spaces H and G are orthogonal according to Corollary 2.5). Due to Lemma 3.2 we have

$$\langle (u \cdot \nabla)u, u \rangle = 0$$

and so in fact

$$\frac{1}{2}\frac{d}{dt}\|u\|^2 + \|\nabla u\|^2 = 0.$$

Integrating in time we obtain

$$\frac{1}{2}\|u(t)\|^2 + \int_0^t \|\nabla u(s)\|^2 \, ds = \frac{1}{2}\|u(0)\|^2 \tag{3.2}$$

for any positive t. It is thus natural to require that for any weak solution, for all $T > 0$ the quantities

$$\operatorname*{ess\,sup}_{0 \le t \le T} \|u(t)\| \quad \text{and} \quad \int_0^T \|\nabla u(s)\|^2 \, ds$$

should be finite. In other words, we should look for weak solutions in the inter-section of the spaces

$$L^\infty(0, T; H) \qquad \text{and} \qquad L^2(0, T; V).$$

(On the torus this definition requires that weak solutions have zero mean for all times; see Exercise 3.8.)

This leads us to the following definition[3] (cf. Heywood, 1988) which is valid on all the domains considered in this book.

Definition 3.3 We say that a function u is a weak solution of the Navier–Stokes equations satisfying the initial condition $u_0 \in H$ if

(i) $u \in L^\infty(0, T; H) \cap L^2(0, T; V)$ for all $T > 0$ and
(ii) u satisfies the equation

$$\int_0^s -\langle u, \partial_t \varphi \rangle + \int_0^s \langle \nabla u, \nabla \varphi \rangle + \int_0^s \langle (u \cdot \nabla) u, \varphi \rangle = \langle u_0, \varphi(0) \rangle - \langle u(s), \varphi(s) \rangle$$

(3.3)

for all test functions $\varphi \in \mathcal{D}_\sigma$ and almost every $s > 0$.

Since a 'function' in $L^\infty(0, T; H) \cap L^2(0, T; V)$ is strictly an equivalence class of functions equal almost everywhere, any such function can be modified on a set of measure zero without changing it in any essential way. Therefore the regularity condition (i) alone is not sufficient to explain how a weak solution u fulfils the initial condition $u(0) = u_0$. In fact condition (ii) of Definition 3.3 makes sure that u is L^2-weakly continuous in time and, in consequence, impos-ing the initial condition makes sense. We prove this result in the next section, together with some other useful auxiliary results.

3.2 Basic properties of weak solutions

In this section we study simple consequences of conditions (i) and (ii) of Defi-nition 3.3. First we will show that from the integrability condition (i) it follows that for every weak solution u the term $(u \cdot \nabla)u$ is integrable, and belongs to $L^r(0, T; L^s)$ for some exponents r and s. The result can be formulated in a gen-eral way for a range of exponents, see (5.16), but we will most frequently use it in the simple form below, where $r = 4/3$ and $s = 6/5$ (this particular choice of exponents is very convenient because $L^{6/5} \subset H^{-1} \subset V^*$).

---

[3] Sometimes it is convenient to say that u is a weak solution of the Navier–Stokes equations on a finite time interval $[0, T]$: by this we will mean that $u \in L^\infty(0, T; H) \cap L^2(0, T; V)$ and u satisfies (3.3) for almost all $0 \le s \le T$.

Lemma 3.4 *If $u \in L^\infty(0, T; H) \cap L^2(0, T; V)$ then*

$$(u \cdot \nabla)u \in L^{4/3}(0, T; L^{6/5}).$$

Proof From the Hölder inequality with exponents $5/3$ and $5/2$ we have

$$\|(u \cdot \nabla)u\|_{L^{6/5}} \leq \left(\int |\nabla u|^{6/5} |u|^{6/5} \right)^{5/6}$$

$$\leq \left(\int |\nabla u|^2 \right)^{1/2} \left(\int |u|^3 \right)^{1/3}$$

$$= \|\nabla u\| \|u\|_{L^3}$$

$$\leq \|\nabla u\| \|u\|_{L^2}^{1/2} \|u\|_{L^6}^{1/2}$$

$$\leq c \|u\|_{H^1}^{3/2} \|u\|^{1/2}.$$

Hence

$$\int_0^T \|(u \cdot \nabla)u\|_{L^{6/5}}^{4/3} \leq c \int_0^T \|u\|_{H^1}^2 \|u\|^{2/3} < \infty,$$

because $u \in L^\infty(0, T; H) \cap L^2(0, T; H^1)$. $\qquad\square$

Let us notice that from Lemma 3.4 and the Hölder inequality we can deduce that

$$\int_0^T \langle (u \cdot \nabla)u, v \rangle < \infty \quad \text{for all} \quad v \in L^4(0, T; L^6).$$

However, from the definition of a weak solution and the Sobolev embedding $H^1 \subset L^6$ it follows that a weak solution u belongs only to the space $L^2(0, T; L^6)$ and is not known to belong to $L^4(0, T; L^6)$. This is one of the reasons why functions with only the regularity of weak solutions cannot be used as test functions in a weak formulation of the 3D Navier–Stokes equations (see also Exercise 3.3).

We also note for further use that a simple interpolation argument gives space–time integrability of any weak solution in a range of spaces.

Lemma 3.5 *If $u \in L^\infty(0, T; H) \cap L^2(0, T; V)$ then*

$$u \in L^r(0, T; L^s(\Omega)),$$

where

$$\frac{2}{r} + \frac{3}{s} = \frac{3}{2}, \qquad 2 \leq s \leq 6.$$

In particular

$$u \in L^{10/3}((0, T) \times \Omega).$$

Proof We interpolate the L^s norm between L^2 and L^6

$$\|u\|_{L^s} \leq \|u\|_{L^2}^{(6-s)/2s} \|u\|_{L^6}^{(3s-6)/2s}.$$

It follows that

$$\int_0^T \|u\|_{L^s}^r \leq \int_0^T \|u\|_{L^2}^{(6-s)r/2s} \|u\|_{L^6}^{(3s-6)r/2s}.$$

Since $u \in L^\infty(0, T; L^2) \cap L^2(0, T; L^6)$ the integral on the right-hand side is bounded when $(3s - 6)r/2s \leq 2$, that is when

$$\frac{2}{r} + \frac{3}{s} \geq \frac{3}{2}. \qquad \square$$

The assertion of Lemma 3.5 measures to some extent how far weak solutions (whose existence is assured) are from the classical solutions we would like to have. In fact we will prove in Chapter 8 that if a weak solution u (satisfying the energy inequality) belongs to the space $L^r(0, T; L^s(\Omega))$ with

$$\frac{2}{r} + \frac{3}{s} = 1,$$

then u is smooth.

We will now focus on consequences of condition (ii) of Definition 3.3. As we mentioned before, this condition provides weak solutions with some form of continuity in time (typically rather weak). The first result in this direction is the following lemma.

Lemma 3.6 *If u is a weak solution of the Navier–Stokes equations then*

$$\int_{t_1}^{t_2} \langle \nabla u, \nabla \phi \rangle + \int_{t_1}^{t_2} \langle (u \cdot \nabla)u, \phi \rangle = \langle u(t_1), \phi \rangle - \langle u(t_2), \phi \rangle \qquad (3.4)$$

for all $\phi \in C_{c,\sigma}^\infty(\Omega)$ and almost every $t_2 \geq t_1$, for almost every $t_1 \geq 0$ including $t_1 = 0$.

Proof Take any $\phi \in C_{c,\sigma}^\infty(\Omega)$ and choose $\alpha \in C_c^\infty[0, \infty)$ with $\alpha(t) = 1$ for all $t \in [t_1, t_2]$. Then $\tilde{\varphi}$ given by

$$\tilde{\varphi}(x, t) = \alpha(t)\phi(x)$$

belongs to the space of test functions $\mathcal{D}_\sigma$. Taking the difference of (3.3) with $\varphi = \tilde{\varphi}$ for $s = t_2$ and $s = t_1$ yields (3.4). $\qquad \square$

It turns out that the converse of Lemma 3.6 is also true, that is if u has the regularity of a weak solution and satisfies (3.4) then u is a weak solution of the Navier–Stokes equations. For the proof see Exercise 3.4.

From Lemma 3.6 one can deduce that a weak solution u has a well-defined time derivative[4] belonging to some, unfortunately very large, Bochner space.

Lemma 3.7 *If u is a weak solution of the Navier–Stokes equations then*

$$\partial_t u \in L^{4/3}(0, T; V^*).$$

In particular, one can modify u on the set of measure zero in such a way that

$$u \in C([0, T]; V^*)$$

and (3.3) and (3.4) are satisfied for all $s > 0$ and all $t_1, t_2 > 0$ respectively.

Proof From Lemma 3.6 we have, for almost all $t > 0$ and for every choice of $\varphi \in C^\infty_{c,\sigma}(\Omega)$,

$$\langle u(t), \varphi \rangle - \langle u_0, \varphi \rangle = \int_0^t \langle g, \varphi \rangle_{V^* \times V},$$

where g is given by

$$\langle g, \varphi \rangle_{V^* \times V} = -\langle \nabla u, \nabla \varphi \rangle - \langle (u \cdot \nabla)u, \varphi \rangle.$$

From our assumptions on the integrability of u, it follows that g is a bounded linear functional on V for almost all s, because

$$|\langle \nabla u, \nabla \varphi \rangle| \leq \|\nabla u\| \|\nabla \varphi\| \qquad \text{and}$$
$$|\langle (u \cdot \nabla)u, \varphi \rangle| \leq \|(u \cdot \nabla)u\|_{L^{6/5}} \|\varphi\|_{L^6}$$
$$\leq c \|(u \cdot \nabla)u\|_{L^{6/5}} \|\nabla \varphi\|,$$

where we used the estimate in the proof of Lemma 3.4 and the Sobolev embedding $H^1 \subset L^6$. Hence Lemma 3.4 yields

$$\|g\|_{V^*}^{4/3} \leq c \|(u \cdot \nabla)u\|_{L^{6/5}}^{4/3} + c \|\nabla u\|^{4/3}$$
$$\leq c \|(u \cdot \nabla)u\|_{L^{6/5}}^{4/3} + c(1 + \|\nabla u\|^2) \in L^1(0, T).$$

It follows that $g \in L^{4/3}(0, T; V^*)$ and using Corollary 1.32 we obtain

$$\partial_t u \in L^{4/3}(0, T; V^*).$$

From Lemma 1.31 we deduce that u is equal almost everywhere to an absolutely continuous function from $[0, T]$ into V^*. Therefore we can modify u on a set of measure zero to obtain

$$u \in C([0, T]; V^*).$$

---

[4] Notice that $\partial_t u$ does not appear in the weak formulation of the Navier–Stokes equations.

To prove that for such a particular representative u equation (3.4) holds for all $t_1, t_2 > 0$ we notice that for all $t_1, t_2 \in (0, T]$ we can find sequences of times $s_n \to t_1$ and $z_n \to t_2$ such that

$$\int_{s_n}^{z_n} \langle \nabla u, \nabla \phi \rangle + \int_{s_n}^{z_n} \langle (u \cdot \nabla) u, \phi \rangle = \langle u(s_n), \phi \rangle - \langle u(z_n), \phi \rangle.$$

The left-hand side is continuous with respect to s_n and z_n, while

$$\langle u(s_n), \phi \rangle \to \langle u(t_1), \phi \rangle \quad \text{and} \quad \langle u(z_n), \phi \rangle \to \langle u(t_2), \phi \rangle,$$

as $n \to \infty$ because $u \in C([0, T]; V^*)$. Passing to the limit as $n \to \infty$, we deduce that (3.4) is satisfied for all $t_1, t_2 > 0$. The proof that (3.3) is satisfied for all $s > 0$ is similar[5] and we omit it. □

In the proof above the function φ had to belong to the space V and could not be taken from the larger space H_0^1, because in the weak formulation of the Navier–Stokes equations we restrict the space of test functions to only those that are divergence free. As a result the question whether $\partial_t u$ belongs to $L^{4/3}(0, T; H^{-1})$ when u is a general weak solution of the Navier–Stokes equations still remains open. (It is worth noticing that H^{-1} is a proper subset of V^*, i.e. if $u \in V^*$ then it does not follow that necessarily $u \in H^{-1}$. Indeed, if $u \in V^*$ then $\langle u, v \rangle$ is bounded for all those $v \in H_0^1$ that are divergence free, but not necessarily for those whose divergence is not zero.)

We also note that the regularity of $\partial_t u$ that is obtained in Lemma 3.7 is not sufficient to make sense of the expression

$$\int_0^T \langle u(s), \partial_t u(s) \rangle \, ds$$

when $u \in L^\infty(0, T; H) \cap L^2(0, T; V)$. This is the main reason why weak solutions are not known to be sufficiently regular to be used as test functions in the weak formulation of the Navier–Stokes equations in 3D.

We are now ready to prove the L^2-weak continuity of weak solutions. We emphasise that we are only able to prove *weak* continuity; continuity remains unknown because it is still open whether the function $t \mapsto \|u(t)\|$ is continuous even for the 'best' weak solutions of the Navier–Stokes equations that we are able to construct.[6]

We also note that in light of Lemma 3.7 we will assume without spelling it out each time that every weak solution we consider has been modified on the

---

[5] The only difference is that the test functions depend on time, so to show that $\langle u(t), \varphi(t) \rangle$ is continuous we notice that every $\varphi \in \mathcal{D}_\sigma$ can be approximated in the norm of V by the sum of divergence-free, smooth functions of the form $\alpha(t) \beta(x)$.

[6] We recall that if $u_n \rightharpoonup u$ and $\|u_n\| \to \|u\|$ then $u_n \to u$ strongly, see Lemma A.20.

set of measure zero so that $u \in C([0, T]; V^*)$ and (3.3) and (3.4) are satisfied for all $s > 0$ and all $t_1, t_2 > 0$, respectively. This assumption will simplify the statement of many results (including the following theorem).

Theorem 3.8 *Weak solutions of the Navier–Stokes equations are L^2-weakly continuous in time, i.e.*

$$\lim_{t \to t_0} \langle u(t), v \rangle = \langle u(t_0), v \rangle \quad \text{for all} \quad v \in L^2(\Omega).$$

In particular, $u(t) \rightharpoonup u_0$ weakly in L^2 as $t \to 0^+$.

Proof We take $v \in L^2$ if Ω is the whole space or a bounded subset of $\mathbb{R}^3$; and we take $v \in \dot{L}^2$ for the torus, since on $\mathbb{T}^3$ we can assume that $v \in \dot{L}^2$, because u then has zero average and in consequence $\langle u(t), c \rangle = 0$ for every $c \in \mathbb{R}^3$ and every $t \geq 0$. Using the Helmholtz–Weyl decomposition v can be written as

$$v = h + \nabla g,$$

where $h \in H$ and $\nabla g \in G$. Using the orthogonality of $u(t) \in H$ and ∇g we obtain

$$\langle u(t), v \rangle = \langle u(t), h + \nabla g \rangle = \langle u(t), h \rangle.$$

Therefore we only need to prove that

$$\lim_{t \to t_0} \langle u(t), h \rangle = \langle u(t_0), h \rangle \quad \text{for all} \quad h \in H(\Omega).$$

Since $C_{c,\sigma}^\infty$ is dense in H we can assume that $h \in C_{c,\sigma}^\infty(\Omega)$. For such h, Lemma 3.6 implies that

$$|\langle u(t) - u(t_0), h \rangle| \leq \left| \int_t^{t_0} \langle \nabla u(t), \nabla h \rangle \right| + \left| \int_t^{t_0} \langle (u \cdot \nabla) u, h \rangle \right| \to 0,$$

when $t \to t_0$, because the functions under the integral signs are integrable due to the integrability assumptions on u and Lemma 3.4. $\qquad \square$

From Theorem 3.8 it follows that for a given weak solution u the value of $u(t)$ is uniquely determined for every positive time (it does not imply that weak solutions are unique, of course).

Corollary 3.9 *Let u be a weak solution of the Navier–Stokes equations. Then for every $T > 0$ the value of $u(T)$ is uniquely determined as the weak limit in the space L^2:*

$$u(t) \rightharpoonup u(T) \quad \text{weakly in } L^2 \text{ when } t \to T.$$

In particular, for any given weak solution u for every $T > 0$ the value of u(T) is uniquely determined by the values of u(t) on the time interval $(0, T)$.

The definition of a weak solution adopted in this chapter seems to be rather restrictive. However, in the next two sections we will show that in order to check that condition (ii) of Definition 3.3 is satisfied it is enough to check some apparently less demanding conditions. This will simplify our proof of the existence of weak solutions in Chapter 4.

3.3 Alternative spaces of test functions

The space $\mathcal{D}_\sigma$ seems to be a very natural choice as the space of test functions. However, in many situations it is easier to use different spaces of test functions. For example, for the torus $\mathbb{T}^3$ or for a bounded subset of $\mathbb{R}^3$ with a smooth boundary it is convenient to define the space $\tilde{\mathcal{D}}_\sigma$.

Definition 3.10 Let Ω be the torus $\mathbb{T}^3$ or a smooth bounded domain in $\mathbb{R}^3$. The space $\tilde{\mathcal{D}}_\sigma(\Omega)$ is given by

$$\left\{ \varphi : \varphi = \sum_{k=1}^{N} \alpha_k(t) a_k(x), \quad \alpha_k \in C_c^1([0, \infty)), \ a_k \in \mathcal{N}, N \in \mathbb{N} \right\},$$

where $\mathcal{N}$ is the basis of H consisting of eigenfunctions of the Stokes operator, described in Theorem 2.24.

The functions in the space $\tilde{\mathcal{D}}_\sigma$ are slightly less regular in time than those in $\mathcal{D}_\sigma$. On a bounded domain they vanish on the boundary, but they usually do not have compact support in Ω, so their derivatives on the boundary may not be zero. However, functions in $\tilde{\mathcal{D}}_\sigma$ have the significant advantage that their dependence on space and time variables is separated, and that they are directly related to the Stokes operator. In fact it turns out that in order to check that a given candidate for a weak solution u satisfies Definition 3.3 it is usually more convenient to check that (3.3) holds for all $\varphi \in \tilde{\mathcal{D}}_\sigma$ and then use the following lemma (cf. Lemma 2.3 in Galdi, 2000).

Lemma 3.11 *Let Ω be the torus $\mathbb{T}^3$ or a smooth bounded domain in $\mathbb{R}^3$. If $u \in L^\infty(0, T; H(\Omega)) \cap L^2(0, T; V(\Omega))$ for all $T > 0$ then the following two statements are equivalent:*

(i) *u satisfies (3.3) for all $\varphi \in \mathcal{D}_\sigma(\Omega)$*
(ii) *u satisfies (3.3) for all $\varphi \in \tilde{\mathcal{D}}_\sigma(\Omega)$.*

Proof   (i) ⇒ (ii). It is clear that we only need to show that the equality (3.3) holds for every fixed time $s > 0$ and every $\phi \in \tilde{\mathcal{D}}_\sigma$ given by

$$\phi(x, t) = \alpha(t)a(x),$$

where $a \in \mathcal{N}$ and $\alpha \in C_c^1([0, \infty))$.

First, we notice that we can find a sequence of smooth functions α_n with compact support in $[0, s + 1)$ such that

$$\alpha_n \to \alpha \quad \text{in } C^1([0, s]).$$

Since functions from $C_{c,\sigma}^\infty(\Omega)$ are dense in V we can also find a sequence of functions $\varphi_n \in C_{c,\sigma}^\infty(\Omega)$ such that

$$\varphi_n \to a \quad \text{in } H^1(\Omega).$$

Then for each n the function ψ_n given by

$$\psi_n(x, t) = \alpha_n(t)\varphi_n(x)$$

is an element of $\mathcal{D}_\sigma$, so from (i) we know that (3.3),

$$\int_0^s -\langle u, \partial_t \varphi \rangle + \int_0^s \langle \nabla u, \nabla \varphi \rangle + \int_0^s \langle (u \cdot \nabla)u, \varphi \rangle = \langle u_0, \varphi(0) \rangle - \langle u(s), \varphi(s) \rangle,$$

is satisfied with $\varphi = \psi_n$ for every n. Furthermore

$$\psi_n \to \phi \quad \text{in } C([0, s]; V) \quad \text{and} \tag{3.5}$$
$$\partial_t \psi_n \to \partial_t \phi \quad \text{in } L^2(0, T; L^2) \tag{3.6}$$

as $n \to \infty$. Now we pass to the limit as $n \to \infty$ in (3.3) with $\varphi = \psi_n$. Using (3.5) and (3.6) we obtain

$$\int_0^s \langle u, \partial_t \psi_n \rangle \;\to\; \int_0^s \langle u, \partial_t \phi \rangle,$$
$$\int_0^s \langle \nabla u, \nabla \psi_n \rangle \;\to\; \int_0^s \langle \nabla u, \nabla \phi \rangle,$$
$$\langle u(0), \psi_n(0) \rangle \;\to\; \langle u(0), \phi(0) \rangle, \quad \text{and}$$
$$\langle u(s), \psi_n(s) \rangle \;\to\; \langle u(s), \phi(s) \rangle.$$

To pass to the limit in the nonlinear term it is enough to notice that, due to the Sobolev embedding $H^1 \subset L^6$, we have $\psi_n \to \phi$ in $C([0, s]; L^6)$, and consequently $\psi_n \to \phi$ in $L^4(0, s; L^6)$. Hence from Lemma 3.4 we obtain

$$\left| \int_0^s \langle (u \cdot \nabla)u, (\psi_n - \phi) \rangle \right| \leq \|(u \cdot \nabla)u\|_{L^{4/3}(0,s;L^{6/5})} \|\psi_n - \phi\|_{L^4(0,s;L^6)} \to 0.$$

(ii) $\Rightarrow$ (i). Let $\phi \in \mathcal{D}_\sigma$. Then $\phi(t) \in H^k \cap V$ for every k and $t > 0$, so we can express $\phi(t)$ in terms of the Stokes eigenfunctions:

$$\phi(t, x) = \sum_{k=1}^{\infty} c_k(t) a_k(x).$$

Define $\psi_n := \sum_{k=1}^{n} c_k(t) a_k(x)$. Then $\psi_n \in \tilde{\mathcal{D}}_\sigma$ and

$$\psi_n \to \phi \quad \text{in} \quad C([0, s], V).$$

Indeed, we have

$$\sup_{0 \le t \le s} \|\phi(t) - \psi_n(t)\|_V^2 = \sup_{0 \le t \le s} \left\| \sum_{k=n+1}^{\infty} c_k(t) a_k(x) \right\|_V^2$$

$$\le \sup_{0 \le t \le s} \sum_{k=n+1}^{\infty} \lambda_k c_k^2(t)$$

$$\le \sup_{0 \le t \le s} \sum_{k=n+1}^{\infty} \frac{\lambda_k^2 c_k^2(t)}{\lambda_n}$$

$$\le \frac{1}{\lambda_n} \sup_{0 \le t \le s} \|A\phi_n(t)\|^2$$

$$\le \frac{1}{\lambda_n} \sup_{0 \le t \le s} \|\phi(t)\|_{H^2}^2 \to 0 \quad \text{as} \quad n \to \infty.$$

Furthermore,

$$\partial_t \psi_n \to \partial_t \phi \quad \text{in} \quad L^2(0, T; L^2),$$

so we can follow the reasoning in the proof of implication (i) $\Rightarrow$ (ii), which was based on the convergence (3.5)–(3.6). $\qquad\square$

The functions in $\tilde{\mathcal{D}}_\sigma$ are C^1 functions of time, but this assumption can be further relaxed.

Lemma 3.12 *Let u be a weak solution of the Navier–Stokes equations, take $\alpha \in H^1(0, T)$ and $a \in \mathcal{N}$. Then (3.3) holds for*

$$\varphi(x, t) = \alpha(t) a(x)$$

for almost all $s \in [0, T]$.

Proof From the assumption we know that there is $g = \dot{\alpha} \in L^2(0, T)$ (by $\dot{\alpha}$ we denote the time derivative) and

$$\alpha(t) = c_0 + \int_0^t g(s) \, ds.$$

Now let

$$\alpha_\varepsilon(t) := c_0 + \int_0^t g_\varepsilon(s)\,\mathrm{d}s,$$

where g_ε denotes a mollified g (extended by zero outside $(0, T)$). Define

$$\varphi_\varepsilon(t, x) := \alpha_\varepsilon(t)a(x).$$

Then $\varphi_\varepsilon \in \tilde{\mathcal{D}}_\sigma$ and

$$\dot{\alpha}_\varepsilon - \dot{\alpha} = g_\varepsilon - g \to 0 \quad \text{in} \quad L^2(0, T)$$

with

$$\sup_{0 \le t \le T} |\alpha_\varepsilon(t) - \alpha(t)| = \sup_{0 \le t \le T} \left| \int_0^t (g_\varepsilon - g) \right| \le \int_0^T |g_\varepsilon - g| \to 0.$$

Hence $\varphi_\varepsilon \to \varphi$ in $C([0, T]; H^1)$ and we conclude the proof following the reasoning in the proof of implication (i) $\Rightarrow$ (ii) from Lemma 3.11. $\square$

3.4 Equivalent weak formulation

Checking condition (ii) of Definition 3.3 for all positive s could be rather inconvenient in the proof of existence of weak solutions. Therefore we will introduce an equivalent but simpler condition. To this end let us first make an easy observation.

Proposition 3.13 *If u is a weak solution of the Navier–Stokes equations then*

$$\int_0^\infty -\langle u, \partial_t\varphi \rangle + \int_0^\infty \langle \nabla u, \nabla\varphi \rangle + \int_0^\infty \langle (u \cdot \nabla)u, \varphi \rangle = \langle u_0, \varphi(0) \rangle$$

for all $\varphi \in \tilde{\mathcal{D}}_\sigma$.

Proof Take any $\varphi \in \tilde{\mathcal{D}}_\sigma$. From Lemma 3.11 it follows that (3.3) is valid for φ for almost all $s > 0$. Since $\varphi(s) = 0$ for large s it follows that the equation above is satisfied. $\square$

The following lemma shows that the converse implication is also true (the result is due to Prodi, 1959, and Serrin, 1963; we use the idea from Lemma 1 in Prodi, 1959).

Lemma 3.14 *Let Ω be a smooth bounded domain in $\mathbb{R}^3$ or the torus $\mathbb{T}^3$. Suppose that $u \in L^\infty(0, T; H(\Omega)) \cap L^2(0, T; V(\Omega))$ for all $T > 0$ and satisfies*

$$\int_0^\infty -\langle u, \partial_t\varphi \rangle + \int_0^\infty \langle \nabla u, \nabla\varphi \rangle + \int_0^\infty \langle (u \cdot \nabla)u, \varphi \rangle = \langle u_0, \varphi(0) \rangle \quad (3.7)$$

for all test functions $\varphi \in \tilde{\mathcal{D}}_\sigma$. Then u is a weak solution of the Navier–Stokes equation with initial condition u_0.

Proof First we will show that (3.3) is valid for almost all $s > 0$ for all $\varphi \in \tilde{\mathcal{D}}_\sigma$. So let us fix $s > 0$ and let $\varphi \in \tilde{\mathcal{D}}_\sigma$ be given by

$$\varphi(x, t) = \alpha(t)a_k(x),$$

where $a_k \in \mathcal{N}$ is one of the eigenfunctions of the Stokes operator from Theorem 2.24. (It is clear that if (3.3) is valid for every φ of this form then it is valid for every $\varphi \in \tilde{\mathcal{D}}_\sigma$.) Then for $h > 0$ we define a piecewise C^1 function

$$\alpha_h(t) = \begin{cases} \alpha(t) & t \in [0, s] \\ \alpha(s)\left[1 - \frac{t-s}{h}\right] & t \in (s, s + h) \\ 0 & t \geq s + h \end{cases}$$

(see Figure 3.1) and a corresponding sequence of functions φ_n by

$$\varphi_n(x, t) = \alpha_{1/n}(t)a_k(x).$$

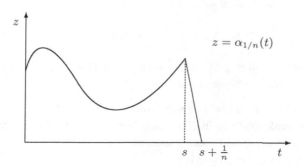

Figure 3.1. The approximating function α_n.

It follows from Lemma 3.12 that (3.7) is valid for φ replaced by φ_n for all n. We want to pass to the limit in this equation. First, we notice that

$$\left| \int_s^\infty \langle \nabla u, \nabla \varphi_n \rangle \right| \leq \left(\int_s^{s+1/n} \|\nabla u\|^2 \right)^{1/2} \left(\int_s^{s+1/n} \|\nabla \varphi_n\|^2 \right)^{1/2}$$

$$\leq \|u\|_{L^2(0,s+1;H^1)}|\alpha(s)|\|\nabla a_k\|n^{-1/2} \to 0,$$

as $n \to \infty$. Since $\varphi_n = \varphi$ on $[0, s]$ we obtain

$$\int_0^\infty \langle \nabla u, \nabla \varphi_n \rangle \to \int_0^s \langle \nabla u, \nabla \varphi \rangle.$$

Similarly, see Exercise 3.5,

$$\int_0^\infty \langle (u \cdot \nabla)u, \varphi_n \rangle \to \int_0^s \langle (u \cdot \nabla)u, \varphi \rangle.$$

The convergence of the term involving the time derivative is more complicated and can be shown as follows:

$$
\begin{aligned}
\int_0^\infty \langle u(t), \partial_t \varphi_n(t) \rangle \, dt &= \int_0^s \langle u(t), \partial_t \varphi(t) \rangle \, dt + \int_s^{s+\frac{1}{n}} \langle u(t), \partial_t \varphi_n(t) \rangle \, dt \\
&= \int_0^s \langle u(t), \partial_t \varphi(t) \rangle \, dt - \frac{\alpha(s)}{1/n} \int_s^{s+\frac{1}{n}} \langle u(t), a_k \rangle \, dt \\
&\to \int_0^s \langle u(t), \partial_t \varphi(t) \rangle \, dt - \alpha(s) \langle u(s), a_k \rangle \\
&= \int_0^s \langle u(t), \partial_t \varphi(t) \rangle \, dt - \langle u(s), \varphi(s) \rangle.
\end{aligned}
$$

This convergence is valid – due to the Lebesgue Differentiation Theorem (Theorem A.2) – whenever s is a Lebesgue point of the function $\langle u(t), a_k \rangle$ and does not depend on the choice of a test function. Since there are only countably many functions $\langle u(t), a_k \rangle$ it follows that

$$\int_0^s -\langle u, \partial_t \varphi \rangle + \int_0^s \langle \nabla u, \nabla \varphi \rangle + \int_0^s \langle (u \cdot \nabla)u, \varphi \rangle = \langle u_0, \varphi(0) \rangle - \langle u(s), \varphi(s) \rangle$$

for almost all s and all $\varphi \in \tilde{\mathcal{D}}_\sigma$. So from Lemma 3.11 it follows that u is a weak solution of the Navier–Stokes equations. $\qquad\square$

Lemmas 3.11 and 3.14 will be very useful in Chapter 4 where we show the existence of weak solutions in dimension three. However, before we give this proof of existence, we will make a short digression to discuss the uniqueness of weak solutions in dimension two.

3.5 Uniqueness of weak solutions in dimension two

We will now prove that weak solutions of the Navier–Stokes equations are unique in space dimension two. The situation is very different from the three-dimensional case, since we have 'better' exponents in the Sobolev embedding theorem, and in consequence we can prove that

$$\partial_t u \in L^2(0, T; V^*)$$

(see Exercise 3.7). Such regularity makes it possible to use weak solutions of the 2D Navier–Stokes equations as test functions and obtain their uniqueness in a relatively easy way, as we now show. This result is due to Lions & Prodi (1959) and is based on an inequality due to Ladyzhenskaya (1959).

Theorem 3.15 *Let $\Omega \subset \mathbb{R}^2$ be a smooth bounded domain, and let u and v be two weak solutions corresponding to the same initial condition $u_0 \in H(\Omega)$. Then $u(t) = v(t)$ for all $t > 0$.*

Proof Let $w = u - v$. We substitute $\varphi = w$ as a test function in the weak form of the equation (see (3.3)) to obtain

$$\int_0^s -\langle u, \partial_t w \rangle + \int_0^s \langle \nabla u, \nabla w \rangle + \int_0^s \langle (u \cdot \nabla) u, w \rangle = \langle u_0, w(0) \rangle - \langle u(s), w(s) \rangle.$$

It is easy to see that in dimension two all functions that appear under the integral signs in (3.3) are integrable (see Exercise 3.7), so the functions $\langle u(s), w(s) \rangle$ and $\langle v(s), w(s) \rangle$ must be absolutely continuous, hence differentiable for almost all $s > 0$. Differentiating the above equality with respect to time we obtain

$$-\langle u, \partial_t w \rangle + \langle \nabla u, \nabla w \rangle + \langle (u \cdot \nabla) u, w \rangle = -\frac{\mathrm{d}}{\mathrm{d}t} \langle u, w \rangle.$$

Similarly, we have

$$-\langle v, \partial_t w \rangle + \langle \nabla v, \nabla w \rangle + \langle (v \cdot \nabla) v, w \rangle = -\frac{\mathrm{d}}{\mathrm{d}t} \langle v, w \rangle.$$

Taking the difference of the two equations we obtain

$$\frac{1}{2} \frac{\mathrm{d}}{\mathrm{d}t} \|w\|^2 + \|\nabla w\|^2 \leq |\langle (u \cdot \nabla) u - (v \cdot \nabla) v, w \rangle|$$

$$= |\langle (u \cdot \nabla) u, w \rangle - \langle ((u - w) \cdot \nabla)(u - w), w \rangle|$$

$$= |\langle (u \cdot \nabla) w, w \rangle + \langle (w \cdot \nabla) u, w \rangle - \langle (w \cdot \nabla) w, w \rangle|.$$

The first and the last term on the right-hand side vanish, according to Lemma 3.2, so

$$\frac{1}{2} \frac{\mathrm{d}}{\mathrm{d}t} \|w\|^2 + \|\nabla w\|^2 \leq |\langle (w \cdot \nabla) u, w \rangle|.$$

Since $w \in H_0^1(\Omega)$ we can use the two-dimensional version of the Ladyzhenskaya inequality,

$$\|w\|_{L^4} \leq c \|w\|^{1/2} \|\nabla w\|^{1/2}$$

(see Exercise 3.6), to obtain

$$\frac{1}{2}\frac{\mathrm{d}}{\mathrm{d}t}\|w\|^2 + \|\nabla w\|^2 \le \|w\|_{L^4}^2\|\nabla u\|$$

$$\le c\|w\|\|\nabla w\|\|\nabla u\|$$

$$\le c\|\nabla u\|^2\|w\|^2 + \frac{1}{2}\|\nabla w\|^2.$$

Therefore

$$\frac{\mathrm{d}}{\mathrm{d}t}\|w\|^2 \le c\|\nabla u\|^2\|w\|^2,$$

and from the Gronwall Lemma (Lemma A.24) we have

$$\|w(t)\|^2 \le \|w(0)\|^2\exp\left(c\int_0^t \|\nabla u(s)\|^2\,\mathrm{d}s\right).$$

Since $u \in L^2(0, T; V)$ and $w(0) = 0$ it follows that $w(t) = 0$ for $t > 0$. $\square$

Notes

The weak solutions of the Navier–Stokes equations, on which we focus throughout the book, were introduced by Leray in 1934. The definition of a 'weak solution' adopted in this chapter is taken from Heywood (1988). Many authors (see Galdi, 2000, for example) define weak solutions in a very similar way, but they include in the condition (ii) of Definition 3.3 equation (3.7) rather than (3.3). In all these cases the regularity of the time derivative, $\partial_t u \in L^{4/3}(0, T; V^*)$, can be deduced from the properties of a weak solution as in Lemma 3.7. The definition in Chemin et al. (2006) is also similar, but they choose to include the condition $u \in C([0, T]; V^*)$ from the start.

In a slightly different approach one can define a weak solution as a function $u \in L^2(0, T; V)$ that satisfies the Navier–Stokes equations as an equality in the dual space V^*, that is

$$\langle \partial_t u, v \rangle + \langle \nabla u, \nabla v \rangle + \langle (u \cdot \nabla)u, v \rangle = 0$$

for almost all times $t \in [0, T]$ and all $v \in V$. Then one can show (see Temam, 1983; cf. Constantin & Foias, 1988) that u is almost everywhere continuous from $[0, T]$ into V^* so that the initial condition $u(0) = u_0$ makes sense.

The pressure is usually absent in the definition of a weak solution, but there are some exceptions. For example, Lemarié-Rieusset (2002) defines a weak solution (on the whole space) as a divergence-free and locally square-integrable vector field that satisfies the Navier–Stokes equations (including the pressure) in the distributional sense (see also Chapter 5).

The proofs of Lemmas 3.11 and 3.14 are based on ideas from Prodi (1959), Serrin (1963), and Galdi (2000) whose exposition we roughly follow in this chapter.

Exercises

3.1 Suppose that

$$u(x, t) = a(t)\nabla\Psi(x),$$

where Ψ is harmonic and a is integrable but discontinuous. Show that u satisfies equation (3.1).

3.2 Prove that the trilinear form $\langle(u \cdot \nabla)v, w\rangle\colon H^1 \times H^1 \times H^1 \to \mathbb{R}$ is continuous.

3.3 Prove that if u is a weak solution of the Navier–Stokes equations then

$$\int_0^T \langle(u \cdot \nabla u), u\rangle = 0,$$

but there exist functions $u, v \in L^\infty(0, T; H(\mathbb{R}^3)) \cap L^2(0, T; V(\mathbb{R}^3))$ such that

$$\int_0^T \langle(u \cdot \nabla)u, v\rangle = \infty.$$

3.4 Prove that if $u \in L^\infty(0, T; L^2) \cap L^2(0, T; H^1)$ and satisfies

$$\int_{t_1}^{t_2} \langle\nabla u, \nabla\phi\rangle + \int_{t_1}^{t_2} \langle(u \cdot \nabla)u, \phi\rangle = \langle u(t_1), \phi\rangle - \langle u(t_2), \phi\rangle$$

for all $\phi \in C_{c,\sigma}^\infty(\Omega)$ and almost every $t_2 \geq t_1$, for almost every t_1 including $t_1 = 0$ then u is a weak solution of the Navier–Stokes equations on $[0, T]$.

3.5 Let φ_n be given as in the proof of Lemma 3.14. Prove that

$$\left|\int_s^{s+1/n} \langle(u \cdot \nabla)u, \varphi_n\rangle\right| \to 0.$$

3.6 In 2D use the Sobolev embedding $H^{1/2} \subset L^4$ and Sobolev interpolation to prove the Ladyzhenskaya inequality

$$\|u\|_{L^4} \leq c\|u\|^{1/2}\|\nabla u\|^{1/2} \qquad \text{for all} \qquad u \in V.$$

Use a similar idea in 3D to show that

$$\|u\|_{L^4} \le c\|u\|^{1/4}\|\nabla u\|^{3/4} \qquad \text{for all} \qquad u \in V.$$

3.7 Prove that if Ω is a smooth bounded domain in 2D and u is a weak solution of the Navier–Stokes equations on Ω then the time derivative of u satisfies $\partial_t u \in L^2(0, T; V^*(\Omega))$.

3.8 Let u be a classical solution of the Navier–Stokes equations on the torus. Prove that if $\int_{\mathbb{T}^3} u(0) = 0$ then $\int_{\mathbb{T}^3} u(t) = 0$ for all $t > 0$.

4

Existence of weak solutions

In this chapter we prove the existence of global-in-time weak solutions of the 3D Navier–Stokes equations. Although we can prove that such weak solutions exist, we are unable to prove their uniqueness.

In the construction of weak solutions we use the Galerkin method, whose general outline we recall in Section 4.1. We prove the existence of weak solutions in bounded domains in Section 4.2, using a simple version of the Aubin–Lions Compactness Lemma (whose proof we postpone until Section 4.5). We then show that our weak solutions satisfy the strong energy inequality and sketch one method for proving the existence of weak solutions on the whole space.

4.1 The Galerkin method

Our definition of a weak solution u in a bounded domain $\Omega \subset \mathbb{R}^3$ (and in $\Omega = \mathbb{T}^3$) guarantees that at each time $t > 0$ the function $u(t)$ is an element of the infinite-dimensional space H spanned by the eigenfunctions of the Stokes operator, as defined in Chapter 2. The Galerkin method allows us to construct a weak solution u as the limit of a sequence of approximate solutions u_n that for each time $t > 0$ belong to the finite-dimensional space $P_n H$ spanned by the first n eigenfunctions,

$$P_n H := \operatorname{span}(a_1, a_2, \ldots, a_n), \quad a_i \in \mathcal{N}.$$

Here $\mathcal{N}$ is the basis given in Theorem 2.24 and by P_n we denote the projection operator $P_n : L^2 \to H$ defined by

$$P_n u = \sum_{i=1}^{n} \langle u, a_i \rangle a_i, \quad \text{where} \quad a_i \in \mathcal{N}. \tag{4.1}$$

Since on the torus and on a smooth bounded domain the a_k are smooth we have $P_n H \subset H^k$ for every $k \in \mathbb{N}$. Usually, we will consider P_n with domain L^2

$$P_n \colon L^2 \to P_n H,$$

but clearly (4.1) makes sense also for every $u \in H^{-1}$, so P_n is also well defined as an operator from H^{-1} to $P_n H$. It is easy to see that

$$\langle P_n u, v \rangle = \langle u, P_n v \rangle,$$

for $u, v \in L^2$, and that $P_n u$ approximate $u \in D(A^{s/2})$ arbitrarily well in the norm of H^s as $n \to \infty$.

Lemma 4.1 *Let Ω be the torus $\mathbb{T}^3$ or a smooth bounded domain in $\mathbb{R}^3$. If $u \in D(A^{s/2})$, $s > 0$, then $P_n u \to u$ in $H^s(\Omega)$ and*

$$\|P_n u\|_{H^s} \leq c \|u\|_{H^s}.$$

Proof See Exercise 4.1 for the case of the torus (the case of a bounded domain is similar). $\qquad\qquad\qquad\qquad\qquad\qquad\qquad\qquad\qquad\qquad\qquad\square$

We will now define a sequence of functions u_n from which we will later choose a subsequence converging to a weak solution u. The functions u_n are smooth in time and space and satisfy an equation (called the Galerkin equation) which can be thought of as the projection of the Navier–Stokes equations onto the space $P_n H$.

Definition 4.2 The nth-order Galerkin equation corresponding to the Navier–Stokes equations is given by

$$\partial_t u_n + A u_n + P_n[(u_n \cdot \nabla)u_n] = 0, \quad u_n(0) = P_n u_0,$$

where A is the Stokes operator (Definition 2.21). We call u_n the Galerkin approximations.

Since the Galerkin equations are finite dimensional, we would expect them to have solutions, at least for a short time. In Section 4.2 we will show that in fact solutions of the Galerkin equation exist globally in time.

If we take the dot product of the Galerkin equation with $\varphi \in \tilde{\mathcal{D}}_\sigma$, where

$$\varphi(x, t) = \alpha(t) a_k(x), \qquad 1 \leq k \leq n, \qquad \alpha \in C_c^1[0, \infty)$$

and integrate in space, then integrating the second term by parts yields

$$\langle \partial_t u_n, \varphi \rangle + \langle \nabla u_n, \nabla \varphi \rangle + \langle (u_n \cdot \nabla)u_n, \varphi \rangle = 0.$$

Integrating in time and using integration by parts in the first term we obtain

$$-\int_0^\infty \langle u_n, \partial_t \varphi \rangle + \int_0^\infty \langle \nabla u_n, \nabla \varphi \rangle + \int_0^\infty \langle (u_n \cdot \nabla) u_n, \varphi \rangle = \langle u_0, \varphi(0) \rangle. \quad (4.2)$$

It follows that u_n satisfies the weak formulation (3.7) of the equations for a subset of the space of test functions $\tilde{\mathcal{D}}_\sigma$, namely those test functions that belong to $P_n H$ for all times. Passing to the limit as $n \to \infty$ in equation (4.2) will enable us to remove this restriction on the choice of φ.

Let us fix $\varphi \in \tilde{\mathcal{D}}_\sigma$. Choose $T > 0$ large enough that the support of φ belongs to $\Omega \times [0, T]$ and $N \in \mathbb{N}$ such that $\varphi(t) \in P_N H$ for all t; it follows that (4.2) is satisfied for all $n \geq N$. In Section 4.2 we will prove that for any $T > 0$ the Galerkin approximations u_n satisfy the uniform bound

$$\sup_{0 \leq t \leq T} \|u_n(t)\|^2 + \int_0^T \|\nabla u_n(s)\|^2 \, ds \leq C.$$

This ensures the weak convergence of a subsequence of u_n (which we relabel) to some $u \in L^2(0, T; H) \cap L^2(0, T; V)$:

$$u_n \rightharpoonup u, \quad \nabla u_n \rightharpoonup \nabla u \quad \text{weakly in } L^2(0, T; L^2).$$

This weak convergence allows us to pass to the limit in the first two terms in equation (4.2) but is not enough to deal with the nonlinear term.

In order to overcome this difficulty we will prove that we can choose a subsequence of u_n (which we again relabel) such that

$$u_n \to u \quad \text{strongly in } L^2(0, T; H)$$

for every $T > 0$. The proof of this fact is based on the following simple version of the Aubin–Lions Lemma, whose proof we postpone until Section 4.5 at the end of this chapter.

Theorem 4.3 (Aubin–Lions Lemma; simple version) *Let H be $H(\mathbb{T}^3)$ or $H(\Omega)$, where Ω is a smooth bounded subset of $\mathbb{R}^3$. Assume that $p, q > 1$ and*

$$\|u_n\|_{L^q(0,T;V)} + \|\partial_t u_n\|_{L^p(0,T;V^*)} \leq C \qquad \text{for all } n \in \mathbb{N}.$$

Then there exists $u \in L^q(0, T; H)$ and a subsequence of u_n such that

$$u_{n_j} \to u \quad \text{strongly in} \quad L^q(0, T; H).$$

This additional strong convergence in L^2 is enough to pass to the limit in (4.2) for any fixed $\varphi \in \tilde{\mathcal{D}}_\sigma$ and thereby conclude that the limit function u satisfies the weak formulation of the Navier–Stokes equations for all $\varphi \in \tilde{\mathcal{D}}_\sigma$. From Lemma 3.14 it follows that u satisfies the weak formulation for all test functions in $\mathcal{D}_\sigma$, and we can conclude that u is indeed a weak solution of the Navier–Stokes equations.

4.2 Existence of weak solutions on bounded domains

Weak solutions of the Navier–Stokes equations can be constructed for each initial condition $u_0 \in H(\Omega)$ on an arbitrary open set in $\mathbb{R}^3$. This result was proved by Hopf (1951) as a generalisation of a previous existence theorem due to Leray (1934) for the whole space $\mathbb{R}^3$ (Hopf's method of proof was very different from Leray's).

In the proof below we assume that the domain $\Omega = \mathbb{T}^3$ or Ω is an open bounded subset of $\mathbb{R}^3$ with smooth boundary. This assumption allows us to use the set $\mathcal{N}$ in the construction of the Galerkin approximations. This simplifies the exposition, but essentially the same computations can be carried out for an arbitrary open set if we replace $\mathcal{N}$ with an appropriate basis of L^2 consisting of smooth compactly supported functions as was done in the original paper by Hopf. The case of $\Omega = \mathbb{R}^3$, which is due to Leray, will be treated later in Section 4.4 (but with a proof different from that of Leray).

Theorem 4.4 (Hopf) *Let Ω be the torus $\mathbb{T}^3$ or a smooth bounded domain $\Omega \subset \mathbb{R}^3$. For every $u_0 \in H(\Omega)$ there exists at least one global-in-time weak solution u of the Navier–Stokes equations satisfying the initial condition u_0.*

Proof
Step 1. We prove that solutions of the Galerkin equations exist at least locally in time

We consider the problem

$$\partial_t u_n + A u_n + P_n[(u_n \cdot \nabla) u_n] = 0, \tag{4.3}$$

$$u_n(0) = P_n u_0, \tag{4.4}$$

for a function u_n of the form

$$u_n(x, t) = \sum_{k=1}^{n} c_k^n(t) a_k(x), \quad a_k \in \mathcal{N},$$

where the $\{c_k^n\}$ are scalar functions of time. It is easy to see that

$$\|P_n u_0 - u_0\| \to 0,$$

see Lemma 4.1. To determine u_n we need to find the (time-dependent) functions c_k^n. To this end we take the inner product of (4.3) in L^2 with a_k,

$k = 1, 2, \ldots, n$. Since $\langle P_n v, a_k \rangle = \langle v, a_k \rangle$ for every $v \in L^2$ and $1 \le k \le n$ this yields[1]

$$\sum_{j=1}^{n} \langle \dot{c}_j^n(t) a_j, a_k \rangle + \sum_{j=1}^{n} \langle c_j^n(t) A a_j, a_k \rangle + \sum_{i,j=1}^{n} \langle (c_i^n(t) a_i \cdot \nabla) c_j^n(t) a_j, a_k \rangle = 0,$$

where $\dot{c}_k^n = \mathrm{d} c_k^n / \mathrm{d} t$. Using the fact that the a_k are eigenfunctions of the Stokes operator and are orthonormal in L^2 we therefore obtain a system of n ODEs,

$$\dot{c}_k^n(t) = -\lambda_k \, c_k^n(t) - \sum_{i,j=1}^{n} B_{ijk} \, c_i^n(t) c_j^n(t), \qquad B_{ijk} = \langle (a_i \cdot \nabla) a_j, a_k \rangle, \quad (4.5)$$

for $k = 1, 2, \ldots, n$. To find initial conditions for c_k^n we take the inner product of (4.4) with a_k, which yields

$$c_k^n(0) = \langle u_0, a_k \rangle.$$

The existence and uniqueness of solutions to this system of equations (the coefficients (c_k^n)), and hence of the corresponding function u_n, on some time interval $[0, T_n)$ is immediate from the classical theory of ODEs (see Hartman, 1973), since the right-hand side of (4.5) is continuous and locally Lipschitz. However, since we have quadratic terms on the right-hand side it is not immediately clear whether the functions c_k^n can be defined globally or whether they blow up at some time $T_n < \infty$.

Step 2. We obtain uniform estimates on the solutions u_n and hence show that they exist globally in time

We already know that u_n exists at least on some time interval $[0, T_n)$. Let $s \in (0, T_n)$. We take the inner product of (4.3) with $u_n(s)$ to get

$$\langle \partial_t u_n(s), u_n(s) \rangle + \langle A u_n(s), u_n(s) \rangle + \langle P_n(u_n(s) \cdot \nabla) u_n(s), u_n(s) \rangle = 0.$$

We notice that

$$\langle \partial_t u_n(s), u_n(s) \rangle = \frac{1}{2} \frac{\mathrm{d}}{\mathrm{d} t} \| u_n(s) \|^2$$

and

$$\langle A u_n(s), u_n(s) \rangle = \langle -\mathbb{P} \Delta u_n, u_n \rangle = \langle -\Delta u_n, \mathbb{P} u_n \rangle$$
$$= \langle -\Delta u_n, u_n \rangle = \| \nabla u_n(s) \|^2.$$

---

[1] We recall that functions a_k are smooth and bounded in every H^k, so all the computations that we do in this section are allowed.

The nonlinear term vanishes using Lemma 3.2:

$$\langle P_n[(u_n \cdot \nabla)u_n], u_n \rangle = \langle (u_n \cdot \nabla)u_n, P_n u_n \rangle = \langle (u_n \cdot \nabla)u_n, u_n \rangle = 0.$$

Therefore for all $s > 0$ we have

$$\frac{1}{2}\frac{d}{dt}\|u_n(s)\|^2 + \|\nabla u_n(s)\|^2 = 0. \tag{4.6}$$

It follows that the L^2 norm of u_n is non-increasing in time. Since

$$\sum_{k=1}^{n} |c_k^n(s)|^2 = \|u_n(s)\|^2$$

it follows that the c_k^n do not blow up in finite time and hence $T_n = \infty$. Moreover, integrating (4.6) in time between 0 and any $t \in [0, \infty)$ we obtain

$$\frac{1}{2}\|u_n(t)\|^2 + \int_0^t \|\nabla u_n(s)\|^2 \, ds = \frac{1}{2}\|u_n(0)\|^2 \le \frac{1}{2}\|u_0\|^2, \tag{4.7}$$

so

$$\sup_{t\ge 0} \|u_n(t)\|^2 \le \|u_0\|^2 \quad \text{and} \quad \int_0^\infty \|\nabla u_n(s)\|^2 \, ds \le \frac{1}{2}\|u_0\|^2.$$

Therefore we have shown that u_n is bounded uniformly (in n) in the norm of $L^\infty(0, \infty; H)$ and the norm of $L^2(0, \infty; V)$ with

$$\sup_{t\ge 0} \|u_n(t)\|^2 + \int_0^\infty \|\nabla u_n(s)\|^2 \, ds \le 2\|u_0\|^2. \tag{4.8}$$

Step 3. In order to use the Aubin–Lions Lemma (Theorem 4.3) we find uniform bounds on $\partial_t u_n$ in $L^{4/3}(0, T; V^*)$ for all $T > 0$

We recall that we have already proved in Lemma 3.7 that weak solutions (when they exist) have their time derivative in $L^{4/3}(0, T; V^*)$ so the bound we want to prove here is quite natural.

Assume that $\varphi \in V$. We take the L^2 inner product of the Galerkin equation (4.3) with φ to obtain

$$\langle \partial_t u_n, \varphi \rangle = -\langle A u_n, \varphi \rangle - \langle P_n[(u_n \cdot \nabla)u_n], \varphi \rangle$$
$$= -\langle A u_n, \varphi \rangle - \langle (u_n \cdot \nabla)u_n, P_n \varphi \rangle.$$

We will now estimate the norm $\|\partial_t u_n\|_{V^*}$ by estimating the right-hand side of the above equality. The first term is easy to estimate: we have

$$
\begin{aligned}
|\langle Au_n, \varphi \rangle| = |\langle -\mathbb{P}\Delta u_n, \varphi \rangle| &\le |\langle -\Delta u_n, \mathbb{P}\varphi \rangle| \\
&= |\langle -\Delta u_n, \varphi \rangle| = |\langle \nabla u_n, \nabla \varphi \rangle| \\
&\le \|\nabla u_n\| \|\varphi\|_V,
\end{aligned}
$$

where we used the fact that functions in V vanish on the boundary (or are periodic in the case of the torus) and that for $\varphi \in V$ we have $\mathbb{P}\varphi = \varphi$.

For the second term we use the standard L^3-L^2-L^6 Hölder inequality (1.6), then the Lebesgue interpolation inequality (1.8)

$$
\|u\|_{L^3} \le \|u\|_{L^2}^{1/2} \|u\|_{L^6}^{1/2},
$$

and finally the Sobolev embedding $H_0^1 \subset L^6$ (or $\dot{H}^1(\mathbb{T}^3) \subset L^6(\mathbb{T}^3)$):

$$
\begin{aligned}
|\langle (u_n \cdot \nabla) u_n, P_n \varphi \rangle| &\le \|u_n\|_{L^3} \|\nabla u_n\|_{L^2} \|P_n \varphi\|_{L^6} \\
&\le c \|u_n\|_{L^2}^{1/2} \|u_n\|_{L^6}^{1/2} \|\nabla u_n\| \|P_n \varphi\|_V \\
&\le c \|u_n\|_{L^2}^{1/2} \|\nabla u_n\|^{3/2} \|\varphi\|_V.
\end{aligned}
$$

Therefore we obtain[2]

$$
\|\partial_t u_n\|_{V^*} \le \|\nabla u_n\| + c \|u_n\|^{1/2} \|\nabla u_n\|^{3/2}.
$$

It follows that

$$
\begin{aligned}
\int_0^T \|\partial_t u_n(s)\|_{V^*}^{4/3} \, ds \\
&\le c \int_0^T \|\nabla u_n(s)\|^{4/3} \, ds + c \int_0^T \|u_n(s)\|^{2/3} \|\nabla u_n(s)\|^2 \, ds \\
&\le c T^{1/3} \|u_n\|_{L^2(0,T;V)}^{4/3} + c \|u_n\|_{L^\infty(0,T;H)}^{2/3} \|u_n(s)\|_{L^2(0,T;V)}^2 \\
&\le c T^{1/3} \|u_0\|^{4/3} + c \|u_0\|^{8/3},
\end{aligned}
$$

where in the last inequality we used the bounds obtained in Step 2. Notice that the right-hand side depends on the time T, so the uniform bound

$$
\|\partial_t u_n\|_{L^{4/3}(0,T;V^*)} \le C_T \quad \text{for all} \quad n \in \mathbb{N} \tag{4.9}
$$

is valid for any $T < \infty$ but not for $T = \infty$.

---

[2] Note that the constant c does not depend on n or Ω.

Step 4. We extract a convergent subsequence and pass to the limit in the equation

Using the Banach–Alaoglu Theorem (Theorem A.22 and Corollary A.23) and Theorem 4.3, we now apply a standard diagonal argument[3] to choose a subsequence of the (u_n) (which we relabel) such that for every time $T > 0$ we have

$$u_n \to u \text{ strongly in } L^2(0, T; H(\Omega)), \tag{4.10}$$

$$u_n \overset{*}{\rightharpoonup} u \text{ weakly-}* \text{ in } L^\infty(0, T; H(\Omega)), \quad \text{and} \tag{4.11}$$

$$\nabla u_n \rightharpoonup \nabla u \text{ weakly in } L^2(0, T; L^2(\Omega)) \tag{4.12}$$

(the same function u occurs as the limit in all of (4.10)–(4.12), see Exercise 4.5). Furthermore, since it will be useful later (see Section 4.3), we use the strong convergence (4.10) to choose u_n in such a way that

$$u_n(t) \to u(t) \text{ in } L^2(\Omega) \quad \text{for almost every } t \in [0, \infty). \tag{4.13}$$

We also note (though it is not needed to conclude this proof) that

$$\partial_t u_n \rightharpoonup \partial_t u \quad \text{in} \quad L^{4/3}(0, T; V^*),$$

see Exercise 4.6.

We now show that u is a weak solution of the Navier–Stokes equations, i.e. that u satisfies (3.3). From Lemma 3.11 and Lemma 3.14 it is enough to check that for any fixed $\varphi \in \tilde{\mathcal{D}}_\sigma$ with $\varphi(x, t) = \alpha(t) a_N(x)$ we have

$$-\int_0^\infty \langle u, \partial_t \varphi \rangle + \int_0^\infty \langle \nabla u, \nabla \varphi \rangle + \int_0^\infty \langle (u \cdot \nabla) u, \varphi \rangle = \langle u_0, \varphi(0) \rangle. \tag{4.14}$$

We choose $T > 0$ such that the support of α is a subset of $[0, T)$. Hence we only need to prove that

$$-\int_0^T \langle u, \partial_t \varphi \rangle + \int_0^T \langle \nabla u, \nabla \varphi \rangle + \int_0^T \langle (u \cdot \nabla) u, \varphi \rangle = \langle u_0, \varphi(0) \rangle.$$

We know, see (4.2), that for all $n \geq N$ the Galerkin approximations u_n satisfy

$$-\int_0^T \langle u_n, \partial_t \varphi \rangle + \int_0^T \langle \nabla u_n, \nabla \varphi \rangle + \int_0^T \langle (u_n \cdot \nabla) u_n, \varphi \rangle = \langle u_0, \varphi(0) \rangle. \tag{4.15}$$

We pass to the limit in (4.15), as $n \to \infty$. From the convergence (4.10) and (4.12) we have

$$\int_0^T \langle u_n, \partial_t \varphi \rangle \to \int_0^T \langle u, \partial_t \varphi \rangle \quad \text{and} \quad \int_0^T \langle \nabla u_n, \nabla \varphi \rangle \to \int_0^T \langle \nabla u, \nabla \varphi \rangle.$$

---

[3] For more details see Exercise 4.4.

To prove that

$$\int_0^T \langle (u_n \cdot \nabla)u_n, \varphi \rangle \to \int_0^T \langle (u \cdot \nabla)u, \varphi \rangle$$

we notice that

$$(u_n \cdot \nabla)u_n - (u \cdot \nabla)u = ((u_n - u) \cdot \nabla)u_n + (u \cdot \nabla)(u_n - u).$$

Hence we need to show that

$$\int_0^T \langle ((u_n - u) \cdot \nabla)u_n, \varphi \rangle \to 0 \quad \text{and} \quad \int_0^T \langle u \cdot \nabla(u_n - u), \varphi \rangle \to 0.$$

From the bound (4.8) and the strong convergence (4.10) it follows that

$$\left| \int_0^T \langle ((u_n - u) \cdot \nabla)u_n, \ \varphi \rangle \right| \leq C_\varphi \int_0^T \|u_n - u\| \|\nabla u_n\|$$

$$\leq C_\varphi \left(\int_0^T \|u_n - u\|^2 \right)^{1/2} \left(\int_0^T \|\nabla u_n\|^2 \right)^{1/2}$$

$$\to 0.$$

Moreover, from the weak convergence of gradients in (4.12) it follows that

$$\int_0^T \langle \partial_j (u_n - u)_i, u_j \varphi_i \rangle \to 0$$

as $n \to \infty$ for all $1 \leq i, j \leq 3$, because $u_j \varphi_i \in L^2$. Hence

$$\int_0^T \langle u \cdot \nabla(u_n - u), \varphi \rangle \to 0.$$

Therefore we obtain (4.14).

It follows (from Lemma 3.14) that u satisfies the weak form of the Navier–Stokes equations for all $\varphi \in \mathcal{D}_\sigma$, as required for a weak solution. Moreover from the fact that u_n is weakly convergent in $L^2(0, T; V)$ and weakly-$*$ convergent in $L^\infty(0, T; H)$ for all $T > 0$ it follows that u belongs to $L^\infty(0, T; H)$ and $L^2(0, T; V)$ for all $T > 0$. Therefore u is a global-in-time weak solution of the Navier–Stokes equations as claimed. $\qquad \square$

Finally, we notice that the same procedure in the case of the torus gives us a stronger bound on the time derivative of a weak solution: we show a bound in a smaller space $H^{-1} \subset V^*$.

Lemma 4.5 *If $u_0 \in H(\mathbb{T}^3)$ then there exists a weak solution of the Navier–Stokes equations corresponding to u_0 and such that*

$$\partial_t u \in L^{4/3}(0, T; H^{-1}).$$

Proof   The proof is exactly the same as before but differs in Step 3 where we can prove a stronger uniform bound on $\partial_t u_n$:

$$\|\partial_t u_n\|_{L^{4/3}(0,T;H^{-1})} \leq C \quad \text{for all} \quad n \in \mathbb{N}.$$

Indeed, in the case of the torus $\mathbb{T}^3$ if $\varphi \in \dot{H}^1$ then we still have the identity

$$\langle \partial_t u_n, \varphi \rangle = -\langle A u_n, \varphi \rangle - \langle (u_n \cdot \nabla) u_n, P_n \varphi \rangle$$

and the estimate

$$|\langle (u_n \cdot \nabla) u_n, P_n \varphi \rangle| \leq c \|u_n\|_{L^2}^{1/2} \|\nabla u_n\|_{L^2}^{3/2} \|\varphi\|_{\dot{H}^1}.$$

However, we can prove that

$$|\langle A u_n, \varphi \rangle| = |\langle -\mathbb{P}\Delta u_n, \varphi \rangle| = |\langle -\Delta u_n, \varphi \rangle|$$
$$\leq |\langle \nabla u_n, \nabla \varphi \rangle| \quad \leq \|u_n\|_{H^1} \|\varphi\|_{\dot{H}^1},$$

where we used the fact that on the torus $\mathbb{P}\Delta u_n = \Delta \mathbb{P} u_n = \Delta u_n$, because $\mathbb{P}$ commutes with derivatives on $\mathbb{T}^3$. Combining this new estimate with the previous bound for the nonlinear term completes the proof. $\square$

4.3 The strong energy inequality and Leray–Hopf weak solutions

The particular method that we have used to construct weak solutions allows us to prove an additional property that to date is unknown for general weak solutions (i.e. functions u that are only known to satisfy Definition 3.3). We have already seen in (3.2) that classical solutions satisfy the energy equality

$$\frac{1}{2}\|u(t)\|^2 + \int_0^t \|\nabla u\|^2 \, ds = \frac{1}{2}\|u_0\|^2 \qquad \text{for all } t \geq 0.$$

We now show that the weak solutions constructed above satisfy a very similar energy *inequality*. This observation in the case of an arbitrary open domain in $\mathbb{R}^3$ is due to Hopf (1951); here we prove a stronger result due to Ladyzhenskaya (1969).[4]

Theorem 4.6 *Let Ω be a bounded domain in $\mathbb{R}^3$ (or the torus $\mathbb{T}^3$). The weak solution u of the Navier–Stokes equations constructed in Theorem 4.4 satisfies*

---

[4] Hopf showed that solutions satisfy (4.17) while Ladyzhensakya proved (4.16). Earlier, Leray (1934) had obtained (4.16) for the weak solutions that he constructed on the whole space (see Section 14.5).

the strong energy inequality:

$$\frac{1}{2}\|u(t)\|^2 + \int_s^t \|\nabla u\|^2 \le \frac{1}{2}\|u(s)\|^2 \quad \text{for all } t > s \qquad (4.16)$$

for almost all times $s \in [0, \infty)$, including $s = 0$.
In particular, u satisfies the energy inequality

$$\frac{1}{2}\|u(t)\|^2 + \int_0^t \|\nabla u\|^2 \le \frac{1}{2}\|u(0)\|^2 \quad \text{for all } t > 0. \qquad (4.17)$$

Proof It follows from (4.13) that for almost all $s, t \in [0, \infty)$ we have

$$u_n(s) \to u(s) \quad \text{and} \quad u_n(t) \to u(t)$$

strongly in $L^2(\Omega)$, so in particular

$$\|u_n(s)\|^2 \to \|u(s)\|^2 \quad \text{and} \quad \|u_n(t)\|^2 \to \|u(t)\|^2. \qquad (4.18)$$

Notice that we always have

$$\|u_n(0)\|^2 \to \|u(0)\|^2,$$

because $u_n(0) = P_n u_0$.

For all s and t for which (4.18) holds we take the lim inf of both sides of the equation

$$\frac{1}{2}\|u_n(t)\|^2 + \int_s^t \|\nabla u_n\|^2 = \frac{1}{2}\|u_n(s)\|^2 \qquad (4.19)$$

(this equation follows immediately from (4.7)). Since $\nabla u_n \rightharpoonup \nabla u$ weakly in $L^2(s, t; L^2(\Omega))$ it follows that

$$\int_s^t \|\nabla u\|^2 \le \liminf_{n \to \infty} \int_s^t \|\nabla u_n\|^2.$$

Therefore passing to the lim inf as $n \to \infty$ on both sides of (4.19) we obtain, using (4.18),

$$\frac{1}{2}\|u(t)\|^2 + \int_s^t \|\nabla u\|^2 \le \frac{1}{2}\|u(s)\|^2$$

for almost all s and almost all $t > s$. We want to prove that this inequality holds in fact for almost all s and *all* $t > s$. To this end let us fix some $s > 0$ for which

$$\frac{1}{2}\|u(t)\|^2 + \int_s^t \|\nabla u\|^2 \le \frac{1}{2}\|u(s)\|^2$$

for almost all $t > s$. Now we choose any $t > s$. We can find a sequence $t' \to t$ such that

$$\frac{1}{2}\|u(t')\|^2 + \int_s^{t'} \|\nabla u\|^2 \leq \frac{1}{2}\|u(s)\|^2.$$

Obviously

$$\int_s^{t'} \|\nabla u\|^2 \quad \to \quad \int_s^t \|\nabla u\|^2.$$

Since u is L^2-weakly continuous, it follows that

$$\|u(t)\|^2 \leq \liminf_{t' \to t} \|u(t')\|^2.$$

Hence

$$\frac{1}{2}\|u(t)\|^2 + \int_s^t \|\nabla u\|^2 \leq \liminf_{t' \to t} \left(\frac{1}{2}\|u(t')\|^2 + \int_s^{t'} \|\nabla u\|^2 \right)$$

$$\leq \frac{1}{2}\|u(s)\|^2. \qquad \qquad \square$$

A consequence of the energy inequality is that the weak solutions that we have constructed attain their initial condition in a strong sense. This fact was shown by Hopf (1951) for the weak solutions he constructed by the Galerkin method in arbitrary domains. We emphasise again that this property is not known for a general weak solution.

Corollary 4.7 *For the weak solutions constructed in Theorem 4.4 we have*

$$u(t) \to u(0) \quad strongly\ in \quad L^2(\Omega) \quad as \quad t \to 0^+.$$

Proof From the energy inequality in (4.17),

$$\|u(t)\|^2 + 2\int_0^t \|\nabla u(s)\|^2\,\mathrm{d}s \leq \|u(0)\|^2,$$

we obtain

$$\limsup_{t \to 0^+} \|u(t)\|^2 \leq \|u(0)\|^2.$$

On the other hand $u(0)$ is the weak limit of $u(t)$ as $t \to 0^+$ so

$$\|u(0)\| \leq \liminf_{t \to 0^+} \|u(t)\|.$$

Hence

$$u(t) \rightharpoonup u(0) \quad \text{in} \quad L^2(\Omega) \quad \text{and} \quad \|u(t)\| \to \|u(0)\| \quad \text{as} \quad t \to 0^+.$$

It follows (see Lemma A.20) that

$$u(t) \to u(0) \quad \text{strongly in} \quad L^2(\Omega) \quad \text{as} \quad t \to 0^+. \qquad \square$$

Note that we have not proved the strong L^2-continuity of weak solutions at times other than $t = 0$; it is not known whether this property holds, even for the solutions that we have constructed.

Corollary 4.8 *If s is a time such that (4.16) holds then $\|u(\cdot)\|$ is right continuous at s.*

Proof Passing to the limit as $t \to s^+$ in (4.16) gives

$$\limsup_{t \to s^+} \|u(t)\| \le \|u(s)\|.$$

Since $u(s)$ is a weak limit of $u(t)$ when $t \to s^+$ it follows that

$$\|u(s)\| \le \liminf_{t \to s^+} \|u(t)\|.$$

Hence $\|u(t)\| \to \|u(s)\|$ when $t \to s^+$, and $u(t) \rightharpoonup u(s)$ as $t \to s^+$. Now the assertion follows from Lemma A.20. $\qquad \square$

Weak solutions of the Navier–Stokes equations satisfying the strong energy inequality have so many good properties that it will be useful to distinguish them from more general weak solutions (which, as far as we know, may not satisfy the strong energy inequality).

Definition 4.9 We say that u is a *Leray–Hopf weak solution* of the Navier–Stokes equations if u is a weak solution of the Navier–Stokes equations that satisfies the strong energy inequality:

$$\frac{1}{2}\|u(t)\|^2 + \int_s^t \|\nabla u\|^2 \le \frac{1}{2}\|u(s)\|^2 \quad \text{for all} \ t > s$$

for $s = 0$ and almost all times $s \in (0, \infty)$.

Note that without a uniqueness result for weak solutions (which is not available) it is possible that there are Leray–Hopf weak solutions that do not coincide with the weak solutions that we have constructed using the Galerkin method.

4.4 Existence of weak solutions on the whole space

In Theorem 4.4 we assumed that Ω is bounded, but the existence of weak solutions can be proved in a similar way on any sufficiently regular open domain.

In fact the first proof of existence of weak solutions of the Navier–Stokes equations is due to Leray (1934) who showed their existence in the case of the whole space.

In this section we will sketch one method of constructing weak solutions on the whole space (a similar method works for more general unbounded domains). The main idea of this proof – due to Heywood (1988) – is to use solutions of the Navier–Stokes equations on a sequence of expanding bounded domains as a sequence of approximate solutions on the whole space and show convergence of such solutions in appropriate function spaces. The resulting solution satisfies the energy inequality but is not known to satisfy the strong energy inequality. We notice however that weak solutions satisfying the strong energy inequality (hence Leray–Hopf weak solutions) can be constructed on the whole space, as proved by Leray in 1934. We will sketch the key ingredient of Leray's proof in Section 14.5.

Theorem 4.10 *For each $u_0 \in H(\mathbb{R}^3)$ there exists at least one weak solution of the Navier–Stokes equations on the whole space.*

Proof Given $u_0 \in H(\mathbb{R}^3)$ we choose a sequence $u_{0,n} \in C_{c,\sigma}^\infty(B_n)$ where B_n is a ball of radius R_n centred at zero, R_n is an increasing sequence with $R_n \to \infty$, and

$$u_{0,n} \to u_0 \text{ in } H(\mathbb{R}^3) \text{ with } \|u_{0,n}\| \le \|u_0\|.$$

Using Theorem 4.4, for each n we can find a weak solution $\tilde{u}_n$ of the Navier–Stokes equations with the zero Dirichlet boundary condition in B_n and with initial condition $u_{0,n}$. We extend the functions $\tilde{u}_n$ by zero outside B_n and denote such extended functions by u_n. (Note that due to the zero boundary conditions these extended functions belong to $V(\mathbb{R}^3)$ for almost every time t and $\|u_n(t)\|_{H^1(\mathbb{R}^3)} = \|u(t)\|_{H^1(B_n)}$.) The sequence u_n of weak solutions share many properties with the sequence of Galerkin approximations considered in the proof of Theorem 4.4. In particular, it follows from our method of construction that we have a uniform bound

$$\frac{1}{2}\|u_n(t)\|_{L^2(\mathbb{R}^3)}^2 + \int_0^t \|\nabla u_n\|_{L^2(\mathbb{R}^3)}^2 \le \frac{1}{2}\|u_n(0)\|^2 \le \frac{1}{2}\|u_0\|^2.$$

From this bound we conclude that for some subsequence of u_n (which we relabel)

$$\nabla u_n \rightharpoonup \nabla u \text{ weakly in } L^2(0, T; L^2(\mathbb{R}^3))$$

for some $u \in L^2(0, T; V(\mathbb{R}^3))$.

Furthermore, for all $n \in \mathbb{N}$ from the estimates on the time derivative in Section 4.2 we have

$$\|\partial_t u_n\|_{L^{4/3}(0,T;V^*(B_n))} \leq cT^{1/4}\|u_n(0)\| + c\|u_n(0)\|^2$$
$$\leq cT^{1/4}\|u_0\| + c\|u_0\|^2 \leq C,$$

for some $C > 0$ since c does not depend on n. Since $V^*(B_n) \subset V^*(B_M)$ for all $n \geq M$ with

$$\| \cdot \|_{V^*(B_M)} \leq \| \cdot \|_{V^*(B_n)},$$

it follows that for each $M \in \mathbb{N}$ we have for all $n \geq M$:

$$\|\partial_t u_n\|_{L^{4/3}(0,T;V^*(B_M))} + \|u_n\|_{L^2(0,T;H^1(B_M))} \leq C.$$

Thus for every $T > 0$ and every $M \in \mathbb{N}$ we can use the 'abstract' version of the Aubin–Lions Lemma (Theorem 4.12 in the next section) with

$$G = H^1(B_M), \qquad H = L^2(B_M), \qquad \text{and} \qquad K = V^*(B_M)$$

to find a subsequence of u_n that converges strongly in $L^2(0, T; L^2(B_M))$. Using the standard diagonal argument we can choose a subsequence of u_n, that we relabel, such that

$$u_n \to u \quad \text{strongly in} \quad L^2(0, T; L^2(B_M))$$

for every $M = 1, 2, 3, \ldots$

It remains to show that the limit function u is a weak solution of the Navier–Stokes equations on the whole space. To do this, given any test function $\phi \in \mathcal{D}_\sigma(\mathbb{R}^3)$ we let M be large enough so that the support of ϕ is contained in $B_M \times [0, T)$. Then for all $n > M$ we have

$$-\int_0^\infty \langle u_n, \partial_t \phi \rangle + \int_0^\infty \langle \nabla u_n, \nabla \phi \rangle + \int_0^\infty \langle (u_n \cdot \nabla) u_n, \phi \rangle = \langle u_0, \phi(0) \rangle$$

since u_n is a weak solution of the Navier–Stokes equations in $B_n \times [0, T)$ and $\phi \in \mathcal{D}_\sigma(B_n)$, because $B_M \subset B_n$ for $n > M$. We pass to the limit as $n \to \infty$, using the weak convergence of gradients and strong convergence of u_n in $L^2(0, T; L^2(B_M))$, and prove that u is a weak solution of the Navier–Stokes equations with initial condition u_0. $\qquad\square$

4.5 The Aubin–Lions Lemma

In this section we give a proof of a simple version of a compactness theorem due to Aubin (1963) and Lions (1969) (for optimal results see Simon, 1987), which we use in the proof of Theorem 4.4 to find a subsequence of the solutions of the Galerkin equations that converges strongly in $L^2(0, T; H)$. We will use ideas from Chemin et al. (2006). We note that (u_n) in the theorem below is an 'abstract' sequence that does not have to be related to the Galerkin approximations.

Theorem 4.11 (Aubin–Lions Lemma; simple version) *Let Ω be the torus $\mathbb{T}^3$ or a smooth bounded domain in $\mathbb{R}^3$. If the sequence (u_n) is uniformly bounded in $L^q(0, T; V)$, $q > 1$, and the partial derivatives $\partial_t u_n$ are uniformly bounded in $L^p(0, T; V^*)$, $p > 1$, then there exist $u \in L^q(0, T; H)$ and a subsequence of (u_n) that converges strongly to u in $L^q(0, T; H)$.*

Proof We will first obtain bounds on finite-dimensional projections of the sequence (u_n). To this end let us consider $\langle u_n, a_j \rangle$, where $a_j \in \mathcal{N}$ is an eigenfunction of the Stokes operator corresponding to the eigenvalue λ_j. From Lemma 1.31 we can deduce that $\langle u_n, a_j \rangle$ is absolutely continuous with $\frac{\mathrm{d}}{\mathrm{d}t} \langle u_n, a_j \rangle = \langle \partial_t u_n, a_j \rangle$ almost everywhere on $[0, T]$, so

$$\langle u_n(s), a_j \rangle = \langle u_n(s^*), a_j \rangle + \int_{s^*}^s \langle \partial_t u_n, a_j \rangle$$

for almost every $s \in [0, T]$. Using an idea from Lemma 10.4 in Renardy & Rogers (2004), there exists a time $s^* \in [0, T]$ such that

$$\langle u_n(s^*), a_j \rangle = \frac{1}{T} \int_0^T \langle u_n, a_j \rangle$$

and we can estimate $\langle u_n(s), a_j \rangle$ as follows:

$$\sup_{0 \leq s \leq T} |\langle u_n(s), a_j \rangle| \leq |\langle u_n(s^*), a_j \rangle| + \left| \int_{s^*}^s \langle \partial_t u_n, a_j \rangle \right|$$

$$\leq \frac{1}{T} \int_0^T \|u_n\| \|a_j\| + \int_0^T \|\partial_t u_n\|_{V^*} \|a_j\|_V$$

$$\leq \frac{1}{T} \|u_n\|_{L^q(0,T;H)} T^{1-1/q} + \|\partial_t u_n\|_{L^p(0,T;V^*)} T^{1-1/p} \lambda_j^{1/2}$$

$$\leq C_1 + C_2 \lambda_j^{1/2}.$$

(We used Hölder's inequality and the equality $\|a_j\|_V^2 = \|\nabla a_j\|^2 = \lambda_j$.) Therefore if we define

$$P_k u_n = \sum_{j=1}^k \langle u_n, a_j \rangle a_j$$

then $P_k u_n \in C([0, T]; H)$ and

$$\sup_{0 \le s \le T} \|P_k u_n(s)\| \le \sum_{j=1}^{k} (C_1 + C_2 \lambda_j^{1/2}) \le k(C_1 + C_2 \lambda_k^{1/2}) \qquad (4.20)$$

because (λ_j) is an increasing sequence.

We claim that for each fixed k the sequence $(P_k u_n)$ has a subsequence that converges in $C([0, T]; P_k H)$. So let $k \in \mathbb{N}$ be fixed. From (4.20) we know that the sequence $(P_k u_n)$ is uniformly bounded with values in a finite-dimensional space $P_k H$. Hence there is a compact set $K \subset P_k H$ such that $P_k u_n : [0, T] \to K$ for all $n \in \mathbb{N}$.

Moreover, the sequence $(P_k u_n)$ is equicontinuous; to prove this we use the fact that $P_k H$ is finite dimensional, and so the norms $\| \cdot \|_{L^2}$ and $\| \cdot \|_{V^*}$ are equivalent on $P_k H$. We can therefore prove the equicontinuity of $(P_k u_n)$ as follows:

$$\|P_k u_n(t_2) - P_k u_n(t_1)\|_{L^2} = \left\| \int_{t_1}^{t_2} \partial_t P_k u_n(s) \, ds \right\|_{L^2}$$

$$\le \int_{t_1}^{t_2} \|\partial_t P_k u_n(s)\|_{L^2} \, ds$$

$$\le c_k \int_{t_1}^{t_2} \|\partial_t P_k u_n(s)\|_{V^*} \, ds$$

$$\le c_k \left(\int_{t_1}^{t_2} \|\partial_t P_k u_n(s)\|_{V^*}^{p} \, ds \right)^{1/p} \left(\int_{t_1}^{t_2} ds \right)^{1-1/p}$$

$$\le \tilde{c}_k |t_2 - t_1|^{1-1/p}.$$

An application of the Arzelà–Ascoli Theorem now provides a convergent subsequence in $C([0, T]; P_k H)$ as claimed.

We are now ready to show that (u_n) has a Cauchy subsequence in $L^q(0, T; H)$. Using the diagonal method we now find a subsequence of u_n (which we relabel) such that for every k the sequence $(P_k u_n)$ is convergent in $L^q(0, T; H)$. We want to prove that this subsequence is a Cauchy sequence in $L^q(0, T; H)$, i.e. that for each $\varepsilon > 0$ there is N such that

$$\int_0^T \|u_n(s) - u_m(s)\|^q \, ds < \varepsilon \quad \text{for all} \quad n, m \ge N. \qquad (4.21)$$

So let $\varepsilon > 0$ be fixed. We notice that for any fixed number $\delta > 0$ we can find a natural number k such that

$$\int_0^T \|P_k u_n(s) - u_n(s)\|^q \, ds < \delta \quad \text{for all} \quad n \ge k. \qquad (4.22)$$

This is possible because, according to our assumption (i) and the fact that $\|u_n\|_V = \|A^{1/2}u_n\|$, we have

$$C \geq \int_0^T \|\nabla u_n(s)\|^q \, ds$$

$$= \int_0^T \left(\sum_{j=1}^{\infty} \lambda_j |\langle u_n(s), a_j \rangle|^2 \right)^{q/2} \, ds$$

$$\geq \int_0^T \left(\sum_{j=k+1}^{\infty} \lambda_j |\langle u_n(s), a_j \rangle|^2 \right)^{q/2} \, ds.$$

So using the fact that the λ_k form an increasing sequence, we obtain

$$C \geq \lambda_{k+1}^{q/2} \int_0^T \left(\sum_{j=k+1}^{\infty} |\langle u_n(s), a_j \rangle|^2 \right)^{q/2} \, ds$$

$$\geq \lambda_{k+1}^{q/2} \int_0^T \|P_k u_n(s) - u_n(s)\|^q \, ds.$$

Since $\lambda_k \to \infty$ as $k \to \infty$ we can choose k large enough that $C\lambda_{k+1}^{-q/2} < \delta$, and then (4.22) is satisfied for all $n > k$.

Now fix k such that (4.22) holds. Using the fact that $(P_k u_n)$ converges in $C([0, T]; H)$ we can find $N > k$ such that

$$\int_0^T \|P_k u_n(s) - P_k u_m(s)\|^q \, ds < \delta$$

for all $n, m \geq N$. To prove (4.21) we use the triangle inequality and obtain

$$\|u_n - u_m\|_{L^q(0,T;H)} \leq \|u_n - P_k u_n\|_{L^q(0,T;H)} + \|P_k u_n - P_k u_m\|_{L^q(0,T;H)}$$
$$+ \|P_k u_m - u_m\|_{L^q(0,T;H)} < 3\delta^{1/q}.$$

Choosing $\delta = \varepsilon^q/3^q$ we obtain (4.21). It follows that (u_n) is a Cauchy sequence in $L^q(0, T; H)$ so it must be strongly convergent to some v in the space $L^q(0, T; H)$. □

It is easy to see that the result of Theorem 4.3 does not have to be restricted to the case where the spaces H and V are related to the Stokes operator. Indeed, the proof can be generalised to the case of a more abstract operator $A \colon D(A) \to H$ where A^{-1} is a compact positive self-adjoint operator and $V = D(A^{1/2})$. In fact Theorem 4.3 is a simplified version of a much more general result that we now recall (see Simon, 1987).

Theorem 4.12 *Suppose that $G \subset H \subset K$ where G, H, and K are reflexive Banach spaces and the embedding $G \subset H$ is compact. If the sequence u_n is bounded in $L^q(0, T; G)$, $q \geq 1$, and $\partial_t u_n$ is bounded in $L^p(0, T; K)$, $p \geq 1$, then there exists a subsequence of u_n that is strongly convergent in $L^q(0, T; H)$.*

We notice that on a bounded domain the embedding $H^1 \subset L^2$ is compact due to the Rellich–Kondrachov Theorem (Theorem 1.8), so one can apply the Aubin–Lions Lemma to the triple of spaces

$$G = H^1, \qquad H = L^2, \qquad K = V^*.$$

This is the version of the result that we used in the proof of existence on $\mathbb{R}^3$.

Notes

The proof of the existence of weak solutions of the Navier–Stokes equations on the whole space was first given by Leray (1934). In Leray's proof weak solutions are obtained as weak limits of solutions of a regularised version of the Navier–Stokes equations, where the nonlinear term is replaced with a mollified version, $(u \cdot \nabla)u \mapsto (\tilde{u} \cdot \nabla)u$, see (14.2) in Chapter 14. The weak solutions constructed by Leray satisfy the strong energy inequality. Leray suspected that the weak solutions he constructed might not be regular and in fact he called them 'turbulent solutions'. He also conjectured the existence of self-similar solutions that blow up in a finite time; this conjecture was essentially refuted by Nečas, Růžička, & Šverák (1996).

The existence of weak solutions (satisfying the energy inequality, but not necessarily the strong energy inequality) in arbitrary open domains is due to Hopf (1951). In his paper he constructs the Galerkin approximations and then uses the Arzelà–Ascoli Theorem and Friedrich's Lemma to obtain convergence of these approximate solutions (see Exercises 4.2–4.9).

Ladyzhenskaya (1969) proved that the solutions constructed by Hopf for bounded domains satisfy the strong energy inequality (Hopf did not show this property). Existence of weak solutions satisfying the strong energy inequality is in fact known for any domain with a uniformly C^2 boundary (see Farwig, Kozono, & Sohr, 2005; see also Galdi & Maremonti, 1986, for the case of exterior domains).

The proof of the existence of weak solutions on $\mathbb{T}^3$ and on bounded domains presented in this chapter follows the lines of the proof in Constantin & Foias (1988), while the proof of the simplified version of the Aubin–Lions Lemma is based on Chemin et al. (2006). The proof of the existence of weak solutions in unbounded domains that we sketch in Section 4.4 is based on Heywood (1988).

One can obtain different definitions of the Galerkin projections P_n by replacing in (4.1) the first eigenfunctions of the Stokes operator $a_k \in \mathcal{N}$ with some other appropriate set of functions forming a basis in H. In fact, as we mentioned before, Hopf used as a basis (a_k) a set of compactly supported smooth functions that allowed him to carry out his proof in arbitrary open domains, including unbounded ones.

While uniqueness of weak solutions of the Navier–Stokes equations is an open problem, a similar question is settled for the incompressible Euler equations. It turns out that weak solutions of the Euler equations are not unique. Scheffer (1993) constructed an example of a nontrivial weak solution of the Euler equations compactly supported in space–time (see also Shnirelman, 1997 & 2000 for a simpler proof). De Lellis & Székelyhidi (2010) showed that there exist bounded, compactly supported initial velocity fields for which there exist infinitely many weak solutions satisfying the energy inequality. There are no similar results known for the Navier–Stokes equations.

Exercises

4.1 Show that if $u \in H \cap \dot{H}^s(\mathbb{T}^3)$ then $P_n u \to u$ in $\dot{H}^s$ as $n \to \infty$ and that

$$\|P_n u\|_{\dot{H}^s} \leq \|u\|_{\dot{H}^s}.$$

4.2 Prove that for every fixed $k \in \mathbb{N}$ the functions c_k^n defined in (4.5) and given by

$$c_k^n(t) = \langle u_n(t), a_k \rangle,$$

where u_n are the Galerkin approximations, are equicontinuous.

4.3 Prove that there is a subsequence u_n of the Galerkin approximations of the Navier–Stokes equations such that $u_n(t)$ tends weakly in L^2 to $u(t)$ uniformly on each time interval $[0, T]$, that is, show that for each fixed $\varphi \in L^2(\Omega)$ and each $\varepsilon > 0$ we can find N such that for all $n > N$

$$|\langle u_n(t) - u(t), \varphi \rangle| \leq \varepsilon$$

for all $t \in [0, T]$.

4.4 Suppose that (u_n) is bounded in $L^2(0, T; L^2)$ for every $T > 0$. Use the diagonal method to show that there is a subsequence of (u_n) that is weakly convergent in $L^2(0, T; L^2)$ for every $T > 0$.

4.5   Prove that if

$$u_n \overset{*}{\rightharpoonup} v_1 \quad \text{in } L^\infty(0, T; L^2),$$

$$u_n \rightharpoonup v_2 \quad \text{in } L^2(0, T; L^2), \qquad \text{and}$$

$$\nabla u_n \rightharpoonup g \quad \text{in } L^2(0, T; L^2)$$

then $v_1 = v_2$ and $g = \nabla v_2$.

4.6   Prove that if (u_n) is a subsequence of Galerkin approximations such that $u_n(t) \to u(t)$ in $L^2(\Omega)$ for almost all $t \in (0, T)$ and $\partial_t u_n \rightharpoonup g$ weakly in $L^{4/3}(0, T; V^*(\Omega))$, then $g = \partial_t u$.

4.7   Use the Poincaré inequality to show that there is an absolute constant C such that

$$\left\| u - \frac{1}{a^3} \int_Q u \right\| \le Ca \|\nabla u\|,$$

for all $u \in H^1(Q)$, where Q is a cube with side of length a.

4.8   Let Q be a cube with side of length a and let

$$Q = \bigcup_{i=1}^{M^3} Q_i,$$

where the Q_i are equal cubes with sides of length a/M. Prove Friedrich's inequality: if $u \in H^1(Q)$ then

$$\|u\|_{L^2(Q)}^2 \le C \frac{a^2}{M^2} \|\nabla u\|^2 + 2\frac{M^3}{a^3} \sum_{i=1}^{M^3} \langle \chi_{Q_i}, u \rangle^2,$$

where C is an absolute constant and χ_{Q_i} is the characteristic function of the cube Q_i.

4.9   Let u_n be a sequence of functions satisfying the estimate

$$\sup_{0 \le t \le T} \|u_n(t)\|^2 + 2 \int_0^T \|\nabla u_n\|^2 \le C_1$$

and such that for each $\psi \in L^2(\mathbb{R}^3)$

$$\langle u_n, \psi \rangle \to \langle u, \psi \rangle \quad \text{in } C([0, T]).$$

Use Friedrich's inequality to show that

$$u_n \to u \quad \text{strongly in} \quad L^2(0, T; L^2(K))$$

for every compact set $K \subset \mathbb{R}^3$.

4.10 Suppose that α_k are C^1 functions and u_n given by

$$u_n(x, t) = \sum_{k=1}^{n} \alpha_k(t) a_k(x)$$

satisfies the weak formulation of the Navier–Stokes equations for every $\varphi = \xi(t) a_k(x) \in \tilde{\mathcal{D}}$, $k \leq n$. Prove that u_n is a solution of the Galerkin equation.

4.11 Suppose that u is a weak solution of the Navier–Stokes equations on $[0, T_1]$ and v is a weak solution of the Navier–Stokes equations on $[T_1, T]$. Prove that if $u(T_1) = v(T_1)$ then w given by $w = u$ on $[0, T_1]$ and $w = v$ on $[T_1, T]$ is a weak solution of the Navier–Stokes equations on $[0, T]$.

5

The pressure

In this chapter we will show how to associate a pressure p with a weak solution u so that the Navier–Stokes equations

$$\partial_t u - \Delta u + (u \cdot \nabla)u + \nabla p = 0 \qquad (5.1)$$

and

$$\nabla \cdot u = 0 \qquad (5.2)$$

hold in the sense of distributions.

We will obtain the pressure by making the following simple observation. If (u, p) are smooth and solve (5.1)–(5.2) then if we take the divergence of (5.1) we obtain

$$\nabla \cdot [(u \cdot \nabla)u] + \Delta p = 0,$$

since u is divergence free. It follows that p is a solution of the equation

$$-\Delta p = \partial_i(u_j\partial_j u_i) = \partial_i\partial_j(u_i u_j), \qquad (5.3)$$

where we have used the fact that u is divergence free to rewrite the right-hand side.

Any two solutions of (5.3) can only differ by a harmonic function. On the torus or the whole space, the requirements of periodicity and zero average or integrability, respectively, are enough to ensure that there is a unique solution of (5.3). In a bounded domain the situation is complicated by the fact that there is no natural boundary condition for p, and we cannot use (5.3) directly; we give sketches of the required indirect argument in this case in Section 5.4.

Given that p is related to u via (5.3) we can write (at least formally)

$$p = (-\Delta)^{-1}\partial_i\partial_j(u_i u_j). \qquad (5.4)$$

111

It is therefore reasonable to expect that p will have similar integrability properties to $u_i u_j$, and using the Calderón–Zygmund Theorem from the theory of singular integral operators (see Appendix B) we will indeed be able to show that

$$\|p\|_{L^r} \leq C_r \|u\|_{L^{2r}}^2, \qquad 1 < r < \infty. \tag{5.5}$$

Once we have a candidate pressure, we will be able to show that the pair (u, p) satisfies (5.1) in the sense of distributions.

5.1 Solving for the pressure on $\mathbb{T}^3$ and $\mathbb{R}^3$

In the absence of boundaries (i.e. on $\mathbb{T}^3$ and $\mathbb{R}^3$) we can obtain estimates on the pressure given knowledge of u using the Calderón–Zygmund theory of singular integral operators; we provide a proof of the relevant results in Appendix B.

Lemma 5.1 *On the domain $\mathbb{T}^3$ or $\mathbb{R}^3$ suppose that*

$$-\Delta p = \partial_i \partial_j (u_i u_j); \tag{5.6}$$

to ensure uniqueness we impose an additional condition: on the torus we require $\int_{\mathbb{T}^3} p = 0$ and on the whole space we require p to be integrable (to some power). Then, for any $1 < r < \infty$,

$$\|p\|_{L^r} \leq C_r \|u\|_{L^{2r}}^2 \tag{5.7}$$

and

$$\|\nabla p\|_{L^r} \leq C_r \|(u \cdot \nabla)u\|_{L^r}. \tag{5.8}$$

Proof We concentrate on the case of the whole space, which is the simplest situation, given that we require the Calderón–Zygmund Theorem. As we have already observed, it follows from (5.6) that

$$p = (-\Delta)^{-1} \partial_i \partial_j (u_i u_j),$$

where $(-\Delta)^{-1} f$ can be interpreted as the integral operator

$$(-\Delta)^{-1} f = \frac{1}{4\pi} \int_{\mathbb{R}^3} \frac{1}{|x - y|} f(y) \, dy,$$

see Appendix B. If we consider $(-\Delta)^{-1}$ and ∂_i in Fourier space they correspond to multiplication by $(2\pi |\xi|)^{-2}$ and $2\pi i\xi_i$ respectively; it is clear that they commute, so we can write

$$p = \partial_i \partial_j (-\Delta)^{-1} (u_i u_j). \tag{5.9}$$

We can now appeal to the Calderón–Zygmund Theorem (Theorem B.3), which guarantees that for every i, j the operator $T_{ij} = \partial_i \partial_j (-\Delta)^{-1}$ is a bounded linear map from $L^r(\mathbb{R}^3)$ to $L^r(\mathbb{R}^3)$ for all $1 < r < \infty$ (see Theorem B.6 in Appendix B). A similar result is available when the domain is a torus, i.e. the corresponding map T_{ij} is a bounded map from $L^r(\mathbb{T}^3)$ to $L^r(\mathbb{T}^3)$ for all $1 < r < \infty$, see the discussion at the end of Appendix B.

It follows from (5.9) and the fact that $T_{ij} \colon L^r \to L^r$ is bounded ($1 < r < \infty$) that if $u \in L^{2r}(\mathbb{R}^3)$ then

$$\|p\|_{L^r(\mathbb{R}^3)} \le C \sum_{i,j} \|u_i u_j\|_{L^r(\mathbb{R}^3)} \le C\|u\|_{L^{2r}(\mathbb{R}^3)}^2. \tag{5.10}$$

If rather than estimates on p we want estimates on ∇p then we can take the gradient in (5.9) to yield

$$\partial_k p = \partial_k \partial_j (-\Delta)^{-1}(u_i \partial_i u_j),$$

after commuting the derivatives and using the fact that u is divergence free. Therefore for each k we can again appeal to the Calderón–Zygmund Theorem to obtain

$$\|\partial_k p\|_{L^r} = \|T_{kj}(u_i \partial_i u_j)\|_{L^r} \le C_r \sum_j \|(u \cdot \nabla)u_j\|_{L^r}$$

and so $\|\nabla p\|_{L^r} \le C_r\|(u \cdot \nabla)u\|_{L^r}$ as claimed.   □

It follows that, in the absence of boundaries, space–time estimates on the regularity of the pressure follow directly from space–time estimates on the regularity of the velocity. Recall (see Lemma 3.5) that for any weak solution u we have

$$u \in L^r(0, T; L^s) \qquad \text{for} \qquad \frac{2}{r} + \frac{3}{s} = \frac{3}{2}, \tag{5.11}$$

and in particular if we equalise the space and time integrability of u then we obtain

$$u \in L^{10/3}(\Omega \times (0, T)). \tag{5.12}$$

Corollary 5.2 *If u is a weak solution in the absence of boundaries then the corresponding pressure p satisfies*[1]

$$p \in L^r(0, T; L^s) \qquad \text{for} \qquad 2/r + 3/s = 3, \quad s > 1 \tag{5.13}$$

---

[1] Equivalently $p \in L^{2s/3(s-1)}(0, T; L^s)$ for $1 < s \le 3$ and $\nabla p \in L^{2s/(4s-3)}(0, T; L^s)$ for $1 < s \le 3/2$.

and in particular

$$p \in L^{5/3}(\mathbb{T}^3 \times (0, T)) \qquad or \qquad p \in L^{5/3}(\mathbb{R}^3 \times (0, T)). \qquad (5.14)$$

Furthermore

$$\nabla p \in L^r(0, T; L^s) \qquad for \qquad 2/r + 3/s = 4, \quad s > 1.$$

Proof It is easy to see that (5.13) follows from (5.11) together with the bound $\|p\|_{L^s} \leq C_s \|u\|_{L^{2s}}^2$ (equation (5.7)). (Note that we must exclude $s = 1$ since the Calderón–Zygmund estimate in (5.10) is not valid in L^1.)

To bound $\|\nabla p\|_{L^s}$ using (5.8) we need to estimate $\|(u \cdot \nabla)u\|_{L^s}$, which we can do using Hölder's inequality (cf. Lemma 3.4). Given that u is a weak solution we cannot control any higher power of $|\nabla u|$ than $|\nabla u|^2$, so to begin with we restrict the range of s to $1 \leq s < 2$. Then, using Hölder's inequality with exponents $2/s$ and $2/(2 - s)$,

$$\|(u \cdot \nabla)u\|_{L^s} \leq \left(\int |u|^s |\nabla u|^s \right)^{1/s}$$

$$\leq \left(\int |\nabla u|^2 \right)^{1/2} \left(\int |u|^{2s/(2-s)} \right)^{(2-s)/2s}$$

$$= \|\nabla u\|_{L^2} \|u\|_{L^{2s/(2-s)}}.$$

In order to control $\|u\|_{L^{2s/(2-s)}}$ we need $2s/(2 - s) \leq 6$, which further restricts the range of s to $1 \leq s \leq 3/2$; then, using Lebesgue interpolation (Theorem 1.5),

$$\|(u \cdot \nabla)u\|_{L^s} \leq \|\nabla u\|_{L^2} \|u\|_{L^2}^{(3-2s)/s} \|u\|_{L^6}^{3(s-1)/s}$$

$$\leq c \|\nabla u\|_{L^2}^{(4s-3)/s} \|u\|_{L^2}^{(3-2s)/s}. \qquad (5.15)$$

Since $u \in L^\infty(0, T; L^2)$ the time integrability of $\|(u \cdot \nabla)u\|_{L^s}$ is determined by the time integrability of the term $\|\nabla u\|_{L^2}^{(4s-3)/s}$; it follows that

$$(u \cdot \nabla)u \in L^{2s/(4s-3)}(0, T; L^s), \qquad 1 \leq s \leq 3/2. \qquad (5.16)$$

This estimate immediately translates to one for ∇p using (5.8):

$$\nabla p \in L^{2s/(4s-3)}(0, T; L^s), \qquad 1 < s \leq 3/2,$$

which is equivalent to (5.2). (Note that once again $s = 1$ is excluded from the possible range of s in (5.8) since we need to use the Calderón–Zygmund Theorem.) □

Similar estimates are available for bounded domains, but the arguments are more involved and the techniques are significantly different; a sketch of the proof is given in Section 5.4.

5.2 Distributional solutions in the absence of boundaries

We now show that given a weak solution u (in the sense of Definition 3.3) and the corresponding pressure constructed as in Lemma 5.1, the Navier–Stokes equations (5.1) hold in the sense of distributions.

We give the proof on the torus, since this is slightly simpler than the argument on $\mathbb{R}^3$ but still illustrates the main ideas. In the statement of the following lemma, $u \otimes u$ is understood as the matrix with components $u_i u_j$; similarly $(\nabla \phi)_{ij} = \partial_i \phi_j$, so that

$$\langle u \otimes u, \nabla \phi \rangle = \int_{\mathbb{T}^3} u_i u_j \partial_i \phi_j.$$

Proposition 5.3 *If u is a weak solution of the Navier–Stokes equations on $\mathbb{T}^3$ and for almost every t we set*

$$-\Delta p = \partial_i \partial_j (u_i u_j)$$

then the Navier–Stokes equations hold in the sense of (space–time) distributions on $\mathbb{T}^3 \times (0, \infty)$, i.e.

$$\int_0^\infty \langle u, \partial_t \phi \rangle + \int_0^\infty \langle u, \Delta \phi \rangle + \int_0^\infty \langle u \otimes u, \nabla \phi \rangle + \int_0^\infty \langle p, \nabla \cdot \phi \rangle = 0 \quad (5.17)$$

for every $\phi \in [C_c^\infty(\mathbb{T}^3 \times (0, \infty))]^3$ and

$$\int_0^\infty \langle u, \nabla \psi \rangle = 0 \quad (5.18)$$

for every $\psi \in C_c^\infty(\mathbb{T}^3 \times (0, \infty))$.

Proof First note that (5.18) is immediate, since u is weakly divergence free (i.e. for almost every t, $\langle u(t), \nabla \eta \rangle = 0$ for every $\eta \in C^\infty(\mathbb{T}^3)$, see Lemma 2.3).

Given $\phi \in [C_c^\infty(\mathbb{T}^3 \times (0, \infty))]^3$, we use the Helmholtz–Weyl Decomposition from Theorem 2.6 to write

$$\phi = \varphi + \nabla \psi,$$

where $\varphi \in [C_c^\infty(\mathbb{T}^3 \times (0, \infty))]^3$ is a divergence-free function of x for every t and $\psi \in C_c^\infty(\mathbb{T}^3 \times (0, \infty))$.

Since φ is divergence free it is now an allowable test function in the definition of a weak solution (Definition 3.3), and so

$$\int_0^s -\langle u, \partial_t \varphi \rangle + \int_0^s \langle \nabla u, \nabla \varphi \rangle + \int_0^s \langle (u \cdot \nabla) u, \varphi \rangle = \langle u_0, \varphi(0) \rangle - \langle u(s), \varphi(s) \rangle$$

for almost every $s > 0$. Since φ is supported on $\mathbb{T}^3 \times [a, b]$, where $[a, b]$ is a compact subset of $(0, \infty)$, it follows that if we take s sufficiently large then the two terms on the right-hand side of this equation are zero, and we can extend the range of integration to $(0, \infty)$ without affecting the left-hand side. Therefore

$$\int_0^\infty -\langle u, \partial_t \varphi \rangle + \int_0^\infty \langle \nabla u, \nabla \varphi \rangle + \int_0^\infty \langle (u \cdot \nabla) u, \varphi \rangle = 0$$

(cf. Lemma 3.14). Since φ is smooth and has compact support we can use the definition of the weak space derivative to write

$$\int_0^\infty \langle u, \partial_t \varphi \rangle + \int_0^\infty \langle u, \Delta \varphi \rangle + \int_0^\infty \langle u \otimes u, \nabla \varphi \rangle = 0. \qquad (5.19)$$

Now, since u is divergence free, if we set $\varphi = \phi - \nabla \psi$ then equation (5.19) becomes

$$\int_0^\infty \langle u, \partial_t \phi \rangle + \int_0^\infty \langle u, \Delta \phi \rangle + \int_0^\infty \langle u_i u_j, \partial_i \phi_j \rangle - \int_0^\infty \langle u_i u_j, \partial_i \partial_j \psi \rangle = 0.$$

Now we observe that since for almost every t

$$-\Delta p = \partial_i \partial_j (u_i u_j)$$

it follows since $\psi(t) \in C_c^\infty(\mathbb{T}^3)$ that we have

$$\langle -\Delta p, \psi(t) \rangle = \langle \partial_i \partial_j (u_i u_j), \psi(t) \rangle$$

for almost every t. Using the definition of the weak derivative and integrating in time we have

$$\int_0^\infty \langle p, -\Delta \psi \rangle = \int_0^\infty \langle u_i u_j, \partial_i \partial_j \psi \rangle;$$

finally, noting that $\nabla \cdot \phi = \nabla \cdot (\varphi + \nabla \psi) = \Delta \psi$, we obtain (5.17). $\square$

A very similar argument works on the whole space: although the Helmholtz–Weyl Decomposition does not preserve the property of compact support in space (which makes $C_c^\infty(\mathbb{R}^3)$ an awkward space of test functions) it does respect the Schwartz space $\mathscr{S}$ in the sense that $\phi \in [\mathscr{S}(\mathbb{R}^3)]^3$ has a decomposition

$$\phi = \varphi + \nabla \psi,$$

where $\varphi \in \mathscr{S}(\mathbb{R}^3)$ with $\nabla \cdot \varphi = 0$ and $\psi \in \mathscr{S}(\mathbb{R}^3)$ (see Exercise 5.2). It is also relatively straightforward to show that smooth functions with compact support in time that are divergence-free elements of $\mathscr{S}(\mathbb{R}^3)$ in the space variables can be used as test functions in the weak formulation of the equations (see Exercise 5.3).

5.3 Additional estimates on weak solutions

In the absence of boundaries we have been able to show that the equations hold in the sense of distributions in a relatively straightforward way. However, on a bounded domain we cannot proceed so directly and instead have to obtain some additional estimates on $\partial_t u$ and Δu. In this section we show how to obtain these estimates in the case of $\mathbb{T}^3$, since such estimates are of interest in themselves, and the arguments are once again simpler in the periodic case.

In a bounded domain these estimates require 'maximal regularity' results for the Stokes equations; in the absence of boundaries results for the heat equation are sufficient, and we give a proof of these results in Appendix D.5. We hope that by presenting the argument in detail in this simpler setting the results for bounded domains (whose proofs we only sketch) become more convincing.

We will use a slight modification of the method from the proof of the existence of weak solutions in Section 4.2 in order to construct a weak solution satisfying better bounds on its time derivative and its Laplacian. These bounds provide another way to show that this weak solution satisfies the Navier–Stokes equations in the sense of distributions, which is also applicable in the case of a bounded domain.

We will need three ingredients: the Helmholtz–Weyl decomposition in L^q (Theorem 2.28), the convergence of 'square Fourier sums' in L^q (Theorem 1.6), and the following maximal regularity result for the heat equation, which states that the solution of (5.20) is as regular as the right-hand side will allow (cf. Theorem D.12). For a simple L^2-based version of this result see Exercise 5.4.

Theorem 5.4 *Let $f \in L^p(0, T; L^q(\mathbb{T}^3))$ with $1 < p, q < \infty$. There is a constant C such that if v is the solution of*

$$\partial_t v - \Delta v = f, \qquad v(0) = 0, \tag{5.20}$$

then

$$\|\partial_t v\|_{L^p(0,T;L^q)} + \|\Delta v\|_{L^p(0,T;L^q)} \le C \|f\|_{L^p(0,T;L^q)}.$$

We are now ready to show how one can obtain further estimates on the time derivative of a weak solution u on the torus. Recall that we showed in Lemma 4.5 that in this case $\partial_t u \in L^{4/3}(0, T; H^{-1})$. We now improve this slightly and obtain $\partial_t u \in L^{4/3}(\varepsilon, T; L^{6/5})$. Since $L^{6/5} \subset H^{-1}$ (see (1.26)) this seems a marginal improvement (at the cost of decreasing the time interval), but it means that $\partial_t u$ can now be viewed a function rather than only a linear functional.

Theorem 5.5 *Given $u_0 \in H$ there exists a Leray–Hopf weak solution u of the Navier–Stokes equations on the torus $\mathbb{T}^3$ with initial data u_0 such that for every $0 < \varepsilon < T < \infty$ we have*

$$\partial_t u \in L^{4/3}(\varepsilon, T; L^{6/5}) \quad \text{and} \quad \Delta u \in L^{4/3}(\varepsilon, T; L^{6/5}).$$

Proof We construct a weak solution u using the Galerkin approximations as we did in Section 4.2, but we choose a special order of the sequence of eigenfunctions of the Stokes operator that corresponds to the 'square sums' of Fourier modes. More precisely, we look for approximate solutions u_N such that for all $t > 0$

$$u_N(t) \in H_N,$$

where

$$H_N := \text{span}\{a_j \in \mathcal{N} : Aa_j = |l|^2 a_j, \ l \in \dot{\mathbb{Z}}^3 \cap [-N, N]^3\}.$$

Before we formulate a modified version of the Galerkin problem for u_N let us note that if $u_0 \in H$ then clearly

$$S_N(u_0) \in H_N,$$

where S_N is defined in Theorem 1.6. Furthermore, if $f \in L^q$ for some $q > 1$ then

$$\mathbb{P}_q S_N(f) \in H_N,$$

where $\mathbb{P}_q$ is the Leray projector on $L^q(\mathbb{T}^3)$, see Theorem 2.28. Now we are ready to define u_N: we look for a u_N of the form

$$u_N = \sum_j c_j^N(t) a_j(x), \quad a_j \in \mathcal{N} \cap H_N,$$

that satisfies the Galerkin equation

$$\partial_t u_N - \Delta u_N + \mathbb{P}S_N[(u_N \cdot \nabla)u_N] = 0,$$
$$u_N(0) = S_N(u_0).$$

Note that the projection $\mathbb{P}S_N(\cdot)$ is just the projection onto the space H_N, so we can omit the q subscript.

It is easy to see that all the uniform bounds that we have obtained before remain valid. In particular, for every $n \in \mathbb{N}$ we have

$$\sup_{0 \le t \le T} \|u_N\| \le \|u_0\| \quad \text{and} \quad \int_0^T \|\nabla u_N\|^2 \le \frac{1}{2}\|u_0\|^2. \tag{5.21}$$

From these bounds we can easily obtain a uniform bound on $(u_N \cdot \nabla)u_N$ in the space $L^{4/3}(0, T; L^{6/5})$, since from (5.15) we have

$$\int_0^T \|(u_N \cdot \nabla)u_N\|_{L^{6/5}}^{4/3} \leq c \int_0^T \|u_N\|^{2/3} \|\nabla u_N\|^2$$
$$\leq c\|u_0\|^{8/3} \tag{5.22}$$

using (5.21).

Next we use the boundedness of the Leray projector and the truncation S_N in $L^{6/5}$ (Theorem 2.28 and Theorem 1.6, respectively) to obtain uniform bounds on $\mathbb{P}S_N[(u_N \cdot \nabla)u_N]$ in $L^{4/3}(0, T; L^{6/5})$:

$$\int_0^T \|\mathbb{P}S_N[(u_N \cdot \nabla)u_N]\|_{L^{6/5}}^{4/3} \leq \tilde{C}_{6/5} \int_0^T \|(u_N \cdot \nabla)u_N\|_{L^{6/5}}^{4/3}$$
$$\leq C\|u_0\|^{8/3}, \tag{5.23}$$

from (5.22).

A uniform bound on $\partial_t u_N$ (and Δu_N) in the space $L^{4/3}(0, T; L^{6/5})$ now follows from the maximal regularity result for the heat equation. We write u_N as

$$u_N = v_N + w_N,$$

where v_N solves the problem

$$\partial_t v_N - \Delta v_N = f_N, \qquad v_N(0) = 0$$

with

$$f_N := \mathbb{P}S_N[(u_N \cdot \nabla)u_N],$$

and w_N solves the heat equation with initial condition $S_N(u_0)$

$$\partial_t w_N - \Delta w_N = 0, \qquad w_N(0) = S_N(u_0).$$

Let us consider first the equation for v_N. From Theorem 5.4 (maximal regularity) and Theorem 1.6 (L^q boundedness of S_N) we obtain

$$\|\partial_t v_N\|_{L^{4/3}(0,T;L^{6/5})} + \|\Delta v_N\|_{L^{4/3}(0,T;L^{6/5})} \leq C\|f_N\|_{L^{4/3}(0,T;L^{6/5})} \leq C \tag{5.24}$$

using (5.23).

Obtaining a similar bound on w_N requires a little more effort, since we only have a uniform bound $\|S_N(u_0)\| \leq \|u_0\|$. First we take the L^2 inner product of $\partial_t w_N - \Delta w_N = 0$ with w_N and integrate in time between 0 and ε. This yields

$$\int_0^\varepsilon \|\nabla w_N\|^2 \leq \frac{1}{2}\|S_N(u_0)\|^2.$$

It follows that there exists $s \in (0, \varepsilon)$ such that

$$\varepsilon \|\nabla w_N(s)\|^2 \le \frac{1}{2} \|S_N(u_0)\|^2 \le \frac{1}{2} \|u_0\|^2.$$

Now if we take the L^2 inner product of equation $\partial_t w_N - \Delta w_N = 0$ with $-\Delta w_N$ and integrate in time between s and T we obtain

$$\int_s^T \|\Delta w_N\|^2 \le \frac{1}{2} \|\nabla w_N(s)\|^2 \le \varepsilon^{-1} \|u_0\|^2;$$

in particular, since $s < \varepsilon$,

$$\int_\varepsilon^T \|\Delta w_N\|^2 \le \varepsilon^{-1} \|u_0\|^2$$

for every N. Therefore, using the equality $\|\partial_t w_N\| = \|\Delta w_N\|$, we obtain

$$\int_\varepsilon^T \|\partial_t w_N\|^2 + \|\Delta w_N\|^2 \le c.$$

Since $\|f\|_{L^{6/5}} \le c\|f\|$ and $4/3 < 2$ we certainly have

$$\int_\varepsilon^T \|\partial_t w_N\|_{L^{6/5}}^{4/3} + \|\Delta w_N\|_{L^{6/5}}^{4/3} \le c \tag{5.25}$$

uniformly in N.

The bounds on v_N in (5.24) and w_N in (5.25) give us a uniform bound on $\partial_t u_N$ and Δu_N in $L^{4/3}(\varepsilon, T; L^{6/5})$. It follows that we can choose a subsequence of u_N that tends to a weak solution u such that both $\partial_t u$ and Δu belong to $L^{4/3}(\varepsilon, T; L^{6/5})$ for every $\varepsilon > 0$. A diagonal argument provides the integrability for any $T > 0$.

We prove the validity of the strong energy inequality as in the proof of Theorem 4.6.   □

Given these estimates on $\partial_t u$ and Δu, we can now use the equation to deduce bounds on ∇p in the same spaces. It will follow that the equations hold almost everywhere, and hence in the sense of distributions. We give this alternative argument in detail (although it is almost redundant in the light of Proposition 5.3) since this is the approach that we have to follow in a bounded domain.

Theorem 5.6 *For every initial condition $u_0 \in H(\mathbb{T}^3)$ there exists a weak solution u that satisfies the Navier–Stokes equations in the distributional sense: there exists a $p \in L_{\mathrm{loc}}^{4/3}(0, \infty; W^{1,6/5}(\mathbb{T}^3))$ such that*

$$\int_0^\infty \langle u, \partial_t \varphi \rangle + \int_0^\infty \langle u, \Delta \varphi \rangle + \int_0^\infty \langle u \otimes u, \nabla \varphi \rangle - \int_0^\infty \langle \nabla p, \varphi \rangle = 0 \tag{5.26}$$

for every $\varphi \in C_c^\infty(\mathbb{T}^3 \times (0, \infty))$.

Proof Let u be a weak solution constructed in the proof of Theorem 5.5. Choose $T > 0$ and take $\varphi \in \mathcal{D}_\sigma$ with compact support in $\mathbb{T}^3 \times (0, T)$. Then

$$-\int_0^T \langle u, \partial_t \varphi \rangle + \int_0^T \langle \nabla u, \nabla \varphi \rangle + \int_0^T \langle (u \cdot \nabla) u, \varphi \rangle = 0.$$

Since for every $0 < \varepsilon < T < \infty$ we have $\partial_t u \in L^{4/3}(\varepsilon, T; L^{6/5})$ we can integrate by parts in the first term on the left-hand side, and using $\varphi(0) = \varphi(T) = 0$ we obtain

$$\int_0^T \langle u, \partial_t \varphi \rangle = -\int_0^T \langle \partial_t u, \varphi \rangle.$$

Similarly, since $\Delta u \in L^{4/3}(\varepsilon, T; L^{6/5})$ it follows that

$$\int_0^T \langle \nabla u, \nabla \varphi \rangle = -\int_0^T \langle \Delta u, \varphi \rangle.$$

Hence

$$\int_0^T \langle \partial_t u - \Delta u + (u \cdot \nabla) u, \varphi \rangle = 0 \tag{5.27}$$

for all $\varphi \in \mathcal{D}_\sigma$ with compact support in $\mathbb{T}^3 \times (0, T)$. Since

$$\partial_t u - \Delta u + (u \cdot \nabla) u \in L^{4/3}(\varepsilon, T; L^{6/5}),$$

from the Helmholtz–Weyl decomposition of $L^{6/5}$ it follows that for almost every time in $(0, T)$ we can find $h \in H_{6/5}$ and $g \in W^{1,6/5}(\mathbb{T}^3)$ such that

$$\partial_t u - \Delta u + (u \cdot \nabla) u = h + \nabla g.$$

From (5.27) we deduce that

$$h = \mathbb{P}_{6/5}[\partial_t u - \Delta u + (u \cdot \nabla) u] = 0,$$

so setting $g = -p$ we deduce that there exists $\nabla p \in L^{4/3}(\varepsilon, T; L^{6/5})$ such that

$$\partial_t u - \Delta u + (u \cdot \nabla) u = -\nabla p,$$

where the equality is understood in the space $L^{4/3}(\varepsilon, T; L^{6/5})$. Therefore

$$\partial_t u - \Delta u + (u \cdot \nabla) u + \nabla p = 0$$

almost everywhere in $\mathbb{T}^3 \times (0, T)$. Since $T > 0$ we arbitrary the same equality holds almost everywhere in $\mathbb{T}^3 \times (0, \infty)$. Now if we multiply this equation by any $\varphi \in C_c^\infty(\mathbb{T}^3 \times (0, \infty))$, integrate over $\mathbb{T}^3$ and time, and then integrate by parts we obtain (5.26). $\qquad\square$

5.4 Pressure in a bounded domain

In a bounded domain we cannot estimate p directly from the equation

$$-\Delta p = \partial_i \partial_j (u_i u_j)$$

(see (5.6)) since there are no boundary conditions for p. To circumvent this problem we show that the estimate for ∇p in (5.8),

$$\|\nabla p\|_{L^r(\Omega)} \le C_r \|(u \cdot \nabla)u\|_{L^r(\Omega)},$$

still holds. We then choose p such that for each time $\int_\Omega p = 0$ and use the 3D Sobolev embedding $W^{1,r}(\Omega) \subset L^{3r/(3-r)}(\Omega)$ to deduce bounds on p. Here we briefly sketch the proof, due to von Wahl (1982) and Sohr & von Wahl (1986), which uses maximal regularity results for the time-dependent Stokes equations and elliptic regularity in $L^r(\Omega)$ for the Stokes equations (see also Seregin, 2014).

Theorem 5.7 *If $\Omega \subset \mathbb{R}^3$ is a smooth bounded domain and u is a weak solution then there exists a corresponding pressure p such that for every $\varepsilon > 0$*

$$p \in L^r(\varepsilon, T; L^s) \quad for \quad 2/r + 3/s = 3, \quad s > 1$$

and

$$\nabla p \in L^r(\varepsilon, T; L^s) \quad for \quad 2/r + 3/s = 4, \quad s > 1.$$

Proof (Sketch)

If we apply $\mathbb{P}_r$, the Leray projector in L^r (see Section 2.4), to every term in the Navier–Stokes equations then we obtain

$$\partial_t u - \mathbb{P}_r \Delta u = -\mathbb{P}_r[(u \cdot \nabla)u]. \qquad (5.28)$$

The projector $\mathbb{P}_r$ eliminates the pressure, but in the case of a bounded domain does not commute with the Laplacian so $\mathbb{P}_r \Delta u \ne \Delta u$. Instead, we define the Stokes operator A_r on $L^r(\Omega)$, setting $A_r = -\mathbb{P}_r \Delta$ (compare with the discussion in Section 2.4) and rewrite the equation as

$$\partial_t u + A_r u = -\mathbb{P}_r[(u \cdot \nabla)u] \qquad (5.29)$$

(which is no more than a change of notation).

We now appeal to 'maximal regularity' results for the linear Stokes equations

$$\partial_t u + A_r u = f(x, t), \qquad u(0) = 0,$$

which guarantee that $\partial_t u$ and $A_r u$ are as regular as f allows:

$$\|\partial_t u\|_{L^q(0,T;L^r)} + \|A_r u\|_{L^q(0,T;L^r)} \leq C\|f\|_{L^q(0,T;L^r)}, \tag{5.30}$$

provided that $1 < q, r < \infty$ (cf. Theorem 5.4).

Applying the regularity from (5.30) to our equation (5.29) we have to include the fact that we have non-zero initial data (cf. proof of Theorem 5.5); then since $\mathbb{P}_r \colon L^r \to L^r$ is bounded (by Theorem 2.16), it follows that for any $\varepsilon > 0$ we have

$$\|\partial_t u\|_{L^q(\varepsilon,T;L^r)} + \|A_r u\|_{L^q(\varepsilon,T;L^r)} \leq C\|\mathbb{P}_r[(u \cdot \nabla)u]\|_{L^q(\varepsilon,T;L^r)}$$
$$\leq C\|(u \cdot \nabla)u\|_{L^q(\varepsilon,T;L^r)}.$$

We cannot immediately deduce any information about ∇p from this, since the equation (5.29), which involves $A_r u$, does not contain the pressure term. However, elliptic regularity results in L^r for the stationary Stokes equation

$$-\Delta u + \nabla q = f \quad \text{with} \quad \nabla \cdot u = 0 \qquad \Leftrightarrow \qquad A_r u = \mathbb{P}_r f$$

guarantee that

$$\|u\|_{W^{2,r}(\Omega)} \leq c\|A_r u\|_{L^r(\Omega)}$$

(see Chapter IV of Galdi, 2011, for example; details of the simpler L^2 case can be found in Chapter 3 of Constantin & Foias, 1988). In particular, the $L^q(0,T;L^r)$ estimate on $A_r u$ implies a similar estimate on Δu (without the Leray projector). It follows that

$$\|\partial_t u\|_{L^q(\varepsilon,T;L^r)} + \|\Delta u\|_{L^q(\varepsilon,T;L^r)} \leq C\|(u \cdot \nabla)u\|_{L^q(\varepsilon,T;L^r)}. \tag{5.31}$$

If we now rewrite (5.28) as

$$\mathbb{P}[\partial_t u - \Delta u + (u \cdot \nabla)u] = 0$$

then every term in this equation is an element of $L^q(\varepsilon,T;L^r)$, and so the equation makes sense pointwise almost everywhere. In particular, for almost every t it follows from the L^r Helmholtz decomposition (Theorem 2.28) that for almost every t there exists a p such that

$$\nabla p = -\partial_t u + \Delta u - (u \cdot \nabla)u. \tag{5.32}$$

Using (5.31) it follows that

$$\|\nabla p\|_{L^q(\varepsilon,T;L^r)} \leq C\|(u \cdot \nabla)u\|_{L^q(\varepsilon,T;L^r)}.$$

Since $(u \cdot \nabla)u \in L^{2r/(4r-3)}(\varepsilon, T; L^r)$ for $1 \le r \le 3/2$ (from (5.16)) it follows that

$$\nabla p \in L^{2r/(4r-3)}(\varepsilon, T; L^r), \quad 1 < r \le 3/2,$$

excluding $r = 1$ since the maximal regularity results are not valid in this case. Now using the Sobolev embedding $W^{1,r} \subset L^{3r/(3-r)}$, where we choose p at each time such that $\int_\Omega p = 0$, we obtain

$$p \in L^{2s/3(s-1)}(\varepsilon, T; L^s) \qquad \text{for} \qquad 3/2 < s \le 3. \qquad \square$$

Given this regularity of $\partial_t u$, Δu, and ∇p, it is very easy to show that the equations also hold in the sense of distributions in a bounded domain.

Proposition 5.8 *On a bounded domain Ω every weak solutions satisfies the Navier–Stokes equations in the sense of distributions, i.e.*

$$\int_0^\infty \langle u, \partial_t \varphi \rangle + \int_0^\infty \langle u, \Delta \varphi \rangle + \int_0^\infty \langle u \otimes u, \nabla \varphi \rangle + \int_0^\infty \langle p, \nabla \cdot \varphi \rangle = 0 \quad (5.33)$$

for every $\varphi \in C_c^\infty(\Omega \times (0, T))$.

Proof We have shown that (5.32) holds almost everywhere in $\Omega \times (0, T)$. Multiplying this equation by $\varphi \in C_c^\infty(\Omega \times (0, T))$, integrating over the space–time domain $\Omega \times (0, T)$, and finally integrating by parts we get (5.33) as in the proof of Theorem 5.6. $\qquad \square$

5.5 Applications of pressure estimates

Much of the analysis in Part I of this book, and some of the analysis in Part II, will in fact proceed without any explicit reference to the pressure. Either we will use divergence-free test functions, which will allow us to 'decouple' the velocity from the pressure, or we will use the vorticity form of the equation, which eliminates the pressure in a different way (Chapters 12 and 13).

The first direct application of the results of this chapter will be in Chapter 11, in which we will prove a local existence result for initial data in L^3; estimates on the pressure in terms of u will be crucial to this argument. We will also see in Chapter 13 that without estimates on the pressure, a solution can be smooth in space at every time t, but not have any time regularity; however, given the pressure estimates we have obtained in this chapter, we can deduce Hölder continuity in time given spatial regularity (see Proposition 13.10).

Perhaps most tellingly, in Chapters 15 and 16 we will develop a local regularity theory, central to which is a localised form of the energy inequality in which the pressure cannot be eliminated. If we want to obtain a localised form of the energy inequality, then we naturally take the L^2 inner product with φu, where φ is a C_c^∞ function that localises the estimate; in this case the contribution from the pressure term is

$$\int \varphi u \cdot \nabla p = -\int p(u \cdot \nabla)\varphi,$$

which need not vanish. In order to develop this local regularity theory we will therefore need some knowledge of the pressure, in particular we will require local estimates on the pressure that can be seen as a generalisation of Lemma 5.1.

Notes

In the context of what is to follow in Chapters 15 and 16 it is interesting that the bounds on p for bounded domains obtained in Theorem 5.7 were not known when Caffarelli, Kohn, & Nirenberg wrote their 1982 paper on partial regularity of the Navier–Stokes equations. They assumed only that $p \in L^{5/4}(\Omega \times (0, T))$ (a result due to von Wahl, 1980) which follows from a similar 'maximal regularity' argument limited to the case $q = r$.

Notice that in general one can only show that a weak solution satisfies the condition $\partial_t u \in L^{4/3}(0, T; V^*)$ and only for the case of the torus have we have proved that $\partial_t u \in L^{4/3}(0, T; H^{-1})$. However since $L^{6/5} \subset H^{-1}$ and $r = 4/3$, $s = 6/5$ satisfy the condition $3/s + 2/r = 4$, it follows from (5.31) that for any Leray–Hopf weak solution corresponding to $u_0 \in V$ we must have $\partial_t u \in L^{4/3}(\varepsilon, T; H^{-1})$ for all $T > 0$.

A treatment of the pressure similar to that given here can be found in the paper by Nečas, Růžička, & Šverák (1996). For more on the semigroup approach see the books by Sohr (2001) or von Wahl (1985).

Exercises

5.1 Using the Calderón–Zygmund Theorem in the form of Theorem B.6 show that the Leray projector on $\mathbb{R}^3$ is bounded from L^r into L^r for $1 < r < \infty$.

5.2 Suppose that $\phi \in [\mathscr{S}(\mathbb{R}^3)]^3$. Show that there exists a $\varphi \in [\mathscr{S}(\mathbb{R}^3)]^3$ and a $\psi \in \mathscr{S}(\mathbb{R}^3)$ such that

$$\phi = \varphi + \nabla\psi, \qquad \nabla \cdot \varphi = 0.$$

5.3 Show that if u is a weak solution of the Navier–Stokes equations on $\mathbb{R}^3$ then

$$\int_0^\infty \langle u, \partial_t \varphi \rangle + \int_0^\infty \langle u, \Delta \varphi \rangle + \int_0^\infty \langle u \otimes u, \nabla \varphi \rangle = 0 \qquad (5.34)$$

for all test functions $\varphi \in C^\infty(\mathbb{R}^3 \times (0, \infty))$ such that $\nabla \cdot \varphi(\cdot, t) = 0$, $\varphi(\cdot, t) \in \mathscr{S}(\mathbb{R}^3)$ for every t, and $\varphi(\cdot, t) = 0$ for t outside a compact subset of $(0, \infty)$.

5.4 Take $f \in L^2(0, T; L^2(\mathbb{T}^3))$ and consider the solution of the heat equation

$$\partial_t u - \Delta u = f(t), \qquad u(0) = 0,$$

which can be written as

$$u(t) = \sum_{k \in \mathbb{Z}^3} \left(\int_0^t e^{-|k|^2(t-s)} \hat{f}_k(t)\, ds \right) e^{ik \cdot x}.$$

Show that $u \in L^2(0, T; \dot{H}^2)$. [Hint: use Young's inequality for convolutions to bound the integral in time component-by-component.]

6

Existence of strong solutions

In this chapter we prove that when the initial condition u_0 belongs to the space V then on some time interval $[0, T]$ one can construct a unique 'strong solution' (see below) of the Navier–Stokes equations. Strong solutions are crucial for the global regularity problem since they immediately become smooth (we will prove this in the next chapter). Moreover, they are unique in the class of all weak solutions satisfying the energy inequality. However, we are only able to prove their existence locally in time (for t small).

Definition 6.1 We say that u is a *strong solution* of the Navier–Stokes equations on the time interval $[0, T]$ if u is a weak solution of the Navier–Stokes equations and in addition

$$u \in L^\infty(0, T; H^1) \cap L^2(0, T; H^2).$$

The definition of a strong solution has so many consequences that we devote the next section solely to the study of its implications. In particular, the resulting regularity of the time derivative of u (in Lemma 6.2 we show that $\partial_t u \in L^2(0, T; L^2)$) means that all the terms appearing in the Navier–Stokes equations can be understood as square integrable functions in the space–time domain. This is very different from the case of a weak solution where the time derivative $\partial_t u$ is only known in general to belong to $L^{4/3}(0, T; V^*)$ so it is not clear whether it can be identified with some 'true' function defined on the space–time domain. Moreover, weak solutions are only known to be L^2-weakly continuous in time, while in the case of strong solutions it follows from the two facts that $u \in L^2(0, T; H^2)$ and $\partial_t u \in L^2(0, T; L^2)$ that

$$u \in C([0, T]; H^1)$$

127

(see Proposition 1.35). It is not surprising, therefore, that strong solutions possess many other desirable properties.

6.1 General properties of strong solutions

We begin with the most obvious implication of Definition 6.1: all terms in the Navier–Stokes equations are defined for a strong solution u as L^2 functions in the space–time domain. We will first consider the terms involving the velocity.

Lemma 6.2 *Suppose that u is a strong solution of the Navier–Stokes equations on $[0, T]$. Then*

$$\partial_t u, \quad \Delta u, \quad and \quad (u \cdot \nabla)u$$

are all elements of $L^2(0, T; L^2)$.

Proof We give a proof valid on all domains considered in this book. It follows immediately from Definition 6.1 that

$$\Delta u \in L^2(0, T; L^2).$$

To prove that $(u \cdot \nabla)u \in L^2(0, T; L^2)$ we notice that

$$
\begin{aligned}
\int_0^T \|(u(s) \cdot \nabla)u(s)\|^2 \, ds &\leq \int_0^T \int |u(x, s)|^2 |\nabla u(x, s)|^2 \, dx \, ds \\
&\leq \int_0^T \int |\nabla u(x, s)|^2 \, dx \, \|u(s)\|_\infty^2 \, ds \\
&\leq \|u\|_{L^\infty(0,T;H^1)}^2 \int_0^T \|u(s)\|_\infty^2 \, ds \\
&\leq \|u\|_{L^\infty(0,T;H^1)}^2 \int_0^T \|u(s)\|_{H^2}^2 \, ds,
\end{aligned}
$$

where we have used the embedding of H^2 into L^∞.

Finally, we show that $\partial_t u$ is square integrable. From Lemma 3.6 and the fact that $\Delta u \in L^2$ we obtain (integrating by parts in the second term)

$$
\begin{aligned}
\langle u(t), \varphi \rangle &= \langle u_0, \varphi \rangle - \int_0^t \langle \nabla u, \nabla \varphi \rangle - \int_0^t \langle (u \cdot \nabla)u, \varphi \rangle \\
&= \langle u_0, \varphi \rangle + \int_0^t \langle \Delta u, \varphi \rangle - \int_0^t \langle (u \cdot \nabla)u, \varphi \rangle
\end{aligned}
$$

for all $t > 0$ and all $\varphi \in C_{c,\sigma}^\infty(\Omega)$. Since H is the completion of $C_{c,\sigma}^\infty$ in the L^2 norm it follows that the above equality holds for all $\varphi \in H$. Differentiating this equality with respect to time we obtain

$$\frac{\mathrm{d}}{\mathrm{d}t} \langle u(t), \varphi \rangle = \langle \Delta u - (u \cdot \nabla)u, \varphi \rangle$$

for all $\varphi \in H$. Now take any $w \in L^2$ and consider its Helmholtz–Weyl decomposition $w = \varphi + \nabla v$. We have for almost all t

$$\frac{\mathrm{d}}{\mathrm{d}t} \langle u(t), w \rangle = \frac{\mathrm{d}}{\mathrm{d}t} \langle u(t), \mathbb{P}w \rangle = \langle \Delta u - (u \cdot \nabla)u, \mathbb{P}w \rangle = \langle g, w \rangle,$$

where $g = \mathbb{P}[\Delta u - (u \cdot \nabla)u]$. Since for a strong solution $\Delta u - (u \cdot \nabla)u$ belongs to $L^2(0, T; L^2)$ we conclude that $g \in L^2(0, T; H)$ and using Lemma 1.31 we obtain $\partial_t u \in L^2(0, T; H) \subset L^2(0, T; L^2)$. □

The next lemma shows that for almost all times the Leray projection of the function $\partial_t u - \Delta u + (u \cdot \nabla)u$ is equal to zero. The result is valid on all three domains considered in this book.

Lemma 6.3 *Suppose that u is a strong solution of the Navier–Stokes equations on $[0, T]$. Then*

$$\int_0^T \langle \partial_t u - \Delta u + (u \cdot \nabla)u, w \rangle = 0 \tag{6.1}$$

for all $w \in L^2(0, T; H)$.

Proof First we use Lemma 1.28 to find a sequence $w_n \in \tilde{\mathcal{D}}_\sigma$ (see Definition 3.10) such that

$$w_n \to w \quad \text{in} \quad L^2(0, T; H).$$

Since $w_n \in \tilde{\mathcal{D}}_\sigma$ we conclude from Lemma 3.11 that

$$\int_0^T -\langle u, \partial_t w_n \rangle + \langle \nabla u, \nabla w_n \rangle + \langle (u \cdot \nabla)u, w_n \rangle \, \mathrm{d}s$$

$$= \langle u(0), w_n(0) \rangle - \langle u(T), w_n(T) \rangle.$$

Since $\partial_t u, \partial_t w_n \in L^2(0, T; L^2) \subset L^2(0, T; H^{-1})$ and $u, w_n \in L^2(0, T; H_0^1)$ we can use Corollary 1.34 and integrate by parts to obtain

$$\int_0^T \langle \partial_t u, w_n \rangle \, \mathrm{d}s = -\int_0^T \langle u, \partial_t w_n \rangle \, \mathrm{d}s + \langle u(T), w_n(T) \rangle - \langle u(0), w_n(0) \rangle.$$

Hence, noticing that $\langle \nabla u, \nabla w_n \rangle = \langle -\Delta u, w_n \rangle$,

$$\int_0^T \langle \partial_t u, w_n \rangle - \langle \Delta u, w_n \rangle + \langle (u \cdot \nabla)u, w_n \rangle = 0.$$

Since $\partial_t u$, Δu, and $(u \cdot \nabla)u$ are all elements of $L^2(0, T; L^2)$ we can pass to the limit as $n \to \infty$ and obtain (6.1).   □

The proof on the whole space is similar but we are not allowed to use the eigenfunctions of the Stokes operator. Instead we use functions w_n of the form $w_n = \sum_{j=1}^{N} c_j(t)v_j(x) \in \mathcal{D}_\sigma(\mathbb{R}^3)$, where c_j are smooth on $[0, T]$, have compact support in $\mathbb{R}$, and $\{v_j\}_{j \in \mathbb{N}}$ is an orthonormal basis of $H(\mathbb{R}^3)$ consisting of smooth divergence-free functions with compact support.

An easy consequence of Lemma 6.3 is that the gradient of the pressure, corresponding to a strong solution, is well defined as an L^2-function in space–time.

Theorem 6.4 *Let Ω be the torus $\mathbb{T}^3$, a smooth bounded domain in $\mathbb{R}^3$ or the whole space $\mathbb{R}^3$. Suppose that u is a strong solution of the Navier–Stokes equations on $[0, T]$. Then there exists a function $p \in L^2(0, T; H^1)$ such that*

$$\partial_t u - \Delta u + (u \cdot \nabla)u + \nabla p = 0$$

for almost all points in $\Omega \times (0, T)$.

Proof Let $\mathbb{P}$ denote the Leray projector from Chapter 2. For almost all times $s \in [0, T]$ we have

$$\partial_t u(s) - \Delta u(s) + (u(s) \cdot \nabla)u(s) \in L^2(\Omega)$$

and we can deduce from Lemma 6.3 that

$$\mathbb{P}\left(\partial_t u(s) - \Delta u(s) + (u(s) \cdot \nabla)u(s)\right) = 0$$

for almost all $s \in [0, T]$. It follows from Theorem 2.16 that for almost all times $s \in [0, T]$ we have

$$\partial_t u(s) - \Delta u(s) + (u(s) \cdot \nabla)u(s) = -\nabla p(s)$$

for some $\nabla p(s) \in L^2(\Omega)$.

Since

$$\|\nabla p\| = \|\partial_t u - \Delta u + (u \cdot \nabla)u\|$$

it follows that $\nabla p \in L^2(0, T; L^2)$ for all domains.

We notice that p is not yet absolutely determined since we can still add a constant to p without changing its gradient. So we need to show that we can choose a function p in such a way that its gradient remains unchanged and $p \in L^2(0, T; L^2)$.

Consider first the case of the torus or a smooth bounded domain. We can modify p by subtracting $\int_\Omega p(t)$ from p for almost every $t \in [0, T]$. Since for a modified (zero-mean) p we have $\|p\| \le c\|\nabla p\|$ according to the Poincaré inequality we deduce that $p \in L^2(0, T; L^2)$.

On the whole space we have $\mathbb{P}\partial_t u = \partial_t u$ and $\mathbb{P}\Delta u = \Delta u$. Therefore p is determined from the Helmholtz–Weyl Decomposition of the nonlinear term:

$$-\nabla p + h = (u \cdot \nabla)u,$$

where h is divergence free. It follows that $-\Delta p = \partial_i \partial_j (u_i u_j)$ in the sense of distributions. Hence from Lemma 5.1 we can deduce that

$$\|p(t)\|_{L^2}^2 \le c\|u(t)\|_{L^4}^4$$

for almost all $t \in [0, T]$. Since $u \in L^4(0, T; L^4)$ when u is a strong solution[1] we have $p \in L^2(0, T; L^2)$. $\qquad\square$

We now show that every strong solution satisfies not only an energy inequality (as is the case with Leray–Hopf weak solutions), but an energy *equality* (as is the case for classical solutions). In particular, every strong solution must also be a Leray–Hopf weak solution (although this is of course a much weaker assertion). This is valid on all three domains considered in this book.

Theorem 6.5 *If u is a strong solution of the Navier–Stokes equations on $[0, T]$ then u satisfies the energy equality*

$$\frac{1}{2}\|u(t_1)\|^2 + \int_{t_0}^{t_1} \|\nabla u(s)\|^2 \, ds = \frac{1}{2}\|u(t_0)\|^2 \tag{6.2}$$

for all $t_0, t_1 \in [0, T], t_0 < t_1$.

Proof Let u be a strong solution and let $\chi_{[t_0, t_1]}$ be the characteristic function of the interval $[t_0, t_1]$, where $0 \le t_0 < t_1 \le T$.

Then $u \chi_{[t_0, t_1]} \in L^2(0, T; H)$ and from Lemma 6.3 it follows that

$$0 = \int_0^T \langle \partial_t u - \Delta u + (u \cdot \nabla)u, u \chi_{[t_0, t_1]} \rangle = \int_{t_0}^{t_1} \langle \partial_t u - \Delta u + (u \cdot \nabla)u, u \rangle.$$

From Lemma 3.2 we have

$$\langle (u \cdot \nabla)u, u \rangle = 0.$$

Moreover, as $\partial_t u \in L^2(0, T; L^2) \subset L^2(0, T; H^{-1})$ and $u \in L^2(0, T; H_0^1)$ we can conclude using Corollary 1.34 that

$$\langle \partial_t u, u \rangle = \frac{1}{2}\frac{d}{dt}\|u\|^2.$$

[1] Notice that a strong solution u belongs to $L^\infty(0, T; L^2) \cap L^\infty(0, T; L^6)$.

Hence for every $0 \leq t_0 < t_1 \leq T$ we obtain

$$\int_{t_0}^{t_1} \frac{1}{2} \frac{d}{dt} \|u\|^2 + \int_{t_0}^{t_1} \|\nabla u(s)\|^2 \, ds = 0. \qquad \square$$

The next lemma shows that a strong solution can be used as a test function in the weak formulation of the Navier–Stokes equations.

Lemma 6.6 *Let u be a weak solution of the Navier–Stokes equations. If*

$$v \in L^2(0, T; H^2 \cap V) \quad and \quad \partial_t v \in L^2(0, T; L^2)$$

then[2] for all $t \in [0, T]$

$$-\int_0^t \langle u, \partial_t v \rangle + \int_0^t \langle \nabla u, \nabla v \rangle + \int_0^t \langle (u \cdot \nabla)u, v \rangle = \langle u(0), v(0) \rangle - \langle u(t), v(t) \rangle.$$

Proof We will give a proof valid for the torus and a smooth bounded domain, but a similar argument works also for the whole space.[3] Define $v_n := P_n(v)$, where P_n is the projection onto the first n eigenfunctions of the Stokes operator. Since v_n is of the form

$$v_n(t, x) = \sum_{k=1}^{n} c_k(t) a_k(x),$$

where $c_k \in L^2(0, T)$, $\dot{c}_k \in L^2(0, T)$, and $a_k \in \mathcal{N}$, we deduce from Lemma 3.12 that v_n can be used as a test function on $[0, T]$. It is easy to see that

$$v_n \to v \quad \text{in } L^2(0, T; H^2) \qquad \text{and}$$

$$\partial_t v_n \to \partial_t v \quad \text{in } L^2(0, T; L^2).$$

Since for any $w \in L^2(0, T; H^2)$ with $\partial_t w \in L^2(0, T; L^2)$ we have from Proposition 1.35

$$\sup_{0 \leq t \leq T} \|w(t)\|_{H^1} \leq C(\|\partial_t w\|_{L^2(0,T;L^2)} + \|w\|_{L^2(0,T;H^2)}) \qquad (6.3)$$

it follows (substituting $w = v_n - v_m$) that (v_n) is a Cauchy sequence in $C([0, T]; H^1)$ and the rest of the proof follows the reasoning from the proof of implication (i) $\Rightarrow$ (ii) in Lemma 3.11. $\qquad \square$

The following result applies to any strong solution of the Navier–Stokes equations and will be very useful in various situations where we will perform

---

[2] As always we assume that in each case we choose a representative of u that belongs to $C([0, T]; V^*)$ and a representative of v that belongs to $C([0, T]; H^1)$.
[3] In the case of $\mathbb{R}^3$ we need to replace $\mathcal{N}$ with a suitable basis of $H^2(\mathbb{R}^3) \cap V$ consisting of functions in $C_{c,\sigma}^\infty(\mathbb{R}^3)$.

computations directly on strong solutions without referring to the Galerkin approximations (cf. Theorem 1.33 and Proposition 1.35).

Lemma 6.7 *Let $u \in L^2(0, T; D(A))$ with $\partial_t u \in L^2(0, T; H)$. Then the function $t \mapsto \|\nabla u\|^2$ is absolutely continuous on $[0, T]$ and*

$$\frac{1}{2}\frac{d}{dt}\|\nabla u\|^2 = \langle \partial_t u, Au \rangle. \tag{6.4}$$

Proof The result is valid on all three domains considered in this book; we give a proof in the case of the torus or a smooth bounded domain. Take a sequence $u_n \in C^1([0, T]; P_n H)$ converging to a strong solution u in the following way:

$$u_n \to u \quad \text{in} \quad L^2(0, T; D(A)) \quad \text{and}$$
$$\partial_t u_n \to \partial_t u \quad \text{in} \quad L^2(0, T; H)$$

(such a sequence exists – see Exercise 6.1). Then we have

$$\langle \partial_t u_n, Au_n \rangle = \langle A^{1/2}\partial_t u_n, A^{1/2}u_n \rangle = \frac{1}{2}\frac{d}{dt}\|\nabla u_n\|^2,$$

so

$$\int_s^t \langle \partial_t u_n, Au_n \rangle = \frac{1}{2}\|\nabla u_n(t)\|^2 - \frac{1}{2}\|\nabla u_n(s)\|^2. \tag{6.5}$$

Using (6.3) for $w = u_n - u_m$ we deduce that $(\|\nabla u_n\|^2)$ is a Cauchy sequence in $C([0, T])$ so we can pass to the limit in (6.5) to obtain

$$\frac{1}{2}\|\nabla u(t)\|^2 = \frac{1}{2}\|\nabla u(s)\|^2 + \int_s^t \langle \partial_t u, Au \rangle,$$

where $\langle \partial_t u, Au \rangle \in L^1(0, T)$, so $t \mapsto \|\nabla u(t)\|^2$ is absolutely continuous and its derivative is given by (6.4). $\qquad\square$

6.2 Local existence of strong solutions

We will now prove that a strong solution exists at least locally in time whenever the initial condition u_0 belongs to the space V. More precisely, we will show that for a given smooth bounded domain (or the torus $\mathbb{T}^3$) we can estimate from below the time of existence of a strong solution arising from u_0. This estimate involves only the H^1 norm of u_0. The same result is valid on the whole space (see Corollary 6.9 and Theorem 6.15). Local-in-time existence of strong solutions on the whole space is due to Leray (1934), while the case of a bounded domain is due to Kiselev & Ladyzhenskaya (1957).

Theorem 6.8 *Let Ω be the torus $\mathbb{T}^3$ or a smooth bounded domain in $\mathbb{R}^3$. There is a constant $c > 0$ (depending on Ω) such that any initial condition $u_0 \in V$ gives rise to a strong solution of the Navier–Stokes equations that exists at least on the time interval $[0, T]$, where*

$$T = c\|\nabla u_0\|^{-4}.$$

Proof Let $\{u_n\}$ be the sequence of Galerkin approximations used in the proof of existence of weak solutions of the Navier–Stokes equations in Theorem 4.4.

We take the inner product of the equation for the Galerkin approximations,

$$\partial_t u_n + A u_n + P_n(u_n \cdot \nabla)u_n = 0,$$

with $A u_n$ to obtain

$$\langle \partial_t u_n, A u_n \rangle + \langle A u_n, A u_n \rangle + \langle (u_n \cdot \nabla)u_n, A u_n \rangle = 0.$$

Hence by Lemma 6.4

$$\frac{1}{2}\frac{\mathrm{d}}{\mathrm{d}t}\|\nabla u_n\|^2 + \|A u_n\|^2 = -\langle (u_n \cdot \nabla)u_n, A u_n \rangle.$$

The main difference between the current situation and the corresponding step in the proof of the existence of weak solutions is that the nonlinear term no longer vanishes. In order to estimate it we use Agmon's inequality (Theorem 1.20) together with the Poincaré inequality and Theorem 2.23 ($\|u\|_{H^2} \le c\|Au\|$)

$$
\begin{aligned}
|\langle (u_n \cdot \nabla)u_n, A u_n \rangle| &\le \|u_n\|_{L^\infty}\|\nabla u_n\|\|A u_n\| \\
&\le c\|u_n\|_{H^1}^{1/2}\|u_n\|_{H^2}^{1/2}\|\nabla u_n\|\|A u_n\| \\
&\le c\|\nabla u_n\|^{3/2}\|A u_n\|^{3/2}.
\end{aligned}
\tag{6.6}
$$

Using Young's inequality with exponents 4 and 4/3 yields

$$|\langle (u_n \cdot \nabla)u_n, A u_n \rangle| \le c\|\nabla u_n\|^6 + \frac{1}{2}\|A u_n\|^2.$$

Therefore we obtain

$$\frac{\mathrm{d}}{\mathrm{d}t}\|\nabla u_n\|^2 + \|A u_n(t)\|^2 \le c\|\nabla u_n(t)\|^6, \tag{6.7}$$

where the constant c does not depend on n.

In particular, we have

$$\frac{\mathrm{d}}{\mathrm{d}t}\|\nabla u_n\|^2 \le c\|\nabla u_n(t)\|^6.$$

Now we compare $\|\nabla u_n\|^2$ with the solution of the problem

$$\frac{dX}{dt} = cX^3, \qquad X(0) = \|\nabla u_0\|^2, \tag{6.8}$$

and deduce (see Exercise 6.2) that for all t such that $0 \le 2ct\|\nabla u_0\|^4 < 1$

$$\|\nabla u_n(t)\|^2 \le X(t) = \frac{\|\nabla u_0\|^2}{\sqrt{1 - 2ct\|\nabla u_0\|^4}}.$$

In particular, if we choose

$$T = \frac{3}{8c\|\nabla u_0\|^4}$$

then for all $t \in [0, T]$ we have the uniform upper bound

$$\sup_{0 \le t \le T} \|\nabla u_n(t)\|^2 \le \sup_{0 \le t \le T} X(t)$$

$$\le X(T) = 2\|\nabla u_0\|^2. \tag{6.9}$$

Now we integrate (6.7) between 0 and T and drop the term $\|\nabla u_n(T)\|^2$ on the left-hand side to obtain

$$\int_0^T \|Au_n(s)\|^2 \, ds \le \|\nabla u_n(0)\|^2 + c \int_0^T \|\nabla u_n(s)\|^6 \, ds$$

$$\le \|\nabla u_0\|^2 + c \int_0^T X^3(s) \, ds,$$

$$\le \|\nabla u_0\|^2 + cTX^3(T)$$

$$\le 4\|\nabla u_0\|^2, \tag{6.10}$$

due to our choice of T. Since $\|v\|_{H^2} \le \tilde{c}\|Av\|$ for all functions $v \in V \cap H^2$ it follows that the functions u_n are uniformly bounded in $L^\infty(0, T; V)$ (from (6.9)) and in $L^2(0, T; H^2)$. We now extract from the subsequence of u_n chosen in the proof of Theorem 4.4 a further subsequence such that (in addition to previous estimates) u_n tends to u weakly-$*$ in $L^\infty(0, T; V)$ and weakly in $L^2(0, T; H^2)$. It follows that u is a weak solution of the Navier–Stokes equations that now also belongs to the space

$$L^2(0, T; H^2) \cap L^\infty(0, T; V),$$

so u is a strong solution on $[0, T]$ as claimed. $\qquad\square$

The following corollary states that the estimates (6.9) and (6.10) are valid for all strong solutions.[4] However, the key observation that we are about to make is

---

[4] We use here the fact that strong solutions arising from an initial condition $u_0 \in V$ are unique in their class (see Exercise 6.5 or Section 6.3 for stronger results).

that the time interval $[0, T]$ on which (6.9) and (6.10) hold depends on the size of the initial data in V and may depend on the shape of Ω but not on its size. In particular, if B_R is the ball centred at zero and of radius $R \geq 1$, and $u_0 \in V(B_R)$, then the guaranteed time of local existence of a strong solution arising from $u_0 \in V(B_R)$ depends only on $\|\nabla u_0\|$ but does not depend on R.

Corollary 6.9 *Let Ω be a smooth bounded domain in $\mathbb{R}^3$, $\lambda \geq 1$ and*

$$\Omega_\lambda := \{x \in \mathbb{R}^3 : x = \lambda y, \ y \in \Omega\}.$$

There exists a constant $c > 0$, independent of λ, with the following property. If u is a strong solution arising from $u_0 \in V(\Omega_\lambda)$ then

$$\|\nabla u_0\|_{L^2(\Omega_\lambda)}^2 \leq M \Rightarrow \sup_{0 \leq s \leq cM^{-2}} \|\nabla u(s)\|_{L^2(\Omega_\lambda)}^2 \leq 2M \quad and$$

$$\int_0^{cM^{-2}} \|Au(s)\|_{L^2(\Omega_\lambda)}^2 \, ds \leq 4M.$$

Proof Our task is to show that c does not depend on λ. So take any choice of $v_0 \in V(\Omega_\lambda)$, and let $\|\nabla v_0\|_{L^2(\Omega_\lambda)}^2 = M$. Define

$$u_0(x) := \lambda v_0(\lambda x) \in V(\Omega).$$

Then $\|\nabla u_0\|_{L^2(\Omega)}^2 = \lambda M$, so u_0 gives rise to a strong solution u such that

$$\sup_{0 \leq s \leq c(\lambda M)^{-2}} \|\nabla u(s)\|_{L^2(\Omega)}^2 \leq 2\lambda M$$

and

$$\int_0^{c(\lambda M)^{-2}} \|Au(s)\|_{L^2(\Omega)}^2 \, ds \leq 4\lambda M,$$

where c is the constant corresponding to the fixed domain Ω (existence of such a constant follows from estimates (6.9) and (6.10)). We will show that the same c can be used on Ω_λ. To this end we note that if v is given by

$$v(x, t) = \frac{1}{\lambda} u(x/\lambda, t/\lambda^2)$$

then v is a strong solution of the Navier–Stokes equations on Ω_λ that arises from v_0 and satisfies the bound

$$\sup_{0 \leq s \leq cM^{-2}} \|\nabla v(s)\|_{L^2(\Omega_\lambda)}^2 = \frac{1}{\lambda} \sup_{0 \leq s \leq c(\lambda M)^{-2}} \|\nabla u(s)\|_{L^2(\Omega)}^2 \leq 2M.$$

Furthermore

$$\int_0^{cM^{-2}} \|Av(s)\|^2_{L^2(\Omega_\lambda)} \, ds = \frac{1}{\lambda^3} \int_0^{cM^{-2}} \|Au(s/\lambda^2)\|^2_{L^2(\Omega)} \, ds$$

$$= \frac{1}{\lambda} \int_0^{c(\lambda M)^{-2}} \|Au(\tau)\|^2_{L^2(\Omega)} \, d\tau \le 4M.$$

To conclude the proof we note that every strong solution arising from v_0 must coincide with v given above, according to the uniqueness of strong solutions in their class (see Exercise 6.5). □

Finally let us notice that since we define strong solutions on closed finite time intervals $[0, T]$ on which the H^1 norm is uniformly bounded it follows from the results we have just proved that the strong solution can always be continued after the time T (but may blow up at some time $t > T$).

6.3 Weak–strong uniqueness and blowup

It is relatively straightforward (see Exercise 6.5) to show that strong solutions are unique, i.e. that there can only be one strong solution corresponding to a given initial condition $u_0 \in V$. But something much stronger is true: strong solutions are unique in the larger class of weak solutions that satisfy the energy inequality (4.17). In particular, strong solutions are unique in the class of Leray–Hopf weak solutions. This result has many far-reaching consequences and is the key to many of the properties of Leray–Hopf weak solutions that we will prove in Chapter 8. The proof below is based on the idea due to Sather and Serrin, see Serrin (1963), whose method allows for the comparison of two weak solutions, where one has some additional regularity.

Theorem 6.10 (Weak–strong uniqueness) *Suppose that u is a strong solution of the Navier–Stokes equations on the time interval $[0, T]$ and that v is any weak solution on $[0, T]$, with the same initial data, that satisfies the energy inequality*

$$\frac{1}{2}\|v(t)\|^2 + \int_0^t \|\nabla v\|^2 \le \frac{1}{2}\|v(0)\|^2 \quad \text{for all } t \in [0, T].$$

Then $u = v$ on $[0, T]$.

Proof From Lemmas 6.3 and 6.6, for all $t \in (0, T)$ we have

$$\int_0^t \langle \partial_t u, v \rangle + \int_0^t \langle \nabla u, \nabla v \rangle + \int_0^t \langle (u \cdot \nabla)u, v \rangle = 0$$

and

$$-\int_0^t \langle v, \partial_t u \rangle + \int_0^t \langle \nabla v, \nabla u \rangle + \int_0^t \langle (v \cdot \nabla)v, u \rangle = \langle v(0), u(0) \rangle - \langle v(t), u(t) \rangle.$$

We add the two equations to eliminate the time derivatives and obtain

$$2 \int_0^t \langle \nabla v, \nabla u \rangle + \int_0^t \langle (v \cdot \nabla)v, u \rangle + \int_0^t \langle (u \cdot \nabla)u, v \rangle = \|u_0\|^2 - \langle v(t), u(t) \rangle$$

since $u_0 = v_0$. Our aim is to derive a differential inequality for the L^2 norm of the difference $w = v - u$. To this end we use the identity

$$\|a - b\|^2 = \|a\|^2 - 2\langle a, b \rangle + \|b\|^2$$

to write

$$2\langle \nabla u, \nabla v \rangle = \|\nabla u\|^2 + \|\nabla v\|^2 - \|\nabla w\|^2,$$

$$\langle u(t), v(t) \rangle = \frac{1}{2}\|u(t)\|^2 + \frac{1}{2}\|v(t)\|^2 - \frac{1}{2}\|w(t)\|^2.$$

Standard computations involving the nonlinear term (see Exercise 6.4) give

$$\int_0^t \langle (u \cdot \nabla)u, v \rangle + \int_0^t \langle (v \cdot \nabla)v, u \rangle = \int_0^t \langle (w \cdot \nabla)w, u \rangle.$$

Hence

$$-\int_0^t \|\nabla w\|^2 + \int_0^t \|\nabla u\|^2 + \int_0^t \|\nabla v\|^2 + \int_0^t \langle (w \cdot \nabla)w, u \rangle$$
$$= \|u_0\|^2 + \frac{1}{2}\|w(t)\|^2 - \frac{1}{2}\|u(t)\|^2 - \frac{1}{2}\|v(t)\|^2.$$

After rearranging the terms we obtain

$$\frac{1}{2}\|w(t)\|^2 + \int_0^t \|\nabla w\|^2 - \int_0^t \langle (w \cdot \nabla w), u \rangle = I_1 + I_2,$$

where

$$I_1 = \frac{1}{2}\|v(t)\|^2 + \int_0^t \|\nabla v(s)\|^2 \, ds - \frac{1}{2}\|v(0)\|^2 \le 0$$

(this is true by assumption) and

$$I_2 = \frac{1}{2}\|u(t)\|^2 + \int_0^t \|\nabla u\|^2 \, ds - \frac{1}{2}\|u(0)\|^2 = 0$$

due to the energy equality for u (see Lemma 6.5). Therefore

$$\frac{1}{2}\|w(t)\|^2 + \int_0^t \|\nabla w(s)\|^2 \, ds \leq \left| \int_0^t \langle (w \cdot \nabla)w, u \rangle \right|$$

$$\leq \int_0^t \|u\|_\infty \|w\| \|\nabla w\|$$

$$\leq \frac{1}{2} \int_0^t \|u\|_\infty^2 \|w\|^2 + \frac{1}{2} \int_0^t \|\nabla w\|^2,$$

and we obtain

$$\|w(t)\|^2 + \int_0^t \|\nabla w(s)\|^2 \, ds \leq \int_0^t \|u(s)\|_\infty^2 \|w(s)\|^2 \, ds$$

$$\leq c \int_0^t \|u(s)\|_{H^2}^2 \|w(s)\|^2 \, ds.$$

Since $u \in L^2(0, T; H^2)$ it follows from the Gronwall inequality for integrals (Lemma A.25) that $w(t) = 0$ for all $t \in (0, T]$. $\qquad\square$

We note that in the proof of Theorem 6.10 it is crucial to avoid using w as a test function in the weak formulation of equations. This is in fact the main reason why uniqueness of strong solutions is known only in the class of weak solutions satisfying the energy inequality and not in the class of all weak solutions.[5]

We now consider a global-in-time Leray–Hopf weak solution u corresponding to an initial condition $u_0 \in V$. There are two possibilities: either u is strong for all positive times, and in this case $\|\nabla u(s)\|$ remains finite for all $s > 0$; or there exists a time T^* when $\|\nabla u\|$ 'blows up', i.e. $\|\nabla u(t)\|$ is not essentially bounded in any (left) neighbourhood of T^*. In the following lemma we estimate the rate at which the L^2 norm of ∇u must increase if u blows up at time T^*. The result is valid on all domains considered in this book (in the case of the whole space it follows from Theorem 6.15).

Lemma 6.11 *There exists a constant $c' > 0$ (depending only on the domain) such that if u is a Leray–Hopf weak solution of the Navier–Stokes equations for some initial data $u_0 \in V$ that blows up at time $T^* > 0$ then*

$$\|\nabla u(t)\|^2 \geq \frac{c'}{\sqrt{T^* - t}}$$

for all $0 \leq t < T^*$.

---

[5] In fact it would be tempting to define $w = u - v$ and substitute this function in the weak formulation of the Navier–Stokes for u and for v. This, however, is not allowed and leads to an invalid proof.

Proof We know that $u(t) \in V$ and the strong energy inequality is satisfied from $s = t$ for all $t < T^*$, see Theorem 6.5. For all such t, using Theorem 6.8, we can construct a strong solution v with initial condition $v(0) = u(t)$, whose maximal time of existence is at least $c\|\nabla u(t)\|^{-4}$. Using weak–strong uniqueness $u(t + s) = v(s)$, and so $T^* > t + c\|\nabla u(t)\|^{-4}$ for all positive $t \le T^*$, otherwise the solution would not blow up at time T^*. It follows that

$$\frac{c}{\|\nabla u(t)\|^4} \le T^* - t$$

for all $t \in (0, T^*)$ and the lemma follows immediately. $\qquad\square$

6.4 Global existence for small data in V

From the proof of Theorem 6.8 we can deduce that for an initial condition that is sufficiently small in H^1 one can construct strong solutions that exist globally in time. We emphasise that we are able to prove the assertion of Theorem 6.12 only on a bounded domain. If Theorem 6.12 was true on $\mathbb{R}^3$ it would immediately imply the global existence of strong solutions, because any $u_0 \in V(\mathbb{R}^3)$ can be rescaled to satisfy the condition below (cf. Lemma 6.13 where such a rescaling is impossible).

Theorem 6.12 *Let Ω be the torus $\mathbb{T}^3$ or a smooth bounded domain in $\mathbb{R}^3$. There exists a constant $C > 0$ (depending on Ω) such that if*

$$\|\nabla u_0\| < C$$

then the strong solution corresponding to $u_0 \in V$ exists globally in time.

Proof Instead of working with the Galerkin approximations we will make estimates directly for u. We know that u_0 gives rise to a strong solution that exists at least on a certain time interval $[0, T)$. On this time interval for each time $t \in (0, T)$ we take the L^2 inner product of the Navier–Stokes equations with Au, and since (6.4) is valid for any strong solution u (see Lemma 6.7), we obtain

$$\frac{\mathrm{d}}{\mathrm{d}t}\|\nabla u\|^2 \le c_1\|\nabla u\|^6 - \|Au\|^2$$

in a similar way as we obtained (6.7). Since we are considering the problem on a bounded domain we have the Poincaré-like inequality

$$\|\nabla u\| \le c_2\|Au\|$$

(see Theorem 2.23 or Exercise 2.11). It follows that

$$\frac{\mathrm{d}}{\mathrm{d}t}\|\nabla u\|^2 \le c_1\|\nabla u\|^6 - \frac{1}{c_2^2}\|\nabla u\|^2$$

$$\le c_1\|\nabla u\|^2\left(\|\nabla u\|^4 - \frac{1}{c_1 c_2^2}\right).$$

Hence if $\|\nabla u_0\| < C := (c_1 c_2^2)^{-1/4}$ then

$$\|\nabla u(t)\|^4 < \frac{1}{c_1 c_2^2} \quad \text{for all} \ t > 0.$$

Therefore $\|\nabla u(t)\|$ is bounded for all $t > 0$ and the strong solution is defined globally in time. □

The global existence of regular solutions can be proved for small data in larger spaces than H^1. In particular, small initial conditions in $H^{1/2}$ (see Chapter 10) and in L^3 (see Chapter 11) also give rise to global-in-time regular solutions. Lemma 6.13 is another 'small data' result, essentially due to Leray (1934). It provides a heuristic indication that we might expect global existence for initial data that is small in $H^{1/2}$, since the norm in $H^{1/2}$ can be interpolated between L^2 and H^1 (see Lemma 1.15) with

$$\|u\|_{H^{1/2}} \le \|u\|_{L^2}^{1/2}\|u\|_{H^1}^{1/2}.$$

Note that it has the significant advantage over Theorem 6.12 that we do not use the Poincaré inequality, so the proof is also valid on the whole space. We also note that the quantity $\|u_0\|\|\nabla u_0\|$ from Lemma 6.13 is scale invariant (see Section 10.1).

Lemma 6.13 *There is a constant $c > 0$ such that if $u_0 \in V$ with*

$$\|u_0\|\|\nabla u_0\| \le c$$

then u_0 gives rise to a global-in-time strong solution of the Navier–Stokes equations.

Proof We estimate the nonlinear term as we did before in (6.6) to obtain

$$\frac{1}{2}\frac{\mathrm{d}}{\mathrm{d}t}\|\nabla u\|^2 + \|Au\|^2 \le \int |u||\nabla u||Au|$$

$$\le c\|u\|_{L^6}\|\nabla u\|_{L^3}\|Au\|$$

$$\le c\|\nabla u\|\|\nabla u\|^{1/2}\|\nabla u\|_{L^6}^{1/2}\|Au\|.$$

Then we use the inequality

$$\|\nabla u\|_{L^6}^{1/2} \le c\|Au\|^{1/2},$$

which follows from the Sobolev embedding[6] and Theorem 2.23 (on a bounded domain $\|u\|_{H^2} \le c\|Au\|$, on the whole space $\|u\|_{\dot{H}^2} \le c\|Au\|$). Hence

$$\frac{1}{2}\frac{d}{dt}\|\nabla u\|^2 + \|Au\|^2 \le c\|\nabla u\|^{3/2}\|Au\|^{3/2}.$$

Now we use the interpolation of the H^1 norm between the L^2 and H^2 norms (on $\mathbb{R}^3$ we use interpolation inequality between the homogenous Sobolev spaces):

$$\|\nabla u\| \le C\|u\|^{1/2}\|Au\|^{1/2}.$$

This interpolation inequality gives us

$$\|\nabla u\|^{3/2}\|Au\|^{3/2} \le c\|u\|^{1/4}\|Au\|^{1/4}\|\nabla u\|\|Au\|^{3/2}$$
$$\le c\|u\|^2\|\nabla u\|^8 + \frac{1}{2}\|Au\|^2$$
$$\le c\|u\|^4\|\nabla u\|^4\|Au\|^2 + \frac{1}{2}\|Au\|^2,$$

where we have used Young's inequality with exponents 8 and 8/7. Hence

$$\frac{d}{dt}\|\nabla u\|^2 \le \|Au\|^2 \left(c\|u\|^4\|\nabla u\|^4 - 1\right). \tag{6.11}$$

Since (from the energy equality)

$$\frac{1}{2}\frac{d}{dt}\|u\|^2 = -\|\nabla u\|^2 \tag{6.12}$$

we know that $\|u\|$ is a decreasing function of time.

Combining (6.11) and (6.12) it follows that

$$\frac{d}{dt}\|u\|^2\|\nabla u\|^2 \le \|u\|^2\|Au\|^2(c\|u\|^4\|\nabla u\|^4 - 1) - \|\nabla u\|^4,$$

and so if $\|u_0\|\|\nabla u_0\| < c^{-1/4}$ the combination $\|u(t)\|\|\nabla u(t)\|$ is decreasing. It then follows from (6.11) and (6.12) that the H^1 norm of u is decreasing, and we obtain the existence of a global solution. □

---

[6] On a bounded domain we use $\|\nabla u\|_{L^6} \le c\|\nabla u\|_{H^1}$, on the whole space $\|\nabla u\|_{L^6} \le c\|\nabla u\|_{\dot{H}^1}$.

6.5 Global strong solutions in the two-dimensional case

Let us now briefly explain why strong solutions are global in time in the two-dimensional case. We will only derive a priori estimates, but the argument can be made rigorous with the use of Galerkin approximations.

Theorem 6.14 (Ladyzhenskaya) *Let $u_0 \in V(\Omega)$, where Ω is a smooth domain in $\mathbb{R}^2$. Then u_0 gives rise to a strong global-in-time solution of the Navier–Stokes equations.*

Proof (sketch) Taking the L^2 inner product of the 2D Navier–Stokes equations with Au, we then integrate by parts and obtain, just as in the 3D case, the inequality

$$\frac{1}{2}\frac{\mathrm{d}}{\mathrm{d}t}\|\nabla u\|^2 + \|Au\|^2 \le |\langle(u\cdot\nabla)u, Au\rangle|.$$

The difference is that in two and three dimensions we have different Sobolev exponents in the Sobolev embedding theorem. As a consequence, while estimating the nonlinear term, we can use the 2D Ladyzhenskaya inequality

$$\|u\|_{L^4} \le c\|u\|^{1/2}\|\nabla u\|^{1/2} \quad u \in H_0^1(\Omega), \quad \Omega \subset \mathbb{R}^2$$

(see Exercise 3.6). We therefore have

$$\begin{aligned}
\frac{1}{2}\frac{\mathrm{d}}{\mathrm{d}t}\|\nabla u\|^2 + \|Au\|^2 &\le |\langle(u\cdot\nabla)u, Au\rangle| \\
&\le \|u\|_{L^4}\|\nabla u\|_{L^4}\|Au\| \\
&\le c\|u\|^{1/2}\|\nabla u\|\|Au\|^{3/2} \\
&\le c\|u\|^2\|\nabla u\|^4 + \frac{1}{2}\|Au\|^2,
\end{aligned}$$

and so

$$\frac{\mathrm{d}}{\mathrm{d}t}\|\nabla u\|^2 + \|Au\|^2 \le \left(c\|u\|^2\|\nabla u\|^2\right)\|\nabla u\|^2.$$

We know that $c\|u\|^2\|\nabla u\|^2$ is integrable in time, since

$$\int_0^T c\|u\|^2\|\nabla u\|^2 \le c\left(\operatorname*{ess\,sup}_{0\le t\le T}\|u(t)\|\right)^2 \int_0^T \|\nabla u(s)\|^2\,\mathrm{d}s \, < \infty$$

because $u \in L^\infty(0, T; H) \cap L^2(0, T; V)$. We can now use the Gronwall Lemma (Lemma A.24) to deduce that

$$\operatorname*{ess\,sup}_{0\le t\le T}\|\nabla u(t)\|^2 + \int_0^T \|Au(s)\|^2\,\mathrm{d}s < \infty$$

for any $T > 0$. In particular, we have global existence of strong solutions. $\square$

The reason why the same argument fails in $\mathbb{R}^3$ is that the exponents in the Ladyzhenskaya inequality are different; in three dimensions we have

$$\|u\|_{L^4} \le c\|u\|^{1/4}\|\nabla u\|^{3/4} \quad u \in H_0^1(\Omega), \quad \Omega \subset \mathbb{R}^3;$$

see Exercise 3.6.

6.6 Strong solutions on the whole space

We will now sketch a proof of the existence of strong solutions on the whole space.

Theorem 6.15 *There exists a constant $c > 0$ such that every initial condition $u_0 \in H^1(\mathbb{R}^3)$ gives rise to a strong solution of the Navier–Stokes equations on the time interval $[0, c\|\nabla u_0\|^{-4}]$.*

Proof We will give a proof corresponding to the construction of weak solutions in Section 4.4.

Let $B_n = B(0, R_n)$ denote the ball centred at zero and of radius R_n, where $R_n \to \infty$, as $n \to \infty$. For a given $u_0 \in V(\mathbb{R}^3)$ we choose a sequence of initial conditions $u_{0,n} \in C_{c,\sigma}^\infty(B_n)$ such that

$$u_{0,n} \to u_0 \quad \text{strongly} \quad \text{in} \quad V(\mathbb{R}^3).$$

Let u_n be a corresponding sequence of Leray–Hopf weak solutions of the Navier–Stokes initial boundary value problem in B_n with the initial condition $u_{0,n}$. Then there is $T > 0$ such that every u_n is strong on $[0, T]$ and

$$\sup_{0 \le s \le T} \|\nabla u_n(s)\|_{L^2(B_n)}^2 + \int_0^T \|Au_n\|_{L^2(B_n)}^2 \le 6\|\nabla u_{0,n}\|^2 \le C\|\nabla u_0\|^2.$$

This estimate holds uniformly for all n, according to Corollary 6.9. Using the bound

$$\|u\|_{\dot{H}^2(B_n)} \le c\|Au\|_{L^2(B_n)},$$

where c is a scale-invariant constant,[7] we deduce that

$$\sup_{0 \le s \le T} \|\nabla u_n(s)\|_{L^2(B_n)}^2 + \int_0^T \|u_n\|_{\dot{H}^2(B_n)}^2 \le C.$$

Since $\|u_n(t)\| \le \|u_{0,n}\| \le c\|u_0\|$ for all t it follows that

$$\sup_{0 \le s \le T} \|u_n(s)\|_{H^1(B_n)}^2 + \int_0^T \|u_n\|_{\dot{H}^2(B_n)}^2 \le C,$$

---

[7] The constant c is scale-invariant because $\|Au\|$ and $\|u\|_{\dot{H}^2}$ scale in the same way.

where C does not depend on n. Let $E_{2,n}$ denote the extension operator

$$E_{2,n} : H^2(B_n) \to H^2(\mathbb{R}^3),$$

described in Theorem 1.16. We have

$$\sup_{0 \le s \le T} \|E_{2,n} u_n(s)\|^2_{H^1(\mathbb{R}^3)} \le \sup_{0 \le s \le T} c_2 \|u_n(s)\|^2_{H^1(B_n)} \le C,$$

and

$$\int_0^T \|E_{2,n} u_n\|^2_{H^2(\mathbb{R}^3)} \le \int_0^T c_2 \|u_n\|^2_{H^2(B_n)}$$

$$= c_2 \int_0^T \|u_n\|^2_{H^2(B_n)} + c_2 \int_0^T \|u_n\|^2_{H^1(B_n)} \le C,$$

where c_2 and C do not depend on n (see Theorem 1.16).

Therefore for some $u \in L^2(0, T; H^2(\mathbb{R}^3)) \cap L^\infty(0, T; V(\mathbb{R}^3))$ and some subsequence of $E_{2,n} u_n$ we have

$$E_{2,n}(u_n) \rightharpoonup u \quad \text{weakly in} \quad L^2(0, T; H^2(\mathbb{R}^3))$$

$$E_{2,n}(u_n) \overset{*}{\rightharpoonup} u \quad \text{weakly-} * \quad \text{in} \quad L^\infty(0, T; H^1(\mathbb{R}^3)).$$

Reasoning as in the proof of Theorem 4.10 we can show that u satisfies (3.3) for $0 \le s \le T$, so u is a weak solution of the Navier–Stokes equations with additional regularity. Hence u is a strong solution of the Navier–Stokes equations on $\mathbb{R}^3$ arising from u_0 and defined on the time interval $[0, T]$. $\qquad\square$

Notes

The local existence of strong solutions of the Navier–Stokes equations was proved by Leray in 1934 for the whole space using a method of successive approximations. Leray also showed that strong (or 'regular', as he called them) solutions are unique in the class of regular solutions and that they exist globally in time if

$$\|\nabla u_0\|^2 \|u_0\|^2 < \delta$$

for sufficiently small $\delta > 0$, see Lemma 6.13.

The existence and uniqueness of global-in-time strong solutions of the Navier–Stokes equations in dimension $n = 2$ is due to Ladyzhenskaya (1959). Global existence and uniqueness of solutions of the Navier–Stokes equations is also known in some cases where the flow has some additional symmetry or when the flow is 'close' to a two-dimensional flow. For example, Ladyzhen-skaya (1970) proved that axisymmetric flows for which the initial condition has zero azimuthal component are globally well posed, that is, unique solutions exist for arbitrary long times (cf. Seregin & Zajaczkowski, 2007, where a simple sufficient condition for regularity of axisymmetric flow can be found). Mahalov, Titi, & Leibovich (1990) proved global well-posedness of flows with helical symmetry.

For flows in thin domains $\Omega_\varepsilon = (0, 1)^2 \times (0, \varepsilon)$ there exists a large set of initial conditions giving rise to global, unique, and strong solutions of the Navier–Stokes equations (see Raugel & Sell, 1993, for example).

Another construction of strong solutions is due to Fujita & Kato (1964) for bounded domains in $\mathbb{R}^3$ and to Kato (1984) for unbounded domains including the whole space. The method uses fractional powers of the Stokes operator and the theory of semigroups. First, the Navier–Stokes equations are written in the abstract form

$$\frac{d}{dt}u = -Au - \mathbb{P}[(u \cdot \nabla)u]$$

and then a solution is given in the integral form, using 'the variation of constants formula'/'Duhamel principle'

$$u(t) = e^{-tA}u_0 + \int_0^t e^{-(t-s)A}\mathbb{P}(u(s) \cdot \nabla)u(s)\,ds.$$

The solution is obtained by a fixed-point argument. In Fujita & Kato (1964) the solution is required to be continuous from $[0, T)$ into H and continuous from $(0, T)$ into the domain of $A^{1/2}$. The uniqueness in this class is obtained for initial conditions in H and existence for initial conditions $u_0 \in D(A^{1/4})$. Kato (1984) constructs solutions that are elements of $C([0, T); \mathbb{P}L^3)$ and shows that they coincide with Leray weak solutions for an initial condition $u_0 \in L^2 \cap L^3$. Such solutions satisfying an integral version of the Navier–Stokes equations and belonging to $C([0, T); X)$, where X is appropriately chosen function space, are usually called 'mild' solutions of the Navier–Stokes equations. An interesting review of this subject can be found in Cannone (2003) and Lemarié-Rieusset (2002).

Exercises

6.1   Let Ω be a smooth bounded domain. Prove that if u belongs to $L^2(0, T; D(A))$ with $\partial_t u \in L^2(0, T; H)$ then there exists a sequence

$$(u_n) \in C^1([0, T]; P_n H)$$

such that

$$u_n \to u \text{ in } L^2(0, T; D(A)) \quad \text{with} \quad \partial_t u_n \to \partial_t u \text{ in } L^2(0, T; H).$$

6.2 Let Y be the solution of the problem

$$Y'(t) = Y^3(t), \quad Y(0) = a > 0.$$

Show that if $X \geq 0$ satisfies

$$X'(t) \leq X^3, \quad X(0) = b,$$

where $0 \leq b \leq a$ then $X(t) \leq Y(t)$ for all non-negative times before Y blows up.

6.3 Prove that if u is a weak solution and v is a strong solution of the Navier–Stokes equations on $[0, T]$ then

$$\int_0^T \langle (u \cdot \nabla)u, v \rangle < \infty.$$

6.4 Prove that if u is a strong solution of the Navier–Stokes equations and v is a weak solution of the Navier–Stokes equations then for $w = v - u$ we have

$$\int_0^t \langle (u \cdot \nabla)u, v \rangle + \int_0^t \langle (v \cdot \nabla)v, u \rangle = \int_0^t \langle (w \cdot \nabla w), u \rangle.$$

6.5 Use the following identity (which holds in L^2 for a strong solution u)

$$\partial_t u + Au + \mathbb{P}(u \cdot \nabla)u = 0$$

to show that if u and v are two strong solutions on $[0, T]$ with the same initial condition then $u = v$ on $[0, T]$.

6.6 Let (u_n) be a sequence of Galerkin approximations converging to a strong solution u on the time interval $[0, T]$. Show that $\partial_t u_n$ are uniformly bounded in $L^2(0, T; L^2)$.

6.7 Suppose that u_0 belongs to H but not to H^1. Let u be a weak solution of the Navier–Stokes equations arising from u_0. Prove that

$$\lim_{t \to 0^+} \|u(t)\|_{H^1} = \infty.$$

7

Regularity of strong solutions

In this chapter we prove that strong solutions of the Navier–Stokes equations immediately become smooth. This is a very powerful result that also gives us a lot of insight into the behaviour of Leray–Hopf weak solutions. We will investigate this further in the next chapter.

Before going into details let us sketch the main idea of the proof. Suppose that we have a strong solution $u \in L^\infty(0, T; V) \cap L^2(0, T; H^2)$, and assume that the initial condition u_0 belongs to $V \cap H^m$ for some $m \geq 2$. We want to prove that in this case

$$u \in L^\infty(0, T; H^m) \cap L^2(0, T; H^{m+1}).$$

To this end for every $k = 2, 3, .., m$ we take the H^k inner product of the equation with u (this is in fact just a formal argument that can be made rigorous when we consider Galerkin approximations) to obtain

$$\frac{1}{2}\frac{d}{dt}\|u\|_{H^k}^2 + \|\nabla u\|_{H^k}^2 + \langle (u \cdot \nabla)u, u \rangle_{H^k} = 0.$$

The key point is that we can use the fact that H^k is a Banach algebra when $k > 3/2$, see Theorem 1.18, to obtain the following estimate for the nonlinear term:

$$
\begin{aligned}
|\langle (u \cdot \nabla)u, u \rangle_{H^k}| &\leq \|(u \cdot \nabla)u\|_{H^k}\|u\|_{H^k} \\
&\leq \tilde{c}_k \|u\|_{H^k}\|\nabla u\|_{H^k}\|u\|_{H^k} \\
&\leq \frac{1}{2}c_k\|u\|_{H^k}^4 + \frac{1}{2}\|\nabla u\|_{H^k}^2.
\end{aligned}
\tag{7.1}
$$

Hence

$$\frac{d}{dt}\|u\|_{H^k}^2 + \|\nabla u\|_{H^k}^2 \leq (c_k\|u\|_{H^k}^2)\|u\|_{H^k}^2.$$

We can now use the Gronwall Lemma to deduce that the condition

$$u \in L^2(0, T; H^k) \quad \text{and} \quad u_0 \in H^k,$$

where $k \geq 2$, implies that $u \in L^\infty(0, T; H^k) \cap L^2(0, T; H^{k+1})$. This is a crucial step in the proof of regularity of strong solutions in the space variables, which can be concluded using a simple bootstrap method; for details see the proof of Theorem 7.3.

As we mentioned before, the rigorous proof concerns Galerkin approximations rather than the strong solution u itself, because even to start our bootstrap method we need to make sense of the H^2 inner product of $\partial_t u$ and u, for which the regularity of a strong solution that we have shown so far is not sufficient. However, uniform bounds on the Galerkin approximations can be easily translated to regularity of the limit function u.

7.1 Regularity in space

Let us recall that in Theorem 6.8 we showed that for every $u_0 \in V$ we can find a time $T > 0$ that depends only on the norm of u_0 in H^1, such that the Galerkin approximations u_n are uniformly bounded in $L^\infty(0, T; H^1) \cap L^2(0, T; H^2)$. Our aim now is to show that if u_0 has the additional regularity $u_0 \in V \cap H^m$ then this sequence of Galerkin approximations is uniformly bounded in H^m on the same time interval $[0, T]$ and

$$\sup_{0 \leq t \leq T} \|u_n\|_{H^m}^2 + \int_0^T \|u_n(s)\|_{H^{m+1}}^2 \, ds \leq C_m.$$

It follows that there is a subsequence u_n that converges weakly to a strong solution

$$u \in L^\infty(0, T; H^m) \cap L^2(0, T; H^{m+1}).$$

The proof of regularity in the space variables that we present below follows the lines of the proof in Constantin & Foias (1988). We give the proof in the case of the torus, but since H^k, $k = 2, 3, \ldots$ is a Banach algebra also on the whole space and on any smooth bounded domain in $\mathbb{R}^3$, minor modifications lead to proofs on every domain considered in this book.

Theorem 7.1 *If u is a strong solution of the Navier–Stokes equations on a time interval $[0, T]$ with initial data $u_0 \in V \cap H^m$, then*

$$u \in L^\infty(0, T; H^m) \cap L^2(0, T; H^{m+1}).$$

Proof Without loss of generality we can assume (see Exercise 7.1) that the strong solution u is obtained as the limit of a sequence of Galerkin approximations[1] such that

$$\sup_{0 \le s \le T} \|u_n(s)\|_{H^1}^2 + \int_0^T \|u_n(s)\|_{H^2}^2 \, ds \le C_1. \tag{7.2}$$

We claim that for every $1 \le k \le m$ there is a constant C_k such that

$$\sup_{0 \le s \le T} \|u_n(s)\|_{H^k}^2 + \int_0^T \|u_n(s)\|_{H^{k+1}}^2 \, ds \le C_k.$$

For $k = 1$ the statement follows from our assumption. Take $2 \le k \le m$ and suppose that the statement is true for $k - 1$, i.e.

$$\sup_{0 \le s \le T} \|u_n(s)\|_{H^{k-1}}^2 + \int_0^T \|u_n(s)\|_{H^k}^2 \, ds \le C_{k-1}.$$

We take the H^k inner product of the equation

$$\partial_t u_n + A u_n + P_n(u_n \cdot \nabla)u_n = 0$$

with u_n and use the fact that H^k is an algebra (since $k > 3/2$) to obtain

$$\frac{1}{2}\frac{d}{dt}\|u_n\|_{H^k}^2 + \|\nabla u_n\|_{H^k}^2 \le |\langle P_n(u_n \cdot \nabla)u_n, u_n\rangle_{H^k}|$$

$$= |\langle (u_n \cdot \nabla)u_n, u_n\rangle_{H^k}|$$

$$\le \|(u_n \cdot \nabla)u_n\|_{H^k}\|u_n\|_{H^k}$$

$$\le c\|u_n\|_{H^k}\|\nabla u_n\|_{H^k}\|u_n\|_{H^k}$$

$$\le c\|u_n\|_{H^k}^4 + \frac{1}{2}\|\nabla u_n\|_{H^k}^2.$$

It follows that

$$\frac{d}{dt}\|u_n\|_{H^k}^2 + \|\nabla u_n\|_{H^k}^2 \le (c\|u_n\|_{H^k}^2)\|u_n\|_{H^k}^2, \tag{7.3}$$

while from our assumption we know that

$$\int_0^T \|u_n(s)\|_{H^k}^2 \, ds \le C_{k-1}.$$

---

[1] In Chapter 9 in Theorem 9.3 we will show that whenever u is a strong solution of the Navier–Stokes equations on the time interval $[0, T]$ then the Galerkin approximations of u converge strongly to u in $L^\infty(0, T; H^1) \cap L^2(0, T; H^2)$, which clearly implies (7.2). However, for the current proof the less elaborate argument from Exercise 7.1 is sufficient.

Therefore the Gronwall inequality (Lemma A.24) yields

$$\|u_n(t)\|_{H^k}^2 \le \|u_n(0)\|_{H^k}^2 \exp\left(c\int_0^t \|u_n(s)\|_{H^k}^2 \, ds\right)$$

$$\le \|u_0\|_{H^k}^2 \exp\left(c\int_0^T \|u_n(s)\|_{H^k}^2 \, ds\right) \le e^{cC_{k-1}}\|u_0\|_{H^k}^2.$$

Moreover, from (7.3) we have

$$\int_0^T \|\nabla u_n(t)\|_{H^k}^2 \, dt \le \|u_n(0)\|_{H^k}^2 + c\int_0^T \|u_n(s)\|_{H^k}^4 \, ds$$

$$\le \|u_0\|_{H^k}^2 + cTe^{2cC_{k-1}}\|u_0\|_{H^k}^4,$$

so it easily follows that

$$\int_0^T \|u_n(s)\|_{H^{k+1}}^2 \, ds \le D_k$$

for some constant D_k independent of n. All of these estimates are valid as long as $2 \le k \le m$. To conclude the proof we notice that from the uniform bounds we have obtained it follows that there must be a weakly convergent subsequence of u_n that converges to a limit u with $u \in L^2(0, T; H^{m+1}) \cap L^\infty(0, T; H^m)$. $\quad\square$

Corollary 7.2 *For every $u_0 \in V \cap H^k$, $k \ge 2$ there exists a local-in-time strong solution $u \in L^\infty(0, T; H^k) \cap L^2(0, T; H^{k+1})$.*

We can now use Theorem 7.1 to prove the smoothness of strong solutions in the space variables on the time interval $(0, T)$. We require no more regularity of the initial condition than $u_0 \in V$. (The proof is based on that of Theorem 10.6 in Constantin & Foias, 1988.)

Theorem 7.3 *Let u be a strong solution of the Navier–Stokes equations [2] on the time interval $[0, T]$, with initial condition $u_0 \in V$. Then for all $0 < \varepsilon < T$ we have*

$$u \in C([\varepsilon, T]; H^m) \quad \text{for all} \quad m \in \mathbb{N}.$$

In particular, for all $t \in (0, T]$ the function $u(t)$ is smooth with respect to the space variables.

Proof Take any $\varepsilon > 0$. From Theorem 7.1 it follows that for almost all $s \in (0, \varepsilon)$ we have $u(s) \in H^2$, so we can choose an $s_1 \in (0, \varepsilon)$ such that

---

[2] This theorem is valid on $\mathbb{T}^3$, $\mathbb{R}^3$, and on a smooth bounded domain. We give the proof in the periodic case.

$u(s_1) \in V \cap H^2$. From Theorem 7.1 it now follows (using the uniqueness of strong solutions) that

$$u \in L^\infty(s_1, T; H^2) \cap L^2(s_1, T; H^3).$$

We can therefore find an $s_2 \in (s_1, \varepsilon)$ such that $u(s_2) \in H^3 \cap H$ and use Theorem 7.1 again to conclude that

$$u \in L^\infty(s_2, T; H^3) \cap L^2(s_2, T; H^4).$$

Continuing in this vein we get

$$u \in L^\infty(\varepsilon, T; H^m) \cap L^2(\varepsilon, T; H^{m+1})$$

for all $m \in \mathbb{N}$. Hence for almost all $t \in (0, T)$ the function $u(t)$ belongs to all Sobolev spaces H^m, so $u(t)$ is smooth for almost all times $t \in (0, T)$.

To get the result for all $t \in (0, T]$ we notice that for each $\varepsilon \in (0, T)$ from the fact that $u \in L^\infty(\varepsilon, T; H^m) \cap L^2(\varepsilon, T; H^{m+1})$ we can deduce (see Exercise 7.2) that

$$\partial_t u \in L^2(\varepsilon, T; H^{m-1})$$

and so (using Corollary 1.35) u belongs to $C([\varepsilon, T]; H^m)$. $\qquad\square$

We have proved the regularity of strong solutions in the case of the torus, but essentially the same idea gives regularity in other (sufficiently smooth) domains. So let us now briefly sketch how regularity of strong solutions can be proved in a smooth bounded domain. Rather than taking the H^s inner product of the Galerkin equation with u_n, we can take the L^2 inner product of the Galerkin equation with $A^s u_n$, $s = 1, 2, 3, \ldots$ We obtain the inequality

$$\frac{1}{2}\frac{d}{dt}\|A^{s/2}u_n\|^2 + \|A^{(s+1)/2}u_n\|^2 \leq |\langle P_n[(u_n \cdot \nabla)u_n], A^s u_n\rangle|.$$

From this inequality we can deduce (see Exercise 7.6) that

$$\frac{d}{dt}\|A^{s/2}u_n\|^2 + \|A^{(s+1)/2}u_n\|^2 \leq c\|u_n\|_{H^s}^2\|A^{s/2}u_n\|^2.$$

Using induction and the Gronwall Lemma we obtain in a similar way as before that

$$\sup_{\varepsilon \leq t \leq T}\|A^{s/2}u_n(t)\|^2 \leq c_s$$

uniformly in n. This finishes the proof because for every $k = 1, 2, 3, \ldots$ we have (Exercise 2.13)

$$\|u\|_{H^k} \le c_k \|A^{k/2} u\|$$

for all $u \in D(A^{k/2})$.

The proof on the whole space requires only minor modifications, since the constants appearing in the inequalities are scale invariant, that is do not change when Ω is transformed by dilation or translation.

7.2 Regularity in space–time

Theorem 7.3 gives smoothness of strong solutions with respect to the space variables. We will now prove that strong solutions are also smooth with respect to time. To this end we will prove that

$$\partial_t^k u \in L^\infty(\varepsilon, T; H^m)$$

for all $k, m \in \mathbb{N}$, and all $\varepsilon > 0$. This is enough to show smoothness, because this assertion implies that $u \in H^k(\Omega \times (\varepsilon, T))$ for all $k \in \mathbb{N}$, so u must be smooth, according to Theorem 1.21. In the proof below we follow Temam (1987).

Lemma 7.4 *Let u be a strong solution of the Navier–Stokes equations on the time interval $[0, T]$. Then for every $\varepsilon > 0$, and all $j, k \in \mathbb{N}$ we have*

$$\partial_t^j u \in L^\infty(\varepsilon, T; H^k).$$

Proof We have shown that if $u_0 \in V$ then for every $\varepsilon \in (0, T)$ the Galerkin approximations u_n are uniformly bounded in $L^\infty(\varepsilon, T; H^k)$ for all $k \in \mathbb{N}$. From the equality

$$\partial_t u_n = -Au_n - P_n[(u_n \cdot \nabla)u_n] \tag{7.4}$$

valid for all positive t we deduce that $\partial_t u_n \in L^\infty(\varepsilon, T; H^k)$ for every k.

Indeed, the term Au_n is easy to handle and for the nonlinear term the bound

$$\|P_n[(u_n \cdot \nabla)u_n]\|_{H^k} \le \|(u_n \cdot \nabla)u_n\|_{H^k}$$
$$\le c\|u_n\|_{H^{k+2}} \|u_n\|_{H^{k+3}}$$

is sufficient (we have used the fact that P_n is a bounded operator from H^k to itself for every $k \in \mathbb{N}$, see Exercise 4.1). Therefore

$$\sup_{\varepsilon \le s \le T} \|P_n[(u_n(s) \cdot \nabla)u_n(s)]\| \le c \sup_{\varepsilon \le s \le T} \|u_n\|_{H^{k+2}} \sup_{\varepsilon \le s \le T} \|u_n\|_{H^{k+3}} \le C.$$

To show that $\partial_t^j u_n$ is uniformly bounded (w.r.t. n) in $L^\infty(\varepsilon, T; H^k)$ for all $j \in \mathbb{N}$ and all $k \in \mathbb{N}$ we can use induction: assume that $\partial_t^l u_n$ is uniformly bounded in every $L^\infty(\varepsilon, T; H^k)$ for all $0 \le l \le j - 1$. Differentiating equation (7.4) $j - 1$ times with respect to time yields

$$\partial_t^j u_n = -A\partial_t^{j-1}u - \sum_{k=0}^{j-1} \binom{j-1}{k} P_n[(\partial_t^k u_n \cdot \nabla)\partial_t^{j-1-k} u_n].$$

Using similar arguments as we did for $\partial_t u_n$ the result follows. □

From Theorem 7.3 and Lemma 7.4 it follows that if u is a strong solution on the time interval $[0, T]$ then for every $\varepsilon > 0$ we have

$$u \in H^k(\Omega \times (\varepsilon, T))$$

for every $k \in \mathbb{N}$, and as a consequence we have the following regularity theorem.

Theorem 7.5 *If u is a strong solution of the Navier–Stokes equations on the time interval $[0, T]$ then $u \in C^\infty(\overline{\Omega} \times [\varepsilon, T])$ for every $0 < \varepsilon < T$.*

Theorem 7.5 is valid on all domains considered in this book and has many interesting consequences, which we will study in the next chapter. However, we can make one important observation immediately. If u_0 gives rise to a solution that blows up for the first time at some $T^* < \infty$, then the solution $u(t)$ is nevertheless smooth on the time interval $(0, T^*)$. We can therefore restrict our attention to classical solutions if we are interested in the phenomenon of blowup.

Let us also remark that when we consider the case of a bounded domain the condition $u_0 \in V \cap H^m$ does not automatically imply[3] that the strong solution u is continuous from $[0, T]$ into H^m with $\partial_t u$ in $C([0, T]; H^{m-2})$, i.e. that the smoothness in Theorem 7.5 extends all the way to $t = 0$ if $u_0 \in H^m \cap V$ for every $m \in \mathbb{N}$ (that is when u_0 is smooth). In fact for such smoothness to hold u_0 must satisfy some additional compatibility conditions. To see this let us consider $u_0 \in H^3 \cap V$. In such a case Au and $\mathbb{P}(u \cdot \nabla)u$ are well defined on the boundary and $\partial_t u \in L^2(0, T; V)$ (see Exercise 7.7). It follows that on the time interval $(0, T)$ the time derivative $\partial_t u$ belongs to V; hence it vanishes on the boundary $\partial\Omega$. At the same time we have

$$\partial_t u = \mathbb{P}\Delta u - \mathbb{P}[(u \cdot \nabla)u]$$

so $\mathbb{P}\Delta u(t) - \mathbb{P}[(u(t) \cdot \nabla)u(t)]$ vanishes on the boundary for all $t > 0$. Hence if $u \in C([0, T]; H^3)$ then we must have

$$\mathbb{P}\Delta u_0 - \mathbb{P}[(u_0 \cdot \nabla)u_0] = 0 \qquad \text{on} \quad \partial\Omega.$$

---

[3] Such an implication is true in the case of the torus.

For more details and the general case see Temam (1982) and Theorems 6.1 and 6.2 in Temam (1987).

Finally, let us note that there are, of course, much stronger upper bounds on the nonlinear term than the inequality

$$|\langle (u \cdot \nabla)u, u \rangle_{H^k}| \leq c \|\nabla u\|_{H^k} \|u\|_{H^k}^2,$$

which we have so frequently used in this chapter. Taking into account the divergence-free property of solutions of the Navier–Stokes equations one can show sharper estimates. For example, the nonlinear term can be bounded in the following way:

$$|\langle (u \cdot \nabla)u, u \rangle_{H^k}| \leq c \|\nabla u\|_{\infty} \|u\|_{H^k}^2,$$

provided that $k \geq 3$. This inequality is usually called Kato's inequality (see Exercise 7.8). One of its simpler versions can be shown in a very elementary way (see Exercise 7.3) and then be used to show a simple criterion for regularity of a solution of the Navier–Stokes equations.

Theorem 7.6 *Let u be a Leray–Hopf weak solution of the Navier–Stokes equations corresponding to an initial condition $u_0 \in V$. If*

$$\int_0^T \|\nabla u(s)\|_{\infty} \, ds < \infty$$

then u is strong on $[0, T]$.

Proof See Exercise 7.4. □

In Chapter 12 we will weaken the assumption made in Theorem 7.6 and show that the weaker condition

$$\int_0^T \|\text{curl } u(s)\|_{\infty} \, ds < \infty$$

still implies that u is strong on $(0, T]$, a result due to Beale, Kato, & Majda (1984).

Notes

The proof of Theorem 7.3 is taken from Constantin & Foias (1988). For a corresponding proof of regularity in the case of a bounded domain we refer the reader to Galdi (2000) who gives a very accessible proof of the regularity of solutions satisfying the Serrin condition $u \in L^r(0, T; L^s)$, where $2/r + 3/s \leq 1$, $s > 3$ (see Chapter 8).

The 'mild solutions' of the Navier–Stokes equations constructed by Kato are also locally smooth in time. This follows from Theorem 1 in Kato (1984), where he states that his solutions satisfy the additional property that

$$u \in L^r(0, T_1; L^s), \quad \text{where} \quad 2/r + 3/s = 1, \quad 3 < s < 9$$

for some $0 < T_1 \le T$. Therefore the regularity of such solutions follows from Theorem 8.17 in the next chapter. Moreover, Kato shows that his mild solutions satisfy

$$(\sqrt{t})^{1-3/q} u \in BC([0, T); \mathbb{P}L^q)$$

and

$$t^{1-3/2q} \nabla u \in BC([0, T); \mathbb{P}L^q)$$

for $3 \le q \le \infty$, where BC denotes bounded continuous functions.

Further regularity results show that a strong solution of the Navier–Stokes equations is in fact real analytic and the radius of analyticity can be bounded from below. The key estimate in this direction is due to Foias & Temam (1989) who proved that strong solutions are analytic in time and they have values in the Gevrey class of functions with,

$$\|A^{1/2} e^{tA^{1/2}} u(t)\|^2 \le C(1 + \|\nabla u_0\|^2)$$

for all $t \in (0, \tau)$, where τ depends only on the norm of u_0 in V. Here

$$e^{tA^{1/2}} = \sum_{n=0}^{\infty} \frac{t^n}{n!} A^{n/2},$$

see also Kahane (1969), Masuda (1967), Komatsu (1979), or Giga (1983).

Exercises

7.1 Explain why in the proof of Theorem 7.3 we can assume that u is obtained as the limit of Galerkin approximations.

7.2 Let u be a strong solution of the Navier–Stokes equations on the torus. Prove that if $u \in L^\infty(\varepsilon, T; H^m) \cap L^2(\varepsilon, T; H^{m+1})$ for some $m \ge 2$ then

$$\partial_t u \in L^2(\varepsilon, T; H^{m-1}).$$

7.3 Let u be a strong solution on the torus. Prove that

$$\frac{\mathrm{d}}{\mathrm{d}t} \|u\|_{\dot{H}^2}^2 + \|u\|_{\dot{H}^3}^2 \le c \|\nabla u\|_\infty \|u\|_{\dot{H}^2}^2.$$

7.4 Let $u_0 \in V(\mathbb{T}^3)$. Using Exercise 7.3, prove that if

$$\int_0^T \|\nabla u(s)\|_\infty \, ds < \infty$$

then u is strong on $[0, T]$.

7.5 Suppose that $u_0 \in V(\mathbb{T}^3)$ and u blows up at time $T > 0$. Show that there is a constant $c_2 > 0$ such that

$$\|u(T - t)\|_{H^2} \geq c_2 t^{-3/4}.$$

7.6 Let u_n be a Galerkin approximation of the strong solution u of the Navier–Stokes equations on a smooth bounded domain. Prove that for $m \geq 2$

$$\frac{d}{dt} \|A^{m/2} u_n\|^2 + \|A^{(m+1)/2} u_n\|^2 \leq c \|u_n\|_{H^m}^2 \|A^{m/2} u_n\|^2.$$

7.7 Let $\Omega \subset \mathbb{R}^3$ be a smooth bounded domain and suppose that the initial condition $u_0 \in H^2(\Omega) \cap V(\Omega)$ gives rise to a strong solution u on time interval $[0, T]$. Prove that $\partial_t u \in L^2(0, T; V)$.

7.8 Prove the estimate

$$\left| \langle (u \cdot \nabla) u, u \rangle_{\dot{H}^k(\mathbb{R}^3)} \right| \leq c \|\nabla u\|_{L^\infty} \|u\|_{\dot{H}^k}^2,$$

for $k \in \mathbb{N}$ and $u \in H^k(\mathbb{R}^3)$ divergence free.

Hint: Consider $\left| \int_{\mathbb{R}^3} [\partial^\alpha ((u \cdot \nabla) u) - (u \cdot \nabla) \partial^\alpha u] \cdot \partial^\alpha u \, dx \right|$, where $|\alpha| = k$. Then use the Leibniz formula for the derivative of the product and the Gagliardo–Nirenberg inequality

$$\|\partial^s f\|_{L^{2l/s}} \leq c \|f\|_{L^\infty}^{1-s/l} \|f\|_{\dot{H}^l}^{s/l} \qquad 0 \leq s \leq l.$$

8

Epochs of regularity and Serrin's condition

In this chapter we present several applications of the smoothness of strong solutions proved in Theorem 7.5. We will often use the fact that a Leray–Hopf weak solution must coincide with a local-in-time strong solution whenever its H^1 norm is essentially bounded and therefore must be smooth for almost all times.[1] Moreover, we use the lower bound on the H^1 norm near any 'blowup time' from Lemma 6.11 to restrict the size of the set of potential singular times of a Leray–Hopf weak solution. As a last application of Theorem 7.5 in this chapter we derive the 'Serrin condition' that provides smoothness of a weak solution u provided that u satisfies the simple integrability condition

$$u \in L^r(0, T; L^s(\Omega)), \qquad \frac{2}{r} + \frac{3}{s} = 1.$$

It should be noted that for the weak solutions that we are able to construct we only have

$$u \in L^r(0, T; L^s(\Omega)), \qquad \frac{2}{r} + \frac{3}{s} = \frac{3}{2},$$

according to Lemma 3.5.

8.1 The putative set of singular times

We can now use the strong energy inequality (4.16) and weak–strong uniqueness (Theorem 6.10) to show that Leray–Hopf weak solutions are smooth for almost all times and that the set of singular times is either empty or very small in terms of the Hausdorff and the box-counting dimensions. The key point

---

[1] This does not rule out the possibility of blowup.

is that whenever we are given the norm of an initial condition in V we can always estimate from below the maximal existence time of the strong solution $(T \geq c\|\nabla u_0\|^{-4})$. These estimates combined with the energy inequality restrict the size of the set of singular times. However, it should be noticed that we restrict throughout this chapter to Leray–Hopf weak solutions, since the validity of the strong energy inequality is crucial to ensure that our weak solutions coincide locally with strong solutions, as we now explain in more detail.

Suppose that we are given a weak solution u on the time interval $[0, T)$ and we know that at a certain time $t_0 \in (0, T)$ our solution $u(t)$ belongs to V. We can deduce from this fact that u is actually smooth on some time interval $(t_0, t_0 + \varepsilon)$ if we know that u coincides on this interval with the strong solution corresponding to the initial condition $u(t_0)$. This is guaranteed by Theorem 6.10 under the assumption that

$$\frac{1}{2}\|u(t_1)\|^2 + \int_{t_0}^{t_1} \|\nabla u(s)\|^2 \, ds \; \leq \; \frac{1}{2}\|u(t_0)\|^2$$

for all $t_1 > t_0$. To make sure that this inequality is satisfied (at least for almost all times $t_0 \in (0, T)$) it is enough to assume that u satisfies the strong energy inequality, or in other words that u is a Leray–Hopf weak solution.

We begin with a simple but striking observation initially due to Leray (1934).

Theorem 8.1 *Any global-in-time Leray–Hopf weak solution u is eventually strong: there exists a $T^* > 0$ such that $u \in C^\infty(\overline{\Omega} \times (T^*, \infty))$.*

Proof From Theorem 6.13 we know that there is a $C > 0$ such that

$$\|u_0\|\|\nabla u_0\| \leq C \qquad \Rightarrow \qquad u \text{ is strong on } (0, \infty).$$

From the energy inequality (4.17) we deduce that

$$\|u(s)\| \leq \|u_0\| \quad \text{for all} \quad s \geq 0.$$

From the bound

$$\int_0^T \|\nabla u(s)\|^2 \, ds \; \leq \; \frac{1}{2}\|u_0\|^2 \qquad \text{for all} \qquad T > 0$$

it follows that the Lebesgue measure of the set

$$\{s > 0 : \|\nabla u(s)\| \leq C/\|u_0\|\}$$

is infinite. Hence we can easily choose $T^* > 0$ such that

$$\|u(T^*)\|\|\nabla u(T^*)\| \leq \|u_0\|\frac{C}{\|u_0\|} = C.$$

Since the strong energy inequality is satisfied for almost every $s > 0$ without loss of generality we can assume that it is satisfied for T^*, i.e.

$$\frac{1}{2}\|u(T)\|^2 + \int_{T^*}^{T} \|\nabla u(s)\|^2 \, ds \leq \frac{1}{2}\|u(T^*)\|^2 \quad \text{for all} \quad T > T^*.$$

From the weak–strong uniqueness (Theorem 6.10) it follows that

$$u(T^* + s) = v(s),$$

where v is a strong solution arising from initial data $u(T^*)$. Since

$$\|u(T^*)\| \, \|\nabla u(T^*)\| \leq C$$

the strong solution v is global in time, according to Theorem 6.13.   □

We now give a precise definition of regular and singular times of a weak solution u.

Definition 8.2 Let u be a weak Leray–Hopf solution with initial condition u_0. We say that t_0 is a *regular time* of the weak solution u if $\|\nabla u(t)\|$ is essentially bounded[2] in some neighbourhood of t_0; t_0 is a *singular time* if it is not regular. We denote the set of all strictly positive regular times by $\mathcal{R}$ and the set of all singular times by $\mathcal{T}$.

Note that if the initial condition belongs to V then any corresponding Leray–Hopf weak solution is initially smooth and either never loses its regularity or blows up for the first time at some $T^* > 0$.

We begin with what is essentially a rephrasing of Theorem 8.1.

Lemma 8.3 *The set $\mathcal{T}$ of singular times of a Leray–Hopf weak solution is compact.*

Proof From Theorem 8.1 we know that $\mathcal{T}$ must be bounded. It is also easy to see that $\mathcal{T}$ is closed, because it follows from the definition of singular points that every accumulation point of $\mathcal{T}$ must be singular.   □

Now let us put together what we know so far about the putative singular times of a global-in-time Leray–Hopf weak solution u arising from an initial condition $u_0 \in V$. Such an initial condition u_0 gives rise to a strong solution on a maximal interval $[0, T_1)$; due to weak–strong uniqueness, all weak solutions satisfying the energy inequality coincide on $[0, T_1)$. If $T_1 = \infty$ then a Leray–Hopf weak solution u is strong for all times and is unique. If $T_1 < \infty$, then

---

[2] The time $t = 0$ is regular if and only if $u_0 \in V$, see Exercise 6.7.

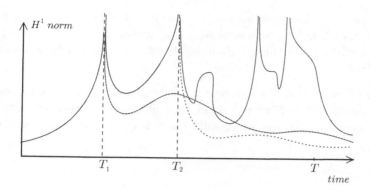

Figure 8.1. If a blowup takes place we can no longer guarantee the uniqueness of solutions and there may be many Leray–Hopf weak solutions, some of which blow up again, and some of which do not. However, there is a time $T > 0$ such that all Leray–Hopf weak solutions arising from u_0 are regular for $t > T$.

all Leray–Hopf weak solutions arising from u_0 must simultaneously blow up at T_1. The value of $u(T_1)$ – as an element of the space L^2 – is the same for every Leray–Hopf weak solution arising from u_0, since $u(t) \rightharpoonup u(T_1)$ in L^2 as $t \to T_1^-$ (see Corollary 3.9). Once a blowup occurs we can no longer guarantee the uniqueness of solutions (see Figure 8.1). If the uniqueness is lost then some of the weak solutions (all corresponding to u_0) may blow up again, while others may remain smooth. Eventually there is some time T after which all these weak solutions are smooth (Theorem 8.1), but there is no reason for the resulting solutions to coincide once the uniqueness has been lost[3] at time T_2.

We will now investigate the size of the set of singular times $\mathcal{T}$ and how a Leray–Hopf weak solution behaves at regular times. Note that it is an immediate consequence of Lemma 8.3 that $\mathcal{R}$ is open; on this set of times the solution is not only bounded but is in fact smooth, as we now show.

Lemma 8.4 *A Leray–Hopf weak solution is smooth on the open set $\Omega \times \mathcal{R}$.*

Proof By definition, for any $t_0 \in \mathcal{R}$ there is some $\delta > 0$ and $M > 0$ such that

$$\|u(s)\|_{H^1} \le M \quad \text{for almost all} \quad s \in (t_0 - \delta, t_0 + \delta).$$

It follows from Theorems 6.8 and 7.5 that there exists a $\delta_1 > 0$ such that for every initial condition v_0 with $\|v_0\|_{H^1} \le M$ there is a corresponding strong solution v that is smooth on the time interval $(0, \delta_1)$. Take any $t' \in (t_0 - \delta, t_0)$ such that $t_0 - t' < \delta_1/2$ and such that the energy inequality holds for t' and

---

[3] To the best of our knowledge we cannot exclude the possibility that at the blowup time T_1 the norm of $u(T_1)$ in H^1 is finite. This case is not shown in Figure 8.1.

$\|u(t')\|_{H^1} \leq M$. Take $v_0 = u(t')$ and let v be the (unique) strong solution arising from v_0. It follows from weak–strong uniqueness that $u(t' + s) = v(s)$ for every $s \in (0, \delta_1)$. Thus there exists a neighbourhood I of t_0 such that u is smooth on $\Omega \times I$. $\square$

We will now prove that $\mathcal{T}$ (if not empty) must be very small. To give restrictions on the size of $\mathcal{T}$ we will recall the definition of the Hausdorff and the box-counting dimensions, which will allow us to quantify the smallness of this set.

8.2 The box-counting and Hausdorff dimensions

Roughly speaking, a compact set X has (upper) box-counting dimension d if the number $N(X, \varepsilon)$ of balls of radius ε needed to cover X is given for small ε by $N(X, \varepsilon) \sim \varepsilon^{-d}$. More precisely we define the (upper) box-counting dimension[4] as follows.

Definition 8.5 The (upper) box-counting dimension[5] of a compact set $X \subset \mathbb{R}^n$ is given by

$$\dim_B(X) = \limsup_{\varepsilon \to 0^+} \left(-\log_\varepsilon N(X, \varepsilon) \right)$$

$$= \limsup_{\varepsilon \to 0^+} \frac{\log N(X, \varepsilon)}{-\log \varepsilon}.$$

For example, the box-counting dimension of the d-dimensional unit cube $[0, 1]^d \subset \mathbb{R}^d$ is equal to d. For $k = 1, 2, 3, \ldots$ the box-counting dimension of the set

$$S_k = \left\{ n^{-k} : n = 1, 2, 3, \ldots \right\} \tag{8.1}$$

is equal to $\frac{1}{k+1}$ (see Exercise 8.1).

For our purposes it turns out to be more convenient to define the box-counting dimension in a slightly different (but equivalent) way.

Lemma 8.6 *For a compact set $X \subset \mathbb{R}^n$ we can also define the box-counting dimension as*

$$\dim_B(X) = \limsup_{\varepsilon \to 0^+} \frac{\log M(X, \varepsilon)}{-\log \varepsilon}, \tag{8.2}$$

----

[4] Also known as the upper Minkowski dimension.
[5] One can also define a lower box-counting dimension by replacing the supremum with the infimum. The box-counting dimension is defined when both the lower and the upper box-counting dimensions coincide, by replacing the limit supremum (or infimum) with just the limit.

where $M(X, \varepsilon)$ denotes the maximal number of disjoint balls of radius ε centred at points of X (see Figure 8.2).

Figure 8.2. The box-counting dimension can be defined as the number d such that $M(X, \varepsilon) \sim \varepsilon^{-d}$, where $M(X, \varepsilon)$ denotes either the maximal number of disjoint balls of radius ε centred at points of X (left); or for which $N(X, \varepsilon) \sim \varepsilon^{-d}$, where $N(X, \varepsilon)$ is the minimal number of such balls needed to cover X (right).

Proof First we show that

$$M(X, \varepsilon) \leq N(X, \varepsilon). \tag{8.3}$$

To prove (8.3) suppose that $c_i \in X$ are the centres of disjoint balls of radius $\varepsilon > 0, i = 1, 2, \ldots, M(X, \varepsilon)$. Since no ball of radius ε can cover two different points c_i and c_j it follows that we need at least $M(X, \varepsilon)$ balls of radius ε to cover the set $\{c_1, c_2, \ldots, c_{M(X,\varepsilon)}\} \subset X$. So (8.3) is proved.

Now we show that

$$N(X, \varepsilon) \leq M(X, \varepsilon/3). \tag{8.4}$$

Suppose that

$$\mathcal{B}_{\mathrm{MAX}} = \{B(c_i, \varepsilon/3) : \ i = 1, 2, \ldots, M(X, \varepsilon/3)\}$$

is a maximal family of disjoint balls of radii $\varepsilon/3$ with centres in X. Then the family

$$\mathcal{B} = \{B(c_i, \varepsilon) : \ i = 1, 2, \ldots, M(X, \varepsilon/3)\}$$

covers the set X (see Figure 8.3).

Indeed, if there was $x \in X$ that did not belong to any ball $B(c_j, \varepsilon)$ then the ball $B(x, \varepsilon/3)$ would be disjoint from every ball $B(c_j, \varepsilon/3)$, for $j = 1, 2, \ldots, M(X, \varepsilon/3)$, and we would be able to find a family of disjoint balls with centres in X that is larger than $\mathcal{B}_{\mathrm{MAX}}$. Thus $\mathcal{B}$ must cover X and (8.4) easily follows.

Using (8.3) and (8.4), for small positive ε we obtain

$$-\log_\varepsilon M(X, \varepsilon) \leq -\log_\varepsilon N(X, \varepsilon) \leq -\log_\varepsilon M(X, \varepsilon/3).$$

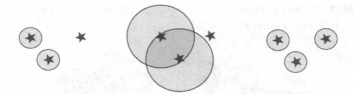

Figure 8.3. From a maximal family $\mathcal{B}_{\text{MAX}}$ of disjoint balls of radius $r > 0$ and with centres in X we can always construct a cover of X by enlarging the radius three times. The reason is simple: if some parts of X are left outside the sum of the enlarged balls (left and middle picture) then $\mathcal{B}_{\text{MAX}}$ is not the largest family of disjoint balls (right picture).

Since

$$\limsup_{\varepsilon \to 0^+} \log_\varepsilon M(X, \varepsilon/3) = \limsup_{\varepsilon \to 0^+} \left(1 - \log_\varepsilon 3\right) \log_{\varepsilon/3} M(X, \varepsilon/3)$$

$$= \limsup_{\varepsilon \to 0^+} \log_\varepsilon M(X, \varepsilon)$$

it follows that (8.2) holds. $\qquad\square$

The following simple lemma will prove very useful.

Lemma 8.7 *Suppose that $K(\varepsilon) = N(X, \varepsilon)$ or $M(X, \varepsilon)$ and define*

$$d = \limsup_{\varepsilon \to 0^+} \frac{\log K(\varepsilon)}{-\log \varepsilon}. \tag{8.5}$$

Then

(i) *for every $0 < d' < d$ there is a sequence $\varepsilon_j \to 0^+$ such that*

$$K(\varepsilon_j) \geq \varepsilon_j^{-d'} \quad \text{for every } j;$$

(ii) *for every $d' > d$ there is $\varepsilon_0 > 0$ such that*

$$K(\varepsilon) \leq \varepsilon^{-d'} \quad \text{for all } 0 < \varepsilon < \varepsilon_0.$$

Proof (i) It follows from (8.5) that there exists a sequence $\varepsilon_j \to 0^+$ such that

$$\log_{\varepsilon_j} K(\varepsilon_j) \to -d,$$

so for $\delta = d - d' > 0$ we can find N such that for all $j > N$ we must have

$$\log_{\varepsilon_j} K(\varepsilon_j) \leq -d + \delta = -d'.$$

The assertion follows. For part (ii) see Exercise 8.2. $\qquad\square$

In addition to the box-counting dimension we will also consider the Hausdorff dimension and the s-dimensional Hausdorff measure.

Definition 8.8 Let

$$\mu_{s,\delta}(X) := \inf \left\{ \sum_{k=1}^{\infty} d_k^s \right\},$$

where the infimum is taken over all coverings of X by open sets $\{U_i\}$ such that diam $U_i = d_i \leq \delta$ for all i, and by diam U_i we denote the diameter of U_i. We can now define the s-dimensional Hausdorff measure as

$$\mathcal{H}^s(X) = \lim_{\delta \to 0^+} \mu_{s,\delta}(X).$$

We note that in $\mathbb{R}^n$ the n-dimensional Hausdorff measure $\mathcal{H}^n$ is proportional to n-dimensional Lebesgue measure (see Theorem 1.12 in Falconer, 1985).

The following lemma, which follows immediately from Definition 8.8, can be used to define the sets of Hausdorff measure zero.

Lemma 8.9 *The set $X \subset \mathbb{R}^n$ has s-dimensional Hausdorff measure zero, $\mathcal{H}^s(X) = 0$, if for every $\varepsilon > 0$ one can cover X by a countable family of balls*[6] *of radii $r_1, r_2, r_3, \ldots$ such that*

$$\sum_{i=1}^{\infty} r_i^s < \varepsilon.$$

Definition 8.10 The Hausdorff dimension of a set $X \subset \mathbb{R}^n$ is given by

$$\dim_{\mathrm{H}}(X) = \inf\{s \geq 0 : \mathcal{H}^s(X) = 0\}.$$

Note that every countable set X has s-dimensional Hausdorff measure zero for all $s > 0$. It follows that every countable set has Hausdorff dimension zero. In particular, for all the sets S_k defined in (8.1) we have

$$\dim_{\mathrm{H}}(S_k) = 0 < \frac{1}{1+k} = \dim_{\mathrm{B}}(S_k),$$

which shows that the Hausdorff dimension can be strictly less than the box-counting dimension. In fact the Hausdorff dimension is never greater than the box-counting dimension, as we now show.

---

[6] It does not matter here whether the balls are closed or open.

Lemma 8.11 *For any compact set $X \subset \mathbb{R}^n$ we have*

$$\dim_H(X) \le \dim_B(X).$$

Proof Suppose that $\dim_B(X) = d$ and take any $s > d$. We want to prove that for every $\delta > 0$ it is possible to cover the set X with a countable family of balls of radii $r_1, r_2, r_3, \ldots$ such that $\sum_{i=1}^{\infty} r_i^s < \delta$. Choose z such that $d < z < s$. Then from Lemma 8.7 we know that we can cover X with $N(X, \varepsilon) \le \varepsilon^{-z}$ balls of radius ε for all sufficiently small ε. Setting $r_i = \varepsilon$ for $i = 1, 2, 3, \ldots, N(X, \varepsilon)$ and $r_i = 0$ for $i > N(X, \varepsilon)$ we get a cover such that

$$\sum_{i=1}^{\infty} r_i^s = N(X, \varepsilon)\varepsilon^s \le \varepsilon^{s-z} \to 0 \qquad \text{as} \qquad \varepsilon \to 0.$$

It follows that $\mathcal{H}^s(X) = 0$ for all $s > \dim_B(X)$, and hence $\dim_H(X) \le \dim_B(X)$. $\qquad\square$

Corollary 8.12 *If $X \subset \mathbb{R}^n$ is compact and $\dim_B(X) < n$ then X has zero Lebesgue measure.*

Proof By Lemma 8.11 it follows that $\dim_H(X) \le \dim_B(X) < n$, and hence $\mathcal{H}^n(X) = 0$. Since $\mathcal{H}^n$ is proportional to the n-dimensional Lebesgue measure the result follows. $\qquad\square$

8.3 Epochs of regularity

We will now use the result of Lemma 6.11, which gives a lower bound on the H^1 norm of any smooth solution that blows up in a finite time, to obtain restrictions on the size of the set of singular times.

Theorem 8.13 *If u is a Leray–Hopf weak solution of the Navier–Stokes equations then the box-counting dimension of the set T of singular times corresponding to u is no greater than $1/2$.*

Proof Fix some small $\varepsilon > 0$. Let F be a maximal family of disjoint intervals of radius ε centred at points of the compact (see Lemma 8.3) set T. Let $t_0, t_1, t_2, \ldots, t_{M(T,\varepsilon)}$ be the centres of the intervals in F. Clearly we can assume that $t_0 < t_1 < \cdots < t_{M(T,\varepsilon)}$, so that $t_1 \ge \varepsilon$. We know from Lemma 6.11 that for each $i = 1, 2, \ldots, M(T, \varepsilon)$ and for almost all times $0 \le t < t_i$ the inequality

$$\|\nabla u(t)\|^2 \ge \frac{c}{\sqrt{t_i - t}}$$

holds. So, if we denote by T a time such that $u(t)$ is regular for all $t \geq T$ then we have

$$\int_0^T \|\nabla u(s)\|^2 \geq \sum_{i=1}^{M(T,\varepsilon)} \int_{t_i - \varepsilon}^{t_i} \|\nabla u(s)\|^2 \, ds$$

$$\geq \sum_{i=1}^{M(T,\varepsilon)} \int_{t_i - \varepsilon}^{t_i} \frac{c}{\sqrt{t_i - s}} \, ds$$

$$= \sum_{i=1}^{M(T,\varepsilon)} \int_0^\varepsilon c s^{-1/2} \, ds$$

$$= 2cM(T, \varepsilon)\sqrt{\varepsilon}.$$

If the box-counting dimension of T was equal to $d > 1/2$ then, according to Lemma 8.7, for any $d' \in (1/2, d)$ we would be able to find a sequence $\varepsilon_j \to 0^+$ such that $M(T, \varepsilon_j) \geq \varepsilon_j^{-d'}$. This would imply that for each $j \in \mathbb{N}$

$$\int_0^T \|\nabla u(s)\|^2 \, ds \geq 2cM(T, \varepsilon_j)\sqrt{\varepsilon_j}$$

$$\geq 2c\varepsilon_j^{-(d'-1/2)},$$

but this is unbounded, since $\varepsilon_j \to 0^+$ as $j \to \infty$, which is impossible. It follows that $\dim_B(T) \leq 1/2$ as claimed. $\square$

Since the Hausdorff dimension of T cannot be greater than the box-counting dimension of T it follows immediately from Theorem 8.13 that the Hausdorff dimension of T is no greater than one half. Therefore from what we have proved so far we can deduce that the s-dimensional Hausdorff measure of T is zero for all $s > 1/2$. The following theorem goes a little further and guarantees that the 1/2-dimensional Hausdorff measure of T is zero.

Theorem 8.14 (Epochs of regularity) *If u is a Leray–Hopf weak solution of the Navier–Stokes equations then u is smooth on an open set of times*

$$\mathcal{R} = \bigcup_{i=1}^\infty (a_i, b_i)$$

that is of full measure. On every $[a, b] \subset (a_i, b_i)$ *the weak solution u is smooth in both space and time. The set* T *of singular times of u has box-counting dimension no greater than one half and* $1/2$-*dimensional Hausdorff measure zero.*

Proof The assertion about R follows from Lemma 8.4, so we only need to prove that the $1/2$-dimensional Hausdorff measure of T is zero.

Choose any $\delta > 0$. Since, according to previous results, the one-dimensional Hausdorff measure of the set T is zero we can cover the set T with a countable family F of intervals,

$$F = \{I_k : I_k = [t_k, t_k + \varepsilon_k], k = 1, 2, 3, \ldots\},$$

where ε_k are non-negative numbers such that

$$\sum_{k=1}^{\infty} \varepsilon_k < \delta.$$

We can assume that $t_k + \varepsilon_k \in T$ for all k: indeed if this is not the case then the set $X_k = [t_k, t_k + \varepsilon_k] \cap T$ is compact (we can assume that X_k is non-empty) and there exists $t' \in X_k$ such that $t_k + \varepsilon_k - t' = \mathrm{dist}(X_k, t_k + \varepsilon_k)$. Since the interval $(t', t_k + \varepsilon_k)$ does not contain any singular times we can replace the time interval $[t_k, t_k + \varepsilon_k]$ with $[t_k, t']$, where $t' \in T$. Thus, according to Lemma 6.11, if the interval $I_k \in F$ then for almost all $s \in I_k$ we have

$$\|\nabla u(s)\|^2 \geq \frac{c}{\sqrt{t_k + \varepsilon_k - s}}$$

for some constant c. Note that without loss of generality we can also assume that the intersection of any two different intervals in F does not contain more than one point. Therefore we have

$$\int_{\bigcup_k I_k} \|\nabla u(s)\|^2 \, ds \geq \sum_{k=1}^{\infty} \int_{t_k}^{t_k+\varepsilon_k} \frac{c}{\sqrt{t_k + \varepsilon_k - s}} \, ds$$

$$\geq \sum_{k=1}^{\infty} 2c\sqrt{\varepsilon_k}.$$

Now it is enough to notice that due to the absolute continuity of integrals, when δ tends to zero so does the integral on the left-hand side. It follows that for every $\varepsilon > 0$ we can choose the family F in such a way that $\sum_{k=1}^{\infty} \sqrt{\varepsilon_k} < \varepsilon$. $\square$

Theorem 8.14 leaves many questions open. In particular, it remains open whether the set of singular times T is at most countable.

8.4 More bounds on weak solutions

We will now use the epochs of regularity result to prove some additional regularity of u. The following result is due to Foias, Guillopé, & Temam (1981) and is a key ingredient in the proof of existence of Lagrangian trajectories associated with weak solutions (see Chapter 17).

Lemma 8.15 *Let u be a Leray–Hopf weak solution of the Navier–Stokes equations on the torus or on a smooth bounded domain Ω. Then[7] for every $T > 0$*

$$\int_0^T \|Au(s)\|^{2/3}\, ds < \infty. \tag{8.6}$$

Consequently we have

$$u \in L^1(0, T; L^\infty). \tag{8.7}$$

Proof It is enough to show (8.6) for some $T > T^*$, where T^* is the last singular time. Without loss of generality we can also assume that $\|\nabla u(t)\| \neq 0$ for all $t \in \mathcal{R}$ (since $\|\nabla u(T_1)\| = 0$ implies that $u = 0$ on $[T_1, \infty)$).

Let $\mathcal{R}$ be the set of regular times. Consider the set

$$(0, T) \cap \mathcal{R} = \bigcup_i (a_i, b_i),$$

where the union is countable or finite, $b_1 = T$, and $b_i \in \mathcal{T}$ for every index $i = 2, 3, 4, \ldots$ Then on every interval of regularity (a_i, b_i) we have, as in the previous computations in (6.7),

$$\frac{d}{dt}\|\nabla u\|^2 + \|Au\|^2 \le c\|\nabla u\|^6.$$

Dividing both sides of the inequality by $\|\nabla u\|^4$ we obtain

$$\frac{1}{\|\nabla u\|^4}\frac{d\|\nabla u\|^2}{dt} + \frac{\|Au\|^2}{\|\nabla u\|^4} = \frac{d}{dt}(-\|\nabla u\|^{-2}) + \frac{\|Au\|^2}{\|\nabla u\|^4} \le c\|\nabla u\|^2.$$

Integrating in time between a_i and b_i and noticing that

$$\lim_{t \to b_i^-} \|\nabla u(t)\|^{-2} = 0$$

for all $i = 2, 3, \ldots$ we get for all $i = 2, 3, \ldots$

$$\frac{1}{\|\nabla u(a_i)\|^2} + \int_{a_i}^{b_i} \frac{\|Au(s)\|^2}{\|\nabla u(s)\|^4}\, ds \le c\int_{a_i}^{b_i} \|\nabla u(s)\|^2\, ds,$$

---

[7] Alternatively we could write $u \in L^{2/3}(0, T; H^2)$, but recall that L^p spaces with $0 < p < 1$ are not Banach spaces.

and for $i = 1$:

$$\frac{1}{\|\nabla u(a_1)\|^2} - \frac{1}{\|\nabla u(T)\|^2} + \int_{a_1}^{T} \frac{\|Au(s)\|^2}{\|\nabla u(s)\|^4}\, ds \le c \int_{a_1}^{T} \|\nabla u(s)\|^2\, ds.$$

Therefore, since $\mathcal{R}$ has full measure in $(0, T)$, we obtain the bound

$$\int_0^T \frac{\|Au(s)\|^2}{\|\nabla u(s)\|^4}\, ds \le c \int_0^T \|\nabla u(s)\|^2\, ds + \frac{1}{\|\nabla u(T)\|^2} < \infty$$

for every $T > T^*$. Now using Hölder's inequality we obtain

$$\int_0^T \|Au(s)\|^{2/3}\, ds = \int_0^T \frac{\|Au(s)\|^{2/3}}{\|\nabla u(s)\|^{4/3}} \|\nabla u(s)\|^{4/3}\, ds$$

$$\le \left(\int_0^T \frac{\|Au(s)\|^2}{\|\nabla u(s)\|^4}\, ds\right)^{1/3} \left(\int_0^T \|\nabla u(s)\|^2\, ds\right)^{2/3}$$

and the assertion follows.

To obtain (8.7) from (8.6) we use the Agmon inequality (Theorem 1.20) and then Young's inequality:

$$\int_0^T \|u(s)\|_\infty\, ds \le c \int_0^T \|\nabla u(s)\|^{1/2} \|Au(s)\|^{1/2}\, ds$$

$$\le c \int_0^T \|\nabla u(s)\|^2 + c \int_0^T \|Au(s)\|^{2/3} < \infty. \qquad \square$$

We will use this result in Chapter 17 to show that for a Leray–Hopf weak solution the Lagrangian trajectories (solutions of $\dot{X} = u(X, t)$) are well defined.

8.5 Serrin's condition

In Chapter 7 we proved that every weak solution u such that

$$\sup_{0 \le s \le T} \|\nabla u(s)\|^2 + \int_0^T \|Au(s)\|^2\, ds < \infty,$$

i.e. a strong solution, is smooth on the time interval $(0, T]$. Our aim now is to derive another condition for smoothness of u, one that concerns only the integrability of u in the space–time domain. More precisely, we will show that if

$$\int_0^T \|u(t)\|_{L^s}^r\, dt < \infty$$

for $s > 3$ and $r \geq 2$ such that

$$2/r + 3/s = 1$$

then u is smooth. We will refer to this condition as 'Serrin's condition' here and in what follows, although similar conditions have been used by many authors, as discussed more fully in the Notes to this chapter. The following lemma is valid on all domains considered in this book.

Lemma 8.16 *Suppose that u is a Leray–Hopf weak solution of the Navier–Stokes equations for some $u_0 \in V$. If $u \in L^r(0, T; L^s)$, where (r, s) satisfy*

$$\frac{2}{r} + \frac{3}{s} = 1, \qquad 2 \leq r < \infty, \tag{8.8}$$

then u is smooth on the time interval $(0, T]$. Furthermore there exists an absolute constant $c > 0$ such that if

$$\|u\|_{L^\infty(0,T;L^3)} < c \tag{8.9}$$

then the same conclusion holds.

It turns out that the result is true for the case $r = \infty$, $s = 3$ without the smallness assumption in (8.9), but the proof is extremely difficult; it was dealt with only relatively recently by Escauriaza, Seregin, & Šverák (2003), see Section 16.4.

Proof First we notice that, even without any additional assumptions, since u_0 belongs to V our weak solution u must be smooth on some time interval $(0, T_1]$. We take an inner product of the Navier–Stokes equations with Au, and using Lemma 6.7 we obtain

$$\frac{1}{2}\frac{\mathrm{d}}{\mathrm{d}t}\|\nabla u\|^2 + \|Au\|^2 \leq |\langle (u \cdot \nabla)u, Au \rangle|.$$

We need to estimate the right-hand side of this inequality. We use the Hölder inequality

$$|\langle (u \cdot \nabla)u, Au \rangle| \leq \|u\|_{L^s}\|\nabla u\|_{L^q}\|Au\|_{L^2}.$$

Here the number q is determined from the equation $q^{-1} + s^{-1} + 2^{-1} = 1$.

First we consider the case $3 < s < \infty$ for which $q = \frac{2s}{s-2} \in (2, 6)$. We eliminate the $L^{2s/(s-2)}$ norm by interpolating between the L^2 norm and the L^6 norm

$$\|\nabla u\|_{L^{2s/(s-2)}} \leq \|\nabla u\|_{L^2}^{(s-3)/s}\|\nabla u\|_{L^6}^{3/s}.$$

Therefore we obtain

$$
\begin{aligned}
|\langle (u \cdot \nabla)u, Au \rangle| &\leq \|u\|_{L^s} \|\nabla u\|_{L^{2s/(s-2)}} \|Au\|_{L^2} \\
&\leq \|u\|_{L^s} \|\nabla u\|_{L^2}^{(s-3)/s} \|\nabla u\|_{L^6}^{3/s} \|Au\| \\
&\leq c \|u\|_{L^s} \|\nabla u\|_{L^2}^{(s-3)/s} \|Au\|^{(s+3)/s} \\
&\leq c \left(\|u\|_{L^s} \|\nabla u\|_{L^2}^{(s-3)/s} \right)^{2s/(s-3)} + \frac{1}{2} \left(\|Au\|^{(s+3)/s} \right)^{2s/(s+3)} \\
&= c \|u\|_{L^s}^r \|\nabla u\|^2 + \frac{1}{2} \|Au\|^2,
\end{aligned}
$$

because $r = 2s/(s - 3)$, according to (8.8). Therefore

$$
\frac{\mathrm{d}}{\mathrm{d}t} \|\nabla u(t)\|^2 + \|Au(t)\|^2 \leq c \|u(t)\|_{L^s}^r \|\nabla u(t)\|^2
$$

for all $t \in (0, T_1)$. The Gronwall Lemma (Lemma A.24) implies that

$$
\|\nabla u(t)\|^2 \leq \|\nabla u(0)\|^2 \exp \left(\int_0^t \|u(s)\|_{L^s}^r \, \mathrm{d}s \right)
$$

for all $t \in (0, T_1)$, and it follows that blowup at time T_1 is impossible for any $T_1 \leq T$.

The case $r = 2$ and $s = \infty$ is similar. From the inequality

$$
\frac{\mathrm{d}}{\mathrm{d}t} \|\nabla u\|^2 + \|Au\|^2 \leq c \|u\|_{L^\infty} \|\nabla u\| \|Au\|
$$
$$
\leq c \|u\|_{L^\infty}^2 \|\nabla u\|^2 + \|Au\|^2
$$

one easily obtains $\nabla u \in L^\infty(0, T; L^2)$.

The other endpoint case, $r = \infty$, $s = 3$, cannot be obtained using the estimates above. Indeed, from the bound on the nonlinear term we only obtain

$$
\frac{\mathrm{d}}{\mathrm{d}t} \|\nabla u\|^2 + \|Au\|^2 \leq c \|u\|_{L^3} \|\nabla u\|_{L^6} \|Au\|
$$
$$
\leq c \|u\|_{L^3} \|Au\|^2
$$

which is not enough to conclude that $\|\nabla u\|$ stays bounded unless we know that u is sufficiently small in $L^\infty(0, T; L^3)$ as in the statement of the lemma. $\qquad \square$

It is easy to see that the assumption that $u_0 \in V$ in Lemma 8.16 is not really needed. Indeed, if $u_0 \in H$ we can find a sequence of initial conditions

$$
u_{0,n} = u(t_n) \in V,
$$

where $t_n \to 0$, such that the energy inequality holds from time t_n, i.e.

$$\frac{1}{2}\|u(t)\|^2 + \int_{t_n}^t \|\nabla u(s)\|^2 \, ds \leq \frac{1}{2}\|u(t_n)\|^2$$

for all $t \in (t_n, T]$. Since $u \in L^r(t_n, T; L^s(\Omega))$ and $u(t_n) \in V$ it follows from Lemma 8.16 that u is smooth on (t_n, T) for every n (we use here the uniqueness of strong solutions in the class of Leray–Hopf weak solutions). Therefore we obtain the following criterion for regularity of weak solutions.

Theorem 8.17 *Let u be a Leray–Hopf weak solution of the Navier–Stokes equations satisfying the 'Serrin condition'*

$$u \in L^r(0, T; L^s), \qquad where \quad \frac{2}{r} + \frac{3}{s} = 1, \quad r \geq 2, s > 3. \quad (8.10)$$

Then u is smooth on the time interval $(0, T]$ and for every $\varepsilon > 0$ is a strong solution of the Navier–Stokes equations on $[\varepsilon, T]$.

We will now prove that Leray–Hopf weak solutions satisfying the Serrin condition (8.10) are unique in the class of all weak solutions satisfying the energy inequality.

Lemma 8.18 *If v is a Leray–Hopf weak solution satisfying the Serrin condition (8.10) on $[0, T]$ then v can be used as a test function in (3.3) for a Leray–Hopf weak solution u.*

Proof Let v be a Leray–Hopf weak solution of the Navier–Stokes equations that satisfies the Serrin condition. Since v is a smooth strong solution on $[\varepsilon, T]$ for every $\varepsilon > 0$ it follows from Lemma 6.6 that

$$-\int_\varepsilon^t \langle u, \partial_t v \rangle = -\int_\varepsilon^t \langle \nabla u, \nabla v \rangle - \int_\varepsilon^t \langle (u \cdot \nabla)u, v \rangle + \langle u(\varepsilon), v(\varepsilon) \rangle - \langle u(t), v(t) \rangle$$

$$(8.11)$$

for all $\varepsilon < t \leq T$.

It is enough to pass to the limit as $\varepsilon \to 0^+$ in this equation. First we pass to the limit on the right-hand side. Since $u(\varepsilon) \to u(0)$ (since the proof of Lemma 4.7 can be applied to any Leray–Hopf weak solutions) and $v(\varepsilon) \to v(0)$ as $\varepsilon \to 0^+$ we have

$$\langle u(\varepsilon), v(\varepsilon) \rangle \to \langle u(0), v(0) \rangle.$$

The first integral term on the right-hand side of the equation is easy to handle. For the second term we only need to show that

$$I = \int_0^T \int |u||\nabla u||v| < \infty.$$

Using the three-term Hölder inequality with exponents $(2s/(s-2), 2, s)$, Lebesgue interpolation, and the inequality $\|u\|_{L^6} \leq c\|\nabla u\|$ we have

$$I \leq \int_0^T \|u\|_{L^{2s/(s-2)}} \|\nabla u\| \|v\|_{L^s}$$

$$\leq c \int_0^T \|u\|^{(s-3)/s} \|\nabla u\|^{(s+3)/s} \|v\|_{L^s}$$

$$\leq c \int_0^T \|u\|^{(s-3)r/(r-1)s} \|\nabla u\|^{(s+3)r/(r-1)s} + \int_0^T \|v\|_{L^s}^r.$$

Since $r/(r-1) = 2s/(s+3)$ we get

$$I \leq \int_0^T \|u\|^{2(s-3)/(s+3)} \|\nabla u\|^2 + \int_0^T \|v\|_{L^s}^r < \infty.$$

Now we can pass to the limit on the right-hand side of (8.11). It follows that we can also pass to the limit on the left-hand side and the assertion follows.   □

Theorem 8.19 *Leray–Hopf weak solutions satisfying the Serrin condition (8.10) are unique in the class of all weak solutions satisfying the energy inequality.*

Proof The proof follows that of Theorem 6.10 (weak–strong uniqueness) with only two differences. First we use Lemma 8.18 instead of Lemma 6.6 to obtain for $w = u - v$

$$\frac{1}{2}\|w(t)\|^2 + \int_0^t \|\nabla w(s)\|^2 \, ds \leq \left| \int_0^t \langle (w \cdot \nabla)w, u \rangle \right|. \qquad (8.12)$$

Then we estimate the nonlinear term in a similar way as we did in the proof of Lemma 8.16

$$|\langle (u \cdot \nabla)w, w \rangle| \leq \|u\|_{L^s} \|w\|_{L^{2s/(s-2)}} \|\nabla w\|$$

$$\leq c\|u\|_{L^s} \|w\|_{L^2}^{(s-3)/s} \|\nabla w\|^{1+3/s}$$

$$\leq \|u\|_{L^s}^r \|w\|^2 + \|\nabla w\|^2.$$

Using this bound on the right-hand side of (8.12) and using the integral version of Gronwall's Lemma (Lemma A.25) concludes the proof.   □

We will use the particular case that $u \in L^4(0, T; L^6)$ implies uniqueness in the analysis of Chapter 10 and the case $u \in L^3(0, T; L^9)$ in Chapter 11.

8.6 Epochs of regularity on the whole space

The proof of the epochs of regularity property is valid for any Leray–Hopf weak solution. Since Leray's construction (see Chapter 14) gives a weak solution satisfying the strong energy inequality (that is a Leray–Hopf weak solution on $\mathbb{R}^3$), we know that the epochs of regularity property is enjoyed at least by one class of weak solutions that we can construct.

However, the method of construction of a weak solution on the whole space that we presented in Chapter 4 does not allow us to conclude in a similar way as we did before that the resulting solution satisfies the strong energy inequality. Nevertheless the solution on the whole space constructed in Section 4.4 can be constructed in such a way that the epochs of regularity property is satisfied.

Proposition 8.20 *The weak solution u constructed in Section 4.4 satisfies the epochs of regularity property in the following sense: there is a countable union of open intervals (a_i, b_i) such that u is strong on every $[\tilde{a}, \tilde{b}] \subset (a_i, b_i)$ and the box-counting dimension of the set $[0, T] \setminus \bigcup_i (a_i, b_i)$ is no greater than one half.*

Proof We will only sketch the argument, for the detailed proof see Heywood (1988), who also proves that the $1/2$-dimensional Hausdorff measure of $\mathcal{T}$ is zero.

Let (u_k) be a subsequence of Leray–Hopf weak solutions of the Navier–Stokes equations on B_k (a ball centred at zero of radius R_k, $R_k \to \infty$) converging to a weak solution u as shown in the construction of weak solutions in Section 4.4. To simplify our exposition we will show the desired property of u on the time interval $[0, 1]$.

In the first step for each $n \in \mathbb{N}$ we divide the interval $[0, 1]$ into 4^n intervals of length 4^{-n} with endpoints at $0, t_1, t_2, \ldots, t_{4^n}$. Then for each fixed n we find a number $M_n = \sqrt{c}2^n$, where c is the constant from Corollary 6.9, so that we have the following implication (for each k)

$$\|\nabla u_k(t)\|^2 \leq M_n \quad \Rightarrow \quad \|\nabla u_k(s)\|^2 \leq 2M_n \quad \text{for every } s \in [t, t + 4^{-n}].$$

In the second step for each $n \in \mathbb{N}$ we define the sets of 'bad times' B_n, and 'good times' G_n:

$$B_n = \{t_j : \|\nabla u_k(t_j)\| \leq M_n \quad \text{for only finitely many } k \in \mathbb{N}\},$$
$$G_n = \{t_j : \|\nabla u_k(t_j)\| \leq M_n \quad \text{for infinitely many } k \in \mathbb{N}\}.$$

Now let us fix $n \in \mathbb{N}$. From our choice of M_n and the definition of G_n it follows that for each $t_j \in G_n$ there exists a subsequence (u_{k_i}) such that

$$t_j \in G_n \quad \Rightarrow \quad \sup_{t_j \le s \le t_{j+1}} \|\nabla u_{k_i}(s)\|^2 \le 2M_n = c_1 2^{n+1}.$$

From this subsequence we can choose a further subsequence that additionally converges to u weakly-$*$ in $L^\infty(t_j, t_{j+1}; H^1)$. Hence we can deduce that the limit function u must satisfy the bound

$$\sup_{s \in [t_j, t_{j+1}]} \|u(s)\|_{H^1} \le C,$$

whenever $t_j \in G_n$ for some $n \in \mathbb{N}$. Therefore if $s \in [0, 1]$ belongs to the set $\bigcup_{t_j \in G_n}(t_j, t_{j+1})$ then s is a regular time of u. It follows that if $s \in [0, 1]$ is a singular time of u then it must belong to the set

$$\tilde{\mathcal{T}} := \bigcap_{n=1}^{\infty} \bigcup_{t_j \in B_n} [t_j, t_{j+1}]. \tag{8.13}$$

In the final step of the proof we find an estimate on the number of elements of B_n. So let $n \in \mathbb{N}$ be fixed. Using Corollary 6.9 again we deduce that

$$t_j \in B_n \quad \Rightarrow \quad \inf_{t_{j-1} \le s \le t_j} \|\nabla u_k(s)\|^2 > \frac{M_n}{2}$$

for all but a finite number of $k \in \mathbb{N}$. Therefore we can find u_N such that

$$\int_{t_{j-1}}^{t_j} \|\nabla u_N(s)\|^2 \, \mathrm{d}s > 4^{-n} M_n / 2 = \tilde{c} 2^{-n} \quad \text{for all } t_j \in B_n.$$

Since we know that $\int_0^1 \|\nabla u_N(s)\|^2 \, \mathrm{d}s \le C$, for some $C > 0$, the number of times t_j that lie in B_n cannot exceed $2^n C / \tilde{c}$. Since the number of elements of B_n is of the order $\sim (4^{-n})^{-1/2}$ or less one can easily show that $\dim_B(\tilde{\mathcal{T}}) \le 1/2$ (Exercise 8.7). $\qquad\square$

Notes

The bound on the Hausdorff dimension of the singular set relies on an observation made by Leray (1934) but the precise modern formulation is due to Scheffer (1976) who showed that the $1/2$-dimensional Hausdorff measure of the set of singular times is zero. The proof we give in this chapter addresses the case of the torus but is in fact valid for any sufficiently regular bounded domain and on the whole space. It works also on unbounded domains with uniform C^2 boundary, since for such domains the existence of weak solutions satisfying the strong energy inequality can be shown, see Farwig, Kozono, & Sohr (2005).

The proof that the box-counting dimension of the set $\mathcal{T}$ of singular times is no greater than one half is taken from Robinson & Sadowski (2007). This result can be strengthened: it was shown by Kukavica (2009b) that the 1/2-dimensional 'box-counting' measure of $\mathcal{T}$ is zero:

$$\mathcal{F}^{1/2}(\mathcal{T}) = 0,$$

where $\mathcal{F}^d(A) = \lim_{\delta \to 0} \mathcal{F}^{d,\delta}(A)$, $\mathcal{F}^{d,\delta}(A) = \sup_{r \in (0,\delta)} \inf nr^d$, and the infimum is taken over all coverings of A by the union $\bigcup_{i=1}^n (t_i - r, t_i + r)$, see Exercise 8.6. More information about the Hausdorff and the box-counting dimensions can be found for example in Falconer (1985, 1990) or Robinson (2011). The proof that a Leray–Hopf weak solution u belongs to $L^1(0, T; L^\infty)$ and $L^{2/3}(0, T; H^2)$ is due to Foias, Guillopé, & Temam (1981), see also Foias et al. (2001).

In this chapter we have shown that the Serrin condition implies regularity for u satisfying the initial-value problem, but in the second part of this book we will show that the Serrin condition also implies spatial regularity when considered only locally, i.e. when u satisfies the Navier–Stokes equations locally in space–time with no boundary conditions (in either space or time).

The question of uniqueness of solutions of the Navier–Stokes equations in various function spaces was studied by Kiselev & Ladyzhenskaya (1957) who proved that in the class $L^\infty(0, T; L^4)$ there could be no more than one solution. Prodi (1959) proved uniqueness of solutions of the Navier–Stokes equations on the whole space under the condition $u \in L^r(0, T; L^s)$, where $2/r + 3/s = 1$ and $s > 3$, with some additional assumptions, hence improving on the result of Kiselev and Ladyzhenskaya (since $L^\infty(0, T; L^4) \subset L^8(0, T; L^4)$). Sather and Serrin (see Serrin, 1963) proved that the condition $u \in L^r(0, T; L^s)$ where $2/r + 3/s = 1$ with $s > 3$ implies that the energy equality is satisfied by u and that u is unique in the class of all weak solutions satisfying the energy inequality. Serrin (1963) also showed an estimate for the nonlinear term needed in the proof of regularity (Lemma 1 in Serrin, 1963), while the whole proof of regularity of solutions satisfying Serrin's condition is due to Ladyzhenskaya (1967). Global results, but in the presence of boundary, were obtained by Sohr (1983) and Giga (1986).

There are also versions of the Serrin condition involving the derivative ∇u. For example, Beirão da Veiga (1995) showed that any Leray–Hopf weak solution that satisfies $\nabla u \in L^r(0, T; L^s(\mathbb{R}^3))$ with $2/r + 3/s = 2$ for $3/2 < s < \infty$ is actually smooth (see also Berselli, 2002). One can also give a criterion for regularity in terms of pressure: if $p \in L^r(0, T; L^s)$ with $2/r + 3/s = 2$, $s > 3/2$ then u is smooth. This result was proved by Berselli & Galdi (2002). There are also some other criteria involving only one component of velocity, see, for example, Cao & Titi (2008) or Zhou & Pokorný (2010).

Finally, let us notice that from the strong energy inequality one can deduce that the L^2 norm of a Leray–Hopf weak solution is decreasing, so it tends to a constant. It is simple to show on $\mathbb{T}^3$ or on a smooth bounded domain that $\|u(t)\| \to 0$ as $t \to \infty$; Leray asked if the same result holds on the whole space. A positive answer was given by Masuda (1984), see also Schonbek (1985). For related questions concerning the large-time behaviour of weak solutions in exterior domains see, for example, Miyakawa & Sohr (1988) or Iftimie, Karch, & Lacave (2014).

Exercises

8.1 Prove that $\dim_B(S_k) = 1/(1 + k)$, where
$$S_k = \left\{ n^{-k} : n = 1, 2, 3 \ldots \right\}.$$

8.2 Let $\dim_B(X) = d$ be the box-counting dimension of a compact set X. Show that for every $d' > d$ there is $\varepsilon_0 > 0$ such that
$$M(X, \varepsilon) \le \varepsilon^{-d'}$$
for all positive $\varepsilon < \varepsilon_0$. Here $M(X, \varepsilon)$ denotes the maximal number of disjoint balls centred at points of X and of radius ε.

8.3 Let Ω be a smooth bounded domain or the whole space $\mathbb{R}^3$. Prove that there exists a constant $c_* > 0$ with the following property. If for some $r > 0$
$$\frac{1}{\sqrt{r}} \int_{t-r}^{t} \|\nabla u(s)\|^2 \, \mathrm{d}s \le c_*$$
then a Leray–Hopf weak solution u is strong on $[t, t + r]$.

8.4 Prove that both quantities
$$\int_0^T \|\nabla u(s)\|_{L^2(\mathbb{R}^3)}^4 \, \mathrm{d}s \quad \text{and} \quad \frac{1}{\sqrt{r}} \int_{t-r}^{t} \|\nabla u(s)\|_{L^2(\mathbb{R}^3)}^2 \, \mathrm{d}s$$
are invariant under the scaling $u_\lambda(x, t) = \lambda u(\lambda x, \lambda^2 t)$.

8.5 Use Exercise 8.3 to give alternative proof of the fact that weak solutions eventually become strong.

8.6 Use Exercise 8.3 to show that
$$\mathcal{F}^{1/2}(\mathcal{T}) = 0;$$

that is prove that for every $\varepsilon > 0$ there is $\delta > 0$ such that

$$\sup_{r \in (0, \delta)} \inf \left\{ n r^{1/2} : \mathcal{T} \subset \bigcup_{i=1}^{n} (t_i - r, t_i + r) \right\} \leq \varepsilon,$$

where $\mathcal{T}$ is the set of singular times.

8.7 Prove that the set $\tilde{\mathcal{T}}$ defined by (8.13) has box-counting dimension no greater than one half.

8.8 Show that if $u_0 \in V$ then

$$\int_0^T \frac{\|u\|_{H^3}^2}{\|u\|_{H^2}^4} \, ds < \infty. \qquad (8.14)$$

Hence deduce that

$$u \in L^{2/3}(0, T; H^3) \qquad \text{and} \qquad \nabla u \in L^1(0, T; L^\infty). \qquad (8.15)$$

9

Robustness of regularity and convergence of Galerkin approximations

In this chapter we will prove two results which require a similar argument: the robustness of strong solutions (i.e. the persistence of regularity under perturbation of the initial condition) and the strong convergence in $L^\infty(0, T; V)$ and $L^2(0, T; H^2)$ of the Galerkin approximations of strong solutions.

9.1 Robustness of strong solutions

First we show that smoothness of solutions of the Navier–Stokes equations on some time interval $[0, T]$ is stable under perturbations to the initial condition that are small with respect to the norm of H^1. In other words, we will show that the set of all initial conditions $u_0 \in V$ that give rise to a strong solution of the Navier–Stokes equations on the time interval $[0, T]$ is open in V (in the topology induced by the H^1 norm). More precisely, let us suppose that

- an initial condition $u_0 \in V$ gives rise to a strong solution u of the Navier–Stokes equations on the time interval $[0, T]$ and
- an initial condition $v_0 \in V$ gives rise to a Leray–Hopf weak solution v of the Navier–Stokes equations on the time interval $[0, T]$.

We will show that if v_0 is sufficiently close to u_0 in V then v is also strong on $[0, T]$, see Figure 9.1.

The condition that v_0 is in the vicinity of u_0 is more exactly expressed by the inequality

$$\|u_0 - v_0\|_{H^1} \le R(u),$$

where $R(u)$ depends on the behaviour of u on the time interval $[0, T]$, or, more precisely, on the norms of $u(s)$ in the spaces H^1 and H^2 for $s \in (0, T]$. The

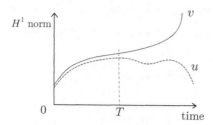

Figure 9.1. Strong solutions with similar initial conditions are close in the H^1 norm on some time interval $[0, T]$.

greater the norms of u in H^1 and H^2 on the time interval $(0, T]$, the smaller the number $R(u)$, and the 'radius of assured regularity' in the space of initial conditions decreases.

The key point in the proof of the robustness of regularity theorem is that we are able to control the H^1 norm of the difference $u - v$ in terms of the norms $\|u\|_{H^1}$ and $\|u\|_{H^2}$ (at least locally in time).

Theorem 9.1 *Let Ω be $\mathbb{T}^3$, $\mathbb{R}^3$, or a smooth bounded domain in $\mathbb{R}^3$. Assume that $u_0 \in V(\Omega)$ gives rise to a strong solution u of the Navier–Stokes equations on the time interval $[0, T]$. If $v_0 \in V$ and*

$$\|\nabla v_0 - \nabla u_0\| \le R$$

then v_0 also gives rise to a strong solution of the Navier–Stokes equations on the time interval $[0, T]$. Here the number R is given by

$$R(u) = \frac{1}{\sqrt{2cT}} \exp\left(-c \int_0^T \|\nabla u(s)\|^4 + \|Au(s)\|\|\nabla u(s)\| \, ds\right), \quad (9.1)$$

where c is some absolute constant (which depends on Ω).

Proof Let v be the strong solution arising from v_0 and denote by w the difference of u and v, i.e.

$$w(t) = u(t) - v(t).$$

Our purpose is to show that the H^1 norm of $w(t)$ cannot blow up on $[0, T]$ when v_0 is sufficiently close to u_0 (this implies that v cannot blow up on $[0, T]$ either). From what we have proved so far it follows that if v loses regularity on the time interval $[0, T]$ then there must be some time $T^* \le T$ such that v is smooth on the time interval $(0, T^*)$ and $\|\nabla v(t)\| \to \infty$ as $t \to T^*$. Clearly, if this happens then we also have $\|\nabla w(t)\| \to \infty$ as $t \to T^*$.

We therefore suppose that there exists such a T^* and deduce a contradiction.

Step 1. We derive a differential inequality for $\|\nabla w\|$ that depends on w and the strong solution u, but not explicitly on v

Since u and v are smooth solutions on $(0, T^*)$ it follows that on the time interval $(0, T^*)$ we have

$$\langle \partial_t u, Aw \rangle + \langle Au, Aw \rangle + \langle (u \cdot \nabla)u, Aw \rangle = 0$$

and

$$\langle \partial_t v, Aw \rangle + \langle Av, Aw \rangle + \langle (v \cdot \nabla)v, Aw \rangle = 0.$$

Subtracting the second of the above equalities from the first yields

$$\frac{1}{2}\frac{d}{dt}\|\nabla w\|^2 + \|Aw\|^2 \leq |\langle (u \cdot \nabla)w, Aw \rangle| + |\langle (w \cdot \nabla)u, Aw \rangle|$$
$$+ |\langle (w \cdot \nabla)w, Aw \rangle|.$$

To bound the terms on the right-hand side we use the Hölder inequality (1.6), then the Lebesgue interpolation (1.8) and the Sobolev embedding $H^1 \subset L^6$, and finally Young's inequality (with exponents 4 and 4/3) to obtain

$$|\langle (u \cdot \nabla)w, Aw \rangle| \leq \|u\|_{L^6}\|\nabla w\|_{L^3}\|Aw\|,$$
$$\leq c\|\nabla u\|\|\nabla w\|^{1/2}\|\nabla w\|_{L^6}^{1/2}\|Aw\|$$
$$\leq c\|\nabla u\|\|\nabla w\|^{1/2}\|Aw\|^{3/2}$$
$$\leq c\|\nabla u\|^4\|\nabla w\|^2 + \frac{1}{3}\|Aw\|^2.$$

Similarly

$$|\langle (w \cdot \nabla)w, Aw \rangle| \leq c\|\nabla w\|^6 + \frac{1}{3}\|Aw\|^2 \quad \text{and}$$
$$|\langle (w \cdot \nabla)u, Aw \rangle| \leq c\|\nabla w\|\|\nabla u\|^{1/2}\|Au\|^{1/2}\|Aw\|$$
$$\leq c\|\nabla u\|\|Au\|\|\nabla w\|^2 + \frac{1}{3}\|Aw\|^2.$$

Combining these estimates yields

$$\frac{d}{dt}\|\nabla w(t)\|^2 \leq c\left(\|\nabla u(t)\|^4 + \|\nabla u(t)\|\|Au(t)\|\right)\|\nabla w(t)\|^2 + c\|\nabla w(t)\|^6,$$

so

$$\frac{d}{dt}\|\nabla w(t)\|^2 \leq \alpha(t)\|\nabla w(t)\|^2 + c\|\nabla w(t)\|^6, \tag{9.2}$$

where

$$\alpha(t) = c \left(\|\nabla u(t)\|^4 + \|\nabla u(t)\| \|Au(t)\| \right). \tag{9.3}$$

Step 2. We use an integrating factor to put the differential inequality into a more manageable form

We write inequality (9.2) as

$$\frac{d}{dt}\|\nabla w\|^2 - \alpha(t)\|\nabla w\|^2 \le c\|\nabla w\|^6$$

and multiply it by $e^{-\int_0^t \alpha(s)\,ds}$. Since

$$e^{-\int_0^t \alpha(s)\,ds} \frac{d}{dt}\|\nabla w(t)\|^2 - \alpha(t)e^{-\int_0^t \alpha(s)\,ds}\|\nabla w(t)\|^2 = \frac{d}{dt}\left(e^{-\int_0^t \alpha(s)\,ds}\|\nabla w(t)\|^2 \right)$$

we obtain

$$\frac{d}{dt}\left(e^{-\int_0^t \alpha(s)\,ds}\|\nabla w(t)\|^2 \right) \le c\, e^{-\int_0^t \alpha(s)\,ds}\|\nabla w(t)\|^6$$

$$= c\, e^{2\int_0^t \alpha(s)\,ds} e^{-3\int_0^t \alpha(s)\,ds}\|\nabla w(t)\|^6$$

$$\le c\, e^{2\int_0^T \alpha(s)\,ds}\left(e^{-\int_0^t \alpha(s)\,ds}\|\nabla w(t)\|^2 \right)^3.$$

Therefore

$$\frac{d}{dt}\left(e^{-\int_0^t \alpha(s)\,ds}\|\nabla w(t)\|^2 \right) \le \beta \left(e^{-\int_0^t \alpha(s)\,ds}\|\nabla w(t)\|^2 \right)^3, \tag{9.4}$$

where α is given by (9.3) and the constant β is given by

$$\beta = c \exp\left(\int_0^T 2\alpha(s)\,ds \right). \tag{9.5}$$

Step 3. We use ODE comparison results to finish the proof

Let us define $X(t) = e^{-\int_0^t \alpha(s)\,ds}\|\nabla w(t)\|^2$. From what we have proved so far it follows that on $(0, T^*)$ the function X satisfies

$$\dot{X} \le \beta X^3, \qquad X(0) = \|\nabla w(0)\|^2,$$

where β is given by (9.5). Therefore $X(t) \le Y(t)$, where Y is the solution of the ODE

$$\dot{Y} = \beta Y^3, \qquad Y(0) = \|\nabla w(0)\|^2,$$

as long as the solution Y exists (see Exercise 6.2). Therefore by direct computation of Y we have

$$e^{-\int_0^t \alpha(s)\,ds}\|\nabla w(t)\|^2 \leq \frac{Y(0)}{\sqrt{1 - 2\beta t Y^2(0)}}$$

as long as

$$Y(0) < \frac{1}{\sqrt{2\beta t}}.$$

So if we know that $Y(0) < 1/\sqrt{2\beta T}$ then clearly for each $t \in (0, T^*)$, where $T^* \leq T$ we have $Y(0) < 1/\sqrt{2\beta t}$ with

$$\|\nabla w(t)\| \leq e^{\int_0^t \alpha(s)\,ds} \frac{Y(0)}{\sqrt{1 - 2\beta t Y^2(0)}}$$

$$\leq e^{\int_0^T \alpha(s)\,ds} \frac{Y(0)}{\sqrt{1 - 2\beta T Y^2(0)}} := C_T.$$

Rewriting the condition $Y(0) < 1/\sqrt{2\beta T}$ in our original variables it follows that if

$$\|\nabla w(0)\| < \sqrt{\frac{c}{2T}} \exp\left(-c \int_0^T \|\nabla u(s)\|^4 + \|\nabla u(s)\|\|Au(s)\|\,ds\right)$$

then $\|\nabla w(t)\| \leq C_T$ for all $t \in [0, T^*)$, so $\|\nabla w(T^*)\| \leq C_T$ and in consequence w does not blow up at any time $T^* \leq T$, so we obtain the assertion of Theorem 9.1. □

9.2 Convergence of Galerkin approximations

Using similar reasoning we will now prove that the Galerkin approximations of a strong solution u of the Navier–Stokes equations, i.e. the solutions of

$$\partial_t u_n + Au_n + P_n[(u_n \cdot \nabla)u_n] = 0, \qquad u_n(0) = P_n u_0$$

(where P_n is the projection onto the space spanned by the first n eigenvectors of the Stokes operator A) converge strongly to u in $L^\infty(0, T; H^1)$ and $L^2(0, T; H^2)$, i.e. both spaces to which u belongs.

It should be emphasised that in the case of the Galerkin approximations of *weak* solutions we do not know whether they converge strongly in $L^2(0, T; H^1)$, even though u itself belongs to $L^2(0, T; H^1)$; in our existence proof we only obtained the weak convergence of a subsequence.

In the case of strong solutions the situation is much better: we will show that any strong solution of the Navier–Stokes equations in a bounded domain is the limit of its Galerkin approximations. The proof does not rely on a priori estimates, since these limit the existence time (as in the proof of Theorem 6.8), but rather makes use of the fact that u is assumed to be strong on the time interval $[0, T]$.

The proof of this is very similar to the proof of Theorem 9.1. In that proof we ended up with the differential inequality

$$\dot{y} \le \beta y^3.$$

Here we will have to deal with a similar inequality,

$$\dot{y} \le \beta y^3 + f(t). \tag{9.6}$$

The following ODE lemma makes precise the idea that functions y satisfying the inequality (9.6) are controlled by their initial data and the 'total forcing' from the $f(t)$ term. Since we will be investigating a sequence of Galerkin approximations the lemma deals with a family of solutions of (9.6) rather than with just a single solution y.

Lemma 9.2 *Take $\alpha > 0$ and let $f_n \ge 0$ be a sequence of non-negative functions on the time interval $[0, T]$. Suppose that (y_n) is a sequence of non-negative functions on $(0, T)$ that satisfy*

$$\frac{\mathrm{d}y_n}{\mathrm{d}t}(t) \le \alpha y_n^3(t) + f_n(t).$$

If

$$\eta_n := y_n(0) + \int_0^T f_n(s)\,\mathrm{d}s \;\to\; 0 \qquad as \quad n \to \infty$$

then $y_n \to 0$ uniformly on $[0, T]$ as $n \to \infty$.

Proof We compare $y_n(t)$ with $Y_n(t)$, the solution of

$$\mathrm{d}Y_n/\mathrm{d}t = \alpha Y_n^3, \qquad Y_n(0) = \eta_n. \tag{9.7}$$

It is easy to see (see Exercise 9.2) that $y_n(t) \le Y_n(t)$ for all t in the interval of existence of Y_n, and one can solve (9.7) to yield

$$Y_n(t) = \frac{\eta_n}{\sqrt{1 - 2\alpha\eta_n^2 t}}.$$

Since $\eta_n \to 0$ the solution Y_n is defined on the whole time interval $[0, T]$ for all n sufficiently large and clearly $Y_n \to 0$ uniformly on $[0, T]$. Since $0 \le y_n \le Y_n$ the assertion follows. $\qquad\square$

We now prove the convergence of Galerkin approximations for strong solutions.

Theorem 9.3 *Let Ω be $\mathbb{T}^3$ or a smooth bounded subset of $\mathbb{R}^3$. If u is the strong solution of the Navier–Stokes equations on the time interval $[0, T]$ corresponding to $u_0 \in V$ and u_n are the Galerkin approximations corresponding to the same initial condition u_0, i.e. the solutions of*

$$\partial_t u_n + Au_n + P_n[(u_n \cdot \nabla)u_n] = 0, \qquad u_n(0) = P_n u_0, \qquad (9.8)$$

then u_n converges strongly to u in both $L^\infty(0, T; H^1)$ and $L^2(0, T; H^2)$.

Proof Let u be the strong solution of the Navier–Stokes equations on $[0, T]$ and u_n be the sequence of Galerkin approximations of u. We want to show that the H^1 norm of the difference

$$w_n = u - u_n$$

tends uniformly to zero on time interval $[0, T]$ as n tends to infinity.

Step 1. We derive a differential inequality for $\|\nabla w_n\|$ that depends on w_n and on the strong solution u, but not explicitly on u_n

Since u is a strong solution, on the time interval $(0, T)$ we have

$$\langle \partial_t u, Aw_n \rangle + \langle Au, Aw_n \rangle + \langle \mathbb{P}(u \cdot \nabla)u, Aw_n \rangle = 0,$$

where $\mathbb{P}$ is the Leray projector. We introduce Q_n via the identity $\mathrm{Id} = P_n + Q_n$ to rewrite this equation as

$$\langle \partial_t u, Aw_n \rangle + \langle Au, Aw_n \rangle + \langle P_n[(u \cdot \nabla)u], Aw_n \rangle = \langle -Q_n\mathbb{P}[(u \cdot \nabla)u], Aw_n \rangle. \tag{9.9}$$

If we take the inner product of (9.8) with Aw_n then we obtain

$$\langle \partial_t u_n, Aw_n \rangle + \langle Au_n, Aw_n \rangle + \langle P_n[(u_n \cdot \nabla)u_n], Aw_n \rangle = 0. \tag{9.10}$$

Taking the difference of (9.9) and (9.10) yields

$$\langle \partial_t w_n, Aw_n \rangle + \langle Aw_n, Aw_n \rangle$$
$$= \langle P_n[(u_n \cdot \nabla)u_n], Aw_n \rangle - \langle P_n[(u \cdot \nabla)u], Aw_n \rangle - \langle Q_n\mathbb{P}[(u \cdot \nabla)u], Aw_n \rangle.$$

We substitute $u_n = u - w_n$ in the first term on the right-hand side, use the identity $\langle P_n w, z \rangle = \langle w, P_n z \rangle$, and after standard computations involving the nonlinear term arrive at

$$\frac{1}{2}\frac{\mathrm{d}}{\mathrm{d}t}\|\nabla w_n\|^2 + \|Aw_n\|^2 \leq |\langle (u \cdot \nabla)w_n, P_nAw_n \rangle| + |\langle (w_n \cdot \nabla)u, P_nAw_n \rangle|$$
$$+ |\langle (w_n \cdot \nabla)w_n, P_nAw_n \rangle| + |\langle Q_n\mathbb{P}[(u \cdot \nabla)u], Aw_n \rangle|.$$

We estimate the terms on the right-hand side using the same methods as in the proof of Theorem 9.1:

$$|\langle (u \cdot \nabla) w_n, P_n A w_n \rangle| \leq c \|\nabla u\|^4 \|\nabla w_n\|^2 + \frac{1}{8} \|P_n A w_n\|^2,$$

$$|\langle (w_n \cdot \nabla) u, P_n A w_n \rangle| \leq c \|\nabla u\| \|Au\| \|\nabla w_n\|^2 + \frac{1}{8} \|P_n A w_n\|^2,$$

$$|\langle (w_n \cdot \nabla) w_n, P_n A w_n \rangle| \leq c \|\nabla w_n\|^6 + \frac{1}{8} \|P_n A w_n\|^2, \quad \text{and}$$

$$|\langle Q_n \mathbb{P}[(u \cdot \nabla) u], A w_n \rangle| \leq c \|Q_n \mathbb{P}[(u \cdot \nabla) u]\|^2 + \frac{1}{8} \|A w_n\|^2.$$

Combining these estimates and noticing that $\|P_n A w_n\| \leq \|A w_n\|$ gives us

$$\frac{\mathrm{d}}{\mathrm{d}t} \|\nabla w_n(t)\|^2 + \|A w_n\|^2 \leq \alpha(t) \|\nabla w_n(t)\|^2 + c \|\nabla w_n(t)\|^6 + f_n(t), \quad (9.11)$$

where

$$\alpha(t) = c \left(\|\nabla u(t)\|^4 + \|\nabla u(t)\| \|Au(t)\| \right), \quad f_n(t) = c \|Q_n \mathbb{P}[(u(t) \cdot \nabla) u(t)]\|^2.$$

Step 2. We use an integrating factor to rewrite (9.11) in a simpler form

We do not present the details here, since they are very similar to the reasoning in Step 2 of the proof of Theorem 9.1. The result is that

$$\frac{\mathrm{d}}{\mathrm{d}t} y_n(t) \leq \beta y_n^3(t) + f_n(t), \quad (9.12)$$

where

$$\beta = c \exp \left(\int_0^T 2\alpha(s) \, \mathrm{d}s \right),$$

$$\alpha(t) = c(\|\nabla u(t)\|^4 + \|\nabla u(t)\| \|Au(t)\|),$$

$$f_n(t) = c \|Q_n \mathbb{P}[(u(t) \cdot \nabla) u(t)]\|^2, \quad \text{and}$$

$$y_n(t) = \|\nabla w_n(t)\|^2 \exp \left(-\int_0^t \alpha(s) \, \mathrm{d}s \right).$$

Step 3. We show that $y_n(0) + \int_0^T f_n(s) \, \mathrm{d}s \to 0$ so that we can use Lemma 9.2

Set

$$\eta_n = \|\nabla w_n(0)\|^2 + \int_0^T \|Q_n \mathbb{P}(u(s) \cdot \nabla) u(s)\|^2 \, \mathrm{d}s;$$

we need to show that $\eta_n \to 0$ as $n \to \infty$.

Clearly

$$\|\nabla w_n(0)\|^2 = \|\nabla u(0) - \nabla P_n u(0)\|^2 \le \|u_0 - P_n u_0\|_{H^1}^2 \to 0,$$

since $u_0 \in D(A^{1/2})$ (we use Lemma 4.1). Now we want to show that

$$\int_0^T \|Q_n \mathbb{P}[(u(s) \cdot \nabla) u(s)]\|^2 \, ds \to 0$$

as $n \to \infty$. We will use the Dominated Convergence Theorem. First, we notice that if u is a strong solution on the time interval $[0, T]$ then – according to Lemma 6.2 – we have $\|(u \cdot \nabla) u\|^2 \in L^1(0, T)$. Hence

$$\|Q_n \mathbb{P}[(u \cdot \nabla) u]\|^2 \le \|(u \cdot \nabla) u\|^2 \in L^1(0, T).$$

Now, since for any strong solution u we have $\mathbb{P}[(u(s) \cdot \nabla) u(s)] \in H(\Omega)$, we deduce that for all $s \in (0, T)$

$$\|Q_n \mathbb{P}[(u \cdot \nabla) u]\|^2 = \|\mathbb{P}[(u \cdot \nabla) u] - P_n \mathbb{P}[(u \cdot \nabla) u]\|^2 \to 0,$$

because a_k form a basis of H. It follows now from the Dominated Convergence Theorem that $\eta_n \to 0$ as required.

Step 4. We use the ODE lemma, Lemma 9.2, to complete the proof

From Lemma 9.2, inequality (9.12), and the fact that $\eta_n \to 0$ as $n \to \infty$, we conclude that

$$\|\nabla w_n\|^2 \to 0 \qquad \text{uniformly on } (0, T).$$

Therefore $u_n \to u$ in $C([0, T]; H^1)$ and also in $L^\infty(0, T; H^1)$. To prove that

$$u_n \to u \quad \text{in} \quad L^2(0, T; H^2)$$

we use the convergence $u_n \to u$ in $C([0, T]; H^1)$ and the inequality (9.12), see Exercise 9.3. □

Notes

Theorem 9.1 is one of many results that show that strong solutions of the Navier–Stokes equations are stable under small perturbations in appropriate function spaces. For example, one of the versions of the robustness of regularity result in Theorem 9.1 was proved by Chernyshenko et al. (2007). It says that if u_0 gives rise to a strong solution u of the Navier–Stokes equations with forcing

f on the time interval $[0, T]$ then for every v_0 and g satisfying the condition

$$\|u_0 - v_0\|_{H^s} + \int_0^T \|f(s) - g(s)\|_{H^s}\, ds \leq c \exp\left(-c \int_0^T \|u(s)\|_{H^{s+1}}\, ds\right),$$

$s \geq 3$, there is a strong solution of the Navier–Stokes equations with forcing g that corresponds to an initial condition v_0.

The paper by Chernyshenko et al. (2007) is closely based on Constantin (1986) who showed that if $u_0 \in H^5$ gives rise to a strong solution of the Euler equations

$$\partial_t u + (u \cdot \nabla) u + \nabla p = f, \quad \operatorname{div} u = 0, \quad u(0) = u_0$$

on the time interval $[0, T]$ and

$$\int_0^T |\operatorname{curl} u(s)|\, ds < \infty$$

then there exists a constant $\nu_0 > 0$ such that for all $\nu \in (0, \nu_0)$ the solution of the Navier–Stokes equations with viscosity ν, initial condition u_0, and forcing f is smooth on $[0, T]$.

The statement of Theorem 9.1 is similar to the one proved in Dashti & Robinson (2008) where the condition for regularity involving forces f and g and initial conditions u_0 and v_0 is of the form

$$\|u_0 - v_0\|_{H^1} + \int_0^T \|f(s) - g(s)\|_{H^1} \leq R(u),$$

where R is given by (9.1). In Chemin et al. (2006) and in Marín-Rubio, Robinson, & Sadowski (2013) the robustness of regularity is studied in the space $\dot{H}^{1/2}$.

The robustness theorem can be used to show that the fact that an initial condition $u_0 \in V$ gives rise to a strong solution on $[0, T]$, where T is given, can be deduced from the properties of Galerkin approximations of sufficiently high order. Indeed if u is a strong solution on $[0, T]$ corresponding to an initial condition u_0 then the Galerkin approximations u_n, where n is sufficiently large, must satisfy a variant of the condition from Theorem 9.1 (see Exercises 9.4–9.6). If this happens we can deduce that all initial conditions in some ball $B(u_{0,n}, R) \subset V$ centred at $u_{0,n}$ and of radius R, depending on u_n, must give rise to strong solutions on $[0, T]$. Since the initial conditions $u_{0,n}$ converge to u_0 in V it is possible to prove that $u_0 \in B(u_{0,n}, R)$ for sufficiently large n (under the assumption that u_0 indeed gives rise to a strong solution).

There are many related results that can be viewed in the context of weak–strong uniqueness, for example there are various results that show that solutions of the Navier–Stokes equations preserving some kind of symmetry are stable in some appropriate function space. In particular, Bardos et al. (2013) proved that two-dimensional flows treated as three-dimensional (that is independent of x_3 variable) are stable in the energy space (for related results see also Gallagher, 1997, and Mucha, 2008). Moreover, they showed that if u is such a flow with an initial condition u_0 and v_0 gives rise to a Leray–Hopf weak solution v then

$$\|u(t) - v(t)\|_{L^2(\Omega)} \le C\left(\|u_0\|_{L^2(\Omega')}\right) \|u_0 - v_0\|_{L^2(\Omega)},$$

where $\Omega' \subset \mathbb{R}^2$ is a smooth bounded domain and $\Omega = \Omega' \times (0, L)$. From this inequality it follows that spontaneous symmetry breaking is not possible for Leray–Hopf weak solutions and that 'two-dimensional' flows are unique in this class of solutions.

Exercises

9.1 Let K be the set of all initial conditions $u_0 \in V(\mathbb{T}^3)$ that give rise to global-in-time strong solutions of the Navier–Stokes equations. Prove that K is open in V.

9.2 Let f be a non-negative function and X a function satisfying the differential inequality

$$\mathrm{d}X/\mathrm{d}t \le \alpha X^3 + f(t), \quad X(0) = a,$$

where $a > 0$ and $\alpha > 0$. Furthermore let Y be the solution of

$$\mathrm{d}Y/\mathrm{d}t = \alpha Y^3, \quad Y(0) = b + \int_0^T f(s)\,\mathrm{d}s.$$

Prove that
  (i) $a < b \Rightarrow X(t) < Y(t)$ and
  (ii) $a \le b \Rightarrow X(t) \le Y(t)$
on the interval of existence of Y.

9.3 Finish the proof of Theorem 9.3, that is show that if u_n is a sequence of Galerkin approximations converging to a strong solution u on the time interval $[0, T]$ then

$$u_n \to u \quad \text{in} \quad L^2(0, T; H^2).$$

9.4 Let u be a strong solution of the Navier–Stokes equations on the torus with forcing $f \in L^2(0, T; \dot{L}^2)$ and an initial condition $u_0 \in V$:

$$\partial_t u - \Delta u + (u \cdot \nabla)u + \nabla p = f, \quad u(0) = u_0.$$

Modify the proof of Theorem 9.1 to show that there is a constant $c > 0$ with the following property: if the initial condition $v_0 \in V$ satisfies

$$\|\nabla u_0 - \nabla v_0\|^2 + \int_0^T \|f(s)\|^2 \, ds \leq \frac{1}{\sqrt{2cT}} e^{-c \int_0^T \|\nabla u\| \|Au\| + \|\nabla u\|^4}$$

then v_0 gives rise to the strong solution of the unforced Navier–Stokes equations on the time interval $[0, T]$.

9.5 Let u_n be a sequence of Galerkin approximations converging to a strong solution u of the Navier–Stokes equations on the torus on the time interval $[0, T]$. The approximations u_n solve the Navier–Stokes equations with forcing f_n:

$$\partial_t u_n - \Delta u_n + (u_n \cdot \nabla) u_n + \nabla p_n = f_n, \quad u_n(0) = P_n u_0,$$

where

$$f_n = Q_n \mathbb{P}[(u_n \cdot \nabla) u_n], \quad Q_n = (\mathrm{Id} - P_n).$$

Prove that $f_n \to 0$ in $L^2(0, T; L^2)$.

9.6 Show that if $u_0 \in V(\mathbb{T}^3)$ gives rise to a strong solution of the Navier–Stokes equations on $[0, T]$, then there exists a natural number n such that the Galerkin approximation u_n corresponding to u_0 satisfies the condition

$$\|u_n(0) - u_0\|^2 + \int_0^T \|f_n(s)\|^2 \, ds \leq \frac{1}{\sqrt{2cT}} e^{-c \int_0^T \|Au_n\| \|\nabla u_n\| + \|\nabla u_n\|^4},$$

where f_n is defined in the previous exercise. (Notice that if this condition is satisfied – and we can check it once u_n and u_0 are given – then it follows that u_0 gives rise to a strong solution on $[0, T]$. Hence one can verify that u arising from u_0 is strong on $[0, T]$ without computing u with exact precision.)

10

Local existence and uniqueness in $\dot{H}^{1/2}$

10.1 Critical spaces

We mentioned in the Introduction that solutions of the Navier–Stokes equations on the whole space have an important scaling property, but we have not yet made significant use of this fact. We now return to this idea and introduce the notion of a 'critical space' as one whose norm is invariant under (a form of) this scaling transformation.

Recall from the discussion in the Introduction that if $u(x, t)$ is a solution of the equations *on the whole space* defined for $t \in [0, T]$ with initial condition u_0, then for any $\lambda \in \mathbb{R}$ the function

$$u_\lambda(x, t) = \lambda u(\lambda x, \lambda^2 t) \tag{10.1}$$

is another solution of the equations,[1] now defined on $[0, T/\lambda^2]$ with initial data $u_{0,\lambda}(x) := \lambda u_0(\lambda x)$. A space whose norm is invariant under the transformation

$$u \mapsto \lambda u(\lambda x)$$

is called *critical* (for the Navier–Stokes equations). Such spaces are in many ways the natural ones in which to study the Navier–Stokes problem, and much recent work has been carried out in this setting. We emphasise, however, that for a space to be critical in terms of this definition we have to treat the problem on the whole of $\mathbb{R}^3$. It is perhaps remarkable that the spaces identified as critical by this scaling argument remain particularly significant even in the case of bounded domains.

In this book we will prove local existence and uniqueness results in two critical spaces, $\dot{H}^{1/2}$ (in this chapter) and L^3 (in the next chapter).

---

[1] Provided that we take $p_\lambda(x, t) = \lambda^2 p(\lambda x, \lambda^2 t)$. In the periodic case the same is true for any $\lambda \in \mathbb{N}$, since such a scaling preserves the periodicity, but this only allows for a 'shrinking of scales' rather than magnification.

We remarked in the Introduction that the importance of such spaces is due to the fact that proving a 'good' local existence result in a critical space, i.e. a result in which the guaranteed existence time depends only on the norm, provides a proof of global existence by rescaling. Unsurprisingly, therefore, we will not prove such a result in this chapter, instead obtaining an existence time that depends on the initial data u_0 in a more involved way.

A related observation is that the idea of 'small initial data' in a space X has no absolute meaning unless X is a critical space. This is relevant in the light of the result of Theorem 6.12 on global existence of a strong solution for initial data that is small in H^1: there we showed that on bounded domains there is a constant ε such that whenever

$$\|u_0\|_{H^1} < \varepsilon$$

the equations have a solution that remains bounded in H^1 (and is therefore strong) for all $t \geq 0$.

A similar result also holds on the whole space (as a corollary of Lemma 6.13), but this would be an unnatural result, since the inhomogeneous H^1 norm,

$$\|u\|_{H^1}^2 = \|\nabla u\|_{L^2}^2 + \|u\|_{L^2}^2,$$

is not invariant under the rescaling transformation $u_0 \mapsto \lambda u_0(\lambda x)$. This means that a careful choice of λ may sometimes lower the norm below the 'smallness' threshold. Indeed, the two 'parts' of the H^1 norm transform differently:

$$\|u_\lambda\|_{L^2}^2 = \int |u_\lambda(x)|^2 \, dx = \int |\lambda u(\lambda x)|^2 \, dx$$
$$= \lambda^2 \int |u(\lambda x)|^2 \, dx = \lambda^2 \int |u(y)|^2 \lambda^{-3} \, dy = \lambda^{-1} \|u\|_{L^2}^2$$

and

$$\|\partial_i u_\lambda\|_{L^2}^2 = \int |\partial_i[\lambda u(\lambda x)]|^2 \, dx = \lambda^2 \int \lambda^2 |(\partial_i u)(\lambda x)|^2 \, dx = \lambda \|\partial_i u\|_{L^2}^2.$$

Therefore

$$\|u_\lambda\|_{H^1}^2 = \lambda^{-1} \|u\|_{L^2}^2 + \lambda \|\nabla u\|^2;$$

we minimise the H^1 norm of u_λ if we take $\lambda = \|u\|_{L^2} / \|\nabla u\|_{L^2}$, and then

$$\|u_\lambda\|_{H^1} = \|u\|_{L^2}^{1/2} \|\nabla u\|_{L^2}^{1/2}. \tag{10.2}$$

The combination of norms on the right-hand side of this equality *is* invariant under the rescaling transformation, and it is perhaps therefore more natural to assume that this combination is 'small' than the H^1 norm itself, as in fact we did in Lemma 6.13.

The expression on the right-hand side of (10.2) is in fact a bound on the $\dot{H}^{1/2}$ norm obtained by interpolating between L^2 and $\dot{H}^1$: see (10.5), below. In this chapter we prove directly a local existence result in $\dot{H}^{1/2}$, which, as we will now see, is the critical Sobolev space for the 3D Navier–Stokes equations.

10.2 Fractional Sobolev spaces and criticality of $\dot{H}^{1/2}$

Recall that in Chapter 1 we defined fractional-order (homogeneous) Sobolev spaces $\dot{H}^s$ on $\mathbb{T}^3$ and $\mathbb{R}^3$ using Fourier series and the Fourier transform, respectively. In the periodic case

$$\dot{H}^s(\mathbb{T}^3) = \left\{ u = \sum_{k \in \dot{\mathbb{Z}}^3} \hat{u}_k e^{ik \cdot x} : \hat{u}_k = \overline{\hat{u}_{-k}}, \quad \sum_{k \in \dot{\mathbb{Z}}^3} |k|^{2s} |\hat{u}_k|^2 < \infty \right\},$$

where the norm is defined by

$$\|u\|_{\dot{H}^s(\mathbb{T}^3)}^2 := (2\pi)^3 \sum_{k \in \dot{\mathbb{Z}}^3} |k|^{2s} |\hat{u}_k|^2;$$

this norm is equal to $\|\Lambda^s u\|$, where

$$\Lambda^s u = \sum_{k \in \dot{\mathbb{Z}}^3} |k|^s \hat{u}_k e^{ik \cdot x} \quad \text{when} \quad u = \sum_{k \in \dot{\mathbb{Z}}^3} \hat{u}_k e^{ik \cdot x}, \tag{10.3}$$

see (1.18). In particular, $\Lambda^2 = -\Delta$.

On the whole space $\dot{H}^s(\mathbb{R}^3)$ is defined as

$$\dot{H}^s(\mathbb{R}^3) = \{ u \in \mathscr{S}' : \hat{u} \in L^1_{\text{loc}}(\mathbb{R}^3), \ |\xi|^s \hat{u}(\xi) \in L^2(\mathbb{R}^3) \},$$

with

$$\|u\|_{\dot{H}^s(\mathbb{R}^3)}^2 := \int_{\mathbb{R}^3} |\xi|^{2s} |\hat{u}(\xi)|^2 \, d\xi.$$

In both cases we can use Sobolev interpolation: whenever $\sigma_1 < s < \sigma_2$ and $u \in \dot{H}^{\sigma_1} \cap \dot{H}^{\sigma_2}$ we have $u \in \dot{H}^s$ and

$$\|u\|_{\dot{H}^s} \le \|u\|_{\dot{H}^{\sigma_1}}^{\alpha} \|u\|_{\dot{H}^{\sigma_2}}^{1-\alpha}, \qquad s = \alpha\sigma_1 + (1-\alpha)\sigma_2 \tag{10.4}$$

(see Lemma 1.15). With $s = 1/2$, $\sigma_1 = 0$, and $\sigma_2 = 1$, this yields

$$\|u\|_{\dot{H}^{1/2}} \le \|u\|_{L^2}^{1/2} \|u\|_{\dot{H}^1}^{1/2} \tag{10.5}$$

(compare this with (10.2)).

We now show that $\dot{H}^{1/2}(\mathbb{R}^3)$ is a critical space, i.e. that the $\dot{H}^{1/2}$ norm is scaling invariant. To do this, we first calculate the Fourier transform of u_λ:

$$\widehat{u_\lambda}(\xi) = \lambda \int e^{-2\pi i \xi \cdot x} u(\lambda x)\, dx$$

$$= \lambda^{-2} \int e^{-2\pi i (\xi/\lambda) \cdot y} u(y)\, dy$$

$$= \lambda^{-2} \hat{u}(\xi/\lambda)$$

(the calculation requires $u \in L^1$ in order to write the Fourier transform in integral form, but the resulting identity can be extended to $u \in L^2$, just as we extend the definition of the Fourier transform itself). Therefore

$$\|u_\lambda\|^2_{\dot{H}^{1/2}(\mathbb{R}^3)} = \int |\xi| |\widehat{u_\lambda}(\xi)|^2\, d\xi = \lambda^{-4} \int |\xi| |\hat{u}(\xi/\lambda)|^2\, d\xi$$

$$= \lambda^{-4} \int |\lambda \xi| |\hat{u}(\xi)|^2\, \lambda^3\, d\xi = \int |\xi| |\hat{u}(\xi)|^2\, d\xi$$

$$= \|u\|^2_{\dot{H}^{1/2}(\mathbb{R}^3)}.$$

For the behaviour of general Sobolev norms and Lebesgue norms under rescaling in $\mathbb{R}^3$ see Exercise 10.1. Exercise 10.2 shows that such rescalings behave very differently for functions defined on $\mathbb{T}^3$.

10.3 Local existence for initial data in $\dot{H}^{1/2}$

In this section we give a simple proof of the local existence of strong solutions for initial data in $\dot{H}^{1/2}(\mathbb{T}^3)$. The argument relies essentially on a priori estimates to show that when $u_0 \in \dot{H}^{1/2}$ the solution must lie in $L^\infty(0, T; \dot{H}^{1/2})$ and $L^2(0, T; \dot{H}^{3/2})$ for some $T > 0$. To make this rigorous we use a Galerkin scheme, which restricts the proof we give to the case of $\mathbb{T}^3$. However, we make no use of the Poincaré inequality, so with a different approximation scheme essentially the same argument yields the result on $\mathbb{R}^3$ (see the Notes).

Note that if $u \in L^\infty(0, T; \dot{H}^{1/2}) \cap L^2(0, T; \dot{H}^{3/2})$ then by Sobolev interpolation (10.4) we have

$$\|u\|^4_{\dot{H}^1} \le \|u\|^2_{\dot{H}^{1/2}} \|u\|^2_{\dot{H}^{3/2}}$$

and hence

$$\int_0^T \|u(t)\|^4_{\dot{H}^1}\, dt \le \int_0^T \|u(t)\|^2_{\dot{H}^{1/2}} \|u(t)\|^2_{\dot{H}^{3/2}}\, dt$$

$$\le \|u\|^2_{L^\infty(0,T;\dot{H}^{1/2})} \|u\|^2_{L^2(0,T;\dot{H}^{3/2})}. \tag{10.6}$$

Since $\dot{H}^1 \subset L^6$ it follows that

$$\|u\|_{L^4(0,T;L^6)} \leq c\|u\|^{1/2}_{L^\infty(0,T;\dot{H}^{1/2})}\|u\|^{1/2}_{L^2(0,T;\dot{H}^{3/2})}. \tag{10.7}$$

Since $2/4 + 3/6 = 1$ it follows that u satisfies the Serrin condition, which guarantees that u is smooth on $(0, T]$ (Theorem 8.17) and unique in the class of Leray–Hopf weak solutions (Theorem 8.19). (For a simpler proof showing that such a solution is unique in the class of Leray–Hopf weak solutions with $u \in L^\infty(0, T; \dot{H}^{1/2}) \cap L^2(0, T; \dot{H}^{3/2})$ see Exercise 10.6.)

In the statement of the following theorem we use the notation $e^{\Delta t} u_0$ to denote the solution at time t of the heat equation with initial data u_0, i.e. $v(t) := e^{\Delta t} u_0$ satisfies

$$\partial_t v - \Delta v = 0, \qquad v(0) = u_0.$$

Theorem 10.1 *There exists an absolute constant $\varepsilon_h > 0$ such that whenever* $u_0 \in \dot{H}^{1/2}(\mathbb{T}^3) \cap H$ *and*

$$\int_0^T \|e^{\Delta s} u_0\|^4_{\dot{H}^1} \, ds < \varepsilon_h \tag{10.8}$$

for some $T > 0$ the Navier–Stokes equations with initial data u_0 have a unique Leray–Hopf weak solution on $[0, T]$, which satisfies

$$u \in L^\infty(0, T; \dot{H}^{1/2}) \cap L^2(0, T; \dot{H}^{3/2})$$

and is therefore smooth on $(0, T]$.

Note that, as we would expect from the discussion at the beginning of the chapter, the existence time T does not depend solely on the norm of u_0 in $\dot{H}^{1/2}$, but on the condition (10.8) which reflects the distribution of energy in the initial condition (see Exercise 10.3). It is striking that this local existence time depends only on properties of solutions of the heat equation; such a result holds in more general critical spaces, see, for example, the nice review article by Cannone (2003).

Proof We consider the Galerkin approximations of u, i.e. the solutions u_n of

$$\partial_t u_n + A u_n + P_n[(u_n \cdot \nabla)u_n] = 0, \qquad u_n(0) = P_n u_0,$$

where P_n is the projector on the space spanned by the first n eigenfunctions of the Stokes operator A (see Chapter 2 for a discussion of the Stokes operator and (4.1) for the definition of P_n).

The arguments used in Chapter 4 guarantee that this Galerkin equation has solutions that exist for all $t > 0$, and that for any choice of $T > 0$ there is a subsequence of the u_n (which we relabel) that converges to a function

$$u \in L^\infty(0, T; H) \cap L^2(0, T; \dot{H}^1)$$

that is a weak solution of the Navier–Stokes equations. To prove that for initial data $u_0 \in \dot{H}^{1/2} \cap H$ we have $u \in L^\infty(0, T; \dot{H}^{1/2}) \cap L^2(0, T; \dot{H}^{3/2})$ we make some further estimates on the Galerkin approximations: we show that there exists a $T > 0$ such that u_n is uniformly bounded (with respect to n) in $L^\infty(0, T; \dot{H}^{1/2}) \cap L^2(0, T; \dot{H}^{3/2})$.

We use a splitting argument, and decompose u_n as $v_n + w_n$, where v_n solves the linear equation

$$\partial_t v_n + A v_n = 0, \qquad v_n(0) = P_n u_0, \tag{10.9}$$

and consequently w_n solves

$$\partial_t w_n + A w_n + P_n[(u_n \cdot \nabla)u_n] = 0, \qquad w_n(0) = 0. \tag{10.10}$$

If we write (10.10) as

$$\partial_t w_n + A w_n + P_n[((v_n + w_n) \cdot \nabla)(v_n + w_n)] = 0, \qquad w_n(0) = 0, \tag{10.11}$$

then this makes it clear that we can solve for v_n and then try to find w_n. The advantage of this splitting is that by dealing with the initial data via the linear equation (10.9), we need only solve the nonlinear equation (10.11) for initial data zero.

While the operator appearing in (10.9) is the Stokes operator A, note that the initial condition u_0 is assumed to be divergence free; since $Au = -\Delta u$ when $u \in D(A) = H \cap \dot{H}^2(\mathbb{T}^3)$, v_n is a solution of heat equation $\partial_t v_n - \Delta v_n = 0$. Now observe that $v_n(t) = P_n v(t)$, where $v(t) = e^{\Delta t} u_0$ is the solution of

$$\partial_t v - \Delta v = 0, \qquad v(0) = u_0 \tag{10.12}$$

(this follows by applying P_n to all the terms in the equation). Since solutions of the heat equation are smooth (see Appendix D) we can act on (10.12) with $\Lambda^{1/2}$ and then take the L^2 inner product with $\Lambda^{1/2} v$ to obtain

$$\langle \partial_t \Lambda^{1/2} v, \Lambda^{1/2} v \rangle + \langle \Lambda^{5/2} v, \Lambda^{1/2} v \rangle = 0,$$

which yields

$$\frac{1}{2} \frac{d}{dt} \|v\|_{\dot{H}^{1/2}}^2 + \|v\|_{\dot{H}^{3/2}}^2 = 0.$$

Therefore

$$\frac{1}{2} \|v(t)\|_{\dot{H}^{1/2}}^2 + \int_0^t \|v(s)\|_{\dot{H}^{3/2}}^2 \, ds = \frac{1}{2} \|u_0\|_{\dot{H}^{1/2}}^2. \tag{10.13}$$

Since $v_n(t) = P_n v(t)$ we have in particular

$$\|v_n(t)\|_{\dot{H}^{1/2}} \le \|u_0\|_{\dot{H}^{1/2}} \qquad (10.14)$$

and

$$\int_0^t \|v_n(s)\|_{\dot{H}^{3/2}}^2 \, ds \le \int_0^t \|v(s)\|_{\dot{H}^{3/2}}^2 \, ds \le \frac{1}{2}\|u_0\|_{\dot{H}^{1/2}}^2 \qquad (10.15)$$

for every n.

We now take the inner product of (10.10) with Λw_n to obtain

$$\frac{1}{2}\|\Lambda^{1/2} w_n\|^2 + \|\Lambda^{3/2} w_n\|^2 = -\langle P_n[(u_n \cdot \nabla)u_n], \Lambda w_n \rangle.$$

We estimate

$$|\langle P_n(u_n \cdot \nabla)u_n, \Lambda w_n \rangle| = |\langle (u_n \cdot \nabla)u_n, \Lambda w_n \rangle|$$
$$\le c\|u_n\|_{L^6}\|\nabla u_n\|_{L^2}\|\Lambda w_n\|_{L^3}$$
$$\le c\|u_n\|_{\dot{H}^1}^2 \|w_n\|_{\dot{H}^{3/2}}, \qquad (10.16)$$

using the Sobolev embeddings $\|f\|_{L^6} \le c\|f\|_{\dot{H}^1}$ and $\|f\|_{L^3} \le c\|f\|_{\dot{H}^{1/2}}$ (see Theorem 1.19). We therefore obtain

$$\frac{1}{2}\frac{d}{dt}\|w_n\|_{\dot{H}^{1/2}}^2 + \|w_n\|_{\dot{H}^{3/2}}^2 \le c\|u_n\|_{\dot{H}^1}^2\|w_n\|_{\dot{H}^{3/2}}$$
$$\le c(\|v_n\|_{\dot{H}^1}^2 + \|w_n\|_{\dot{H}^1}^2)\|w_n\|_{\dot{H}^{3/2}},$$

since $u_n = v_n + w_n$. Using Young's inequality to absorb the $\|w_n\|_{\dot{H}^{3/2}}$ term on the left-hand side yields

$$\frac{d}{dt}\|w_n\|_{\dot{H}^{1/2}}^2 + \|w_n\|_{\dot{H}^{3/2}}^2 \le c_1\|w_n\|_{\dot{H}^1}^4 + c_1\|v_n\|_{\dot{H}^1}^4;$$

we now use the interpolation inequality $\|w_n\|_{\dot{H}^1}^2 \le \|w_n\|_{\dot{H}^{1/2}}\|w_n\|_{\dot{H}^{3/2}}$ (from (10.4)) to end up with

$$\frac{d}{dt}\|w_n\|_{\dot{H}^{1/2}}^2 + \|w_n\|_{\dot{H}^{3/2}}^2 \le c_1\|w_n\|_{\dot{H}^{1/2}}^2\|w_n\|_{\dot{H}^{3/2}}^2 + c_1\|v_n\|_{\dot{H}^1}^4.$$

This is a differential inequality of the form

$$\frac{dx}{dt} + y \le cxy + \delta(t), \qquad \text{with} \qquad x(0) = 0, \qquad (10.17)$$

where

$$x = \|w_n\|_{\dot{H}^{1/2}}^2, \quad y = \|w_n\|_{\dot{H}^{3/2}}^2, \quad c = c_1, \quad \text{and} \quad \delta = c_1\|v_n\|_{\dot{H}^1}^4.$$

Lemma 10.3 shows that if $D := \int_0^T \delta(s)\,ds < 1/4c$ then (10.17) implies that

$$\sup_{0 \le s < T} x(s) \le 2D \qquad \text{and} \qquad \int_0^T y(s)\,ds \le 2D.$$

As a consequence, provided that

$$c_1 \int_0^T \|v_n\|_{\dot{H}^1}^4\,ds < \frac{1}{4c_1} \tag{10.18}$$

we obtain the estimates

$$\sup_{0 \le s \le T} \|w_n(s)\|_{\dot{H}^{1/2}}^2 \le \frac{1}{2c_1} \qquad \text{and} \qquad \int_0^T \|w_n(s)\|_{\dot{H}^{3/2}}^2\,ds \le \frac{1}{2c_1}. \tag{10.19}$$

Since $v_n = P_n v$ we have

$$\int_0^T \|v_n(s)\|_{\dot{H}^1}^4\,ds \le \int_0^T \|v(s)\|_{\dot{H}^1}^4\,ds; \tag{10.20}$$

the estimate in (10.13) shows that $v \in L^\infty(0,T;\dot{H}^{1/2}) \cap L^2(0,T;\dot{H}^{3/2})$, and so, using (10.6), it follows that $v \in L^4(0,T;\dot{H}^1)$. We can therefore choose T sufficiently small that

$$c_1 \int_0^T \|v(s)\|_{\dot{H}^1}^4\,ds < \frac{1}{4c_1},$$

and then (10.20) ensures that (10.18) holds for every n. It follows that for this choice of T we can obtain the estimates in (10.19) uniformly in n.

By combining the estimates in (10.14), (10.15), and (10.19), we obtain a uniform bound in $L^\infty(0,T;\dot{H}^{1/2}) \cap L^2(0,T;\dot{H}^{3/2})$ for $u_n = w_n + v_n$. We take further subsequences (which we once again relabel) so that u_n converges weakly-$*$ in $L^\infty(0,T;\dot{H}^{1/2})$ and weakly in $L^2(0,T;\dot{H}^{3/2})$. These bounds are then preserved in the limit as $n \to \infty$, so that the limit (which we already know is a weak solution) also satisfies

$$u \in L^\infty(0,T;\dot{H}^{1/2}) \cap L^2(0,T;\dot{H}^{3/2}).$$

It now follows from (10.7) that $u \in L^4(0,T;L^6)$, and so u is smooth on $(0,T]$ and is unique in the class of Leray–Hopf weak solutions (Theorems 8.17 and 8.19). $\qquad\square$

It is simple to use this result to prove local existence for all initial data in $\dot{H}^{1/2}$, and global existence for initial data whose $\dot{H}^{1/2}$ norm is small.

Corollary 10.2 *Suppose that $u_0 \in H \cap \dot{H}^{1/2}(\mathbb{T}^3)$. Then*

(i) *there exists a time $T = T(u_0) > 0$ such that the Navier–Stokes equations have a solution*

$$u \in L^\infty(0, T; \dot{H}^{1/2}) \cap L^2(0, T; \dot{H}^{3/2}) \qquad (10.21)$$

that is unique in the class of Leray–Hopf weak solutions and is smooth on $(0, T]$;

(ii) *there exists an absolute constant $\varepsilon_{1/2}$ (which does not depend on u_0) such that if*

$$\|u_0\|_{\dot{H}^{1/2}} < \varepsilon_{1/2}$$

then the solution satisfies (10.21) for every $T > 0$ and hence is unique in the class of Leray–Hopf weak solutions and smooth for all $t > 0$.

Proof   We have shown in (10.13) that the solution $v(t)$ of $\partial_t v + Av = 0$ with $v(0) = u_0$ satisfies

$$\frac{1}{2}\|v(t)\|_{\dot{H}^{1/2}}^2 + \int_0^t \|v(s)\|_{\dot{H}^{3/2}}^2 \, ds \le \frac{1}{2}\|u_0\|_{\dot{H}^{1/2}}^2. \qquad (10.22)$$

Using the Sobolev interpolation $\|v(t)\|_{\dot{H}^1}^4 \le \|v(t)\|_{\dot{H}^{1/2}}^2 \|v(t)\|_{\dot{H}^{3/2}}^2$ we therefore have

$$\int_0^T \|v(t)\|_{\dot{H}^1}^4 \, dt \le \left(\sup_{0 \le t \le T} \|v(t)\|_{\dot{H}^{1/2}}^2 \right) \int_0^T \|v(t)\|_{\dot{H}^{3/2}}^2 \, dt$$

$$\le \|u_0\|_{\dot{H}^{1/2}}^2 \int_0^T \|v(t)\|_{\dot{H}^{3/2}}^2 \, dt.$$

For part (i) note that since $\|v(t)\|_{\dot{H}^{3/2}}^2$ is integrable we can always choose T sufficiently small that the right-hand side of the above inequality is less than ε_h, as required by condition (10.8) in Theorem 10.1. For part (ii) it follows from (10.22) that $\int_0^T \|v(t)\|_{\dot{H}^{3/2}}^2 \, dt \le \|u_0\|_{\dot{H}^{1/2}}^2$ for all $T > 0$, and so in fact

$$\int_0^T \|v(t)\|_{\dot{H}^1}^4 \, dt \le \|u_0\|_{\dot{H}^{1/2}}^4$$

and (10.8) can be satisfied for any $T > 0$ if $\|u_0\|_{\dot{H}^{1/2}} < \varepsilon_{1/2} := \varepsilon_h^{1/4}$. $\square$

10.4 An auxiliary ODE lemma

Finally we give a proof of the ODE lemma that was used to complete the proof of Theorem 10.1, and which we will also use in the proof of Theorem 11.3 in the next chapter. Note that the assumption that $x(0) = 0$ is crucial.

Lemma 10.3 *Suppose that x, y, δ are real-valued, non-negative functions that are continuous on $[0, T)$, x is differentiable, and for some $c > 0$*

$$\frac{dx}{dt} + y \leq cxy + \delta(t), \quad t \in [0, T), \quad and \quad x(0) = 0.$$

If $D := \int_0^T \delta(s)\,ds < 1/4c$ then

$$\sup_{0 \leq s \leq T} x(s) \leq 2D < \frac{1}{2c} \quad and \quad \int_0^T y(s)\,ds \leq 2D < \frac{1}{2c}.$$

Proof   If $D = 0$ then the assertion follows easily from Gronwall's Lemma. So let us assume that $D > 0$ and let

$$Y(t) = \int_0^t y(s)\,ds.$$

The function Y is continuous and non-decreasing with $Y(0) = 0$. To prove that $Y(t) \leq 2D$ for all $t \in [0, T]$ it suffices to show that for all $t \in [0, T)$ we have

$$Y(s) \leq 2D \quad \text{for} \quad s \in [0, t] \quad \Rightarrow \quad Y(t) < 2D.$$

This rules out the possibility that $Y(t) = 2D$ for some $t \in [0, T)$, which implies that $Y(t) \leq 2D$ for all $t \in [0, T]$ since Y is continuous.

So let us fix $t \in [0, T)$, assume that $Y(t) \leq 2D$, and let $s \in [0, t]$. Integrating the differential inequality from 0 to s yields

$$x(s) + \int_0^s y(r)\,dr \leq c\left\{\sup_{0 \leq r \leq s} x(r)\right\}\int_0^s y(r)\,dr + \int_0^s \delta(r)\,dr.$$

So for all $s \in [0, t]$ we have

$$x(s) + Y(s) \leq c\left\{\sup_{0 \leq r \leq s} x(r)\right\}Y(s) + D. \tag{10.23}$$

Since x is continuous on $[0, t]$ there is an $s_0 \in [0, t]$ such that

$$\sup_{0 \leq r \leq t} x(r) = x(s_0).$$

If $Y(s_0) = 0$ then it follows from (10.23) with $s = s_0$ that

$$x(s_0) \leq D < 2D.$$

If $Y(s_0) > 0$ then we can use our assumption: $Y(s_0) \leq Y(t) \leq 2D < \frac{1}{2c}$ to deduce from (10.23) that

$$x(s_0) + Y(s_0) < c\{x(s_0)\}\frac{1}{2c} + D,$$

and so

$$x(s_0) < 2D - 2Y(s_0) < 2D.$$

Using the assumption $Y(t) \leq 2D < 1/2c$ in (10.23) with $s = t$ yields

$$x(t) + Y(t) \leq \frac{1}{2} \sup_{0 \leq r \leq t} x(r) + D,$$

so from the inequality $x(s_0) < 2D$ it follows that $Y(t) < 2D$. Hence $Y(t) < 2D$ for all $t \in [0, T)$ and the bound on $x(t)$ follows from (10.23). $\qquad \square$

Notes

The local existence of solutions in $H^{1/2}$, and global existence for small data, was first proved by Fujita & Kato (1964) for the Dirichlet problem in a bounded domain, using the semigroup approach (which they develop in their paper): in fact they treated initial data in the space $D(A^{1/4})$, where A is the Stokes operator on Ω. They showed that the space

$$\{u \in C^0([0, T]; H) : t^{1/4}\|\nabla u(t)\| \to 0 \text{ as } t \to 0\}$$

is preserved by the mapping

$$u \mapsto e^{-At}u_0 + \int_0^t e^{-A(t-s)}\mathbb{P}[(u(s) \cdot \nabla)u(s)]\,ds \qquad (10.24)$$

when T is sufficiently small and constructed a solution by successive approximations, i.e. repeated applications of this map. In (10.24) the notation $e^{-At}u_0$ denotes the solution at time t of the Stokes equation $\partial_t u + Au = 0$ with initial data u_0. (See also the Notes at the end of Chapter 6.)

Chemin (1992) proved global existence for small data in $\dot{H}^{1/2}(\mathbb{R}^3)$ using energy estimates but did not address the issue of local existence for large data. One can easily see (at least formally) that small $\dot{H}^{1/2}$ data gives rise to a global solution that is bounded in $\dot{H}^{1/2}$ using the estimate

$$\frac{1}{2}\frac{d}{dt}\|u\|_{\dot{H}^{1/2}}^2 + \|u\|_{\dot{H}^{3/2}}^2 \leq c\|u\|_{\dot{H}^1}^2\|u\|_{\dot{H}^{3/2}} \leq c\|u\|_{\dot{H}^{1/2}}\|u\|_{\dot{H}^{3/2}}^2$$

(see (10.16) for the estimate on the nonlinear term). It follows that if the initial condition u_0 satisfies $\|u_0\|_{\dot{H}^{1/2}} < 1/c$ then $\|u(t)\|_{\dot{H}^{1/2}}$ is non-increasing and therefore stays bounded for all time.

The argument used in this chapter follows Marín-Rubio, Robinson, & Sadowski (2013) and offers a simplified version of that in Chemin et al. (2006); the splitting method, which in the context of the Navier–Stokes equations goes as far back as Leray (1934), is also used by Calderón (1990). Both Marín-Rubio, Robinson, & Sadowski (2013) and Chemin et al. (2006) also contain an $\dot{H}^{1/2}$ version of the 'robustness of regularity' result that we proved in Chapter 9 in the case of H^1 initial data.

Proofs of all the results in this chapter for the case of the whole space can be carried out in a very similar way to the analysis for $\mathbb{T}^3$ given here, replacing P_n by the Fourier truncation operator S_n, defined by setting[2]

$$\widehat{S_n f}(\xi) = \chi_{|\xi| \le n} \hat{f}(\xi),$$

and considering the solution of

$$\partial_t u_n + A u_n + S_n[(u_n \cdot \nabla) u_n] = 0 \qquad u_n(0) = S_n u_0.$$

The key point is that once again the solutions of the heat part of the solution can be bounded in a way that does not depend on n. The fact that the resulting convergence of u_n to u is only valid on compact subsets of $\mathbb{R}^3$ can be overcome in a similar way as in the proof of existence of weak solutions on the whole space (see Theorem 4.10). The details are given by Chemin et al. (2006).

The existence of solutions with initial data in critical spaces (based largely on an analysis of the mapping in (10.24)) has been the subject of much research. The review article by Cannone (2003) contains a good overview (see also Cannone, 1995, and many of the results are given in the book by Lemarié-Rieusset, 2002). The largest critical space in which global existence for small data has been obtained (by Koch & Tataru, 2001) is the space BMO^{-1}, where $u_0 \in \mathrm{BMO}^{-1}$ if it can be written as $\sum_i \partial_i f_i$, where each $f_i \in \mathrm{BMO}$ (functions of bounded mean oscillation, see Stein, 1993); equivalently $u_0 \in \mathrm{BMO}^{-1}$ if the solution v of the heat equation with initial data $v(0) = u_0$ satisfies

$$\sup_{x \in \mathbb{R}^3, \, r \ge 0} \left(|B(x, r)|^{-1} \int_{B(x,r)} \int_0^{r^2} |v|^2 \, dt \, dy \right)^{1/2} < \infty.$$

---

[2] In dimensions two and larger this truncation operator is unbounded from L^r into L^r unless $r = 2$, a remarkable fact proved by Fefferman (1971).

Koch and Tataru also obtain local existence for all data in a slightly smaller space that they term $\overline{\mathrm{VMO}}^{-1}$; this is the collection of all u_0 such that

$$\lim_{R \to 0} \sup_{x \in \mathbb{R}^3, \, 0 \le r \le R} \left(|B(x, r)|^{-1} \int_{B(x,r)} \int_0^{r^2} |v|^2 \, \mathrm{d}t \, \mathrm{d}y \right)^{1/2} = 0.$$

(VMO consists of functions of vanishing mean oscillation, see Stein, 1993.)

Exercises

10.1 Show that under the rescaling $u_\lambda(x) = \lambda u(\lambda x)$

$$\|u_\lambda\|_{\dot{H}^s(\mathbb{R}^3)} = \lambda^{s-1/2} \|u\|_{\dot{H}^s(\mathbb{R}^3)}$$

and

$$\|u_\lambda\|_{L^p(\mathbb{R}^3)} = \lambda^{1-(3/p)} \|u\|_{L^p(\mathbb{R}^3)}.$$

10.2 Show that for any $\lambda \in \mathbb{N}$ and any $p > 0$,

$$\int_{\mathbb{T}^3} |f(\lambda x)|^p \, \mathrm{d}x = \int_{\mathbb{T}^3} |f(x)|^p \, \mathrm{d}x.$$

10.3 If

$$u_0 = \sum_{k \in \mathbb{Z}^3} \hat{u}_k e^{ik \cdot x}$$

then the solution $S(t)u_0$ of the heat equation $\partial_t v - \Delta v = 0$ satisfying $v(0) = u_0$ is given by

$$e^{\Delta t} u_0 = \sum_{k \in \mathbb{Z}^3} e^{-|k|^2 t} \hat{u}_k e^{ik \cdot x}.$$

Assuming that $u_0 \in \dot{H}^{1/2}$ evaluate

$$\int_0^T \|e^{\Delta s} u_0\|_{\dot{H}^{3/2}}^2 \, \mathrm{d}s$$

in terms of the Fourier coefficients of u_0, and hence show that there are u_n with $\|u_n\|_{\dot{H}^{1/2}} = 1$ such that

$$\int_0^{T_n} \|e^{\Delta s} u_n\|_{\dot{H}^{3/2}}^2 \, \mathrm{d}s = \varepsilon_0,$$

with $T_n \to 0$, provided that $\varepsilon_0 < 1/2$.

10.4 Repeat the analysis of the previous exercise when $u_0 \in \dot{H}^1$, and show that for any $\varepsilon_0 > 0$ and $T > 0$ there exists a corresponding δ_0 such that

$$\|u_0\|_{\dot{H}^1} \leq \delta_0 \qquad \Rightarrow \qquad \int_0^T \|e^{\Delta s} u_0\|_{\dot{H}^{3/2}}^2 \, ds \leq \varepsilon_0.$$

10.5 Show that if u is a Leray–Hopf weak solution obtained as the limit of a sequence of Galerkin approximations and $u \in L^2(0, T; H^{3/2})$ then we have $u \in L^\infty(0, T; H^{1/2})$.

10.6 Show that for a given initial condition $u_0 \in \dot{H}^{1/2}$ there can be only one weak solution with the regularity $u \in L^\infty(0, T; \dot{H}^{1/2}) \cap L^2(0, T; \dot{H}^{3/2})$.

11

Local existence and uniqueness in L^3

In the previous chapter we gave a proof of local existence in the critical Sobolev space $\dot{H}^{1/2}$. In this chapter we give a similar proof (in the absence of boundaries) of the local existence of solutions with initial data in the space L^3, which is another critical space. Due to the embedding $\dot{H}^{1/2} \subset L^3$, the space L^3 is strictly larger than $\dot{H}^{1/2}$. It is simple to check that on the whole space the L^3 norm is invariant under the scaling transformation $u(x) \mapsto u_\lambda(x) := \lambda u(\lambda x)$:

$$\int_{\mathbb{R}^3} |u_\lambda(x)|^3 \, dx = \lambda^3 \int_{\mathbb{R}^3} |u(\lambda x)|^3 \, dx = \int_{\mathbb{R}^3} |u(y)|^3 \, dy.$$

We give a relatively elementary proof of local existence in L^3, following Robinson & Sadowski (2014), using techniques similar to those employed by Beirão da Veiga (1987). We use the splitting approach that we employed for $u_0 \in \dot{H}^{1/2}$, along with the bound on the pressure

$$\|p\|_{L^r} \leq c_r \|u\|_{L^{2r}}^2, \qquad 1 < r < \infty, \tag{11.1}$$

obtained in Chapter 5. Such a local existence result was stated by Leray (1934) without proof; the first detailed proof was due to Giga (1986) who used the theory of semigroups.

As in the case of $\dot{H}^{1/2}$, the condition we obtain for local existence will be expressed in terms of the solution $v(t)$ of the heat equation with initial condition u_0; it ensures local existence and uniqueness for all $u_0 \in L^3$ and global existence (and uniqueness) for initial data whose L^3 norm is sufficiently small.

Since the proof we present is valid with only minor variations for both $\mathbb{T}^3$ and $\mathbb{R}^3$ we drop the spatial specification from our spaces, i.e. L^3 could denote $L^3(\mathbb{T}^3)$ or $L^3(\mathbb{R}^3)$; we make the distinction only when the two cases have to be treated differently.

11.1 Preliminaries

We will use throughout the algebraic identity, valid for any $\gamma \in \mathbb{R}$,

$$\partial(|u|^\gamma) = \partial(u_k u_k)^{\gamma/2} = \gamma \, u_k(\partial u_k)(u_k u_k)^{\gamma/2-1}$$
$$= \gamma \, u_k(\partial u_k)|u|^{\gamma-2} = \gamma \, (u \cdot \partial u)|u|^{\gamma-2}, \qquad (11.2)$$

where ∂ denotes any partial derivative.

We want to follow the time evolution of $\|u\|_{L^3}^3$, and since

$$\int \partial_t u \cdot u|u| = \frac{1}{3} \frac{d}{dt} \int |u|^3 \qquad (11.3)$$

we will take the inner product of the equation with $u|u|$ and integrate. The following lemma shows that the quantity $\int |u||\nabla u|^2$ will arise naturally in our energy estimates. (All spatial integrals in this chapter are taken over $\mathbb{T}^3$ or $\mathbb{R}^3$ unless noted otherwise, and often we will omit the integration variable dx.)

Lemma 11.1 *If $u \in C^\infty(\mathbb{T}^3)$ or $C_c^\infty(\mathbb{R}^3)$ then*

$$\int -\Delta u \cdot u|u| \geq \int |\nabla u|^2 |u|. \qquad (11.4)$$

For a generalisation suited to an analysis of the L^α norm for $\alpha > 3$ see Exercise 11.1.

Proof Integrating by parts and using (11.2) we obtain

$$-\int (\partial_i \partial_i u_j) u_j |u| = \int (\partial_i u_j)(\partial_i u_j)|u| + \int (\partial_i u_j) u_j u_k (\partial_i u_k)|u|^{-1}$$

$$= \int |\nabla u|^2 |u| + \int \left[u_j(\partial_i u_j)|u|^{-1/2} \right] \left[u_k(\partial_i u_k)|u|^{-1/2} \right]$$

$$= \int |\nabla u|^2 |u| + \int \left[\frac{2}{3} \partial_i (|u|^{3/2}) \right] \left[\frac{2}{3} \partial_i (|u|^{3/2}) \right]$$

$$= \int |\nabla u|^2 |u| + \frac{4}{9} \int |\nabla |u|^{3/2}|^2;$$

since the last term is positive we can drop it to obtain (11.4). $\qquad \square$

We now show that bounds on $\int |\nabla u|^2 |u|$ provide bounds on $\|u\|_{L^9}$, which will be useful later. For a more general variant see Exercise 11.2.

Lemma 11.2 *There exists a constant $c > 0$ such that*

$$\|u\|_{L^9}^3 \leq c \int_{\mathbb{R}^3} |\nabla u|^2 |u| \qquad (11.5)$$

for any $u \in W^{1,9/4}(\mathbb{R}^3)$. The same is valid for functions on the torus provided that $\int_{\mathbb{T}^3} u = 0$.

We take $u \in W^{1,9/4}$ to ensure that the expression on the right-hand side of (11.5) is finite, since

$$\int |u||\nabla u|^2 \leq \left(\int |u|^9\right)^{1/9} \left(\int |\nabla u|^{9/4}\right)^{8/9};$$

the first factor is finite since $u \in L^9$ by the embedding $W^{1,9/4} \subset L^9$ (see Theorem 1.7) and the second factor is finite since $u \in W^{1,9/4}$. However, we obtain $u \in L^9$ whenever the right-hand side of (11.5) is finite (and not only for $u \in W^{1,9/4}$).

Proof We give the proof of (11.5) on $\mathbb{R}^3$, which is simple, deferring the proof on $\mathbb{T}^3$ to the end of the chapter. First take $u \in C_c^\infty(\mathbb{R}^3)$. Certainly $|u|^{3/2} \in \dot{H}^1(\mathbb{R}^3)$, and from the embedding $\dot{H}^1(\mathbb{R}^3) \subset L^6(\mathbb{R}^3)$ we have

$$\|u\|_{L^9}^3 = \left(\int |u|^9\right)^{1/3} = \||u|^{3/2}\|_{L^6}^2 \leq c \int \left|\nabla |u|^{3/2}\right|^2.$$

On the other hand, from (11.2) it follows that

$$\left|\partial_k |u|^{3/2}\right|^2 = \left|\frac{3}{2} u_i |u|^{-1/2} \partial_k u_i\right|^2 \leq \frac{9}{4} |u||\nabla u|^2,$$

so

$$\left|\nabla |u|^{3/2}\right|^2 \leq c|u||\nabla u|^2$$

and (11.5) follows. For $u \in W^{1,9/4}$ we use the density of $C_c^\infty(\mathbb{R}^3)$ in $W^{1,9/4}(\mathbb{R}^3)$.

However, when $\Omega = \mathbb{T}^3$ and $\int_{\mathbb{T}^3} u = 0$ the function $|u|^{3/2}$ does not have zero average, so we cannot apply the embedding $\dot{H}^1(\Omega) \subset L^6(\Omega)$ and we have to argue more carefully. This is done using a proof by contradiction, as in the usual proof of the standard Poincaré inequality for functions with zero average (see Theorem 1 in Section 5.8 in Evans, 1998, for example). For details see Lemma 11.5 at the end of the chapter. $\qquad\square$

11.2 Local existence in L^3

We will use the same method that we used in the last chapter to study the equations with initial data in the critical Sobolev space $\dot{H}^{1/2}$: we split the solution into one part that satisfies the heat equation and deals with the initial data, and another that contains the nonlinear term but has zero initial data.

We first show that if $u_0 \in H^2$ then the solution is smooth on $(0, T)$ for some time T that depends only on the L^3 norm of u_0 and properties of $v(t)$, the solution of (11.7). We then use an approximation argument to prove a local existence result for data in L^3. To show that the solution remains smooth we will show that u must satisfy the Serrin condition $u \in L^3(0, T; L^9)$, which by Theorems 8.17 and 8.19 guarantees that u is smooth and unique on $(0, T]$.

Theorem 11.3 *There is an absolute constant $\varepsilon_3 > 0$ such that if $u_0 \in H^2$ with $\nabla \cdot u_0 = 0$, and for some $T > 0$*

$$\|u_0\|_{L^3}^3 \int_0^T \int |\nabla v(t)|^2 |v(t)| \, dx \, dt < \varepsilon_3, \tag{11.6}$$

where $v(t) = e^{\Delta t} u_0$, then the solution u of the Navier–Stokes equations with initial condition u_0 is unique and smooth on $\Omega \times (0, T]$.

Proof Since $u_0 \in H^2$, by Theorem 7.1 there exists a time $T' > 0$ such that the Navier–Stokes equations have a solution $u \in L^\infty(0, T'; H^2)$, which is smooth on $(0, T']$. We set

$$T^* = \sup\{T' : u \in L^\infty(0, T'; H^2)\},$$

so that u is smooth on $(0, T^*)$ and $\|u(t)\|_{H^2} \to \infty$ as $t \to T^*$. While u is smooth we can calculate with the equations rigorously rather than having to use the weak formulation. We want to prove that if ε_3 in (11.6) is sufficiently small then $T^* \geq T$, where T comes from (11.6). We do this by showing that if $T^* < T$ then $u \in L^3(0, T^*; L^9)$; the global Serrin condition of Theorem 8.17 then guarantees that u must be smooth on $(0, T^*]$. In particular $u(T^*) \in H^2$ and the solution can be extended past T^*, a contradiction.

We use the splitting method from Chapter 10: we let $u = v + w$ where v and w are given by[1]

$$\partial_t v - \Delta v = 0 \qquad v(0) = u_0 \tag{11.7}$$

and

$$\partial_t w - \Delta w + ((v + w) \cdot \nabla)(v + w) + \nabla p = 0 \qquad w(0) = 0.$$

We want to prove that $u \in L^3(0, T^*; L^9)$. If we take the dot product of the v equation with $v|v|$, integrate in space, and use (11.3) along with Lemma 11.1 we obtain

$$\frac{1}{3} \frac{d}{dt} \|v\|_{L^3}^3 + \int |\nabla v|^2 |v| \leq 0.$$

---

[1] Recall that for divergence-free u_0 the solution of (11.7) in the absence of boundaries coincides with the solution of the Stokes system

$$\partial_t v - \Delta v + \nabla q = 0, \quad \nabla \cdot v = 0, \quad v(0) = u_0.$$

Integrating in time this yields

$$\|v(t)\|_{L^3}^3 + 3 \int_0^t \int |\nabla v|^2 |v| \leq \|u_0\|_{L^3}^3. \qquad (11.8)$$

(We note for later use that this inequality does not require that $u_0 \in H^2$, only $u_0 \in L^3$.) Therefore according to Lemma 11.2 (which we can use to guarantee that $\|v\|_{L^9}^3 \leq c \int |\nabla v|^2 |v|$) we will have $v \in L^3(0, T^*; L^9)$. Since $u = v + w$ it will therefore be enough to show that $w \in L^3(0, T^*; L^9)$.

We multiply the w equation by $w|w|$ and integrate in space. Using (11.4) from Lemma 11.1 for the linear term we obtain

$$\frac{1}{3}\frac{d}{dt}\|w\|_{L^3}^3 + \int |\nabla w|^2 |w| \leq -\int [(u \cdot \nabla)u] \cdot w|w| - \int \nabla p \cdot w|w|.$$

We bound the two terms on the right-hand side separately. First

$$-\int [(u \cdot \nabla)u] \cdot w|w| = -\int (u_i \partial_i u_j) w_j |w| = \int u_i u_j \partial_i (w_j |w|)$$

after an integration by parts, using the fact that u is divergence free. Now, using (11.2) this is equal to

$$\int u_i u_j (\partial_i w_j)|w| + \int u_i u_j w_j w_k (\partial_i w_k)|w|^{-1}$$

$$\leq 2\int |u|^2 |\nabla w||w|$$

$$\leq 4\int |w|^3 |\nabla w| + 4\int |\nabla w||w||v|^2,$$

since $|u|^2 \leq 2|w|^2 + 2|v|^2$. For the term involving the pressure we have

$$-\int \nabla p \cdot w|w| = \int p w_j \partial_j(|w|) = \int p w_j w_k (\partial_j w_k)|w|^{-1} \leq \int |p||w||\nabla w|.$$

We now estimate each of the resulting terms on the right-hand side,

$$4\int |w|^3 |\nabla w| + 4\int |\nabla w||w||v|^2 + \int |p||w||\nabla w| =: R_1 + R_2 + R_3,$$

using the notation

$$I(u) := \int |\nabla u|^2 |u|. \qquad (11.9)$$

To bound these three terms we will use Lebesgue interpolation and the Hölder inequality, the inequality from Lemma 11.2, and the pressure estimates from the previous chapter.

For R_1 we use the Lebesgue interpolation $\|w\|_{L^5} \leq \|w\|_{L^3}^{2/5} \|w\|_{L^9}^{3/5}$:

$$R_1 = 4 \int |w|^3 |\nabla w| \leq 4 \left(\int |w||\nabla w|^2 \right)^{1/2} \left(\int |w|^5 \right)^{1/2}$$
$$\leq 4I(w)^{1/2} \|w\|_{L^3} \|w\|_{L^9}^{3/2}$$
$$\leq cI(w)\|w\|_{L^3},$$

using (11.5) in the form $\|w\|_{L^9}^3 \leq cI(w)$. For R_2 we also use Hölder's inequality with exponents $(2, 18, 9/4)$, along with the Lebesgue interpolation inequality $\|v\|_{L^{9/2}} \leq \|v\|_{L^3}^{1/2} \|v\|_{L^9}^{1/2}$:

$$R_2 = 4 \int |\nabla w||w||v|^2$$
$$\leq 4 \left(\int |w||\nabla w|^2 \right)^{1/2} \left(\int |w|^9 \right)^{1/18} \left(\int |v|^{9/2} \right)^{4/9}$$
$$\leq cI(w)^{2/3} \|v\|_{L^{9/2}}^2$$
$$\leq \frac{1}{4}I(w) + c\|v\|_{L^{9/2}}^6$$
$$\leq \frac{1}{4}I(w) + c\|v\|_{L^3}^3 \|v\|_{L^9}^3$$
$$\leq \frac{1}{4}I(w) + c\|v\|_{L^3}^3 I(v).$$

(Recall that we use c to denote a numerical constant, and that its exact value can change from line to line.) Finally for R_3 we use the estimate $\|p\|_{L^{9/4}} \leq c\|u\|_{L^{9/2}}^2$ (see (11.1)):

$$R_3 = \int |p||w||\nabla w|$$
$$\leq \left(\int |w||\nabla w|^2 \right)^{1/2} \left(\int |w|^9 \right)^{1/18} \left(\int |p|^{9/4} \right)^{4/9}$$
$$\leq cI(w)^{2/3} \|u\|_{L^{9/2}}^2$$
$$\leq \frac{1}{4}I(w) + c\|u\|_{L^{9/2}}^6$$
$$\leq \frac{1}{4}I(w) + c\|v\|_{L^3}^3 \|v\|_{L^9}^3 + c\|w\|_{L^3}^3 \|w\|_{L^9}^3$$
$$\leq \frac{1}{4}I(w) + c\|v\|_{L^3}^3 I(v) + c\|w\|_{L^3}^3 I(w),$$

where we have used the fact that $u = v + w$.

Combining these estimates leads to the inequality

$$\frac{1}{3}\frac{\mathrm{d}}{\mathrm{d}t}\|w\|_{L^3}^3 + \frac{1}{2}I(w) \le cI(w)\|w\|_{L^3} + c\|v\|_{L^3}^3 I(v) + c\|w\|_{L^3}^3 I(w).$$

Since

$$cI(w)\|w\|_{L^3} = cI(w)^{2/3}I(w)^{1/3}\|w\|_{L^3} \le \frac{1}{6}I(w) + cI(w)\|w\|_{L^3}^3,$$

we obtain

$$\frac{\mathrm{d}}{\mathrm{d}t}\|w\|_{L^3}^3 + I(w) \le c_*\|w\|_{L^3}^3 I(w) + c_*\|v\|_{L^3}^3 I(v),$$

for some particular numerical value c_*.

This is of the form required by the ODE result of Lemma 10.3:

$$\frac{\mathrm{d}x}{\mathrm{d}t} + y \le cxy + \delta(t), \qquad \text{with} \qquad x(0) = 0,$$

where

$$x = \|w\|_{L^3}^3, \quad y = I(w), \quad c = c_*, \quad \text{and} \quad \delta = c_*\|v\|_{L^3}^3 I(v).$$

Applying Lemma 10.3 we can deduce, since $T^* < T$ by assumption, that if

$$c_* \int_0^T \|v(s)\|_{L^3}^3 I(v(s))\,\mathrm{d}s < 1/4c_* \tag{11.10}$$

then

$$\sup_{0 \le s \le T^*} \|w(s)\|_{L^3}^3 + \int_0^{T^*}\int_{\mathbb{R}^3} |\nabla w(x,s)|^2|w(x,s)|\,\mathrm{d}x\,\mathrm{d}s \le \frac{1}{c_*}.$$

$$\tag{11.11}$$

Therefore $w \in L^3(0, T^*; L^9)$. Since we have shown that $v \in L^3(0, T^*; L^9)$ it follows that $u = v + w \in L^3(0, T^*; L^9)$ satisfies the Serrin condition of Theorem 8.17 and so is smooth on $(0, T^*]$, and we obtain the required contradiction. □

We will now relax our assumption on u_0 to obtain smooth (finite-energy) solutions for[2] $u_0 \in L^3 \cap L^2$. As with the similar results for $u_0 \in H^{1/2}$ in Chapter 10, without the fact that these solutions are smooth and unique, the implication '$u_0 \in L^3 \implies u \in L^\infty(0, T; L^3)$' could be thought of as a mere mathematical curiosity.

The following theorem is valid on $\mathbb{T}^3$ and $\mathbb{R}^3$.

---

[2] Of course, on $\mathbb{T}^3$ the additional restriction $u_0 \in L^2$ is redundant.

Theorem 11.4 *There is an absolute constant* $\varepsilon_3 > 0$ *with the following property: if* $u_0 \in L^3 \cap L^2$ *is divergence free and* $v(t)$ *is the solution of the heat equation with initial data* u_0 *then whenever*

$$\|u_0\|_{L^3}^3 \int_0^T \int |\nabla v(t)|^2 |v(t)| \, dx \, dt < \varepsilon_3 \tag{11.12}$$

the Navier–Stokes equations have a solution $u \in L^\infty(0, T; L^3)$ *that is smooth on* $(0, T]$. *In particular:*

(i) *if* $u_0 \in L^3 \cap L^2$ *then there exists a time* $T > 0$ *such that the equations have a unique local solution* $u \in L^\infty(0, T; L^3)$ *that is smooth on* $(0, T]$ *and*
(ii) *there exists an absolute constant* $\varepsilon_3 > 0$ *such that if* $u_0 \in L^3 \cap L^2$ *with* $\|u_0\|_{L^3} < \varepsilon_3$ *then the equations have a unique global solution* u, *which remains bounded in* L^3 *for all* $t > 0$ *and is smooth for all* $t > 0$.

Proof   Given $u_0 \in L^3 \cap L^2$, let v be the solution of

$$\partial_t v - \Delta v = 0, \qquad v(0) = u_0$$

and let v_α be the solution of

$$\partial_t v_\alpha - \Delta v_\alpha = 0, \qquad v_\alpha(0) = u_{0,\alpha} := e^{\alpha \Delta} u_0,$$

where we use $e^{t\Delta} u_0$ to denote the solution at time t of the heat equation

$$\partial_t w - \Delta w = 0 \qquad w(0) = u_0.$$

Note that $v_\alpha(t) = v(t + \alpha)$.

Since $u_{0,\alpha} \in H^2$ for all $\alpha > 0$ (see Theorem D.5) we deduce from Theorem 11.3 that for each $\alpha > 0$ there exists a smooth solution u_α of the Navier–Stokes equations on some time interval $[0, T_\alpha)$ with

$$u_\alpha(0) = u_{0,\alpha},$$

where we take T_α such that

$$\|u_{0,\alpha}\|_{L^3}^3 \int_0^{T_\alpha} I(v_\alpha(t)) \, dt < \varepsilon_3,$$

with I as defined in (11.9). Choose any T' in the range $0 < T' < T$, where T is defined in (11.12). Let α be sufficiently small that $T' + \alpha < T$. Since we have $v_\alpha(t) = v(t + \alpha)$ it follows that

$$\int_0^{T'} I(v_\alpha(t)) \, dt = \int_\alpha^{T'+\alpha} I(v(t)) \, dt \le \int_0^T I(v(t)) \, dt.$$

Since

$$u_{0,\alpha} \to u_0 \quad \text{in } L^2 \text{ and in } L^3 \tag{11.13}$$

as $\alpha \to 0$ (due to properties of the heat semigroup, see Theorem D.3) we can choose α_0 such that for all $\alpha < \alpha_0$ we have

$$\|u_{0,\alpha}\|_{L^3}^3 \int_0^{T'} I(v_\alpha(t))\, dt < \varepsilon_3.$$

It follows (since (11.10) implies (11.11)) that for every $0 < T' < T$ we can find a sequence of smooth solutions of the Navier–Stokes equations defined on the time interval $[0, T')$ such that $u_\alpha(0) = u_{0,\alpha}$ and u_α are uniformly bounded in $L^3(0, T'; L^9)$.

It now follows from Theorem 8.19 that u_α is the unique Leray–Hopf weak solution with $u_\alpha(0) = u_{0,\alpha}$; in particular, for sufficiently small α we have

$$\sup_{0 \le t \le T'} \|u_\alpha(t)\|_{L^2}^2 + \int_0^{T'} \|\nabla u_\alpha(t)\|_{L^2}^2\, dt \le 2\|u_{0,\alpha}\|_{L^2}^2$$

$$\le 2\|u_0\|_{L^2}^2 + 1,$$

using (11.13). Hence u_α is uniformly bounded in $L^\infty(0, T'; L^2)$ and ∇u_α is uniformly bounded in $L^2(0, T'; L^2)$. Following a similar argument as for (4.9) we can also show that $\partial_t u_\alpha$ is uniformly bounded in $L^{4/3}(0, T'; V^*)$. Therefore we can use Theorem 4.11 to obtain a sequence of solutions bounded in $L^2(0, T'; H^1)$ that converges strongly in $L^2(0, T'; L^2(\mathbb{T}^3))$ (or, in case of the whole space $\mathbb{R}^3$ in $L^2(0, T'; L^2(\Omega'))$ for any bounded $\Omega' \subset \mathbb{R}^3$).

We now extract a further subsequence that also converges weakly in $L^3(0, T; L^9)$. Since now the limiting function u satisfies $u \in L^3(0, T'; L^9)$ it is a smooth solution of the Navier–Stokes equations defined on the time interval $(0, T')$ that satisfies the initial condition u_0 (Theorem 8.17); it is also the only such solution (Theorem 8.19). Since this holds for any T' with $0 < T' < T$ we obtain the first assertion in the statement of the theorem.

To finish the proof of the theorem (i.e. to prove (i) and (ii)), observe that for any $u_0 \in L^3$ equation (11.8),

$$\|v(t)\|_{L^3}^3 + 3\int_0^t \int |\nabla v|^2 |v| \le \|u_0\|_{L^3}^3, \tag{11.14}$$

shows that $\int |\nabla v|^2 |v|$ is integrable in time, and hence it is possible to choose T sufficiently small that

$$\|u_0\|_{L^3}^3 \int_0^T \int |\nabla v(t)|^2 |v(t)|\, dt < \varepsilon_3,$$

which yields (i). For (ii) we also use (11.14), which shows that

$$\|u_0\|_{L^3}^3 \int_0^T \int |\nabla v(t)|^2 |v(t)| \, dt \le \frac{1}{3} \|u_0\|_{L^3}^6,$$

and so the solution exists on $(0, T)$ for any $T > 0$ when $\|u_0\|_{L^3}^6 < 3\varepsilon_3$. $\qquad\square$

11.3 A proof of Lemma 11.2 on $\mathbb{T}^3$

To finish we give a proof of Lemma 11.2 in the periodic case, which is taken from Robinson & Sadowski (2014).

Lemma 11.5 *There exists a constant $c > 0$ such that if $u \in W^{1,9/4}(\mathbb{T}^3)$ with $\int_{\mathbb{T}^3} u = 0$ then*

$$\|u\|_{L^9}^3 \le c \int_{\mathbb{T}^3} |\nabla u|^2 |u|. \tag{11.15}$$

Proof If we apply the embedding $H^1(\mathbb{T}^3) \subset L^6(\mathbb{T}^3)$ to the function $|u|^{3/2}$, then

$$\|u\|_{L^9}^3 = \left\| |u|^{3/2} \right\|_{L^6}^2 \le C \left\| |u|^{3/2} \right\|_{H^1}^2 = C \left\| \nabla |u|^{3/2} \right\|_{L^2}^2 + C\|u\|_{L^3}^3$$

$$\le C \left\| |u|^{1/2} |\nabla u| \right\|_{L^2}^2 + C\|u\|_{L^3}^3,$$

and so there is a constant C such that

$$\|u\|_{L^9}^3 \le C \int_{\mathbb{T}^3} |u||\nabla u|^2 + C\|u\|_{L^3}^3. \tag{11.16}$$

If (11.15) does not hold then there exists a sequence $(u_n) \in W^{1,9/4}$, $u_n \ne 0$, such that

$$\|u_n\|_{L^9}^3 \ge n \int_{\mathbb{T}^3} |u_n||\nabla u_n|^2 \quad \text{and} \quad \int_{\mathbb{T}^3} u_n = 0.$$

Now we normalise the sequence u_n in L^3, setting $f_n = u_n/\|u_n\|_{L^3}$. Then

$$\|f_n\|_{L^3} = 1$$

and we still have

$$\|f_n\|_{L^9}^3 \ge n \int_{\mathbb{T}^3} |f_n||\nabla f_n|^2 \quad \text{and} \quad \int_{\mathbb{T}^3} f_n = 0. \tag{11.17}$$

Using (11.16) it follows that

$$\|f_n\|_{L^9}^3 \le C \int_{\mathbb{T}^3} |f_n||\nabla f_n|^2 + C\|f_n\|_{L^3}^3$$

$$\le \frac{C}{n} \|f_n\|_{L^9}^3 + C,$$

from which it follows that for all $n > C$

$$\|f_n\|_{L^9}^3 \le Cn/(n - C),$$

i.e. f_n is uniformly bounded in L^9, $\|f_n\|_{L^9} \le M$ for all n (for some $M > 0$). It now follows from (11.2) and (11.17) that

$$\int_{\mathbb{T}^3} |\nabla(f_n|f_n|^{1/2})|^2 \le c \int_{\mathbb{T}^3} |f_n||\nabla f_n|^2 \le \frac{c}{n}\|f_n\|_{L^9}^3 \le \frac{cM}{n} \to 0$$

as $n \to \infty$. In particular, $\|\nabla(f_n|f_n|^{1/2})\|_{L^2} \to 0$ as $n \to \infty$.

Notice also that $f_n|f_n|^{1/2} \in L^2$ with $\|f_n|f_n|^{1/2}\|_{L^2} = \|f_n\|_{L^3}^{3/2} = 1$ for every n. It follows that $g_n := f_n|f_n|^{1/2}$ forms a bounded sequence in H^1 for which $\|\nabla g_n\|_{L^2} \to 0$ as $n \to \infty$. One can now follow the proof of Theorem 1 in Section 5.8 in Evans (1998) to conclude that there is a subsequence such that $g_{n_j} = f_{n_j}|f_{n_j}|^{1/2} \to g$ strongly in L^2, where $\|\nabla g\| = 0$ and so g is a constant. It follows that

$$\|g\|_{L^2} = \lim_{j \to \infty} \|f_{n_j}|f_{n_j}|^{1/2}\|_{L^2} = 1.$$

Now, we find a further subsequence such that $f_{n_j}|f_{n_j}|^{1/2} \to g$ almost everywhere, which implies that $|f_{n_j}|^{3/2} \to |g|$ almost everywhere. It follows that $|f_{n_j}|^{1/2} \to |g|^{1/3}$ a.e., and so $f_{n_j} \to f := g|g|^{-1/3}$ almost everywhere. We also know that $\|f_{n_j}\|_{L^3} = 1$, and $\|f\|_{L^3} = 1$. It follows that $f_{n_j} \to f$ strongly in L^3. The zero average condition is preserved by the convergence, which implies that $f = 0$. But this contradicts the fact that $\|f\|_{L^3} = 1$, and the result follows. $\square$

Notes

The first complete proof of local existence in L^3 was given by Giga (1986), who used semigroup techniques, i.e. an analysis of the mapping

$$u \mapsto e^{-At}u_0 + \int_0^t e^{-A(t-s)}\mathbb{P}[(u(s) \cdot \nabla)u(s)]\,ds$$

discussed in the Notes to Chapter 10 (equation (10.24)). He also considered local existence in L^p for $p > 3$.

A similar approach to that used here (i.e. 'energy estimates') was used by Beirão da Veiga (1987), who considered the behaviour of L^p norms, $p > 3$, for the equations on $\mathbb{R}^3$, using a whole space version of Lemma 11.2 adapted for such Lebesgue spaces; see also Beirão da Veiga & Secchi (1987) and Exercise 11.4. The proof of local existence in L^3 presented here is due to Robinson & Sadowski (2014); the validity of such a proof in the periodic case relies on Lemma 11.5 and the possibility of using a periodic variant of the Calderón–Zygmund Theorem in order to estimate the pressure (as discussed in Appendix B).

Giga (1986) and Robinson & Sadowski (2014) also obtain lower bounds in L^p on solutions under the assumption that the first blowup time is $T > 0$, namely

$$\|u(T - t)\|_{L^p} \geq ct^{-(p-3)/2p}, \tag{11.18}$$

where c is an absolute constant that depends on p but not on the initial data, see Exercise 11.8. (Such lower bounds were stated by Leray without proof; they also follow immediately from the analysis of Beirão da Veiga, 1987, although they are not stated there.)

Exercises

11.1 Show that if $u \in C_c^\infty(\mathbb{R}^3)$ and $\alpha \geq 2$ then

$$\int -\Delta u \cdot u|u|^{\alpha-2} \geq \int |\nabla u|^2 |u|^{\alpha-2}.$$

11.2 Take α with $2 \leq \alpha < \infty$. Show that there exists a constant c_α such that if $u \in W^{1,3\alpha/(1+\alpha)}(\mathbb{R}^3)$ then

$$\|u\|_{L^{3\alpha}}^\alpha \leq c_\alpha \int_{\mathbb{R}^3} |\nabla u|^2 |u|^{\alpha-2} \, dx.$$

(This generalises Lemma 11.2.)

11.3 Show that if u is smooth and $\alpha \geq 0$ then

$$\int (u \cdot \nabla)u \cdot u|u|^\alpha \, dx = 0.$$

11.4 Using the results of Exercises 11.1 and 11.3 show that if u is a smooth solution of the Navier–Stokes equations then for any $\alpha \geq 2$

$$\frac{1}{\alpha}\frac{d}{dt}\|u\|_{L^\alpha}^\alpha + \frac{1}{2}\int |u|^{\alpha-2}|\nabla u|^2 \leq c\int |p|^2 |u|^{\alpha-2}. \tag{11.19}$$

Use the result of Exercise 11.2, along with the estimate $\|p\|_{L^\alpha} \leq c\|u\|_{L^{2\alpha}}^2$ (valid for $1 < \alpha < \infty$), and Lebesgue interpolation to deduce that

$$\frac{1}{\alpha}\frac{d}{dt}\|u\|_{L^\alpha}^\alpha + \frac{1}{2c_\alpha}\|u\|_{L^{3\alpha}}^\alpha \leq c\|u\|_{L^\alpha}^{\alpha-1}\|u\|_{L^{3\alpha}}^3. \tag{11.20}$$

11.5 Deduce from Exercise 11.4 that there exists an absolute constant $\varepsilon > 0$ such that if $\|u_0\|_{L^3} < \varepsilon$ then the solution u remains bounded in L^3 for all $t > 0$.

11.6 Suppose that $X \colon [0, T) \to \mathbb{R}$ satisfies $\dot{X} \le cX^\theta$ for some $\theta > 1$ and that $X(t) \to \infty$ as $t \to T$. Show that there is a $\kappa = \kappa(c, \theta) > 0$ such that

$$X(t) \ge \kappa(T - t)^{-1/(\theta-1)} \qquad \text{for all} \qquad t \in [0, T).$$

11.7 Deduce from Exercise 11.4 that if $\alpha > 3$ then

$$\frac{1}{\alpha}\frac{\mathrm{d}}{\mathrm{d}t}\|u\|_{L^\alpha}^\alpha \le c\|u\|_{L^\alpha}^{\alpha(\alpha-1)/(\alpha-3)} \tag{11.21}$$

and hence show, using the result of Exercise 11.6, that if a solution blows up at time T then (11.18) must hold.

11.8 Show that if there exists an absolute constant c such that whenever u is a regular solution on the whole space that blows up at time T

$$\|u(T - t)\|_{L^p} \ge ct^{-\gamma} \tag{11.22}$$

then necessarily $\gamma = (p - 3)/2p$. [Consider the rescaled solutions $u_\lambda(x, t) = \lambda u(\lambda x, \lambda^2 t)$ and use the result of Exercise 10.1.]

11.9 For $\alpha > 3$, by interpolating the L^α norm on the right-hand side of (11.20) between L^3 and $L^{3\alpha}$, show that

$$\frac{1}{\alpha}\frac{\mathrm{d}}{\mathrm{d}t}\|u\|_{L^\alpha}^\alpha + \frac{1}{2c_\alpha}\|u\|_{L^{3\alpha}}^\alpha \le c\|u\|_{L^3}^2\|u\|_{L^{3\alpha}}^\alpha,$$

and hence deduce, using Exercise 11.5, that if $u_0 \in L^2 \cap L^\infty$ with $\|u_0\|_{L^3} < \varepsilon$ then the L^α norm of u is non-increasing for every $3 \le \alpha < \infty$.

Part II

Local and Partial Regularity

Overview of Part II

In the first part of this book we have constructed global weak solutions for initial data in H (divergence-free L^2), and local strong solutions for initial data in V (divergence-free H_0^1). We showed that strong solutions immediately become smooth, and are unique in the class of Leray–Hopf weak solutions, i.e. those satisfying the strong energy inequality.

One consequence of these results that will be particularly important in what follows is the existence of 'epochs of regularity': any Leray–Hopf weak solution is regular on a collection of time intervals that have full measure (Theorem 8.14). In particular, this means that the spatial derivatives in the equation can be understood classically for a.e. t.

In the second part of this book we focus on local properties of solutions, in particular on 'conditional regularity' results of a local form. We have already seen examples of conditional regularity results on the whole domain: for example, the result of Theorem 8.17 guarantees that a Leray–Hopf weak solution u is smooth on $(0, T)$ if

$$u \in L^r(0, T; L^s(\Omega)), \qquad \frac{2}{r} + \frac{3}{s} = 1, \qquad 3 < s \leq \infty. \qquad (11.23)$$

The first major result we prove in Part II is a local version of this: if u is a weak solution on $\Omega \times (0, T)$, and for some $U \subset \Omega$ and $(a, b) \subset (0, T)$

$$u \in L^r(a, b; L^s(U))$$

(with the same conditions on r and s as in (11.23)) then u is smooth with respect to the spatial variables on $U \times (a, b)$. This result was initially due to Serrin (1962) under the assumption that $2/r + 3/s < 1$ and proved for the case of equality by Fabes, Jones, & Rivière (1972) and Struwe (1988); in Chapter 13 we follow a proof due to Takahashi (1990) which also deals with the case of equality. The result also holds when $u \in L^\infty(0, T; L^3)$, but the proof (due to Escauriaza, Seregin, & Šverák, 2003) is much harder; we give a sketch in Section 16.4.

Throughout most of the first part of the book we used the Helmholtz–Weyl decomposition to eliminate the pressure. However, this is a 'global' approach and cannot be used in local arguments. In the local regularity result described above we eliminate the pressure by analysing the equation for the curl of u: if we define the vorticity $\omega = \text{curl } u$ then the Navier–Stokes equations become

$$\partial_t \omega - \Delta \omega + (u \cdot \nabla)\omega = (\omega \cdot \nabla)u, \qquad \nabla \cdot \omega = 0.$$

The pressure no longer appears, but this is not a closed system of equations for ω unless we have a way of reconstructing the velocity field u from ω. This problem forms the subject of Chapter 12, in which we derive the Biot–Savart Law and hence estimates for u based on estimates for ω.

The other local regularity results that we treat here, due to Caffarelli, Kohn, & Nirenberg (1982), do not have a vorticity-based formulation, and will require us to consider the pressure in some detail. These conditional regularity results are of a different type, known as 'ε-regularity' results. These guarantee local regularity under a smallness condition on certain integral quantities. For example, we will show in Chapter 15 that there is an absolute constant ε_0 such that if

$$\frac{1}{r^2} \int_{Q_r(a,s)} |u|^3 + |p|^{3/2} < \varepsilon_0 \qquad (11.24)$$

then $u \in L^\infty(Q_{r/2}(a, s))$, where $Q_r(a, s)$ is the 'parabolic cylinder' defined as

$$Q_r(a, s) = \{(x, t) : |x - a| < r, \ s - r^2 < t < s\}.$$

There are two things to note about the condition (11.24): the most obvious, that it includes the pressure, we leave aside for a moment. More tellingly, it requires the local smallness of an integral quantity that we know is globally bounded. If u is unbounded in a neighbourhood of (a, s) then this implies a lower bound on the left-hand side of (11.24), and one can then show using a covering argument that the set of all 'singular points' of u cannot be too large: thus a conditional ε-regularity result of this sort can be used to deduce an unconditional 'partial regularity' result. In this case, (11.24) implies that the set of space–time singularities has box-counting dimension no larger than $5/3$. In Chapter 16 we prove another ε-regularity result: there is an absolute constant $\varepsilon_1 > 0$ such that if

$$\limsup_{r \to 0} \frac{1}{r} \int_{Q_r(a,s)} |\nabla u|^2 \leq \varepsilon_1$$

then there exists $\rho > 0$ such that $u \in L^\infty(Q_\rho(a, s))$. The corresponding partial regularity result in this case is that the singular set has one-dimensional Hausdorff measure zero and hence Hausdorff dimension no larger than one (in fact the result is a little stronger, as we will see).

The proof of these local ε-regularity results uses a local form of the energy inequality,

$$2 \int_0^t \int_\Omega |\nabla u|^2 \varphi \, dx \, ds$$
$$\leq \int_0^t \int_\Omega |u|^2 \left\{ \frac{\partial \varphi}{\partial t} + \Delta \varphi \right\} dx \, ds + \int_0^t \int_\Omega (|u|^2 + 2p)(u \cdot \nabla)\varphi \, dx \, ds,$$

valid for any smooth scalar function $\varphi \geq 0$ with compact support in the space–time domain $\Omega \times (0, T)$. While this equation (as an equality) can be derived for classical solutions by multiplying by $u\varphi$ and integrating by parts, its validity for weak solutions must be proved (much as we had to prove the validity of the 'global' energy inequality for a certain class of weak solutions in Chapter 4). The existence of solutions satisfying this inequality, termed 'suitable weak solutions', is established in Chapter 14 by taking limits of a regularised form of the Navier–Stokes equations. Using this method we are also able to prove the validity of the strong energy inequality for weak solutions on the whole space in Theorem 14.4.

In the final chapter (Chapter 17) we consider the existence and uniqueness of Lagrangian trajectories, i.e. solutions of the ordinary differential equation

$$\frac{dX}{dt} = u(X, t) \qquad X(0) = a \in \mathbb{T}^3,$$

where u is a suitable weak solution with initial data in $H \cap \dot{H}^{1/2}$. We prove existence following Foias, Guillopé, & Temam (1985) and then almost everywhere uniqueness (i.e. for almost every $a \in \mathbb{T}^3$) using an argument due to Robinson & Sadowski (2009).

12

Vorticity

In our treatment of solutions of the initial value problem for the Navier–Stokes equations we eliminated the pressure by means of the Leray projector, which we introduced in Chapter 2. In this chapter we present an alternative method for eliminating the pressure, based on studying the curl of the velocity, which is known as the vorticity. The idea behind this approach relies on the fact that in three dimensions the curl of the gradient of any function is zero, so that $\operatorname{curl} \nabla p = 0$.

12.1 The vorticity equation

We will denote the vorticity, $\operatorname{curl} u$, by ω, and obtain an evolution equation for ω. Notice that since u is divergence free, so is ω. By taking the curl of the Navier–Stokes equations,

$$\partial_t u - \Delta u + (u \cdot \nabla)u + \nabla p = 0,$$

we obtain

$$\partial_t \omega - \Delta \omega + \operatorname{curl}((u \cdot \nabla)u) = 0. \tag{12.1}$$

In order to compute the curl of the nonlinear term, we use the vector identity

$$(u \cdot \nabla)u = \frac{1}{2}\nabla(|u|^2) - u \times \omega, \tag{12.2}$$

see Exercise 12.1. Hence, since the curl of a gradient is zero,

$$\operatorname{curl}((u \cdot \nabla)u) = -\operatorname{curl}(u \times \omega) = (u \cdot \nabla)\omega - (\omega \cdot \nabla)u,$$

where for the last equality we have used the vector identity

$$\operatorname{curl}(A \times B) = A(\nabla \cdot B) - B(\nabla \cdot A) + (B \cdot \nabla)A - (A \cdot \nabla)B$$

224

(see Exercise 12.2) and the fact that u and ω are divergence free.

As a result of these elementary calculations we obtain the following evolution equation for ω:

$$\partial_t \omega - \Delta \omega + (u \cdot \nabla)\omega = (\omega \cdot \nabla)u. \tag{12.3}$$

We notice that while for sufficiently smooth u and ω the equation makes perfect sense as is, when considering a weak solution of the Navier–Stokes equations, we do not in principle have enough regularity to write various terms in (12.3).

We now obtain a formulation of equation (12.3) in the distributional sense, and then show that if u is a weak solution of the Navier–Stokes equations then the pair (u, ω) satisfies this distributional formulation of the vorticity equation.

We consider test functions ϕ that belong to $C_c^\infty(\mathbb{R}^n \times (0, \infty))$, as for the moment we are not concerned with the initial value problem (and so the restriction $\phi(x, 0) = 0$ is allowable). Note that such ϕ need not be divergence free.

Taking the dot product of the equation with ϕ, integrating over space–time, and formally integrating by parts we obtain

$$\int_0^\infty \int -\omega \cdot \partial_t \phi - \omega \cdot \Delta \phi - (u \cdot \nabla \phi) \cdot \omega + (\omega \cdot \nabla)\phi \cdot u \, dx \, dt = 0,$$

which we write as

$$\int_0^\infty \langle \omega, \partial_t \phi \rangle + \langle \omega, \Delta \phi \rangle + \langle (u \cdot \nabla)\phi, \omega \rangle - \langle (\omega \cdot \nabla)\phi, u \rangle \, dt = 0. \tag{12.4}$$

Having derived the distributional form of the equation we are ready to show that weak solutions of the Navier–Stokes equation satisfy this formulation of the vorticity equation.

Theorem 12.1 *Let u be a weak solution of the Navier–Stokes equations. Then the pair (u, ω), where $\omega = \operatorname{curl} u$), satisfies (12.4) for every choice of test function $\phi \in C_c^\infty(\mathbb{R}^n \times (0, \infty))$.*

Proof Let u be a weak solution of the Navier–Stokes equations in the sense of Definition 3.3. Given $\phi \in C_c^\infty(\mathbb{R}^n \times (0, \infty))$, we observe that $\operatorname{curl} \phi$ is divergence free and as a consequence can be used as a test function in (3.3). We have (using that $\phi(0) = 0$ and taking $s = \infty$)

$$\int_0^\infty -\langle u, \partial_t \operatorname{curl} \phi \rangle + \langle \nabla u, \nabla \operatorname{curl} \phi \rangle + \langle (u \cdot \nabla)u, \operatorname{curl} \phi \rangle = 0. \tag{12.5}$$

We now integrate by parts in the first two terms and use the fact that $\langle f, \operatorname{curl} g \rangle = \langle \operatorname{curl} f, g \rangle$ (see Exercise 12.3), to obtain

$$\int \langle \omega, \partial_t \phi \rangle + \langle \omega, \Delta \phi \rangle - \langle (u \cdot \nabla)u, \operatorname{curl} \phi \rangle = 0;$$

to complete the proof we need to show that

$$-\langle (u \cdot \nabla)u, \operatorname{curl} \phi \rangle = \langle (u \cdot \nabla)\phi, \omega \rangle - \langle (\omega \cdot \nabla)\phi, u \rangle.$$

Using (12.2), the left-hand side becomes

$$-\langle (u \cdot \nabla)u, \operatorname{curl} \phi \rangle = -\left\langle \frac{1}{2}\nabla(|u|^2) - u \times \omega, \operatorname{curl} \phi \right\rangle = \langle u \times \omega, \operatorname{curl} \phi \rangle.$$

Using tensor notation we can rewrite the right-hand side as

$$\int \epsilon_{ijk} u_j \omega_k \epsilon_{imn} \partial_m \phi_n = \int [\delta_{jm}\delta_{kn} - \delta_{jn}\delta_{km}] u_j \omega_k \partial_m \phi_n$$

$$= \int u_j \omega_k \partial_j \phi_k - u_j \omega_k \partial_k \phi_j$$

$$= \langle (u \cdot \nabla)\phi, \omega \rangle - \langle (\omega \cdot \nabla)\phi, u \rangle,$$

where we have used (1.4). □

12.2 The Biot–Savart Law

Notice that in order to consider equations (12.3) or (12.4) as a closed system involving only the vorticity we need an expression for u in terms of ω: we therefore need to study the system

$$\operatorname{curl} u = \omega, \qquad \operatorname{div} u = 0. \tag{12.6}$$

For us the most useful result will be a local representation of the velocity in terms of the vorticity, which we present in full detail later in this chapter. For now we will focus on the case of solving (12.6) on the whole space $\mathbb{R}^3$, with similar results holding in the periodic case.

We can think of the system (12.6) as a time-independent problem, with no a priori relationship to the Navier–Stokes equations. The first observation is that a necessary condition for (12.6) to have a solution is that $\operatorname{div} \omega = 0$ (take the divergence of the first equation in (12.6)) and so we will assume that this condition holds (as it does for any ω that arises as the curl of a Navier–Stokes solution u).

In order to find an expression for u in terms of ω, we will assume sufficient regularity and decay for the vorticity ($\omega \in C_c^\infty(\mathbb{R}^3)$ for example) to make it possible to justify the following manipulations.

Taking the curl of the first equation in (12.6) and using the vector identity $\Delta u = \nabla(\operatorname{div} u) - \operatorname{curl}(\operatorname{curl} u)$ (see Exercise 2.2), we obtain (since u is divergence free)

$$\operatorname{curl} \omega = \operatorname{curl}\operatorname{curl} u = \nabla(\nabla \cdot u) - \Delta u = -\Delta u,$$

and so it follows that

$$u = (-\Delta)^{-1}\operatorname{curl} \omega.$$

Now, using the fundamental solution of the Laplacian (see Theorem C.4) we obtain

$$u(x) = \frac{1}{4\pi} \int_{\mathbb{R}^3} \frac{1}{|x-y|} \operatorname{curl} \omega(y)\,dy;$$

integrating by parts to move the curl onto the kernel we find (after careful algebraic manipulations in which it is particularly important to pay attention to the signs) that

$$u(x) = -\frac{1}{4\pi} \int_{\mathbb{R}^3} \frac{x-y}{|x-y|^3} \times \omega(y)\,dy, \tag{12.7}$$

a formula known as the Biot–Savart Law. On several occasions it will be useful for us to rewrite this expression as

$$u(x) = -\frac{1}{4\pi} \int_{\mathbb{R}^3} \left(\nabla_y \frac{1}{|x-y|}\right) \times \omega(y)\,dy.$$

As mentioned above it is possible to extend this representation of the velocity to vorticities that satisfy just some L^p integrability condition. More precisely we have the following result.

Theorem 12.2 *Let*

$$T(\omega) := -\frac{1}{4\pi} \int_{\mathbb{R}^3} \frac{x-y}{|x-y|^3} \times \omega(y)\,dy$$

be the operator on the right-hand side of (12.7). Then T is a bounded linear operator from L^p into L^q provided $1 < p < 3$ and $1 + \frac{1}{q} = \frac{2}{3} + \frac{1}{p}$. Hence defining $u = T(\omega)$ we have $\omega = \operatorname{curl} u$ and $\|u\|_{L^q} \le c\|\omega\|_{L^p}$.

Sketch of the proof It is straightforward to see that T is well defined for smooth, compactly supported functions. Since these functions are dense in L^p for $1 < p < 3$, T will extend to L^p provided we show that $\|T(w)\|_{L^q} \le c\|w\|_{L^p}$ for smooth compactly supported functions.

Notice that T is given by the convolution (see Appendix A) of ω with a kernel of order $|x|^{-2}$. In three dimensions this kernel belongs to the space $L^{3/2,\infty}$

(the space weak $L^{3/2}$, see Appendix A), and so applying Young's convolution inequality involving weak L^p spaces (see Theorem A.16) we obtain

$$\|T(\omega)\|_{L^q} \le c \left\| \frac{1}{|x|^2} \right\|_{L^{3/2,\infty}} \|\omega\|_{L^p}, \tag{12.8}$$

provided that $1 < p, q < \infty$ satisfy $1 + \frac{1}{q} = \frac{2}{3} + \frac{1}{p}$. Finally notice that the equality relating p and q further reduces the range of p to $1 < p < 3$, completing the proof. $\qquad\qquad\square$

We note that the relation between the exponents p and q in the previous theorem is the same as for the three-dimensional Sobolev embedding. Recall that in $\mathbb{R}^3$ we have $\|f\|_{L^q} \le c\|\nabla f\|_{L^q}$, where $q = 3p/(3 - p)$, which can be rewritten as $1 + \frac{1}{q} = \frac{2}{3} + \frac{1}{p}$. Theorem 12.2 shows that it is not necessary to know the full gradient of u to reconstruct the velocity and to obtain this estimate, but knowing only the vorticity suffices.

For example, a bound on the vorticity in $L^{6/5}$ would translate into a bound for the velocity u in L^2. Also, the case $\omega \in L^2$ (which corresponds to $p = 2$ in (12.8)) yields a velocity in L^6. In upcoming sections we will obtain various estimates for ∇u in terms of ω in various Lebesgue spaces, by means of equation (12.7).

We have managed to obtain a solution for system (12.6) at a fixed time, under the condition that $\operatorname{div}\omega = 0$. We remark that if this condition is satisfied by the initial data then it is satisfied for all positive times, since if we take the divergence of equation (12.1) and use the fact that $\operatorname{div}\operatorname{curl}A = 0$ for every vector field A then we obtain the equation $\partial_t(\operatorname{div}\omega) = \Delta(\operatorname{div}\omega)$, with initial condition zero, and so (by the uniqueness of solutions of the heat equation, see Appendix D) we have $\operatorname{div}\omega = 0$ for all time.

12.3 The Beale–Kato–Majda blowup criterion

As mentioned before, our main use of the vorticity will be in local arguments, but for the sake of completeness we present here one of the best-known blowup criteria, which is expressed in terms of the maximum of the vorticity.

The result we present here was first shown for the Euler equations (i.e. without the Laplacian term) by Beale, Kato, & Majda (1984); it is probably more significant in that context. We refer the reader to the original work of Beale, Kato, & Majda (1984) or to the book of Majda & Bertozzi (2002), where a detailed exposition of this and other related results is presented. The proof for the Navier–Stokes equations is considerably easier than for the Euler equations.

Theorem 12.3 *Let $u_0 \in V(\mathbb{R}^3)$ and let u be the corresponding solution of the Navier–Stokes equations. If for every time $T > 0$ we have*

$$\int_0^T \|\omega(\cdot, s)\|_{L^\infty} \, ds < \infty \tag{12.9}$$

then the solution exists globally in time as a strong solution. Conversely, if T is the maximal time of existence of the strong solution (that is, the first singularity occurs at time $t = T$) then we necessarily must have

$$\lim_{t \to T} \int_0^t \|\omega(\cdot, s)\|_{L^\infty} \, ds = \infty.$$

We remark that the regularity of the initial condition ($u_0 \in V$) guarantees that the integral in (12.9) makes sense near zero, see Exercise 8.8. The result of Theorem 12.3 also applies to the Euler equations; however, one requires more regularity of the initial data (H^3 suffices) to construct a strong solution for the 3D Euler equations (a notion we have not made precise for this equation).

Finally we notice that Theorem 7.6 guarantees that the solution is globally strong provided that

$$\int_0^T \|\nabla u(\cdot, s)\|_{L^\infty} \, ds < \infty, \tag{12.10}$$

for every $T > 0$. Condition (12.9) weakens (12.10) by replacing ∇u with ω.

Proof It suffices to show that the solution can be continued past the time $t = T$ whenever $\int_0^T \|\omega\|_{L^\infty} \, ds < \infty$, and in order to prove this we show that $\|u(\cdot, T)\|_{H^1} < \infty$ and then use the existence results from Chapter 6 to continue the solution.

First we notice that $\|\nabla u\| = \|\omega\|$; indeed,

$$\int |\omega|^2 = \int \epsilon_{ijk}(\partial_j u_k)\epsilon_{imn}(\partial_m u_n)$$

$$= \int [\delta_{jm}\delta_{kn} - \delta_{jn}\delta_{km}](\partial_j u_k)(\partial_m u_n)$$

$$= \int (\partial_j u_k)(\partial_j u_k) - (\partial_j u_k)(\partial_k u_j) = \int (\partial_j u_k)(\partial_j u_k) = \int |\nabla u|^2,$$

since $\int (\partial_j u_k)(\partial_k u_j) = 0$ due to the incompressibility condition on u (just integrate by parts).

Taking the L^2 inner product of the vorticity equation

$$\partial_t \omega - \Delta \omega + (u \cdot \nabla)\omega = (\omega \cdot \nabla)u$$

with ω and dropping the term $\|\nabla\omega\|^2$ we find that

$$\frac{d}{dt}\|\omega\|_{L^2}^2 \le |\langle(\omega\cdot\nabla)u,\omega\rangle| \le \|\omega\|_{L^\infty}\|\nabla u\|_{L^2}\|\omega\|_{L^2} \le c\|\omega\|_{L^\infty}\|\omega\|_{L^2}^2,$$

$$(12.11)$$

from which we obtain

$$\|\omega(t)\|_{L^2} \le \|\omega_0\|_{L^2}\exp\left(\int_0^t \|\omega(s)\|_{L^\infty}\,ds\right);$$

$$(12.12)$$

this is finite by assumption for all $t \le T$ (combining (12.9) with the fact that $\|\omega_0\| = \|\nabla u_0\| < \infty$). Therefore $u(T)$ is in H^1 and as we have seen in Lemma 6.11 this rules out the existence of a singularity at time T, and so we can continue the solution past this time. □

12.4 The vorticity in two dimensions

In this section we briefly discuss the two-dimensional case, since the vorticity formulation provides a way to see the fundamental differences between the equations in two and three dimensions.

Notice that in the two-dimensional case we could extend the velocity vector $u(x_1, x_2) = (u_1(x_1, x_2), u_2(x_1, x_2))$ to a three-dimensional vector by including an additional zero component (we will underline the three-vectors obtained in this way); the resulting 'three-dimensional velocity' $\underline{u} := (u(x_1, x_2), 0)$ solves the 3D Navier–Stokes equations. In this form, we can now take the curl of the velocity $\underline{u}$ and note that the first and second components of the vorticity (that we will denote by $\underline{\omega}$) vanish identically:

$$\underline{\omega} = \text{curl}\,\underline{u} = (\partial_1 u_2 - \partial_2 u_1)\underline{e}_3 =: \omega\underline{e}_3,$$

where now ω is a scalar, given by $\partial_1 u_2 - \partial_2 u_1$.

In this case the nonlinear term $(\underline{\omega}\cdot\nabla)\underline{u}$ on the right-hand side of the vorticity equation vanishes, since $\underline{\omega}\cdot\nabla = \omega\partial_3$ and $\underline{u}$ depends only on x_1 and x_2. The only non-trivial component of the vorticity equation is now

$$\partial_t\omega - \Delta\omega + (u\cdot\nabla)\omega = 0.$$

$$(12.13)$$

Notice a fundamental difference between the two-dimensional and three-dimensional cases: the so-called 'vortex-stretching' term $(\underline{\omega}\cdot\nabla)\underline{u}$ vanishes in two dimensions.

The vorticity formulation allows us to prove in an elementary manner that solutions of the Navier–Stokes equations in two dimensions are globally strong.

Theorem 12.4 *Let $\omega_0 \in L^2$. Then there exists a unique solution of the 2D Navier–Stokes equation that remains strong for all time.*

Proof Since $\|\omega_0\| = \|\nabla u_0\|$ our existence result guarantees that we have a unique strong solution for a short time. Notice that by taking the inner product of the equation (12.13) with ω we obtain

$$\frac{1}{2}\frac{d}{dt}\|\omega\|^2 + \|\nabla\omega\|^2 = 0,$$

from which it follows that $\|\omega(t)\| \le \|\omega_0\|$ for all $t \ge 0$. Since $\|\nabla u\| = \|\omega\|$, we obtain control on the H^1 norm of u and as a consequence the regularity of the solution; therefore u is a strong solution for all time. $\qquad\square$

12.5 A local version of the Biot–Savart Law

In the next chapter we will focus on the local regularity of solutions of the 3D Navier–Stokes equations, and show that the integrability condition

$$u \in L^r(a, b; L^s(\Omega)), \qquad \frac{2}{r} + \frac{3}{s} = 1,$$

a local version of the Serrin condition from Theorem 8.17, guarantees the smoothness in space of the local solution.

Much of the analysis will be carried out in terms of the vorticity. To translate regularity information for ω into information about the velocity u, we need to obtain a local version of the formula (12.7) that we derived to reconstruct u from ω (i.e. to invert the curl operator) on the whole of $\mathbb{R}^3$. We would expect curl^{-1} to be a nonlocal operator (as in (12.7)), requiring information on the boundary to be uniquely defined, information that unfortunately we do not have. An additional problem in attempting to define u uniquely, given ω, comes from the fact that we are dealing with space–time functions, while curl acts only on the space variables; we could therefore add any function of t alone to u and still satisfy the equation curl $u = \omega$ on the corresponding time interval.

For these reasons our 'local representation' formula, i.e. the reconstruction of u involving ω only on a subdomain, will involve an additional additive term, for which we will not have quantitative estimates, but which we know is harmonic (and therefore smooth in space). More precisely, we have the following result (cf. (12.7)).

Theorem 12.5 (Biot–Savart Law, local version) *Assume that u and ω are defined in an open domain B, satisfy $\omega = $ curl u, and that $u \in L^q(B)$ and*

$\nabla u \in L^r(B)$ *for some* $1 \leq q, r \leq \infty$. *Then for any open set* $\Omega \subset\subset B$, *and for all* $x \in \Omega$ *we have*

$$u(x) = -\frac{1}{4\pi} \int_\Omega \left(\nabla_y \frac{1}{|x-y|} \right) \times \omega(y) \, dy + A(x), \qquad (12.14)$$

where A is a harmonic function.

Proof   Notice that without any requirements on A equation (12.14) is trivially true, taking it as the definition of A. So all we need to prove is that A is harmonic. Notice that it is enough to show that

$$u(x) + \frac{1}{4\pi} \int_\Omega \left(\nabla_y \frac{1}{|x-y|} \right) \times \omega(y) \, dy$$

is weakly harmonic[1], since any weakly harmonic function is in fact harmonic (Theorem C.3 – Weyl's Lemma).

We want to show that for all $\phi \in C_c^\infty(\Omega)$

$$\int \left[u(x) \Delta\phi(x) + \frac{1}{4\pi} \int_\Omega \left(\nabla_y \frac{1}{|x-y|} \right) \times \omega(y) \, dy \, \Delta\phi(x) \right] dx = 0.$$

We start by considering the term involving the Biot–Savart integral. We have

$$\int \frac{1}{4\pi} \int_\Omega \left(\nabla_y \frac{1}{|x-y|} \right) \times \omega(y) \, dy \, \Delta\phi(x) \, dx$$

$$= \int_\Omega \frac{1}{4\pi} \left(\int \left(\nabla_y \frac{1}{|x-y|} \right) \Delta\phi(x) \, dx \right) \times \omega(y) \, dy$$

$$= \int_\Omega \frac{1}{4\pi} \left(\int \left(-\nabla_x \frac{1}{|x-y|} \right) \Delta\phi(x) \, dx \right) \times \omega(y) \, dy$$

$$= \int_\Omega \frac{1}{4\pi} \left(\int \left(\frac{1}{|x-y|} \right) \Delta\nabla\phi(x) \, dx \right) \times \omega(y) \, dy$$

$$= -\int_\Omega \nabla\phi(y) \times \omega(y) \, dy, \qquad (12.15)$$

where to obtain (12.15) we have used the fact that $\frac{1}{4\pi} \frac{1}{|x|}$ is the fundamental solution for the Laplacian in $\mathbb{R}^3$, so that

$$\frac{1}{4\pi} \int_{\mathbb{R}^3} \left(\frac{1}{|x-y|} \right) \Delta f(x) \, dx = -f(y).$$

In order to complete the proof we use the identity

$$\int \omega(y) \times \nabla\phi(y) \, dy = \int \nabla u(y) \cdot \nabla\phi(y) \, dy,$$

----
[1] A function f is weakly harmonic on Ω if $\int_\Omega f \Delta\phi \, dx = 0$ for every $\phi \in C_c^\infty(\Omega)$.

which holds as long as u is divergence free (see Exercise 12.4). Integrating by parts in the right-hand side to move all the derivatives to ϕ yields the desired result. □

We now turn our attention to space–time estimates for the velocity u in terms of ω using this local form of the Biot–Savart Law; we include the assumption that $u \in L^\infty(0, t; L^2(\Omega))$, since this space–time regularity is known for weak solutions. For simplicity we will concentrate on the case that Ω is a ball B_R.

Theorem 12.6 *Suppose that $u \in L^\infty(0, t; L^2(B_R))$ and that ω is the curl of u. Then for any $R' < R$,*

(i) *for $2 \le \beta \le \infty$*

$$\omega \in L^\beta(0, t; L^\infty(B_R)) \quad \Rightarrow \quad u \in L^\beta(0, t; L^\infty(B_{R'}));$$

(ii) *for $0 < \alpha < 1$, $k = 0, 1, 2, \ldots$,*

$$\omega \in L^\infty(0, t; C^{k,\alpha}(B_R)) \quad \Rightarrow \quad u \in L^\infty(0, t; C^{k+1,\alpha'}(B_{R'}))$$

for all $0 < \alpha' < \alpha$.

Proof Choose R'' with $R' < R'' < R$. At each time t we use the representation (12.14) with $\Omega = B_{R''}$.

As pointed out before, an important issue is the time regularity of A, so we start by considering (12.14) as an equation for A, namely

$$A(x) := u(x) + \frac{1}{4\pi} \int_{B_{R''}} \left(\nabla_y \frac{1}{|x - y|} \right) \times \omega(y) \, dy.$$

We have different integrability properties for u and for the integral term above. We are assuming that $u \in L^\infty(0, t; L^2(B_R))$, while we will show that the second term belongs to $L^\beta(0, t; L^\infty(B_{R''}))$. As a consequence u will limit the spatial regularity of A while the integral term will limit the time regularity of A. In order to obtain some space–time regularity for the integral term we notice that since $\nabla \frac{1}{|x|}$ is integrable near the origin in $\mathbb{R}^3$

$$\left\| \frac{1}{4\pi} \int_{B_{R''}} \left(\nabla \frac{1}{|x - y|} \right) \times \omega(y, t) \, dy \right\|_{L^\infty(B_{R''})} \le C(R'') \|\omega\|_{L^\infty(B_{R''})}$$

and so

$$\left\| \frac{1}{4\pi} \int_{B_{R''}} \left(\nabla \frac{1}{|x - y|} \right) \times \omega(y, t) \, dy \right\|_{L^\beta(0, t; L^\infty(B_{R''}))} \le C(R'') \|\omega\|_{L^\beta(0, t; L^\infty(B_{R''}))},$$

from which it follows that

$$A(x) = u(x) + \frac{1}{4\pi} \int_{B_{R''}} \left(\nabla_y \frac{1}{|x-y|} \right) \times \omega(y) \, dy \, \in L^\beta(0, t; L^2(B_{R''})).$$

(12.16)

Since A is harmonic we can use the mean value formula from Lemma C.1: for $x \in B_{R'}$ we choose $\rho = (R'' - R')/2$ so that $B_\rho(x) \subset B_{R''}$ for any $x \in B_{R'}$, and then we can write

$$A(x) = \frac{1}{|B_\rho|} \int_{B_\rho(x)} A(y) \, dy$$

$$\leq \frac{1}{|B_\rho|} \int_{B_\rho(x)} |A(y)| \, dy \leq \frac{1}{|B_\rho|^{1/2}} \|A\|_{L^2(B_\rho)},$$

which yields an L^∞ bound in $B_{R'}$, in terms of $C(R', R'')\|A\|_{L^2(B_{R''})}$. Using (12.16), we obtain $A \in L^\beta(0, t; L^\infty(B_{R'}))$. The proof of statement (i) is now complete as we have shown that both terms in the representation (12.14) of u are in $L^\beta(0, t; L^\infty(B_{R'}))$.

The proof of (ii) is more involved. As before we start by considering A. Notice that the assumption on ω guarantees that $\omega \in L^\infty(0, t; L^\infty(B_R))$, which by part (i) implies that $A \in L^\infty(0, t; L^\infty(B_{R'}))$. Now since A is harmonic, Theorem C.2 allows us to control the L^∞ norm of any derivative of A in a smaller ball in terms of A (and no derivatives) in B_R. As a result we find that $A \in L^\infty(0, t; C^m(B_{R'}))$ for any m and $R' < R$. Now that we have the desired regularity for A, the proof reduces to studying the regularity of the integral term

$$\frac{1}{4\pi} \int_\Omega \left(\nabla_y \frac{1}{|x-y|} \right) \times \omega(y) \, dy$$

in the representation (12.14).

We first observe that it suffices to prove the result for $k = 0$, with the result for higher derivatives being obtained by applying the result to $\partial^k u$. Also, by a translation and rescaling we can assume that ω is $C^{0,\alpha}$ in $B_1(0)$. We want to show that u is $C^{1,\alpha}$ in $B_{R'}$ for any $0 < R' < 1$.

Choose R'' and R''' with $R' < R'' < R''' < 1$, and consider a cutoff function ϕ such that $\phi(x) = 1$ for $x \in B_{R''}$ and supp $\phi \subset B_{R'''}$. Since A is smooth we need only consider the two expressions

$$\int_{B_1} \left(\nabla \frac{1}{|x-y|} \right) \times \phi(y)\omega(y) \, dy$$

and

$$\int_{B_1} \left(\nabla \frac{1}{|x - y|} \right) \times [1 - \phi(y)] \omega(y) \, dy.$$

The second term is easy to handle, as we are considering $x \in B_{R'}$ and the support of $[1 - \phi(y)]\omega(y)$ is contained in $R'' \le |y| \le 1$. Therefore $\frac{1}{|x-y|}$ is smooth in the region that we are considering, and so the corresponding term is actually smooth (we are only trying to show that u is $C^{1,\alpha}$).

For the first term, to simplify the presentation, rather than writing the product $\phi(y)\omega(y)$ every time we ignore the cutoff function and make the extra assumption that $\omega \in C_c^{0,\alpha}(B_1)$.

The derivative of u can be written as a singular integral of the form

$$\partial_k \int_{\mathbb{R}^3} \left(\nabla \frac{1}{|x - y|} \right) \times \omega(y) \, dy = \text{p.v.} \int_{\mathbb{R}^3} K(x - y)\omega(y) \, dy,$$

with K homogeneous of degree -3 (i.e. $K(\lambda x) = \lambda^{-3} K(x)$ for all $\lambda > 0$) and with zero average on every sphere, which in particular implies that $\int_{|y|>\varepsilon} K(y) \, dy = 0$ for every $\varepsilon > 0$. Here p.v. stands for principal value, which by definition means

$$\text{p.v.} \int_{\mathbb{R}^3} K(x - y)\omega(y) \, dy = \lim_{\varepsilon \to 0} \int_{|x-y|>\varepsilon} K(x - y)\omega(y) \, dy.$$

We will not require the precise expression for K, just the fact that it defines a singular integral operator; a detailed derivation can be found in Section 2.4.3 of Majda & Bertozzi (2002) for example.

Obtaining the required Hölder regularity for $\partial_k u$ reduces to showing that

$$\left| \text{p.v.} \int_{\mathbb{R}^3} K(y)\omega(x + h - y) \, dy - \text{p.v.} \int_{\mathbb{R}^3} K(y)\omega(x - y) \, dy \right| \le C|h|^\alpha,$$

where we have used the properties of a convolution to evaluate ω at $x + h - y$ and $x - y$, in order to take advantage of the regularity of ω. Notice that since $\int_{|y|>\varepsilon} K(y) \, dy = 0$ for every $\varepsilon > 0$, we can rewrite the left-hand side as

$$\left| \text{p.v.} \int_{\mathbb{R}^3} K(y)[\omega(x + h - y) - \omega(x + h) - \omega(x - y) + \omega(x)] \, dy \right|$$

$$\le \text{p.v.} \int_{\mathbb{R}^3} |K(y)||\omega(x + h - y) - \omega(x + h) - \omega(x - y) + \omega(x)| \, dy.$$

Notice that since $\omega \in C^{0,\alpha}$ we have

$$|\omega(x + h - y) - \omega(x + h) - \omega(x - y) + \omega(x)| \le C \min\{|y|^\alpha, |h|^\alpha\}. \tag{12.17}$$

This is easy to see by considering the two possible splits for the left-hand side

$$|\omega(x + h - y) - \omega(x + h)| + |-\omega(x - y) + \omega(x)| \le C|y|^\alpha$$

and

$$|\omega(x + h - y) - \omega(x - y)| + |-\omega(x + h) + \omega(x)| \le C|h|^\alpha.$$

We split the domain of integration depending on h as follows:

$$\text{p.v.} \int_{\mathbb{R}^3} |K(y)||\omega(x + h - y) - \omega(x + h) - \omega(x - y) + \omega(x)| \, dy$$

$$= \text{p.v.} \int_{|y|<|h|} \cdots + \int_{1 \ge |y| \ge |h|} \cdots \, .$$

For the first integral we have (using (12.17) with right-hand side $C|y|^\alpha$ in this region)

$$\text{p.v.} \int_{|y|<|h|} \cdots \le C \int_{|y|<|h|} \frac{1}{|y|^3} |y|^\alpha \, dy \le C|h|^\alpha$$

while for the second (using (12.17) with right-hand side $C|h|^\alpha$ in this region)

$$\int_{1 \ge |y| \ge |h|} \cdots \le C \int_{1 \ge |y| \ge |h|} \frac{1}{|y|^3} |h|^\alpha \, dy \le C|h|^\alpha |\log|h|| \le C|h|^{\alpha'},$$

for any $0 < \alpha' < \alpha$, and $|h| < 1$, say. (The presence of the logarithmic term is the reason why we cannot prove that $u \in C^{1,\alpha}$.) $\qquad\square$

Notes

The Biot–Savart Law usually refers to the equation relating the magnetic field to the current that generates it (originally described empirically by Biot & Savart, 1820). In the context of fluid mechanics, the structural similarities mean the same terminology is used.

The Beale–Kato–Majda Theorem presented in Section 12.3 is more relevant in the context of the Euler equations. We refer the reader to Majda & Bertozzi (2002) for a different proof and a more extensive treatment of the inviscid case.

In the case of the Euler equation several extensions of the Beale–Kato–Majda Theorem have been obtained by considering norms other than L^∞. For example, Kozono & Taniuchi (2000) obtained the result for the BMO norm, Kozono, Ogawa, & Taniuchi (2002) for Besov spaces (in particular for $\dot{B}_{\infty,\infty}^{0,0}$) and finally

Planchon (2003) obtained a criterion in terms of the quantity

$$\sup_j \int_0^T \|\Delta_j \omega\|_{L^\infty} \, dt,$$

where $\Delta_j \omega$ here corresponds to the (smooth) localisation of ω around the frequency 2^j.

Exercises

12.1 Prove the vector identity

$$(u \cdot \nabla)u = \frac{1}{2}\nabla(|u|^2) - u \times \omega.$$

12.2 Prove the vector identity

$$\text{curl}\,(A \times B) = A(\nabla \cdot B) - B(\nabla \cdot A) + (B \cdot \nabla)A - (A \cdot \nabla)B.$$

12.3 Prove the identity $\langle f, \text{curl}\, g \rangle = \langle \text{curl}\, f, g \rangle$.

12.4 Prove that if $u \in L^q_{\text{loc}}(\mathbb{R}^3)$ and $\nabla u \in L^r_{\text{loc}}(\mathbb{R}^3)$ and $1 \le q, r \le \infty$ then

$$\int_{\mathbb{R}^3} \omega \times \nabla \phi \, dx = \int_{\mathbb{R}^3} \nabla u \cdot \nabla \phi \, dx$$

for all $\phi \in C_c^\infty(\mathbb{R}^3)$, provided that u is divergence free.

13

The Serrin condition for local regularity

In this chapter we show that if u is a local weak solution of the Navier–Stokes equations on a space–time domain $U \times (t_1, t_2)$ that satisfies

$$u \in L^{q'}(t_1, t_2; L^q(U)), \qquad \frac{2}{q'} + \frac{3}{q} = 1, \qquad (13.1)$$

then u is smooth in the spatial variables. The constraint relating q and q' (Serrin's condition) appeared in Chapter 8, providing smoothness of weak solutions. (Note that here q' does not denote the dual exponent to q; given the many exponents in this chapter it is convenient to use exponents with a prime for the time integrability.) To date it is not known whether or not the weak solutions constructed in Chapter 4 satisfy (13.1), so what we prove is a form of 'conditional regularity'.

13.1 Local weak solutions

We start by describing what we mean by a local weak solution; our definition is satisfied by the restriction of any weak solution to the space–time domain $U \times (t_1, t_2)$, where U is some open subset of $\mathbb{R}^3$, but is more general. We will use the notation $L_t^{r'} L_x^r$ to denote $L^{r'}(t_1, t_2; L^r(U))$.

Definition 13.1 We say that u is a local weak solution of the Navier–Stokes equations in $U \times (t_1, t_2)$ if

(i) $u \in L^\infty(t_1, t_2; L^2(U))$, $\nabla u \in L^2(t_1, t_2; L^2(U))$, and
(ii) u satisfies the equation

$$\int_{t_1}^{t_2} -\langle u, \partial_t \varphi \rangle + \int_{t_1}^{t_2} \langle \nabla u, \nabla \varphi \rangle + \int_{t_1}^{t_2} \langle (u \cdot \nabla)u, \varphi \rangle = 0 \quad (13.2)$$

for all functions $\varphi \in C_c^\infty(U \times (t_1, t_2))$ that are divergence free.

Definition 13.1 differs from Definition 3.3 in Chapter 3, by restricting to a smaller class of test functions (see Definition 3.1 for details on $\mathcal{D}_\sigma$, the space of test functions), since we are not focusing on the initial value problem (indeed zero might not be in (t_0, t_1), and for us it will be convenient to allow t to range over all $\mathbb{R}$) or boundary values. Instead the equation must be satisfied weakly in Ω while retaining the estimates on u and ∇u that we know hold for weak solutions of the initial-value problem.

We start by considering an example due to Serrin (1962) that shows unexpected issues concerning the regularity of local weak solutions. Indeed, even if we assume smoothness in the space variables, we might not see any regularity in time.

Example 13.2 (Serrin's example) Let us consider the following local solution of the Navier–Stokes equations in a smooth bounded domain U in $\mathbb{R}^3$,

$$u(x, t) = a(t)\nabla\Psi(x),$$

where Ψ is a harmonic function and $a \in L^1(0, \infty) \cap L^\infty(0, \infty)$. Theorem C.3 guarantees the smoothness of Ψ, and hence of u, in the space variables.

We will show that u satisfies the local weak formulation of the Navier–Stokes equations from Definition 13.1, because every term in (13.2) vanishes. Indeed, $u(t)$ is the gradient of a smooth function for all t, while $\partial_t\varphi$ is divergence free, and so the first term of the weak formulation vanishes due to the Helmholtz–Weyl decomposition. Moreover, since φ has compact support in Ω we can integrate by parts in the second term and get

$$\langle\nabla u, \nabla\varphi\rangle = -\langle\Delta u, \varphi\rangle = -a\langle\Delta\nabla\Psi, \varphi\rangle = -a\langle\nabla(\Delta\Psi), \varphi\rangle = 0.$$

Finally, we have

$$[(u \cdot \nabla)u] \cdot \varphi = a^2(\partial_i\Psi)(\partial_i\partial_j\Psi)\varphi_j = \frac{a^2}{2}\left[\partial_j\sum_i(\partial_i\Psi)^2\right]\varphi_j = \frac{a^2}{2}\nabla|\nabla\Psi|^2 \cdot \varphi.$$

Therefore

$$\langle(u \cdot \nabla)u, \varphi\rangle = \frac{1}{2}\langle\nabla(|\nabla\Psi|^2), \varphi\rangle = 0,$$

due to the Helmholtz–Weyl decomposition. We emphasise that in the calculations above we have not required any regularity for the function a, and so $a \in L^1(0, \infty) \cap L^\infty(0, \infty)$ suffices for u to be a solution of the Navier–Stokes equations.

The function u is an example of a local weak solution that, while smooth in the space variables, is not sufficiently regular in time to be a classical solution of the Navier–Stokes equations. The reason why it is not possible to deduce time regularity of the solution from its space regularity is the presence of the pressure, which disappears in our weak formulation, and for which we will not be able to make any regularity claims. In the results below we will discuss sufficient conditions (see Theorem 13.11) that guarantee the spatial smoothness of local weak solutions, which, as this example indicates, is the most we can hope for. We note that when considering the initial-value problem, as opposed to the purely local problem discussed in the example above, it is in fact possible to obtain additional Hölder regularity in time. We discuss this result at the end of the chapter in Section 13.5.

The majority of the estimates in this chapter will be local, and we will normally consider domains Ω that are open balls in $\mathbb{R}^3$, the natural choice for a local neighbourhood of a point in $\mathbb{R}^3$. When considering space–time domains, it is natural to consider the product of balls in space with intervals in time (rather than four-dimensional balls), but given the parabolic nature of the Navier–Stokes equation, it becomes appropriate to link the length of the time interval to the radius of the ball in $\mathbb{R}^3$ in a specific way. This leads us to consider the following 'parabolic cylinders' as our fundamental domains.

Definition 13.3 Given $(x_0, t_0) \in \mathbb{R}^3 \times \mathbb{R}$, and $R > 0$, *the centred parabolic cylinder* $Q_R^*(x_0, t_0)$ is the set

$$Q_R^*(x_0, t_0) = B_R(x_0) \times (t_0 - R^2/2, t_0 + R^2/2),$$

where $B_R(x_0)$ denotes the open ball of radius R centred at x_0, see Figure 13.1.

We note that we will encounter other families of parabolic cylinders in later chapters. These will not be centred cylinders, but instead will be defined by $Q_R(x_0, t_0) = B_R(x_0) \times (t_0 - R^2, t_0)$. The cylinders $Q_R(x_0, t_0)$, which do not contain the point (x_0, t_0), are more common in the development of the partial regularity theory for the Navier–Stokes equations, this being the reason why we reserve a simpler notation for them. Also note that for notational convenience we allow for time to be negative, and indeed most of our arguments will be carried out in cylinders of the form $Q_R^*(0, 0)$. For simplicity of notation we will refer to $Q_R^*(0, 0)$ as Q_R^*.

The main reason for using parabolic cylinders (as opposed to balls in four dimensions for example) is dictated by the natural scaling of the equation. As seen in the Introduction and Chapter 10, if u is a solution then so is

$$u_\lambda(x, t) = \lambda u(\lambda x, \lambda^2 t). \tag{13.3}$$

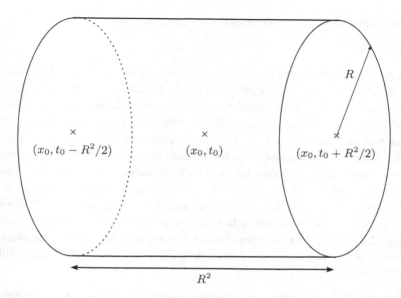

Figure 13.1. The parabolic cylinder $Q_R^*(x_0, t_0)$. The horizontal axis will correspond to time in our pictures. The point (x_0, t_0) corresponds to the 'centre' of the cylinder.

There is a particular family of norms of the form $L_t^q L_x^q$ that are preserved (in an appropriate sense) under this scaling transformation, and it is this family of norms that will be of interest to us. The $L_t^q L_x^q$ norm of the velocity is scale-invariant with respect to the scaling in (13.3) precisely when $\frac{2}{q} + \frac{3}{q} = 1$: if $u(x, t)$ solves the Navier–Stokes equations in the parabolic cylinder $Q_R^*(0, 0)$ then $u_R(x, t)$ solves the Navier–Stokes equations in $Q_1^*(0, 0)$ and

$$\|u_R\|_{L_t^{q'} L_x^q(Q_1^*(0,0))} = \|u\|_{L_t^{q'} L_x^q(Q_R^*(0,0))},$$

provided that (q, q') satisfy the condition $\frac{2}{q'} + \frac{3}{q} = 1$ (see Exercise 13.1).

We now state the main result of this chapter and then make some clarifying remarks.

Theorem 13.4 *Let u be a local weak solution of the Navier–Stokes equations (as in Definition 13.1) on some parabolic cylinder $Q_R^*(x_0, t_0)$, and suppose that*

$$u \in L_t^{q'} L_x^q(Q_R^*(x_0, t_0)), \qquad \frac{2}{q'} + \frac{3}{q} \le 1,$$

where $3 < q \le \infty$ and $2 \le q' < \infty$. Then u is smooth with respect to the space variables in $Q_R^*(x_0, t_0)$.

Note that the case of strict inequality in Theorem 13.4, i.e.

$$\frac{2}{q'} + \frac{3}{q} < 1,$$

is a simple consequence of the result with equality, by an application of Hölder's inequality. This case was originally proved by Serrin (1962), and we will see later that this result is a key ingredient of the proof in the harder case of equality.

Theorem 13.4 guarantees the spatial smoothness of the solution, based on a condition that looks rather weak. However, it says nothing about the time regularity of the solution. Indeed, as we have seen in Example 13.2, it is not possible to show any regularity in time for local weak solutions, unless additional conditions are added on the pressure p. However, by imposing conditions on the pressure it is possible to show that u is Hölder continuous in time. We will show this in Proposition 13.10, and in fact such an assumption on the pressure also allows for some Hölder continuity in the following theorem, due to Escauriaza, Seregin, & Šverák (2003) which deals with the case $q = 3$, $q' = \infty$ excluded in Theorem 13.4.

Theorem 13.5 *Let u be a local weak solution of* (13.2) *on some centred parabolic cylinder $Q_R^* = Q_R^*(x_0, t_0)$, and further assume that $p \in L^{3/2}(Q_R^*)$. If*

$$u \in L_t^\infty L_x^3(Q_R^*)$$

then the function u is smooth with respect to the space variables in Q_R^* and Hölder continuous in time.

We postpone a sketch of the proof of this result to Section 16.4, as it requires the partial regularity results due to Caffarelli, Kohn, & Nirenberg (1982) that we will present in Chapters 15 and 16.

In order to perform a local analysis of the smoothness of u without having to consider the pressure we will use the vorticity formulation of the equation introduced in the previous chapter,

$$\partial_t \omega - \Delta \omega = (\omega \cdot \nabla)u - (u \cdot \nabla)\omega. \tag{13.4}$$

As shown before we can understand this equation in the sense of distributions, or in its weak formulation as

$$\int \langle \omega, \partial_t \phi \rangle + \langle \omega, \Delta \phi \rangle + \langle (u \cdot \nabla)\phi, \omega \rangle - \langle (\omega \cdot \nabla)\phi, u \rangle \, dt = 0$$

for all smooth vector fields ϕ with compact support in the domain of definition of u.

The proof of Theorem 13.4 will rely on two main ideas: first, on obtaining estimates for ω by considering equation (13.4) as a heat equation for ω, with the right-hand side $(\omega \cdot \nabla)u - (u \cdot \nabla)\omega$ which has some known regularity. Smoothing properties of solutions of the heat equation then improve the known regularity of ω. The second idea will be to transfer these improved estimates for ω into improved estimates for u, using the results on the local Biot–Savart Law from the previous chapter.

Over the next few sections we present an essentially self-contained proof of Theorem 13.4. Regarding the more delicate result in Theorem 13.5, we present a sketch of the proof at the end of Chapter 16.

13.2 Main auxiliary theorem: a smallness condition

The main auxiliary result used in the proof of Theorem 13.4 will be presented in Theorem 13.6. We will refer to this theorem as an ε-regularity result, as it deduces the local smoothness of the solution under the assumption that the $L_t^{q'} L_x^q$ norm of u is sufficiently small. We also present a brief discussion of why Theorem 13.6 cannot be used to prove Theorem 13.4 in the case $(q', q) = (\infty, 3)$.

Theorem 13.6 *Let u be a local weak solution of* (13.2) *on some parabolic cylinder $Q_R^*(x_0, t_0)$. Then for every q with $3 \le q \le \infty$ there exists an ε_q such that if*

$$\|u\|_{L_t^{q'} L_x^q(Q_R^*(x_0, t_0))} < \varepsilon_q, \qquad where \qquad \frac{2}{q'} + \frac{3}{q} = 1,$$

then u is spatially smooth function in $Q_{R/2}^*(x_0, t_0)$.

We start by showing how Theorem 13.4 follows from Theorem 13.6. The proof follows via a simple argument using absolute continuity that is only available when $3 < q \le \infty$, $2 \le q' < \infty$, even though we notice that Theorem 13.6 does not have this limitation.

Proof of Theorem 13.4 It suffices to show that for every $(a, s) \in Q_R^*(x_0, t_0)$ the velocity u is smooth in the space variables in a neighbourhood of (a, s). Now, since $q' < \infty$, if $u \in L_t^{q'} L_x^q(Q_R^*(x_0, t_0))$ then there exists a $\rho > 0$ sufficiently small such that if $Q_\rho^*(a, s) \subset Q_R^*(x_0, t_0)$ we have

$$\|u\|_{L_t^{q'} L_x^q(Q_\rho^*(a, s))} < \varepsilon_q,$$

where ε_q comes from Theorem 13.6. Applying Theorem 13.6 to $Q_\rho^*(a, s)$ we obtain the smoothness in the space variables in $Q_{\rho/2}^*(a, s)$, completing the proof. $\square$

Notice that the fact that $u \in L_t^\infty L_x^3(Q_R^*)$ does not guarantee that we can find a smaller cylinder $Q_\rho^*(a, s) \subset Q_R^*$ within which $\|u\|_{L_t^\infty L_x^3(Q_\rho^*(a, s))} < \varepsilon_q$. This is the reason why we cannot use Theorem 13.6 (which does not have any restriction on the exponents) to complete the proof in this endpoint case.

As the proof of Theorem 13.6 is relatively intricate and contains multiple auxiliary results, we will start by sketching the main steps, indicating how they are put together.

13.3 The case $\frac{2}{q'} + \frac{3}{q} < 1$

We devote this section to presenting the main ideas in the proof of Theorem 13.6, with the full proof contained in Section 13.4.2. In order to clarify the main ideas we first sketch the details assuming that $\frac{3}{q} + \frac{2}{q'} < 1$; the result with this more restrictive constraint is a key ingredient in the proof of the result in the case $\frac{3}{q} + \frac{2}{q'} = 1$. More precisely, we will show the following result.

Theorem 13.7 *Let u be a weak solution of* (13.2) *on some centred parabolic cylinder $Q_R^* = Q_R^*(x_0, t_0)$, and suppose that*

$$u \in L_t^{q'} L_x^q(Q_R^*), \qquad \frac{2}{q'} + \frac{3}{q} < 1,$$

where $3 < q \le \infty$ and $2 < q' \le \infty$. Then u is smooth with respect to the space variables in Q_R^*.

Notice that the strict inequality in $\frac{2}{q'} + \frac{3}{q} < 1$ prevents the combination $(3, \infty)$, while still allowing other pairs with $q' = \infty$, hence the range of exponents in the statement.

13.3.1 Sketch of the argument

We will start by sketching the proof Theorem 13.7, originally due to Serrin (1962); we do not follow his proof, and instead use a different localisation procedure due to Takahashi (1990).

As we will see in the proof, under this stronger restriction on the exponents we no longer have to appeal to an auxiliary theorem with a smallness condition like Theorem 13.6.

Sketch of the proof It is sufficient to prove that under the conditions of the theorem u is smooth in $Q_{R/2}^*$: any point $(x, t) \in Q_R^*(x_0, t_0)$ belongs to a cylinder $Q_R^*(x, t)$, such that $Q_{2\rho}^*(x, t) \subset Q_R^*(x_0, t_0)$. Clearly $u \in L_t^{q'} L_x^q(Q_R^*(x_0, t_0))$ implies that $u \in L_t^{q'} L_x^q(Q_{2\rho}^*(x, t))$, and so if we can show that $u \in L_t^{q'} L_x^q(Q_R^*(x, t))$ implies that u is spatially smooth in $Q_{R/2}^*(x, t)$, then by applying the result to $Q^*_{2\rho}(x, t)$ we find that u is spatially smooth on $Q^*_\rho(x, t)$ and hence at (x, t).

Step 1. Localisation of the solution

In view of the remark above, we only need to prove the result in a cylinder of half the diameter of the original one, and so we can (smoothly) localise the solution to the smaller cylinder, in such a way that the localisation vanishes at the boundary of the larger cylinder.

Given the cylinder Q^*_R in the statement of the theorem we consider a cutoff function ϕ that equals 1 in a smaller cylinder, say $Q^*_{\rho_i R}$, and is supported in $Q^*_{\rho_s R}$, with $1/2 < \rho_i < \rho_s < 1$. We do not fix any explicit values of ρ_i and ρ_s at this stage, as we will have to iterate the localisation argument, and in the final step we want to be able to use a cutoff function that equals 1 in $Q^*_{R/2}$, i.e. we want $\rho_i > 1/2$ in the final iteration. The number of iterations depends on the exponents q and q', but it will always be finite, making it possible to choose suitable ρ_i and ρ_s at every stage to guarantee we can prove the final result for $Q^*_{R/2}$.

We denote by W the localisation of ω, that is $W = \phi\omega$, and we consider the evolution of W. We can write the equation for W using the following shorthand,

$$\partial_t W - \Delta W = \partial(Wu) + Wu + \partial W + W,$$

where we use the notation $\partial(Wu)$ to denote terms that can be written as first derivatives of products of components of u with components of W, and that might have as coefficients smooth functions other that u and W, like the cutoff ϕ, or its derivatives. The other terms on the right-hand side are to be interpreted in a similar way.

It is reasonable to believe that the most difficult term to handle will be the first one, given its quadratic nature, and the fact that it also contains one derivative. To simplify the exposition in this sketch of the proof, we will therefore assume that W satisfies

$$\partial_t W - \Delta W = \partial(Wu) \tag{13.5}$$

on Q^*_R, with W vanishing on the boundary of Q^*_R.

Step 2. Use estimates for the heat equation to show that $\omega \in L^\infty_t L^\infty_x$

At this point we view (13.5) as the heat equation $\partial_t W - \Delta W = \partial f$, for a function f that we only assume belongs to some space $L^{d'}_t L^a_x$ (as a consequence of the hypotheses on u and ω). We will write $W = (\partial_t - \Delta)^{-1} \partial f$ and look for any available estimates of the form $\|W\|_{L^r_t L^r_x} \leq c\|f\|_{L^{d'}_t L^a_x}$, hoping to make r and r' as large as possible, to improve on the regularity of W and hence ω.

We will prove all the required estimates for the heat equation in Appendix D; the result we need in this sketch comes from Theorem D.7: if $1 \le a \le r \le \infty$, $1 < a' \le r' \le \infty$ satisfy

$$\frac{2}{a'} + \frac{3}{a} < \frac{2}{r'} + \frac{3}{r} + 1 \qquad (13.6)$$

then there exists a constant c (that might depend on R) such that

$$\|W\|_{L_t^{r'} L_x^r} = \|(\partial_t - \Delta)^{-1} \partial f\|_{L_t^{r'} L_x^r} \le c\|f\|_{L_t^{a'} L_x^a}.$$

Notice that while Theorem D.7 gives estimates over the whole space the support of the functions involved is contained in Q_R^*. Therefore all norms above can be taken over Q_R^*.

Applying this estimate to equation (13.5) and using Hölder's inequality we obtain

$$\|W\|_{L_t^{r'} L_x^r} \le c\|Wu\|_{L_t^{a'} L_x^a}$$
$$\le c\|W\|_{L_t^{m'} L_x^m} \|u\|_{L_t^{q'} L_x^q},$$

where

$$\frac{1}{a} = \frac{1}{q} + \frac{1}{m} \qquad \text{and} \qquad \frac{1}{a'} = \frac{1}{q'} + \frac{1}{m'}.$$

In terms of q, q', m, and m', condition (13.6) becomes

$$\frac{2}{q'} + \frac{2}{m'} + \frac{3}{q} + \frac{3}{m} < \frac{2}{r'} + \frac{3}{r} + 1,$$

which we can rewrite as

$$\frac{2}{m'} - \frac{2}{r'} + \frac{3}{m} - \frac{3}{r} < 1 - \left(\frac{2}{q'} + \frac{3}{q}\right). \qquad (13.7)$$

By assumption, the right-hand side is strictly positive. Notice that if we were studying the case $\frac{2}{q'} + \frac{3}{q} = 1$ then the right-hand side would vanish and the following approach would produce no improvement in the exponents; this is why the case of equality is significantly harder.

We now use the hypothesis that $u \in L_t^{q'} L_x^q(Q_R^*)$ together with the fact that $W \in L_t^{m'} L_x^m(Q_R^*)$ for some m, m' (initially $m = m' = 2$ since we are dealing with a weak solution) to show that in fact W is more regular and belongs to $L_t^{r'} L_x^r$, with $r > m$ and $r' > m'$. By iterating this argument it is possible to show that in fact we have $W \in L_t^\infty L_x^\infty$ in a slightly smaller cylinder. Full details of the iteration scheme and an estimate on the number of steps required are presented in the next section.

Finally we note that since $W = \omega$ in $Q^*_{\rho_i R}$ it follows that $\omega \in L^\infty_t L^\infty_x(Q^*_{\rho_i R})$.

Step 3. Deduce that $u \in L^\infty_t L^\infty_x$ using the local Biot–Savart Law

We showed in the previous chapter that we can reconstruct u from ω using a local version of the Biot–Savart Law (Theorem 12.5). In fact we obtained the expression

$$u(x) = -\frac{1}{4\pi} \int_\Omega \left(\nabla_y \frac{1}{|x-y|} \right) \times \omega(y)\,dy + A(x),$$

where the function A is a harmonic. In Theorem 12.6 we showed that if we have $u \in L^\infty_t L^2_x$ then for any $2 \le \beta \le \infty$

$$\omega \in L^\beta_t L^\infty_x(Q^*_R) \qquad \Rightarrow \qquad u \in L^\beta_t L^\infty_x(Q^*_{R'}),$$

for any $R' < R$, which in our case implies that $u \in L^\infty_t L^\infty_x(Q^*_{R'})$.

Step 4. Higher regularity via parabolic theory

Further regularity is obtained by using general results for parabolic equations, coupled to the result of Theorem 12.6 that allows us to deduce the regularity of u from that of ω. We show in Theorem D.8 that if w solves

$$\partial_t w - \Delta w = \partial f \qquad \text{in } Q^*_R$$

then for $R' < R$ and $0 < \alpha < 1$

$$f \in L^\infty_t L^\infty_x(Q^*_R) \qquad \Rightarrow \qquad w \in L^\infty_t C^{0,\alpha}_x(Q^*_{R'}), \qquad (13.8)$$

$$f \in L^\infty_t C^{0,\alpha}_x(Q^*_R) \qquad \Rightarrow \qquad w \in L^\infty_t C^1_x(Q^*_{R'}). \qquad (13.9)$$

Since in our case $f = Wu$, and we have shown that both u and ω are in $L^\infty_t L^\infty_x$, we find via (13.8) that in fact W is more regular. Using the Biot–Savart Law (Theorem 12.6) we can translate this result into regularity for u. This in turn means that f is more regular, and so we can use (13.9) to gain even more regularity of ω. By repeating this approach we can obtain the smoothness of u and ω. Full details are presented in the next section. $\qquad\qquad\square$

13.3.2 Proof of Theorem 13.7

Recall that we are trying to prove that if u is a weak solution of (13.2) on some parabolic cylinder $Q^*_R(x_0, t_0)$ and

$$u \in L^{q'}_t L^q_x(Q^*_R), \qquad \frac{2}{q'} + \frac{3}{q} < 1, \qquad (13.10)$$

where $3 < q \leq \infty$ and $2 < q' \leq \infty$, then u is smooth with respect to the space variables in Q_R^*.

In the sketch presented above, Steps 1–3 boil down to deducing that under these conditions $u \in L_t^\infty L_x^\infty(Q_{R/2}^*)$. We now follow the above outline, filling in the missing details in the localisation step and including all the terms in the resulting equation for W.

Proof We note that it is enough to solve the problem for the unit cylinder $Q_1^*(0, 0)$.

Step 1. Localisation

Since the main goal is to prove an interior estimate for ω, that is, an $L_t^\infty L_x^\infty$ estimate in the smaller cylinder $Q_{1/2}^*$, we start by localising the equation. As before, we consider a cutoff function ϕ that equals 1 in $Q_{\rho_i}^*$ and vanishes outside $Q_{\rho_s}^*$, with $1/2 < \rho_i < \rho_s < 1$ to be determined later.

As noted in Chapter 12, the vorticity equation needs to be understood in its weak formulation or in the distributional sense. As a consequence, the equations considered in the following discussion need to be understood in a similar fashion. Define $W = \phi\omega$. We start by considering the action of the heat operator on W, that is $(\partial_t - \Delta)W$, in terms of the cutoff ϕ and ω. An elementary calculation shows that (for the ith component)

$$\partial_t W_i - \Delta W_i = \phi(\partial_t \omega_i - \Delta\omega_i) + (\partial_t\phi - \Delta\phi)\omega_i - 2\nabla\phi \cdot \nabla\omega_i, \quad (13.11)$$

with the equation understood in the distributional sense.

We now use the fact that ω is a solution of the vorticity form of the Navier–Stokes equations (13.4) to rewrite the first term on the right-hand side. We rewrite the term $-2\nabla\phi \cdot \nabla\omega_i$ as $-2\partial_j((\partial_j\phi)\omega_i) + 2(\Delta\phi)\omega_i$, where we use the standard convention of summation over repeated indices, i.e. over j in this instance. We obtain

RHS of (13.11)

$$= \phi(\omega_j\partial_j u_i - u_j\partial_j\omega_i) + (\partial_t\phi)\omega_i + (\Delta\phi)\omega_i - 2\partial_j((\partial_j\phi)\omega_i)$$

$$= \partial_j(\phi\omega_j u_i - \phi u_j\omega_i)$$

$$\quad - (\partial_j\phi)(\omega_j u_i - u_j\omega_i) + (\partial_t\phi + \Delta\phi)\omega_i - 2\partial_j((\partial_j\phi)\omega_i).$$

We can rewrite equation (13.11) (using W whenever we see $\phi\omega$) as

$$\partial_t W_i - \Delta W_i = \partial_j(W_j u_i - u_j W_i) - (\partial_j\phi)(\omega_j u_i - u_j\omega_i)$$

$$\quad + (\partial_t\phi + \Delta\phi)\omega_i - 2\partial_j((\partial_j\phi)\omega_i). \quad (13.12)$$

Step 2. Improve on the integrability properties of W

We assume that $W \in L_t^{m'} L_x^m$ and deduce that $W \in L_t^{r'} L_x^r$ for some $r' > m'$, $r > m$.

Via heat estimates we reduce estimating the $L_t^{r'} L_x^r$ norm of W_i to estimating each of the terms on the right-hand side of equation (13.12). We have

$$\|W_i\|_{L_t^{r'} L_x^r} \le c \Big[\|(\partial_t - \Delta)^{-1} (\partial_j (W_j u_i - u_j W_i))\|_{L_t^{r'} L_x^r}$$
$$+ \|(\partial_t - \Delta)^{-1} ((\partial_j \phi)(\omega_j u_i - u_j \omega_i))\|_{L_t^{r'} L_x^r}$$
$$+ \|(\partial_t - \Delta)^{-1} ((\phi_t + \Delta \phi)\omega_i)\|_{L_t^{r'} L_x^r}$$
$$+ \|(\partial_t - \Delta)^{-1} (\partial_j ((\partial_j \phi)\omega_i))\|_{L_t^{r'} L_x^r} \Big]$$
$$\le c \Big[\|(W_j u_i - u_j W_i)\|_{L_t^{a'} L_x^a} + \|(\partial_j \phi)(\omega_j u_i - u_j \omega_i)\|_{L_t^{l'} L_x^l}$$
$$+ \|(\phi_t + \Delta \phi)\omega_i\|_{L_t^{l'} L_x^l} + \|(\partial_j \phi)\omega_i\|_{L_t^{m'} L_x^m} \Big],$$

provided that $1 \le a \le r \le \infty$, $1 < a' \le r' \le \infty$ satisfy

$$\frac{2}{a'} + \frac{3}{a} < \frac{2}{r'} + \frac{3}{r} + 1;$$

$1 \le l \le r \le \infty$, $1 < l' \le r' \le \infty$ satisfy

$$\frac{2}{l'} + \frac{3}{l} < \frac{2}{r'} + \frac{3}{r} + 2;$$

and finally $1 \le m \le r \le \infty$, $1 < m' \le r' \le \infty$ satisfy

$$\frac{2}{m'} + \frac{3}{m} < \frac{2}{r'} + \frac{3}{r} + 1.$$

Here we have used Theorem D.7 for the first and fourth term (yielding the conditions on a, a', m and m' above) and Theorem D.6 for the second and third term (yielding the conditions on l and l').

Given q, q' satisfying the Serrin condition

$$\frac{2}{q'} + \frac{3}{q} < 1 \tag{13.13}$$

and m, m' arising from our assumption at this stage of the induction for ω we define a, a', l, l' by

$$\frac{1}{l} = \frac{1}{a} = \frac{1}{q} + \frac{1}{m} \qquad \text{and} \qquad \frac{1}{l'} = \frac{1}{a'} = \frac{1}{q'} + \frac{1}{m'}, \tag{13.14}$$

and so the constraints above become

$$\frac{2}{q'} + \frac{3}{q} + \frac{2}{m'} + \frac{3}{m} < \frac{2}{r'} + \frac{3}{r} + 1, \tag{13.15}$$

$$\frac{2}{q'} + \frac{3}{q} + \frac{2}{m'} + \frac{3}{m} < \frac{2}{r'} + \frac{3}{r} + 2, \qquad \text{and}$$

$$\frac{2}{m'} + \frac{3}{m} < \frac{2}{r'} + \frac{3}{r} + 1.$$

Notice that the second and third conditions will be satisfied automatically if the first is.

Given the definitions of a, a', l, l' we can use Hölder's inequality to obtain

$$\|W_i\|_{L_t^{l'} L_x^{l}} \le c \Big[\|u\|_{L_t^{q} L_x^{q}} \|W_i\|_{L_t^{m'} L_x^{m}} + \|u\|_{L_t^{q} L_x^{q}} \|\omega\|_{L_t^{m'} L_x^{m}} + \|\omega\|_{L_t^{l'} L_x^{l}} + \|\omega\|_{L_t^{m'} L_x^{m}} \Big].$$

Since $u \in L_t^{q} L_x^{q}$ (and $l \le m$, $l' \le m'$ by definition, see (13.14)) we obtain

$$\|W_i\|_{L_t^{l'} L_x^{l}} \le c \|\omega\|_{L_t^{m'} L_x^{m}},$$

where the exponents must satisfy the only remaining constraint (13.15), which we rewrite as

$$\frac{2}{m'} - \frac{2}{r'} + \frac{3}{m} - \frac{3}{r} < 1 - \frac{2}{q'} - \frac{3}{q}. \tag{13.16}$$

It is clear from the above constraint that since the right-hand side is strictly positive, we can choose r, r' in such a way that $m < r$ and $m' < r'$, hence gaining regularity for W, and since $W = \phi \omega$ and $\phi = 1$ in $Q_{\rho_i}^*$ obtaining a bound on $\|\omega\|_{L_t^{l'} L_x^{l}(Q_{\rho_i}^*)}$.

In order to quantify the gain in regularity, it will be convenient for us to define

$$\chi = \frac{1}{6} \left(1 - \left(\frac{2}{q'} + \frac{3}{q} \right) \right),$$

so that the right-hand side of (13.16) is exactly 6χ. Now, we will explicitly choose r and r', taking

$$r = \begin{cases} \frac{m}{1-m\chi} & m\chi < 1 \\ \infty & m\chi \ge 1 \end{cases} \qquad \text{and} \qquad r' = \begin{cases} \frac{m'}{1-m'\chi} & m'\chi < 1 \\ \infty & m'\chi \ge 1 \end{cases} \tag{13.17}$$

(this is not the only possibility). It is easy to check that

$$\frac{3}{m} - \frac{3}{r} \le 3\chi \qquad \text{and} \qquad \frac{2}{m'} - \frac{2}{r'} \le 2\chi,$$

so that the left-hand side of (13.16) is $\leq 5\chi$, hence verifying the constraint on the exponents.

We have proved that

$$\omega \in L_t^{m'} L_x^m(Q_1^*) \quad \Rightarrow \quad \omega \in L_t^{m'/(1-m'\chi)} L_x^{m/(1-m\chi)}(Q_{\rho_i}^*),$$

assuming that r and r' are finite. Repeating this argument, we obtain

$$\omega \in L_t^{m'/(1-2m'\chi)} L_x^{m/(1-2m\chi)}(Q_{\rho_i^2}^*),$$

assuming that $m'/(1 - 2m'\chi)$ and $m/(1 - 2m\chi)$ are finite. Notice that the exponents are calculated using (13.17), substituting in terms of m the value $m/(1 - m\chi)$ (similarly for m') to obtain

$$\frac{\dfrac{m}{1 - m\chi}}{1 - \dfrac{m}{1 - m\chi}\chi} = \frac{m}{1 - 2m\chi}.$$

After k iterations of this procedure we obtain

$$\omega \in L_t^{m'/(1-km'\chi)} L_x^{m/(1-km\chi)}(Q_{\rho_i^k}^*),$$

assuming that the exponents are finite. We would stop iterating once both exponents are infinity, that is, when

$$\frac{m}{1 - km\chi}\chi \geq 1 \quad \text{and} \quad \frac{m'}{1 - km'\chi}\chi \geq 1,$$

which can be rewritten as $m(k + 1)\chi \geq 1$ and $m'(k + 1)\chi \geq 1$. This means that we have to iterate this procedure only a finite number of times. Assuming that we have to iterate this procedure k times, we can choose ρ_i with $\rho_i^{k+1} > 3/4$, so that we obtain $\omega \in L_t^\infty L_x^\infty(Q_{3R/4})$.

Steps 3 and 4. Regularity of u

Just as in the sketch, it follows using the local version of the Biot–Savart Law and the results of Theorem 12.6 that $\omega \in L_t^\infty L_x^\infty(Q_{3R/4})$ implies that we have $u \in L_t^\infty L_x^\infty(Q_{3R'/4})$, for any $R' < R$.

To deduce the smoothness of u we combine our results for the Biot–Savart Law in Theorem 12.6 with parabolic regularity results (Theorem D.8) precisely as in the sketch above; the additional terms only affect the argument minimally.

Further regularity is obtained by using general results for parabolic equations, coupled to the result of Theorem 12.6 that allows us to deduce the regularity of u from that of ω. Combining Theorem D.8 and Theorem D.9 we know that if W solves

$$\partial_t W - \Delta W = \partial f + g \qquad \text{in } Q_R$$

then for $R' < R, 0 < \alpha < 1$,

$$f, g \in L_t^\infty L_x^\infty(Q_R) \quad \Rightarrow \quad W \in L_t^\infty C_x^{0,\alpha}(Q_{R'}), \qquad (13.18)$$

$$f, g \in L_t^\infty C_x^{0,\alpha}(Q_R) \quad \Rightarrow \quad W \in L_t^\infty C_x^1(Q_{R'}). \qquad (13.19)$$

Let us apply these results to equation (13.12), which we rewrite as

$$\partial_t W_i - \Delta W_i = \partial_j(\phi \omega_j u_i - u_j \phi \omega_i) \qquad (13.20)$$

$$- (\partial_j \phi)(\omega_j u_i - u_j \omega_i) + (\partial_t \phi + \Delta \phi)\omega_i - 2\partial_j((\partial_j \phi)\omega_i). \quad (13.21)$$

On the right-hand side we have rewritten W as $\phi \omega$. We will obtain the additional regularity by applying (13.18) and (13.19), with ∂f given by the right-hand side of (13.20) and g by (13.21). Given that our initial assumption is that $\omega, u \in L_t^\infty L_x^\infty(Q_{3R'/4})$, for any $R' < R$, choosing ϕ appropriately (to guarantee it vanishes outside $Q_{3R'/4}$ and equals 1 in $Q_{3R''/4}$), for any $R'' < R'$ we can use (13.18) to obtain:

$$f, g \in L_t^\infty L_x^\infty(Q_{3R'/4}) \quad \Rightarrow \quad W, \omega \in L_t^\infty C_x^{0,\alpha}(Q_{3R''/4}), \qquad 0 < \alpha < 1.$$

Now we can use our results for u given in terms of ω by the Biot–Savart Law (Theorem 12.6) to obtain further regularity for u on $Q_{3R''/4}, R''' < R''$. We have

$$\omega \in L_t^\infty C_x^{0,\alpha}(Q_{3R''/4}) \quad \Rightarrow \quad u \in L_t^\infty C_x^{1,\alpha'}(Q_{3R'''/4}), \qquad 0 < \alpha' < \alpha.$$

Since now $\omega u \in L_t^\infty C_x^{0,\alpha'}(Q_{3R'''/4})$ we can use (13.19) for any $R'''' < R'''$:

$$f, g \in L_t^\infty C_x^{0,\alpha'}(Q_{3R'''/4}) \quad \Rightarrow \quad W, \omega \in L_t^\infty C_x^1(Q_{3R''''/4}).$$

Smoothness of u and ω on $Q_{R/2}$ now follows by repeating this argument inductively. $\qquad\qquad\square$

13.4 The case $\frac{2}{q'} + \frac{3}{q} = 1$

In this section we prove Theorem 13.6, which we restate here for convenience.

Theorem *Let u be a local weak solution of* (13.2) *on some parabolic cylinder $Q_R^*(x_0, t_0)$. Then for every q with $3 \le q \le \infty$ there exists an ε_q such that if*

$$\|u\|_{L_t^{q'} L_x^q(Q_R^*)} < \varepsilon_q, \qquad \text{where} \qquad \frac{2}{q'} + \frac{3}{q} = 1, \qquad (13.22)$$

then u is a smooth function with respect to the space variables in $Q_{R/2}^*$.

13.4.1 Sketch of the argument

We now sketch the proof of Theorem 13.6, where we also have the smallness assumption (13.22).

Step 1. Localisation

The argument initially proceeds as in the case $\frac{2}{q'} + \frac{3}{q} < 1$, with the same localisation (recall that $W = \phi\omega$) as in Step 1 in Section 13.3.1, yielding the equation

$$\partial_t W - \Delta W = \partial(Wu) + Wu + \partial W + W.$$

This time we will keep two terms in the sketch proof and consider

$$\partial_t W - \Delta W = \partial(Wu) + \partial W. \tag{13.23}$$

Step 2. Use estimates for the heat equation to show that $\omega \in L_t^\beta L_x^\infty$

As before, we now use estimates for the heat equation (Theorems D.6 and D.7) to bound the solution W of (13.23) in terms of the right-hand side. Notice that here we use the fact that the initial data for W is zero, which is achieved by the localisation.

If we try to reproduce the approach of the previous section then, following the same heat equation estimates, we end up requiring the relationship

$$\frac{2}{m'} - \frac{2}{r'} + \frac{3}{m} - \frac{3}{r} < 1 - \left(\frac{2}{q'} + \frac{3}{q}\right) \tag{13.24}$$

between the various exponents, but now the right-hand side vanishes. However, the case $r' < \infty$ of Theorem D.7 allows for equality in (13.24), so we can try to proceed in this way. This is still not enough to obtain any immediate gain in regularity without an additional ingredient; it is here that the smallness condition (13.22) will come to the rescue.

Let us now proceed with the estimates. We want to study equation (13.23) assuming only that $\omega \in L_t^{s'} L_x^s$ and $u \in L_t^{q'} L_x^q$. Using Theorems D.6 and D.7 we obtain

$$\|W\|_{L_t^{r'} L_x^r} \leq c\|Wu\|_{L_t^{m'} L_x^m} + c\|W\|_{L_t^{s'} L_x^s},$$

provided that the exponents satisfy

$$\frac{2}{m'} + \frac{3}{m} \leq \frac{2}{r'} + \frac{3}{r} + 1 \quad \text{and} \quad \frac{2}{s'} + \frac{3}{s} \leq \frac{2}{r'} + \frac{3}{r} + 1.$$

We want to apply Hölder's inequality to the first term on the right-hand side, and so we choose m, m' to satisfy

$$\frac{1}{m} = \frac{1}{q} + \frac{1}{r} \quad \text{and} \quad \frac{1}{m'} = \frac{1}{q'} + \frac{1}{r'}.$$

It follows that

$$\|W\|_{L_t^{r'} L_x^r} \leq C\|u\|_{L_t^{q'} L_x^q} \|W\|_{L_t^{r'} L_x^r} + c\|W\|_{L_t^{s'} L_x^s}. \tag{13.25}$$

We now use the smallness condition on $\|u\|_{L_t^{q'} L_x^q}$: since C is a universal constant, if we assume that $\|u\|_{L_t^{q'} L_x^q}$ is sufficiently small (say $\leq \frac{1}{2C}$) we can absorb this term on the left-hand side and deduce that

$$\|W\|_{L_t^{r'} L_x^r} \leq 2c\|W\|_{L_t^{s'} L_x^s}, \tag{13.26}$$

where we still require $\frac{2}{s'} + \frac{3}{s} \leq \frac{2}{r'} + \frac{3}{r} + 1$. This allows for an improvement in the regularity of W provided that the right-hand side of (13.26) is finite. In terms of ω we obtain the estimate

$$\|\omega\|_{L_t^{r'} L_x^r(Q_{\rho_i R})} \leq 2c\|\omega\|_{L_t^{s'} L_x^s(Q_R)}, \tag{13.27}$$

given that $\phi = 1$ on $Q_{\rho_i R}$ and is bounded above on Q_R.

We should point out that this argument is more delicate than presented here, and has the potentially fatal flaw that we have implicitly assumed that $\|W\|_{L_t^{r'} L_x^r}$ is finite, which a priori we do not know (and is in fact what we set out to prove). Indeed, equation (13.25) would be true even if $\|W\|_{L_t^{r'} L_x^r} = \infty$, in which case we could not use the smallness of $\|u\|_{L_t^{q'} L_x^q}$ to obtain the improved regularity in equation (13.27). In order to ensure that this issue does not arise, in the proof proper we will have to regularise the equation to guarantee that its solution belongs to $L_t^{r'} L_x^r$; we then derive the estimate above for the regularised solution, and show that it holds in the limit for the solution of the original equation. Full details are presented in the next section.

Step 3. Iterate to show that $\omega \in L_t^\beta L_x^\infty$

Just as in the case of strict inequality we now iterate this procedure, replacing Q_R by $Q_{\rho_i R}$ and (s, s') by the higher (r, r'); our eventual conclusion (after a finite number of iterations) is that $\omega \in L_t^\beta L_x^\infty(Q_{3R/4})$ for every $\beta \in [2, \infty)$.

Step 4. Use the Biot–Savart Law to deduce that $u \in L_t^\beta L_x^\infty$

We can now use the Biot–Savart Law and consequent estimates to deduce that $u \in L_t^\beta L_x^\infty(Q_{R/2})$ for every $\beta \in [2, \infty)$. With this additional regularity for u, we notice that the pair of exponents (β, ∞) satisfy the Serrin condition with a strict inequality for $\beta > 2$, that is $\frac{2}{\beta} + \frac{3}{\infty} < 1$, and so the theorem follows directly from Theorem 13.7.

13.4.2 Proof of Theorem 13.6

The sketch proof above shows that it is sufficient to prove that for exponents q, q' such that

$$3 < q \leq \infty, \quad 2 \leq q' < \infty, \quad \frac{2}{q'} + \frac{3}{q} = 1 \qquad (13.28)$$

there exists a constant ε_q such that if u is a weak solution of (13.2) on $Q_R(x_0, t_0)$ and

$$\|u\|_{L_t^{q'} L_x^q(Q_R)} < \varepsilon_q$$

then $\omega \in L_t^\beta L_x^\infty(Q_{R/2})$ for some β with $2 \leq \beta < \infty$.

Proof   As before, we note that it is enough to solve the problem for the unit cylinder: we can rescale and consider a weak solution u on $Q_1(0, 0)$ such that

$$\|u\|_{L_t^{q'} L_x^q(Q_1)} < \varepsilon_q.$$

We now follow the same steps as in the sketch of the proof presented in Section 13.4.1.

Step 1. Localisation

We proceed just as in the proof of Theorem 13.7. Recall that we have chosen ρ_i and ρ_s such that $1/2 < \rho_i < \rho_s < 1$ and set $W = \phi\omega$ where ϕ is a cutoff function that is 1 on Q_{ρ_i} and has support within Q_{ρ_s}. We obtain the equation

$$\partial_t W_i - \Delta W_i = \partial_j(W_j u_i - u_j W_i) - (\partial_j \phi)(\omega_j u_i - u_j \omega_i)$$
$$+ (\partial_t \phi + \Delta\phi)\omega_i - 2\partial_j((\partial_j \phi)\omega_i). \qquad (13.29)$$

In order to make the argument in the sketch proof rigorous we consider a regularised version of this equation. We denote by $u_{i,\varepsilon}$ the convolution in both space and time of u_i with ρ_ε for some $\rho \in C_c^\infty(\mathbb{R}^4)$ that satisfies $\int_{\mathbb{R}^4} \rho = 1$. That is, given $\rho \in C_c^\infty(\mathbb{R}^4)$ with $\rho \geq 0$, we define

$$\rho_\varepsilon(y, s) = \frac{1}{\varepsilon^4} \rho\left(\frac{y}{\varepsilon}, \frac{s}{\varepsilon}\right)$$

and $u_{i,\varepsilon} := u_i ** \rho_\varepsilon$, where

$$f ** g(x,t) := \int_{-\infty}^{\infty} \int_{\mathbb{R}^3} f(x - y, t - s) g(y,s) \, dy \, ds.$$

This procedure is just a space–time version of the mollification technique described in Appendix A; the resulting function $u_{i,\varepsilon}$ is smooth in both variables (see Theorem A.11).

Step 2. Use estimates for the heat equation to show that $\omega \in L_t^\beta L_x^\infty$

We denote by W_i^ε the solution of the regularised equation

$$\partial_t W_i^\varepsilon - \Delta W_i^\varepsilon = \partial_j (W_j^\varepsilon u_{i,\varepsilon} - u_{j,\varepsilon} W_i^\varepsilon) - (\partial_j \phi)(\omega_{j,\varepsilon} u_{i,\varepsilon} - u_{j,\varepsilon} \omega_{i,\varepsilon})$$
$$+ (\partial_t \phi + \Delta \phi) \omega_{i,\varepsilon} - 2 \partial_j ((\partial_j \phi) \omega_{i,\varepsilon}).$$

Since we have used a cutoff function we know that $W_i^\varepsilon(x,t) = 0$ for all $t \in (-1, -1 + \rho_s)$, and so using Theorem D.1 we know that there exists a unique smooth solution of this equation which we can write as

$$W_i^\varepsilon(x,t) = (\partial_t - \Delta)^{-1} (\partial_j (W_j^\varepsilon u_{i,\varepsilon} - u_{j,\varepsilon} W_i^\varepsilon))$$
$$- (\partial_t - \Delta)^{-1} ((\partial_j \phi)(\omega_{j,\varepsilon} u_{i,\varepsilon} - u_{j,\varepsilon} \omega_{i,\varepsilon}))$$
$$+ (\partial_t - \Delta)^{-1} ((\partial_t \phi + \Delta \phi) \omega_{i,\varepsilon})$$
$$- 2(\partial_t - \Delta)^{-1} (\partial_j ((\partial_j \phi) \omega_{i,\varepsilon})).$$

Notice that since we have mollified ω and u on the right-hand side, and we are using a cutoff function, all norms appearing below are finite.

We reduce estimating the $L_t^{r'} L_x^r$ norm of W_i^ε to estimating each of the terms on the right-hand side. We therefore have

$$\|W_i^\varepsilon\|_{L_t^{r'} L_x^r} \le c \Big[\|(\partial_t - \Delta)^{-1}(\partial_j (W_j^\varepsilon u_{i,\varepsilon} - u_{j,\varepsilon} W_i^\varepsilon))\|_{L_t^{r'} L_x^r}$$
$$+ \|(\partial_t - \Delta)^{-1}((\partial_j \phi)(\omega_{j,\varepsilon} u_{i,\varepsilon} - u_{j,\varepsilon} \omega_{i,\varepsilon}))\|_{L_t^{r'} L_x^r}$$
$$+ \|(\partial_t - \Delta)^{-1}((\partial_t \phi + \Delta \phi) \omega_{i,\varepsilon})\|_{L_t^{r'} L_x^r}$$
$$+ 2\|(\partial_t - \Delta)^{-1}(\partial_j ((\partial_j \phi) \omega_{i,\varepsilon}))\|_{L_t^{r'} L_x^r} \Big]$$
$$\le c \Big[\|(W_j^\varepsilon u_{i,\varepsilon} - u_{j,\varepsilon} W_i^\varepsilon)\|_{L_t^{q'} L_x^q}$$
$$+ \|(\partial_j \phi)(\omega_{j,\varepsilon} u_{i,\varepsilon} - u_{j,\varepsilon} \omega_{i,\varepsilon})\|_{L_t^{r'} L_x^l}$$
$$+ \|(\partial_t \phi + \Delta \phi) \omega_{i,\varepsilon}\|_{L_t^{r'} L_x^l} + 2\|(\partial_j \phi) \omega_{i,\varepsilon}\|_{L_t^{m'} L_x^m} \Big],$$

provided that $1 \le a \le r \le \infty$, $1 < a' \le r' < \infty$ satisfy

$$\frac{2}{a'} + \frac{3}{a} \le \frac{2}{r'} + \frac{3}{r} + 1, \tag{13.30}$$

$1 \le l \le r \le \infty$, $1 < l' \le r' < \infty$ satisfy

$$\frac{2}{l'} + \frac{3}{l} \le \frac{2}{r'} + \frac{3}{r} + 2, \tag{13.31}$$

and $1 \le m \le r \le \infty$, $1 < m' \le r' < \infty$ satisfy

$$\frac{2}{m'} + \frac{3}{m} \le \frac{2}{r'} + \frac{3}{r} + 1. \tag{13.32}$$

Here we have used Theorem D.7 for the first and fourth term (yielding the conditions on a, a', m and m' above) and Theorem D.6 for the second and third term (yielding the conditions on l and l'). We define a, a', l, l' in terms of m, m', q, q' as in the sketch (see (13.14)) by

$$\frac{1}{l} = \frac{1}{a} = \frac{1}{q} + \frac{1}{m}, \qquad \text{and} \qquad \frac{1}{l'} = \frac{1}{a'} = \frac{1}{q'} + \frac{1}{m'}. \tag{13.33}$$

Notice that (13.30) and (13.31) follow from (13.32), since we are assuming the Serrin condition $\frac{2}{q'} + \frac{3}{q} = 1$.

Given the definitions of a, a', l and l' we can use Hölder's inequality to obtain

$$\|W_i^\varepsilon\|_{L_t^{l'} L_x^r} \le c \Big[\|u_\varepsilon\|_{L_t^q L_x^p} \|W_i^\varepsilon\|_{L_t^{l'} L_x^r} + \|u_\varepsilon\|_{L_t^q L_x^p} \|\omega_\varepsilon\|_{L_t^{m'} L_x^m} \\ + \|\omega_\varepsilon\|_{L_t^{l'} L_x^l} + \|\omega_\varepsilon\|_{L_t^{m'} L_x^m} \Big].$$

Using the fact that $\|u_\varepsilon\|_{L_t^q L_x^p} \le \|u\|_{L_t^q L_x^p}$ (and similarly for the other mollified terms) and the fact that in a bounded domain L^p spaces are nested, we obtain

$$\|W_i^\varepsilon\|_{L_t^{l'} L_x^r} \le c \|u\|_{L_t^q L_x^p} \|\omega\|_{L_t^{l'} L_x^r} + c \|\omega\|_{L_t^{m'} L_x^m}.$$

We now choose ε_q to ensure that $\|u\|_{L_t^q L_x^p}$ is small enough (e.g. $<1/2c$) that we can absorb the first term on the right-hand side on the left-hand side to obtain

$$\|W_i^\varepsilon\|_{L_t^{l'} L_x^r} \le c \|\omega\|_{L_t^{m'} L_x^m}.$$

Given this estimate, which is uniform in ε, we can find a subsequence (that we denote the same way) and a function $\widetilde{W}_i$ such that W_i^ε converges weakly to $\widetilde{W}_i$ in $L_t^{l'} L_x^r$, for which

$$\|\widetilde{W}_i\|_{L_t^{l'} L_x^r} \le c \|\omega\|_{L_t^{m'} L_x^m}.$$

In particular, since $\omega \in L_t^2 L_x^2$ we have $\|\widetilde{W}_i\|_{L_t^{a'} L_x^2} \le c \|\omega\|_{L_t^2 L_x^2}$ for $2 < a' < \infty$.

We now prove that $\widetilde{W}_i$ solves (13.29). For any $\psi \in C_c^\infty(Q_R^*(x_0, t_0))$ we have

$$-\int \langle W_i^\varepsilon, \partial_t \psi_i \rangle + \langle W_i^\varepsilon, \Delta \psi_i \rangle \, dt = \int -\langle W_j^\varepsilon u_{i,\varepsilon} - u_{j,\varepsilon} W_i^\varepsilon, \partial_j \psi_i \rangle$$
$$- \langle (\partial_j \phi)(\omega_{j,\varepsilon} \partial_j u_{i,\varepsilon} - u_{j,\varepsilon} \partial_j \omega_{i,\varepsilon}), \psi_i \rangle$$
$$+ \langle (\partial_t \phi + \Delta \phi) \omega_{i,\varepsilon}, \psi_i \rangle + 2 \langle ((\partial_j \phi) \omega_{i,\varepsilon}), \partial_j \psi_i \rangle \, dt.$$

Since $\psi_i \in C_c^\infty$ is a test function, and $W_i^\varepsilon \rightharpoonup \widetilde{W}_i$ in $L_t^{\alpha'} L_x^2$ we have immediate convergence for the terms involving $\langle W_i^\varepsilon, \partial_t \psi_i \rangle$, and $\langle W_i^\varepsilon, \Delta \psi_i \rangle$. Since $w_{i,\varepsilon}$ converges in $L_t^2 L_x^2$ and $u \in L_t^q L_x^q$ we have convergence of every term in the second and third lines of the equation.

This leaves only $\langle W_j^\varepsilon u_{i,\varepsilon} - u_{j,\varepsilon} W_i^\varepsilon, \partial_j \psi_i \rangle$. Clearly it is enough to consider one of the terms. We choose the first, and rewrite it as

$$\int \langle W_j^\varepsilon u_i, \partial_j \psi_i \rangle + \langle W_j^\varepsilon (u_{i,\varepsilon} - u_i), \partial_j \psi_i \rangle \, dt;$$

for the second term, we use the uniform boundedness of W_j^ε in $L_t^{\alpha'} L_x^2$ and the fact that $\|u_{i,\varepsilon} - u_i\|_{L_t^{q'} L_x^q} \to 0$ to obtain convergence, while for the first we need only observe that $u_i \in L_t^q L_x^q$ and W_j^ε converges weakly-$*$ in $L_t^{\alpha'} L_x^2$.

The idea now is to bootstrap the resulting inequality

$$\|W\|_{L_t^{r'} L_x^r} \le c \|\omega\|_{L_t^{m'} L_x^m},$$

where

$$\frac{2}{m'} + \frac{3}{m} \le \frac{2}{r'} + \frac{3}{r} + 1,$$

starting from our assumption $\omega \in L_t^2 L_x^2$, i.e. $m = m' = 2$; the additional restriction that we have is $1 < r' < \infty$.

For $m = m' = 2$ we notice that $r = 2$, and $2 \le r' < \infty$ are acceptable. So we have shown that $W \in L_t^\beta L_x^2$ for any $2 \le \beta < \infty$, which shows $\omega \in L_t^\beta L_x^2$ in Q_{ρ_i}.

Step 3. Iterate to show that $\omega \in L_t^\beta L_x^\infty$

We now iterate this construction by choosing a new cutoff function, supported where the previous one was identically 1, and equal to 1 in a slightly smaller cylinder; we can show that $\omega \in L_t^\beta L_x^{r_k}(Q_{\rho_i^k})$ implies $\omega \in L_t^\beta L_x^{r_{k+1}}(Q_{\rho_i^{k+1}})$, provided that

$$\frac{2}{\beta} + \frac{3}{r_k} \le \frac{2}{\beta} + \frac{3}{r_{k+1}} + 1.$$

For example, taking $r_1 = 2$ in the first step we find that $r_2 = 6$, with the option of taking r_3 to be any number in the next iteration. This shows that we only need to iterate the procedure three times to obtain $\omega \in L_t^\beta L_x^\infty(Q_{\rho_i^3})$ for $2 \le \beta < \infty$.

Choosing ρ_i sufficiently large so that $\rho_i^3 > 3/4$, for example, we find that $\omega \in L_t^\beta L_x^\infty(Q_{3/4})$.

Step 4. Use the Biot–Savart Law to deduce that $u \in L_t^\beta L_x^\infty(Q_{R/2})$

It follows using the local version of the Biot–Savart Law and the results of Theorem 12.6 that $\omega \in L_t^\beta L_x^\infty(Q_{3R/4})$ implies that $u \in L_t^\beta L_x^\infty(Q_{3R/4})$.

Notice that at this point we can invoke Serrin's original result, as the pair of exponents ∞, β satisfy the Serrin condition with a strict inequality for $\beta > 2$, that is $\frac{2}{\beta} + \frac{3}{\infty} < 1$, and so the theorem follows directly from Theorem 13.7. $\quad\square$

We postpone the proof of Theorem 13.5 concerning the case $L_t^\infty L_x^3$ until Section 16.4 as it requires the partial regularity theory of Caffarelli, Kohn, & Nirenberg (1982). Instead in the final section of this chapter we consider the regularity in time for solutions that are spatially smooth.

13.5 Local Hölder regularity in time for spatially smooth u

We saw earlier (see the comments following Theorem 13.5) that given a purely local definition of a 'weak solution', smoothness of u in the space variables does not imply any temporal regularity, since the regularity of $\partial_t u$ is limited by that of ∇p. But when we have a solution of the initial-value problem, i.e. of the equations we considered throughout Part I, the analysis of Chapter 5 shows that ∇p does have some regularity; local smoothness of such a solution then implies Hölder continuity in time. This is a useful result, since it implies in particular that if a solution to the Navier–Stokes initial-value problem satisfies the condition of Theorem 13.4, then it is locally Hölder continuous in time.

We will need two lemmas. The first is a simple Sobolev-embedding result.

Lemma 13.8 *Suppose that $U \subset \mathbb{R}^3$ is a smooth bounded domain and that $\partial_t g \in L^q(0, T; L^r(U))$. Then*

$$\|g(t_2) - g(t_1)\|_{L^r(U)} \le C|t_2 - t_1|^{1-1/q}.$$

Proof We have

$$\|g(t_2) - g(t_1)\|_{L^r(U)} \le \int_{t_1}^{t_2} \|\partial_t g(\tau)\|_{L^r(U)}\, d\tau$$

$$\le \left(\int_{t_1}^{t_2} \|\partial_t g(\tau)\|_{L^r(U)}^q\, d\tau \right)^{1/q} |t_2 - t_1|^{1-1/q},$$

as claimed. $\quad\square$

We will also require the following variant of Agmon's inequality (see Theorem 1.20).

Lemma 13.9 *Let $U \subset \mathbb{R}^3$ be a smooth bounded domain and suppose that $f \in H^k(U)$ for some integer $k \geq 2$, and that $1 \leq r \leq 2$. Then*

$$\|f\|_{L^\infty(U)} \leq C_{k,r} \|f\|_{L^r(U)}^\theta \|f\|_{H^k(U)}^{1-\theta}, \tag{13.34}$$

where

$$\theta = \frac{r(k - \frac{3}{2})}{kr + \frac{3}{2}(2 - r)}.$$

Proof If we interpolate the H^1 and H^2 norms between L^2 and H^k using Lemma 1.17,

$$\|f\|_{H^1} \leq c\|f\|_{L^2}^{1-1/k}\|f\|_{H^k}^{1/k}, \qquad \|f\|_{H^2} \leq c\|f\|_{L^2}^{1-2/k}\|f\|_{H^k}^{2/k},$$

then from the standard Agmon inequality (Theorem 1.20) we obtain

$$\|f\|_{L^\infty} \leq c\|f\|_{H^1}^{1/2}\|f\|_{H^2}^{1/2} \leq c\|f\|_{L^2}^{1-3/2k}\|f\|_{H^k}^{3/2k}.$$

Now interpolate L^2 between L^r and L^∞,

$$\|f\|_{L^2} \leq \|f\|_{L^r}^{r/2}\|f\|_{L^\infty}^{1-r/2},$$

which implies that

$$\|f\|_{L^\infty} \leq c\|f\|_{L^r}^{r(1-3/2k)/2}\|f\|_{L^\infty}^{(1-3/2k)(1-r/2)}\|f\|_{H^k}^{3/2k};$$

this yields (13.34). $\qquad\qquad\qquad\qquad\qquad\qquad\qquad\qquad\qquad\qquad\quad\square$

We can now show that solutions of the initial-value problem that are locally smooth in space are also locally Hölder continuous in time, using the bounds we have on ∇p in Corollary 5.2.

Proposition 13.10 *Suppose that u is a weak solution of the Navier–Stokes initial-value problem on any three-dimensional domain Ω (a smooth bounded domain, $\mathbb{R}^3$, or $\mathbb{T}^3$), and let U be a smooth bounded subdomain of Ω. If $u \in L^\infty(0, T; H^k(U))$ for some integer $k \geq 2$ then for any γ in the range*

$$0 < \gamma < \frac{1}{2}\left(\frac{k - \frac{3}{2}}{k + \frac{3}{2}}\right)$$

there exists a constant $C = C_\gamma$ such that

$$|u(x, t) - u(x, s)| \leq C|t - s|^\gamma$$

for all $x \in U$.

Proof Corollary 5.2 or its extension to bounded domains (Theorem 5.7) shows that

$$(u \cdot \nabla)u + \nabla p \in L^{2r/(4r-3)}(0, T; L^r(\Omega)), \qquad 1 < r \le 3/2,$$

for any (global) weak solution of the Navier–Stokes equations. Since by assumption $\Delta u \in L^\infty(0, T; L^2(U))$, it follows that

$$\partial_t u = \Delta u - (u \cdot \nabla)u - \nabla p \in L^{2r/(4r-3)}(0, T; L^r(U)) \qquad 1 < r \le 3/2,$$

where we have used the fact that the equation holds in the sense of distributions (Proposition 5.3 or Theorem 5.6). It now follows from Lemma 13.8 that

$$\|u(\cdot, t) - u(\cdot, s)\|_{L^r} \le C|t - s|^{(3-2r)/2r},$$

and then using Lemma 13.9 we obtain

$$\|u(t) - u(s)\|_{L^\infty} \le C_{k,r}\|u(t) - u(s)\|_{L^r}^\theta, \qquad \theta = \frac{r(k - \frac{3}{2})}{kr + \frac{3}{2}(2 - r)}$$

$$\le C_{k,r}|t - s|^{\theta(3-2r)/2r}.$$

Noting that we can take r arbitrarily close to 1 it follows that

$$|u(x, t) - u(x, s)| \le \|u(t) - u(s)\|_{L^\infty} \le C|t - s|^\gamma$$

for any $\gamma < \frac{1}{2}(k - \frac{3}{2})/(k + \frac{3}{2})$ as claimed. $\qquad \square$

Combining this result with that of Theorem 13.4 and 13.5, we obtain the following local regularity result based on the Serrin criterion.

Theorem 13.11 *Suppose that u is a weak solution of the Navier–Stokes initial-value problem on any three-dimensional domain Ω (a smooth bounded domain, $\mathbb{R}^3$, or $\mathbb{T}^3$). If U is a smooth bounded subdomain of Ω and*

$$u \in L^r(a, b; L^s(U)), \qquad \frac{2}{r} + \frac{3}{s} = 1, \quad s > 3,$$

then on $U \times (a, b)$ the function u is smooth in space and Hölder continuous in time with any exponent strictly less than one half.

Notes

In the local case the results are due to Serrin (1962) for the strict inequality $2/q' + 3/q < 1$, Fabes, Jones, & Rivière (1972) for $2/q' + 3/q = 1$ with $3 < q < \infty$, and Struwe (1988) for $2/q' + 3/q = 1$ with $q > 3$. The approach presented in this text is due to Takahashi (1990), where he obtains the same result as Struwe and covers the case $L_t^\infty L_x^3$ under a smallness assumption.

Uniqueness for the case $L_t^\infty L_x^3$ was known prior to Escauriaza, Seregin, & Šverák (2003). See, for example, Masuda (1984), Sohr & von Wahl (1984), Kozono & Sohr (1996), and Kozono (1998). Some partial regularity results for $L_t^\infty L_x^3$ are due to Neustupa (1999) and Beirão da Veiga (2000).

Exercise

13.1 Show that if (q, q') satisfy the condition $\frac{2}{q'} + \frac{3}{q} = 1$ then

$$\|u_R\|_{L_t^{q'} L_x^q(Q_1^*(0,0))} = \|u\|_{L_t^{q'} L_x^q(Q_R^*(0,0))}.$$

14

The local energy inequality

In Chapters 15 and 16 we will prove two local regularity criteria; a key ingredient in the proofs is a localised form of the energy inequality that forms the subject of this chapter. We will prove the existence of Leray–Hopf weak solutions that also satisfy the following localised form of the energy inequality:

$$2 \int_0^t \int_\Omega |\nabla u|^2 \varphi \, dx \, ds$$

$$\leq \int_0^t \int_\Omega |u|^2 \{\partial_t \varphi + \Delta \varphi\} \, dx \, ds + \int_0^t \int_\Omega (|u|^2 + 2p)(u \cdot \nabla)\varphi \, dx \, ds \tag{14.1}$$

for any smooth scalar function $\varphi \geq 0$ with compact support in the space–time domain $\mathbb{T}^3 \times (0, T)$.

First we give a formal derivation of this local energy inequality, assuming that the solutions are smooth. We then give a rigorous justification of the validity of this inequality (on a periodic domain), i.e. we prove the existence of at least one weak solution for which it is satisfied; such weak solutions are termed 'suitable'. To do this we use a regularised version of the Navier–Stokes equation

$$\partial_t u - \Delta u + [(\psi_\varepsilon * u) \cdot \nabla]u + \nabla p = 0, \qquad \nabla \cdot u = 0, \tag{14.2}$$

where ψ_ε is a mollifier in the space variables. (This regularisation was in fact used by Leray in his 1934 paper to prove the existence of weak solutions on the whole space.) The particular form of (14.2) allows us to derive a local energy inequality for (14.2), and then deduce (14.1) for those weak solutions of the Navier–Stokes equations that arise as the limit of such regularised solutions by taking the limit (along a subsequence) as $\varepsilon \to 0$. It will also allow us to prove the validity of the strong energy inequality for a class of weak solutions defined on $\mathbb{R}^3$.

14.1 Formal derivation of the local energy inequality

In order to give a formal derivation of the local energy inequality, we assume that we have a smooth solution of the Navier–Stokes equations. We then choose[1] a function $\varphi \in C_c^\infty(\mathbb{T}^3 \times (0, T))$ and take the scalar product of the equation with φu to obtain

$$\varphi u \cdot \partial_t u - \varphi u \cdot \Delta u + \varphi u \cdot [(u \cdot \nabla)u] + \varphi u \cdot \nabla p = 0.$$

We now integrate over $\mathbb{T}^3$ and use integration by parts repeatedly. We have

$$-\int_{\mathbb{T}^3} \varphi u \cdot \Delta u \, dx = \int_{\mathbb{T}^3} \varphi |\nabla u|^2 + (\partial_j \varphi)(\partial_j u_i) u_i \, dx,$$

and then noting that

$$\int_{\mathbb{T}^3} (\partial_j \varphi)(\partial_j u_i) u_i \, dx = \frac{1}{2} \int_{\mathbb{T}^3} (\partial_j \varphi) \partial_j |u|^2 \, dx = -\frac{1}{2} \int_{\mathbb{T}^3} (\Delta \varphi)|u|^2 \, dx$$

we obtain

$$-\int_{\mathbb{T}^3} \varphi u \cdot \Delta u \, dx = \int_{\mathbb{T}^3} \varphi |\nabla u|^2 \, dx - \frac{1}{2} \int_{\mathbb{T}^3} (\Delta \varphi)|u|^2 \, dx. \qquad (14.3)$$

Similarly

$$\int_{\mathbb{T}^3} \varphi u_i u_j (\partial_j u_i) \, dx = \frac{1}{2} \int_{\mathbb{T}^3} \varphi u_j \partial_j |u|^2;$$

and so, since u is divergence free,

$$\int_{\mathbb{T}^3} \varphi u_i u_j (\partial_j u_i) \, dx = -\frac{1}{2} \int_{\mathbb{T}^3} [(u \cdot \nabla)\varphi]|u|^2 \, dx. \qquad (14.4)$$

For the final term on the right-hand side a straightforward integration by parts yields

$$\int_{\mathbb{T}^3} \varphi u \cdot \nabla p = -\int_{\mathbb{T}^3} p(u \cdot \nabla)\varphi,$$

again using the fact that u is divergence free.

Now, since $\partial_t(|u|^2 \varphi) = 2\varphi u \cdot \partial_t u + |u|^2 \partial_t \varphi$, if we use (14.3) and (14.4) and integrate from 0 to T then we obtain

$$2 \int_0^T \int_{\mathbb{T}^3} |\nabla u|^2 \varphi \, dx \, ds$$

$$= \int_0^T \int_{\mathbb{T}^3} |u|^2 \{\partial_t \varphi + \Delta \varphi\} \, dx \, ds + \int_0^T \int_{\mathbb{T}^3} (|u|^2 + 2p)(u \cdot \nabla)\varphi \, dx \, ds,$$

$$(14.5)$$

---

[1] In this formal derivation there is no need to restrict to $\varphi \geq 0$.

which is the *local energy equality*. As with the standard energy equality for weak solutions, when we obtain a version of this in a rigorous way via a limiting argument we will obtain an inequality (the $=$ is replaced by $\leq$).

In Section 14.4 we will derive an equivalent form of the local energy inequality which will be more useful later, but for now we aim to prove the validity of the inequality version of (14.5).

14.2 The Leray regularisation

We will treat the equations on a periodic domain, although Theorem 14.1 also holds on the whole space. Suppose that $\psi \in C_c^\infty(\mathbb{R}^3)$ is a standard mollifier, i.e. a function with $0 \leq \psi \leq 1$ such that

$$\text{supp } \psi \subset B_1(0) \qquad \text{and} \qquad \int_{\mathbb{R}^3} \psi(x)\,dx = 1$$

(see Appendix A), and let $\psi_\varepsilon = \varepsilon^{-3}\psi(x/\varepsilon)$. We will consider, for some fixed $\varepsilon > 0$ sufficiently small that the support of ψ_ε is contained wholly within one period, a regularised version of the Navier–Stokes equations,

$$\partial_t u - \Delta u + [(\psi_\varepsilon * u) \cdot \nabla]u + \nabla p = 0, \qquad \nabla \cdot u = 0. \qquad (14.6)$$

Here $*$ denotes the convolution,

$$(\psi_\varepsilon * u)(x) = \int_{\mathbb{R}^3} \psi_\varepsilon(x - y)u(y)\,dy,$$

where we extend u periodically to the whole of $\mathbb{R}^3$ in this integral. By using Young's inequality for convolutions it is easy to prove the standard properties of $\psi_\varepsilon * u$ for $u \in L^p$: boundedness and convergence in L^p,

$$\|\psi_\varepsilon * u\|_{L^p} \leq \|u\|_{L^p}, \qquad \lim_{\varepsilon \to 0} \|\psi_\varepsilon * u - u\|_{L^p} = 0, \qquad (14.7)$$

and the smoothing estimates

$$\|\psi_\varepsilon * u\|_{L^q} \leq \varepsilon^{-(3/p-3/q)}\|u\|_{L^p}, \qquad \|\psi_\varepsilon * u\|_{H^s} \leq \varepsilon^{-s}\|u\|_{L^2} \qquad (14.8)$$

($p \leq q \leq \infty$ and $s \geq 0$); for more details see Appendix A. (Note that we only regularise in the space variable, so that (14.6) makes sense as an equation for the time evolution of u.)

One effect of the regularisation is that solutions of (14.6) behave better than those of the Navier–Stokes equations themselves: for each fixed $\varepsilon > 0$ this equation now has a unique smooth solution for all $t \geq 0$. While for each fixed $\varepsilon > 0$ the solutions are more regular (see (14.9)), the improved bounds are not uniform in ε (for otherwise they would hold for the Navier–Stokes limit); we only obtain uniform bounds that are consistent with what we already know for the Navier–Stokes equations (as in the second paragraph of the statement of the following theorem). However, this is also the case for the Galerkin approxima- tions, but there are certain cancellations that occur for (14.6) when deriving its local energy inequality that do not occur for the Galerkin approximations.

By a weak solution of (14.6) satisfying the initial condition $u_0 \in H$ we mean a function $u \in L^\infty(0, T; H) \cap L^2(0, T; V)$ for all $T > 0$ such that

$$\int_0^s -\langle u, \partial_t \varphi \rangle + \int_0^s \langle \nabla u, \nabla \varphi \rangle + \int_0^s \langle [(\psi_\varepsilon * u) \cdot \nabla] u, \varphi \rangle = \langle u_0, \varphi(0) \rangle - \langle u(s), \varphi(s) \rangle$$

for all $\varphi \in \mathcal{D}_\sigma$ and almost every $s > 0$ (cf. Definition 3.3).

Theorem 14.1 *Fix $\varepsilon > 0$. Then for each $u_0 \in H$ the regularised equation* (14.6) *has a unique weak solution u^ε such that for every $T > 0$*

$$u^\varepsilon \in L^\infty(0, T; H) \cap L^2(0, T; V) \qquad and \qquad \partial_t u^\varepsilon \in L^2(0, T; V^*) \quad (14.9)$$

and, for any $\tau > 0$, $u^\varepsilon \in L^\infty(\tau, T; H^k)$ for every $k \geq 0$.

For a fixed $u_0 \in H$, the solutions u^ε are bounded uniformly (with respect to ε) in $L^\infty(0, T; H) \cap L^2(0, T; V)$ and $\partial_t u^\varepsilon$ is bounded uniformly in $L^{4/3}(0, T; V^*)$, for every $T > 0$.

Furthermore there exists $p^\varepsilon \in L^{5/3}(\mathbb{T}^3 \times (0, \infty))$ such that the equations

$$\partial_t u^\varepsilon - \Delta u^\varepsilon + [(\psi_\varepsilon * u^\varepsilon) \cdot \nabla] u^\varepsilon + \nabla p^\varepsilon = 0, \qquad \nabla \cdot u^\varepsilon = 0 \quad (14.10)$$

hold in the sense of distributions on $\mathbb{T}^3 \times (0, \infty)$.

Leray (1934) proved the existence of regular solutions for (14.6) on $\mathbb{R}^3$ by using an explicit integral formula for the solution of the linear problem

$$\partial_t v - \Delta v + \nabla q = f(x, t), \qquad \nabla \cdot v = 0,$$

combined with an iterative method applied to this equation with the particular choice $f = -[(\psi * u) \cdot \nabla] u$. However, in the case of a periodic domain it is possible to prove the result using the Galerkin method as in Chapter 4, and we give a sketch of this proof here. We will return to the whole space case in Section 14.5.

Sketch proof   Consider the Galerkin approximations

$$\partial_t u_n - \Delta u_n + P_n \{[(\psi_\varepsilon * u_n) \cdot \nabla] u_n\} = 0, \qquad u_n(0) = P_n u_0 \quad (14.11)$$

(recall that P_n denotes the orthogonal projection onto the space spanned by the first n eigenfunctions of the Stokes operator, see (4.1)); this eliminates the ∇p term in (14.6). Note that we now have two parameters, the regularisation parameter ε and the dimension n of the Galerkin approximation; here we will fix ε and take $n \to \infty$. We take the inner product of (14.11) with u_n; since

$$\langle P_n\{[(\psi_\varepsilon * u_n) \cdot \nabla]u_n\}, u_n \rangle = \langle [(\psi_\varepsilon * u_n) \cdot \nabla]u_n, u_n \rangle = 0$$

we obtain

$$\frac{1}{2}\frac{d}{dt}\|u_n\|^2 + \|\nabla u_n\|^2 = 0.$$

Integrating this equation from 0 to t we obtain

$$\frac{1}{2}\|u_n(t)\|^2 + \int_0^t \|\nabla u_n(s)\|^2 \, ds \le \frac{1}{2}\|u_n(0)\|^2 \le \frac{1}{2}\|u_0\|^2,$$

which gives uniform bounds (with respect to n) on u_n in $L^\infty(0, T; L^2)$ and in $L^2(0, T; H^1)$. These are the same bounds that were obtained for the Navier–Stokes equations themselves.

However, with the regularised advection term the bounds on

$$\partial_t u_n = \Delta u_n - P_n\{[(\psi_\varepsilon * u_n) \cdot \nabla]u_n\}$$

are better than without the regularisation: taking the inner product with some $\varphi \in V$ yields

$$|\langle \partial_t u_n, \varphi \rangle| \le |\langle \nabla u_n, \nabla \varphi \rangle| + \left| \int [(\psi_\varepsilon * u_n) \cdot \nabla P_n \varphi] \cdot u_n \, dx \right|$$

$$\le \|\nabla u_n\|\|\nabla \varphi\| + \|\psi_\varepsilon * u_n\|_{L^6}\|u_n\|_{L^3}\|\nabla \varphi\|.$$

Since $\|u_n\|_{L^3} \le \|u_n\|_{L^2}^{1/2}\|u_n\|_{L^6}^{1/2}$ and $\|u_n\|_{L^6} \le c\|u_n\|_{H^1} \le c\|\nabla u_n\|$ it follows that

$$\|\partial_t u_n\|_{V^*} \le \|\nabla u_n\| + c\|\psi_\varepsilon * u_n\|_{H^1}\|u_n\|^{1/2}\|\nabla u_n\|^{1/2}$$

$$\le \|\nabla u_n\| + c\varepsilon^{-1}\|u_n\|^{3/2}\|\nabla u_n\|^{1/2},$$

since $\|\psi_\varepsilon * u\|_{H^1} \le c\varepsilon^{-1}\|u\|$ (see (14.8)). Therefore for each $\varepsilon > 0$ we obtain $\partial_t u_n \in L^2(0, T; V^*)$, although this bound degrades as $\varepsilon \to 0$. Recall that for standard weak solutions of the Navier–Stokes equations we have only $\partial_t u \in L^{4/3}(0, T; V^*)$ (see Lemma 3.7); we can still obtain such a bound here for u_n, and this bound *does* hold uniformly with respect to ε.

Now, as in Chapter 4, we can take the limit (via a subsequence) as $n \to \infty$ to obtain a solution u^ε of the regularised equation (14.6). This solution now has enough regularity to ensure that it is unique (see Exercise 14.1) and that it satisfies the strong energy inequality

$$\frac{1}{2}\|u^\varepsilon(t)\|^2 + \int_s^t \|\nabla u^\varepsilon\|^2 \le \frac{1}{2}\|u^\varepsilon(s)\|^2 \quad \text{for all } t > s.$$

for almost all times $s \in [0, \infty)$, including $s = 0$ (cf. Theorem 4.6). We can deduce further that for each fixed $\varepsilon > 0$ this solution is smooth for $t > 0$. Indeed, note that for almost every $t > 0$, $u(t) \in H \cap H^1$.

The arguments of Chapter 5 yield a $p^\varepsilon \in L^{5/3}(\mathbb{T}^3 \times (0, T))$ such that the equations are satisfied in the sense of distributions, as claimed.

Since solutions are unique, it now suffices to consider the smoothness of solutions that start with $u_0 \in H \cap H^1$; for such initial data, if we take the inner product of (14.11) with $-\Delta u_n$ then we have

$$\frac{1}{2}\frac{\mathrm{d}}{\mathrm{d}t}\|\nabla u_n\|^2 + \|\Delta u_n\|^2 = \int [((\psi_\varepsilon * u_n) \cdot \nabla)u_n] \cdot \Delta u_n \, \mathrm{d}x$$
$$\leq \|\psi_\varepsilon * u_n\|_{L^\infty}\|\nabla u_n\|\|\Delta u_n\|$$
$$\leq c\varepsilon^{-3/2}\|u_n\|\|\nabla u_n\|\|\Delta u_n\|,$$

which, after an application of Young's inequality, yields

$$\frac{\mathrm{d}}{\mathrm{d}t}\|\nabla u_n\|^2 + \|\Delta u_n\|^2 \leq c\varepsilon^{-3}\|u_n\|^2\|\nabla u_n\|^2.$$

Use of Gronwall's inequality now shows that u_n is bounded, uniformly in n, in $L^\infty(0, T; H^1) \cap L^2(0, T; H^2)$ for any $T > 0$. Taking limits as $n \to \infty$ it follows that for $u_0 \in H \cap H^1$, equation (14.6) has a global strong solution, and one can then follow the regularity arguments in Chapter 7 to guarantee that this solution is in fact smooth for all $t > 0$. $\qquad\square$

In the next section we use a limiting argument as $\varepsilon \to 0$ to obtain a class of weak solutions that satisfy the local energy inequality.

14.3 Rigorous derivation of the local energy inequality

In this section we aim to prove, in the absence of boundaries, that there are weak solutions that satisfy the local energy inequality.

Theorem 14.2 *For any $u_0 \in H$ there exists a Leray–Hopf weak solution u of the Navier–Stokes equations that satisfies the local energy inequality*

$$2\int_0^T \int_{\mathbb{T}^3} |\nabla u|^2 \varphi \, \mathrm{d}x \, \mathrm{d}s$$
$$\leq \int_0^T \int_{\mathbb{T}^3} |u|^2 \{\partial_t\varphi + \Delta\varphi\} \, \mathrm{d}x \, \mathrm{d}s + \int_0^T \int_{\mathbb{T}^3} (|u|^2 + 2p)(u \cdot \nabla)\varphi \, \mathrm{d}x \, \mathrm{d}s$$
$$(14.12)$$

for any smooth scalar function $\varphi \geq 0$ with compact support in the space–time domain $\mathbb{T}^3 \times (0, T)$.

The following argument, valid in the absence of boundaries, is due to Biryuk, Craig, & Ibrahim (2007). We use the existence and regularity results for the Leray-regularised equation from Theorem 14.1; although we only proved this theorem (or at least sketched the proof) in the periodic case, the result is also true on the whole space, as shown by Leray (1934). Caffarelli, Kohn, & Nirenberg (1982) showed that Theorem 14.2 is also true in the case of a bounded domain, see the Notes.

Proof   As in the previous section we consider the Leray regularisations

$$\partial_t u^\varepsilon - \Delta u^\varepsilon + [(\psi_\varepsilon * u^\varepsilon) \cdot \nabla]u^\varepsilon + \nabla p^\varepsilon = 0, \qquad \nabla \cdot u^\varepsilon = 0, \qquad (14.13)$$

which (according to Theorem 14.1) have smooth solutions u^ε for all $t > 0$. Since there exists a p^ε such that this equation holds in the sense of distributions and all the other terms in the equation are smooth for all $t > 0$, it follows that ∇p^ε is also smooth for all $t > 0$.

Noting that $\tilde{u}^\varepsilon := \psi_\varepsilon * u^\varepsilon$ is divergence free, we can follow the calculations of Section 14.1 which were formal for the Navier–Stokes equations, but are rigorous here given the smoothness of u^ε, to derive a local form of the energy inequality for (14.13). We need only think afresh for the nonlinear term, for which we have[2]

$$\int_{\mathbb{T}^3} \varphi u_i^\varepsilon \tilde{u}_j^\varepsilon (\partial_j u_i^\varepsilon) \, \mathrm{d}x = -\int_{\mathbb{T}^3} (\partial_j \varphi) \tilde{u}_j^\varepsilon |u^\varepsilon|^2 \, \mathrm{d}x - \int_{\mathbb{T}^3} \varphi (\partial_j u_i^\varepsilon) \tilde{u}_j^\varepsilon u_i^\varepsilon \, \mathrm{d}x,$$

and we obtain the local energy equality for (14.13),

$$2 \int_0^T \int_{\mathbb{T}^3} |\nabla u^\varepsilon|^2 \varphi \, \mathrm{d}x \, \mathrm{d}s = \int_0^T \int_{\mathbb{T}^3} |u^\varepsilon|^2 \{\partial_t \varphi + \Delta \varphi\} \, \mathrm{d}x \, \mathrm{d}s$$
$$+ \int_0^T \int_{\mathbb{T}^3} (|u^\varepsilon|^2 + 2p^\varepsilon)(\tilde{u}^\varepsilon \cdot \nabla)\varphi \, \mathrm{d}x \, \mathrm{d}s. \quad (14.14)$$

Since the solutions u^ε are bounded uniformly with respect to ε both in $L^\infty(0, T; H)$ and in $L^2(0, T; V)$, and $\partial_t u^\varepsilon$ is bounded uniformly with respect

---

[2] This is the step that requires us to use the Leray regularisation rather than a Galerkin approximation. The nonlinear term in the Galerkin approximation (4.3) is $P_n(u_n \cdot \nabla)u_n$; multiplying by φu_n and integrating yields

$$\int (\varphi u_n) \cdot P_n[(u_n \cdot \nabla)u_n] = \int [P_n(\varphi u_n)] \cdot [(u_n \cdot \nabla)u_n].$$

Since $P_n(\varphi u_n) \neq \varphi u_n$ the cancellation that occurs for the Leray-regularised equation does not occur here.

to ε in $L^{4/3}(0, T; V^*)$, the arguments used in the proof of Theorem 4.4 guarantee that

$$u^\varepsilon \text{ converges to } u \text{ weakly in } L^2(0, T; V) \text{ and strongly in } L^2(0, T; L^2),$$

as $\varepsilon \to 0$ through some sequence (ε_j), and this is sufficient to guarantee that u is a weak solution of the Navier–Stokes equations. (Only the argument to show the convergence of the nonlinear term, which includes the regularisation, is slightly different; see Exercise 14.2.) This is sufficient to ensure that u satisfies the strong energy inequality, so is indeed a Leray–Hopf weak solution.

To show that the resulting weak solution u also satisfies the local energy inequality we have to show that all the terms in (14.14) converge to the corresponding limiting terms in (14.12). For simplicity we write $\varepsilon \to 0$ in what follows, but this is always to be understood as taking the limit through the sequence (ε_j).

Since $\nabla u^\varepsilon \rightharpoonup \nabla u$ weakly in $L^2(\mathbb{T}^3 \times (0, T))$ and $\varphi \in C_c^\infty(\mathbb{T}^3 \times (0, T))$, it follows from the definition of weak convergence that $\varphi^{1/2} \nabla u^\varepsilon \rightharpoonup \varphi^{1/2} \nabla u$ in $L^2(\mathbb{T}^3 \times (0, T))$, and so

$$\|\varphi^{1/2} \nabla u\|_{L^2(\mathbb{T}^3 \times (0,T))} \le \liminf_{\varepsilon \to 0} \|\varphi^{1/2} \nabla u^\varepsilon\|_{L^2(\mathbb{T}^3 \times (0,T))},$$

i.e.

$$\int_0^T \int_{\mathbb{T}^3} |\nabla u|^2 \varphi \, \mathrm{d}x \, \mathrm{d}s \le \liminf_{\varepsilon \to 0} \int_0^T \int_{\mathbb{T}^3} |\nabla u^\varepsilon|^2 \varphi \, \mathrm{d}x \, \mathrm{d}s.$$

(Note that this is the only part of the argument that requires us to take $\varphi \ge 0$.) It therefore suffices to show the convergence of the terms on the right-hand side of (14.14) to the corresponding terms without ε.

The first term on the right-hand side converges because $\partial_t \varphi + \Delta \varphi$ is a fixed function in $C_c^\infty(\mathbb{T}^3 \times (0, T))$ and $u^\varepsilon \to u$ strongly in $L^2(0, T; L^2)$. For the nonlinear term $\int |u^\varepsilon|^2 (\tilde{u}^\varepsilon \cdot \nabla)\varphi$, note that

$$|u^\varepsilon|^2 \tilde{u}^\varepsilon - |u|^2 u = |u^\varepsilon - u|^2 \tilde{u}^\varepsilon + 2u \cdot (u^\varepsilon - u)\tilde{u}^\varepsilon + |u|^2(\tilde{u}^\varepsilon - u).$$

Then, since $u^\varepsilon \to u$ in $L^2(\mathbb{T}^3 \times (0, T))$ and u^ε is uniformly bounded in $L^{10/3}(\mathbb{T}^3 \times (0, T))$ (by Lebesgue interpolation, see Lemma 3.5), it follows using the Lebesgue interpolation inequality

$$\|u^\varepsilon - u\|_{L^3} \leq \|u^\varepsilon - u\|_{L^2}^{1/6} \|u^\varepsilon - u\|_{L^{10/3}}^{5/6}$$

that $u^\varepsilon \to u$ in $L^3(\mathbb{T}^3 \times (0, T))$. We also have $\tilde{u}^\varepsilon \to u$ in $L^2(\mathbb{T}^3 \times (0, T))$, as

$$\|\tilde{u}^\varepsilon - u\|_{L^3} \leq \|\tilde{u}^\varepsilon - (\psi_\varepsilon * u)\|_{L^3} + \|(\psi_\varepsilon * u) - u\|_{L^3}.$$

Since $\nabla\varphi \in L^\infty(\mathbb{T}^3 \times (0, T))$ it is sufficient to show that

$$I_\varepsilon := \int_0^T \int_\Omega |u^\varepsilon - u|^2 |\tilde{u}^\varepsilon| + 2|u||u^\varepsilon - u||\tilde{u}^\varepsilon| + |u|^2 |\tilde{u}^\varepsilon - u|$$

tends to zero. Given the basic boundedness and convergence properties of mollification (see (14.7)) and the fact that the L^3 norms of $\tilde{u}^\varepsilon$ and u are all bounded independently of ε, this convergence to zero follows from repeated applications of Hölder's inequality:

$$|I_\varepsilon| \leq \|u^\varepsilon - u\|_{L^3}^2 \|\tilde{u}^\varepsilon\|_{L^3} + 2\|u\|_{L^3} \|u^\varepsilon - u\|_{L^3} \|\tilde{u}^\varepsilon\|_{L^3} + \|u\|_{L^3}^2 \|\tilde{u}^\varepsilon - u\|_{L^3}$$

(recall that $\tilde{f}$ denotes the convolution of f with ψ_ε).

Finally we have to deal with the pressure term $2 \int p^\varepsilon (\tilde{u}^\varepsilon \cdot \nabla)\varphi$. If we take the divergence of (14.13) then p^ε satisfies

$$-\Delta p^\varepsilon = \partial_j(\tilde{u}_k^\varepsilon \partial_k u_j^\varepsilon) = \partial_j \partial_k(\tilde{u}_k^\varepsilon u_j^\varepsilon),$$

using the fact that $\tilde{u}^\varepsilon$ is divergence free. Thus

$$p^\varepsilon = \partial_j \partial_k(-\Delta)^{-1}(\tilde{u}_k^\varepsilon u_j^\varepsilon).$$

For each j, k we know (see Theorem B.6) that $T_{jk} = \partial_j \partial_k(-\Delta)^{-1}$ is a bounded linear map from L^r to L^r (for $1 < r < \infty$). Since u_ε is uniformly bounded in L^3, $\tilde{u}_k^\varepsilon u_j^\varepsilon$ is uniformly bounded in $L^{3/2}$, and therefore p^ε is uniformly bounded in $L^{3/2}(\mathbb{T}^3 \times (0, T))$. Furthermore, since $p = \partial_j \partial_k(-\Delta)^{-1}(u_k u_j)$ it follows that

$$p^\varepsilon - p = T_{jk}(\tilde{u}_k^\varepsilon u_j^\varepsilon - u_k u_j),$$

and so

$$\|p^\varepsilon - p\|_{L^{3/2}} \leq C \sum_{k,j} \|\tilde{u}_k^\varepsilon u_j^\varepsilon - u_k u_j\|_{L^{3/2}}$$

$$\leq C\|\tilde{u}_k^\varepsilon[u_j^\varepsilon - u_j]\|_{L^{3/2}} + C\|[\tilde{u}_k^\varepsilon - u_k]u_j\|_{L^{3/2}}$$

$$\leq C\|\tilde{u}^\varepsilon\|_{L^3}\|u^\varepsilon - u\|_{L^3} + C\|\tilde{u}^\varepsilon - u\|_{L^3}\|u\|_{L^3}.$$

Since $u^\varepsilon \to u$ and $\tilde{u}^\varepsilon \to u$ in L^3 as $\varepsilon \to 0$ it follows that $p^\varepsilon \to p$ in $L^{3/2}(\mathbb{T}^3 \times (0, T))$. Now we can write

$$\left| \int p^\varepsilon (\tilde{u}^\varepsilon \cdot \nabla)\varphi - p(u \cdot \nabla)\varphi \right|$$

$$= \left| \int p^\varepsilon [(\tilde{u}^\varepsilon - u) \cdot \nabla]\varphi + (p^\varepsilon - p)(u \cdot \nabla)\varphi \right|$$

$$\leq \int |p^\varepsilon||\tilde{u}^\varepsilon - u||\nabla\varphi| + |p^\varepsilon - p||u||\nabla\varphi|$$

$$\leq \{ \|p^\varepsilon\|_{L^{3/2}} \|\tilde{u}^\varepsilon - u\|_{L^3} + \|p^\varepsilon - p\|_{L^{3/2}} \|u\|_{L^3} \} \|\nabla\varphi\|_{L^\infty},$$

which shows convergence of the final term. □

We briefly discuss the proof in the case of a bounded domain in the Notes at the end of the chapter.

14.4 Derivation of an alternative local energy inequality

In this section we give the arguments on a general domain Ω, which could be a bounded domain, the torus, or the whole space.

In the following chapter we will require the local energy inequality in a slightly different form, namely

$$\int_\Omega |u(x, t)|^2 \phi \, dx + 2 \int_a^t \int_\Omega |\nabla u|^2 \phi \, dx \, ds$$

$$\leq \int_a^t \int_\Omega |u|^2 (\partial_t \phi + \Delta\phi) \, dx \, ds + \int_a^t \int_\Omega (|u|^2 + 2p)u \cdot \nabla\phi \, dx \, ds \quad (14.15)$$

for any $t \in (a, b)$ and every non-negative $\phi \in C_c^\infty(\Omega \times (a, b))$. This form of the inequality is useful since it will allow us to estimate the spatial integral $\int_\Omega |u|^2 \phi$ at a fixed time t rather than only a space–time integral.

We will now show that this is a consequence of the local energy inequality in the form proved above,

$$2 \int_a^b \int_\Omega |\nabla u|^2 \varphi \, dx \, ds$$

$$\leq \int_a^b \int_\Omega |u|^2 \{\partial_t \varphi + \Delta\varphi\} \, dx \, ds + \int_a^b \int_\Omega (|u|^2 + 2p)u \cdot \nabla\varphi \, dx \, ds,$$

valid for $\varphi \geq 0$ in $C_c^\infty(\Omega \times (a, b))$, using an argument similar to that employed in the proof of Lemma 3.14.

For a given $\phi \in C_c^\infty(\Omega \times (a, b))$ we consider the family of test functions

$$\varphi_\varepsilon(x, s) = \phi(x, s)\chi((t - s)/\varepsilon),$$

where χ is a smooth function such that $0 \leq \chi \leq 1$, $\chi(s) = 0$ for $s \leq 0$, and $\chi(s) = 1$ for $s \geq 1$. Noting that

$$\chi\left(\frac{t-s}{\varepsilon}\right) = \begin{cases} 0 & s \geq t \\ 1 & s \leq t - \varepsilon, \end{cases}$$

that

$$\chi'\left(\frac{t-s}{\varepsilon}\right) = 0 \quad \text{for} \quad s \leq t - \varepsilon, \ s \geq t, \qquad \int_{t-\varepsilon}^t \varepsilon^{-1}\chi'\left(\frac{t-s}{\varepsilon}\right) ds = 1,$$

and that

$$\frac{\partial \varphi_\varepsilon}{\partial s}(x, s) = \frac{\partial \phi}{\partial s}(x, s)\,\chi\left(\frac{t-s}{\varepsilon}\right) - \phi(x, s)\,\varepsilon^{-1}\chi'\left(\frac{t-s}{\varepsilon}\right)$$

we obtain (14.15) by letting $\varepsilon \to 0$. The limit of every term except that involving $\partial_t \varphi$ is easily obtained using the Dominated Convergence Theorem, since, for example,

$$\int_a^b \int_\Omega |\nabla u|^2 \varphi_\varepsilon = \int_a^t \int_\Omega |\nabla u|^2 \varphi_\varepsilon,$$

and for $s \leq t$ we have $|\nabla u|^2 \varphi_\varepsilon \to |\nabla u|^2 \phi$ as $\varepsilon \to 0$ along with the bound $|\nabla u|^2 \varphi_\varepsilon \leq |\nabla u|^2 \phi$, which is integrable on $\Omega \times (a, b)$.

The remaining term arising from the time derivative of φ is

$$\int_{t-\varepsilon}^t \varepsilon^{-1}\chi'\left(\frac{t-s}{\varepsilon}\right) \int_\Omega |u(x, s)|^2 \phi(x, s)\, dx\, ds;$$

after taking an appropriate subsequence, for almost every t this term converges to

$$\int_\Omega |u(x, t)|^2 \phi(x, t)\, dx,$$

since $\varepsilon^{-1}\chi'(t - s/\varepsilon)$ is an approximation of the identity, see Theorem A.9.

14.5 Derivation of the strong energy inequality on the whole space

In Section 4.4 we proved the existence of weak solutions for the equations posed on the whole space, but we were unable to prove the existence of weak solutions satisfying the strong energy inequality. We now show how to construct solutions on the whole space that do satisfy the strong energy inequality, by taking the limit of solutions of the Leray regularisation

$$\partial_t u^\varepsilon - \Delta u^\varepsilon + (\tilde{u}^\varepsilon \cdot \nabla) u^\varepsilon + \nabla p^\varepsilon = 0, \qquad \nabla \cdot u^\varepsilon = 0, \qquad u^\varepsilon(0) = \psi_\varepsilon * u_0.$$
$$(14.16)$$

In this formulation of the Leray regularisation, recall that $\tilde{u}^\varepsilon = \psi_\varepsilon * u^\varepsilon$ and that the superscript ε indexes the solution.

We have already remarked that Leray (1934) proved the existence of solutions of (14.16) on the whole space, although our proof in Theorem 14.1 is specialised to the torus. Given that these solutions are smooth, if we take the L^2 inner product of (14.16) with u^ε and integrate in time between s and t then we obtain the energy equality

$$\frac{1}{2} \|u^\varepsilon(t)\|^2 + \int_s^t \|\nabla u^\varepsilon\|^2 = \frac{1}{2} \|u^\varepsilon(s)\|^2, \qquad (14.17)$$

and in particular the bound

$$\frac{1}{2} \|u^\varepsilon(t)\|^2 + \int_0^t \|\nabla u^\varepsilon\|^2 = \frac{1}{2} \|u^\varepsilon(0)\|^2 \le \frac{1}{2} \|u_0\|^2; \qquad (14.18)$$

so u^ε is uniformly bounded in $L^\infty(0, T; H) \cap L^2(0, T; V)$. As in the proof of Theorem 14.1 it is also straightforward to show that $\partial_t u^\varepsilon$ is uniformly bounded in $L^{4/3}(0, T; V^*)$. It is then possible to extract a subsequence (which we relabel) such that $u^\varepsilon \to u$ strongly in $L^2(0, T; L^2(K))$ for every compact $K \subset \mathbb{R}^3$ and $\nabla u^\varepsilon \to \nabla u$ weakly in $L^2(0, T; L^2(\mathbb{R}^3))$. This is enough to show that u is a weak solution, arguing as we did in Theorem 4.10.

In order to pass to the limit in (14.17) we need to exclude the possibility that energy can 'escape to $|x| = \infty$' by finding uniform bounds on the 'tail' of $|u^\varepsilon|^2$. To do this we follow Section 27 of Leray (1934).

Proposition 14.3 *For all $\eta > 0$, $T > 0$ there exists $R = R(u_0, T, \eta)$, such that*

$$\int_{|x|>R} |u^\varepsilon(x, t)|^2 \le \eta \qquad \text{for every} \quad t \in [0, T].$$

Proof First, choose $R_1 > 0$ sufficiently large that

$$\int_{|x|>R_1} |u^\varepsilon(x, 0)|^2 \le \int_{|x|>R_1-\varepsilon} |u_0(x)|^2 < \frac{\eta}{2}.$$

For any $R > R_1$ define

$$\varrho(x) = \begin{cases} 0 & |x| < R_1 \\ \frac{|x|-R_1}{R-R_1} & R_1 \le |x| \le R \\ 1 & |x| > R \end{cases}$$

and take the L^2 inner product of (14.16) with $u^\varepsilon \varrho$ and integrate by parts so that

$$\frac{1}{2}\frac{\mathrm{d}}{\mathrm{d}t}\int_{\mathbb{R}^3}\varrho|u^\varepsilon|^2 + \int_{\mathbb{R}^3}\varrho|\nabla u^\varepsilon|^2$$

$$= -\int_{\mathbb{R}^3}(\partial_j u_i^\varepsilon)u_i^\varepsilon(\partial_j\varrho) + \int_{\mathbb{R}^3}|u^\varepsilon|^2(\tilde{u}^\varepsilon\cdot\nabla)\varrho + \int_{\mathbb{R}^3}p^\varepsilon(u^\varepsilon\cdot\nabla)\varrho.$$

Now integrate from 0 to t to obtain

$$\frac{1}{2}\int_{\mathbb{R}^3}\varrho|u^\varepsilon(t)|^2 \le \frac{1}{2}\int_{\mathbb{R}^3}\varrho|u^\varepsilon(0)|^2$$

$$+ \int_0^t\int_{\mathbb{R}^3}\left\{|\nabla u^\varepsilon||u^\varepsilon| + |u^\varepsilon|^2|\tilde{u}^\varepsilon| + |p^\varepsilon||u^\varepsilon|\right\}|\nabla\varrho|,$$

where we have dropped the positive term involving $|\nabla u^\varepsilon|^2$ from the left-hand side.

Now we use the facts that $\varrho \in H^1(\mathbb{R}^3)$, $|\nabla\varrho| = \frac{1}{R-R_1}$ for $R_1 < |x| < R$, and $|\nabla\varrho| = 0$ for $|x| < R_1$ and $|x| > r$ to write

$$\frac{1}{2}\int_{|x|>R}|u^\varepsilon(t)|^2 \le \frac{1}{2}\int_{|x|>R_1}|u^\varepsilon(0)|^2$$

$$+ \frac{1}{R-R_1}\int_0^t\int_{\mathbb{R}^3}|\nabla u^\varepsilon||u^\varepsilon| + |u^\varepsilon|^2|\tilde{u}^\varepsilon| + |p^\varepsilon||u^\varepsilon|.$$

The integral in the second term on the right-hand side is bounded by

$$\|u^\varepsilon(0)\|\int_0^t\left(\|\nabla u^\varepsilon\| + \|u^\varepsilon\|_{L^4}^2 + \|p^\varepsilon\|_{L^2}\right) \le \|u^\varepsilon(0)\|\int_0^t\left(\|\nabla u^\varepsilon\| + c\|u^\varepsilon\|_{L^4}^2\right),$$

using the Calderón–Zygmund estimate $\|p^\varepsilon\|_{L^2} \le c\|u^\varepsilon\|_{L^4}^2$ (5.7). Now observe that

$$\int_0^t\|\nabla u^\varepsilon\| \le t^{1/2}\left(\int_0^t\|\nabla u^\varepsilon\|^2\right)^{1/2} \le t^{1/2}\|u^\varepsilon(0)\| \le t^{1/2}\|u_0\|$$

and

$$\int_0^t\|u^\varepsilon\|_{L^4}^2 \le \int_0^t\|u^\varepsilon\|_{L^2}^{1/2}\|u^\varepsilon\|_{L^6}^{3/2}$$

$$\le c\|u^\varepsilon(0)\|^{1/2}\int_0^t\|\nabla u^\varepsilon\|^{3/2} \le c\|u^\varepsilon(0)\|^{1/2}t^{1/4}\left(\int_0^t\|\nabla u^\varepsilon\|^2\right)^{3/4}$$

$$\le ct^{1/4}\|u^\varepsilon(0)\|^2 \le ct^{1/4}\|u_0\|^2.$$

Therefore

$$\frac{1}{2}\int_{|x|>R}|u^\varepsilon(t)|^2 \le \frac{1}{2}\int_{|x|>R_1}|u^\varepsilon(0)|^2 + \frac{c\|u_0\|}{R-R_1}\left\{t^{1/2} + t^{1/4}\|u_0\|\right\}$$

(this is Leray's equation (5.7) in more modern notation). Choosing R so large that

$$\frac{c\|u_0\|}{R - R_1}\left\{T^{1/2} + T^{1/4}\|u_0\|\right\} < \frac{\eta}{4}$$

the result follows.   □

We now use this result to prove the validity of the strong energy inequality for solutions obtained on the whole space as limits of the u^ε.

Theorem 14.4 *Given $u_0 \in H(\mathbb{R}^3)$ there exists a weak solution u of the Navier–Stokes equations that satisfies the strong energy inequality, i.e.*

$$\frac{1}{2}\|u(t)\|^2 + \int_s^t \|\nabla u\|^2 \le \frac{1}{2}\|u(s)\|^2 \qquad \text{for all} \quad t \ge s \qquad (14.19)$$

for $s = 0$ and almost every $s > 0$.

Proof   First observe that if $f \in L^2(\mathbb{R}^3)$,

(i) $f^\varepsilon \to f$ strongly in $L^2(K)$ for every compact $K \subset \mathbb{R}^3$, and
(ii) for every $\eta > 0$ there exists $R(\eta)$ such that

$$\int_{|x| \ge R} |f^\varepsilon|^2 < \eta \qquad \text{for all} \quad \varepsilon > 0$$

then $f^\varepsilon \to f$ in $L^2(\mathbb{R}^3)$. Indeed, given $\delta > 0$, choose R sufficiently large that

$$\int_{|x| \ge R} |f^\varepsilon|^2 < \frac{\delta}{3} \qquad \text{and} \qquad \int_{|x| \ge R} |f|^2 < \frac{\delta}{3}$$

and then choose ε_0 such that

$$\|f^\varepsilon - f\|_{L^2(B_R)} < \frac{\delta}{3} \qquad \text{for every} \qquad 0 < \varepsilon < \varepsilon_0,$$

from which it follows that

$$\|f^\varepsilon - f\|_{L^2(\mathbb{R}^3)} \le \|f^\varepsilon - f\|_{L^2(B_R)} + \int_{|x| \ge R} |f^\varepsilon|^2 + \int_{|x| \ge R} |f|^2 < \delta$$

for every $0 < \varepsilon < \varepsilon_0$, i.e. $f^\varepsilon \to f$ in $L^2(\mathbb{R}^3)$.

We now use this result for the subsequence u^ε that converges to a weak solution u, for which we know that $u^\varepsilon \to u$ strongly in $L^2(0, T; L^2(K))$ for every compact $K \subset \mathbb{R}^3$ and $\nabla u^\varepsilon \to \nabla u$ weakly in $L^2(0, T; L^2(\mathbb{R}^3))$. Since it follows that for almost every $t > 0$

$$u^\varepsilon(t) \to u(t) \quad \text{in} \quad L^2(K)$$

for every compact subset $K \subset \mathbb{R}^3$, we can deduce that $u^{\varepsilon}(t) \to u(t)$ in $L^2(\mathbb{R}^3)$ for almost every $t > 0$.

We now take the limit as $\varepsilon \to 0$ in (14.17) and in (14.18) to obtain

$$\frac{1}{2}\|u(t)\|^2 + \int_s^t \|\nabla u\|^2 \leq \frac{1}{2}\|u(s)\|^2$$

for almost every $t \geq s \geq 0$ (including $s = 0$). Finally, proceeding as in the proof of Theorem 4.6 we use the weak continuity in L^2 to show that this inequality holds for every $t \geq s$. $\qquad\square$

Notes

The proof of the existence of suitable weak solutions given here relies strongly on the Calderón–Zygmund estimates available for the pressure in the absence of boundaries.

Caffarelli, Kohn, & Nirenberg (1982) proved the existence of suitable weak solutions on bounded domains using a different method, adapted from that of Scheffer (1977). To find a solution that satisfies the local energy inequality on $(0, T)$ they take $n > 0$ and set $\delta = T/n$, then find functions u_n and p_n that satisfy

$$\partial_t u_n - \Delta u_n + (\Psi_\delta(u_n) \cdot \nabla)u_n + \nabla p_n = 0, \qquad (14.20)$$

where $\Psi_\delta(u_n)$ is a space–time mollification of u_n that depends only on the values of u at times before $t - \delta$. To do this they set

$$\Psi_\delta(u)(z) := \delta^{-4} \int_{\Omega \times (0,T)} \psi(\zeta/\delta)\tilde{u}(z - \zeta)\,d\zeta,$$

where $\psi \in C_c^{\infty}(\mathbb{R}^4)$ is a standard mollifier whose support is contained in the set $\{|x|^2 < t,\ 1 < t < 2\}, z = (x, t), \zeta = (\xi, \tau)$, and $\tilde{u}(x, t)$ is the extension of u by zero outside $\Omega \times (0, T)$. This means that (14.20) is in fact a linear equation of the form

$$\partial_t u - \Delta u + (w \cdot \nabla)u + \nabla p = 0 \qquad \nabla \cdot u = 0, \qquad (14.21)$$

where $w \in C^{\infty}(\overline{\Omega} \times [0, T]; \mathbb{R}^3)$ is a given divergence-free function, on a sequence of 'time slices' of the form $\Omega \times (m\delta, (m + 1)\delta), 0 \leq m \leq N - 1$.

The existence of solutions of (14.21) and hence of (14.20) can be shown using the Galerkin method, and these solutions are smooth enough to allow for a version of the local energy inequality along the lines of that derived for the Leray approximation as equation (14.14). Caffarelli, Kohn, and Nirenberg then use estimates (due to Solonnikov, 1964) on solutions of the linear equation

(14.21) to give uniform bounds on u_n and p_n. These allow passage to the limit, in much the same way as we achieved above in the proof of Theorem 14.2.

Other proofs of the existence of solutions that satisfy the local energy inequality are given in Lions (1994), which is similar to the method used in this chapter, and by Beirão da Veiga (1985a,b) who first regularised the Navier–Stokes equations by adding an extra bi-Laplacian term and considered

$$\partial_t u + \varepsilon \Delta^2 u - \Delta u + (u \cdot \nabla)u + \nabla p = 0, \qquad \nabla \cdot u = 0,$$

then letting $\varepsilon \to 0$ to obtain the required inequality. A 'Galerkin-like' method, using certain carefully chosen finite-dimensional spaces for the approximations, has been developed by Guermond (2006, 2007). Berselli and Spirito have constructed suitable weak solutions using an artificial compressibility method (Berselli & Spirito, 2016a) and via semi-discretisations (Berselli & Spirito, 2016b).

Exercises

14.1 Show that the weak solution of the regularised equation

$$\partial_t u - \Delta u + [(\psi_\varepsilon * u) \cdot \nabla]u + \nabla p = 0, \qquad \nabla \cdot u = 0,$$

whose existence is shown in Theorem 14.1, is unique.

14.2 Show that if $u^\varepsilon \to u$ strongly in $L^2(0, T; L^2)$ and $u^\varepsilon \rightharpoonup u$ weakly in $L^2(0, T; H^1)$ then

$$\int_0^T \langle [(\psi_\varepsilon * u^\varepsilon) \cdot \nabla]u^\varepsilon, \varphi \rangle \to \int_0^T \langle (u \cdot \nabla)u, \varphi \rangle$$

for every $\varphi \in \mathcal{D}_\sigma$.

15

Partial regularity I: $\dim_B(S) \leq 5/3$

In Chapter 13 we gave a proof of the conditional local regularity result that if u is a weak solution of the Navier–Stokes equations that satisfies the 'Serrin condition' on $U \times (t_1, t_2)$,

$$u \in L^r(t_1, t_2; L^s(U)), \qquad \frac{2}{r} + \frac{3}{s} = 1, \tag{15.1}$$

then u is smooth in U.

The proof there essentially fell into two parts:

(i) local boundedness of ω and hence of u and
(ii) bootstrapping the regularity of ω and u to show that they are smooth in space.

We went on to show that when u is a solution of the Navier–Stokes initial-value problem on $\Omega \times (0, T)$ it is possible to use estimates on the time derivative to show that local spatial regularity of the solution (such as that obtained in (ii) above) can be used to deduce Hölder continuity in time (Proposition 13.10).

In this chapter we give a complete proof of an alternative conditional regularity theorem due to Caffarelli, Kohn, & Nirenberg (1982) (hereafter CKN). The proof leads to the conclusion that u is locally bounded, and hence can be viewed as a variant of step (i) of our previous arguments. By following step (ii) and then using Proposition 13.10 it follows that u is spatially smooth and Hölder continuous in time (provided that we assume that ∇p has the regularity we know holds for solutions of the initial-value problem).

In Chapter 13 we used the parabolic cylinders $Q_r^*(a, s)$ as a natural set within which to localise our solution. In what follows we will also use parabolic cylinders, but it will turn out that it is more convenient to use the cylinders $Q_r(a, s)$

279

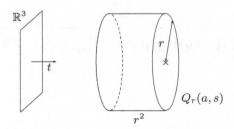

Figure 15.1. A schematic picture of the parabolic cylinder $Q_r(a, s)$ for some $r < 1$. The centre-right point (a, s) is marked by a cross.

defined by

$$Q_r(a, s) = \{(x, t) : |x - a| < r, \ s - r^2 < t < s\}.$$

While we will still refer to (a, s) as the 'centre' of this cylinder, it is now rather the 'centre-right' point, as illustrated in Figure 15.1; since the cylinder is open, the point (a, s) is not contained in the cylinder.

We will show in Theorem 15.4, closely following the original proof of CKN, that there exists an absolute constant $\varepsilon_0 > 0$ such that if for some $r > 0$, for which $Q_r(a, s) \subset \Omega \times (0, T)$, u and p satisfy

$$\frac{1}{r^2} \int_{Q_r(a,s)} |u|^3 + |p|^{3/2} < \varepsilon_0 \tag{15.2}$$

then u is bounded in $Q_{r/2}(a, s)$ (and is therefore smooth in space and Hölder continuous in time when u is a solution of the initial-value problem). In the majority of the integrals in this chapter and the next we will omit the integration variables $(\mathrm{d}x \, \mathrm{d}t)$ unless this causes any ambiguity.

Note that there is a significant difference between this smallness condition and the integrability condition of Serrin (15.1). For the latter we must assume a degree of integrability of the weak solution that we cannot guarantee is satisfied, while in the CKN condition we must assume that an integral that we know is globally bounded is in fact sufficiently small locally (we work with scale-invariant quantities so that requiring 'smallness' is meaningful, see the discussion in Chapter 10 and below). For this reason the CKN condition allows us to go further than the Serrin condition, and to prove an unconditional bound on the size of the set of possible singularities.

Because of the bootstrapping allowed by step (ii) above, it is reasonable to make the following definition of what it means for a point to be 'regular'.

Definition 15.1 A point $(x, t) \in \Omega \times (0, T)$ is a *regular point* of a weak solution u if there is a space–time neighbourhood $\mathcal{U}$ of (x, t) such that $u \in L^\infty(\mathcal{U})$. We call (x, t) a *singular point* if it is not a regular point.

By giving a local condition guaranteeing that (x, t) is a regular point, we will be able to find a bound on the dimension of the space–time 'singular set'

$$S = \{(x, t) \in \Omega \times (0, \infty) : (x, t) \text{ is a singular point}\},$$

i.e. the collection of all $(x, t) \in \Omega \times (0, \infty)$ that are singular points. Note that we exclude $t = 0$ from this definition.

Observe that if (x, t) is singular then negating the regularity condition in (15.2) and noting that $\int_{Q_r(x,t)} 1 = cr^5$, a simple application of Hölder's inequality implies that

$$\frac{1}{r^2} \int_{Q_r(x,t)} |u|^3 \le \frac{1}{r^2} \left(\int_{Q_r(x,t)} |u|^{10/3} \right)^{9/10} \left(\int_{Q_r(x,t)} 1 \right)^{1/10}$$

$$= \frac{c}{r^{3/2}} \left(\int_{Q_r(x,t)} |u|^{10/3} \right)^{9/10},$$

and similarly for the pressure term.[1] Therefore when (x, t) is singular we must have

$$\int_{Q_r(x,t)} |u|^{10/3} + |p|^{5/3} > c\varepsilon_0^{10/9} r^{5/3}$$

for all $r > 0$, where c is an absolute constant. Since we know that for any weak solution $u \in L^{10/3}(\Omega \times (0, T))$ (see (5.12)) and $p \in L^{5/3}(\Omega \times (0, T))$ (see (5.14)), u cannot be singular at 'too many points'. Making this precise, one can show that the box-counting dimension of the singular set must satisfy $\dim_B(S) \le 5/3$ (Theorem 15.8).

While we prove a result for solutions that need only be defined locally (on some domain $U \times (a, b)$) we do assume certain properties of these solutions that are known when they solve the Navier–Stokes initial-value problem. The existence of such 'suitable weak solutions' follows from Theorem 14.2.

Definition 15.2 A pair (u, p) is *suitable* on $U \times (a, b)$ if

(i) $u \in L^\infty(a, b; L^2(U))$, $\nabla u \in L^2(U \times (a, b))$, and $p \in L^{3/2}(U \times (a, b))$;
(ii) $-\Delta p = \partial_i \partial_j (u_i u_j)$ for almost every $t \in (a, b)$; and

---

[1] This is not quite right, since (15.2) implies that $u \in L^\infty(Q_{r/2}(a, s))$ and $(a, s) \notin Q_{r/2}(a, s)$. However, the rigorous argument, with a small shift of the original cylinder, follows this outline very closely.

(iii) the local energy inequality

$$\int_U |u(t)|^2 \phi(t)\,dx + 2 \int_a^t \int_U |\nabla u|^2 \phi\,dx\,ds$$
$$\le \int_a^t \int_U |u|^2 (\phi_t + \Delta\phi)\,dx\,ds + \int_a^t \int_U (|u|^2 + 2p)u \cdot \nabla\phi\,dx\,ds$$

(15.3)

is valid for every $t \in (a, b)$ and for every non-negative scalar test function $\phi \in C_c^\infty(U \times (a, b))$.

Note that the definition does not include the assumption that u is a (Leray–Hopf) weak solution of the Navier–Stokes equations, but only: (i) that (u, p) has the regularity of a weak solution; (ii) that u and p are related as they would be for a weak solution; and (iii) that the local energy inequality holds. It is worth commenting on the assumed regularity of p in (i), since we know that for a solution of the initial-value problem we have more, namely $p \in L^{5/3}(U \times (a, b))$. At this point we can attempt to justify (i) by asserting that additional regularity for p would neither simplify the following proof nor strengthen the result. An indication of why $p \in L^{3/2}$ is the most appropriate assumption can be obtained by considering the final term of the local energy inequality (15.3),

$$\int_{Q_r} (|u|^2 + 2p)u \cdot \nabla\phi\,dx\,ds.$$

This is cubic in u, so we would expect $\int |u|^3$ to play a role in the argument. We can then note that using Hölder's inequality on $\int pu$ (neglecting the $\nabla\phi$ term for now) yields $\|u\|_{L^3}\|p\|_{L^{3/2}}$, making $\int |p|^{3/2}$ the expression in p that we might expect to arise most naturally.

The partial regularity theorems that we will prove are in fact valid for any suitable pair, and not only for suitable weak solutions, since the only ingredients they require are properties (i)–(iii) of Definition 15.2. Scheffer (1987, see also 1985) showed that if we only use these ingredients and no additional properties of weak solutions then the 'headline result' of the next chapter that $\dim_H(S) \le 1$ cannot be improved: he showed that for any $\gamma < 1$ it is possible to construct a pair (u, p) satisfying these three properties such that the singular set of u satisfies $\dim_H(S) \ge \gamma$.

We remark here that throughout the proof we will work with suitable weak solutions on parabolic cylinders. In particular, in the case of a suitable weak solution on $Q_1(0, 0) = B_1 \times (-1, 0)$, the local energy inequality (15.3) becomes

$$\int_{B_1} |u(t)|^2 \phi(t) \, dx + 2 \int_{-1}^{t} \int_{B_1} |\nabla u|^2 \phi \, dx \, ds$$

$$\leq \int_{-1}^{t} \int_{B_1} |u|^2 (\phi_t + \Delta \phi) \, dx \, ds + \int_{-1}^{t} \int_{B_1} (|u|^2 + 2p) u \cdot \nabla \phi \, dx \, ds$$

for all non-negative $\phi \in C_c^\infty(Q_1(0,0))$ and every $t \in (-1, 0)$.

15.1 Scale-invariant quantities

We have already remarked that on the whole space the Navier–Stokes equations have the following scaling property: if $u(x, t)$ is a solution with corresponding pressure $p(x, t)$, then so is $u_\lambda(x, t) = \lambda u(\lambda x, \lambda^2 t)$ with corresponding pressure $\lambda^2 p(\lambda x, \lambda^2 t)$. Such a property is shared by local solutions, in the following sense: if (u, p) is a suitable weak solution on the parabolic cylinder $Q_r(0,0)$, then (u_λ, p_λ) is a suitable weak solution on the parabolic cylinder $Q_{r/\lambda}(0,0)$.

As with 'small data' results for the initial value problem (see Chapter 10), if we want to have a local regularity result based on some quantity being 'small', in the sense that it is less than some absolute constant, then this should be a quantity that is invariant under this rescaling transformation. Otherwise we could use an appropriate rescaling to make it 'small' and apply our result to any solution. For example, if we neglect the pressure for the time being, the condition we will require on u to obtain local regularity will be

$$\frac{1}{r^2} \int_{Q_r(0,0)} |u|^3 < \varepsilon_0 \tag{15.4}$$

for some $r > 0$. If we rescale u to u_λ then

$$\frac{1}{(r/\lambda)^2} \int_{Q_{r/\lambda}(0,0)} |u_\lambda|^3 \, dx \, dt = \frac{\lambda^2}{r^2} \int_{Q_{r/\lambda}(0,0)} \lambda^3 |u(\lambda x, \lambda^2 t)|^3 \, dx \, dt$$

$$= \frac{1}{r^2} \int_{Q_r(0,0)} |u(y, s)|^3 \, dy \, ds,$$

substituting $y = \lambda x$ and $s = \lambda^2 t$. So we cannot change the value of the quantity on the left-hand side of (15.4) via a rescaling of u.

Since the arguments in the proofs of the local regularity theorems of both this chapter and the next proceed by estimating quantities at one scale in terms of quantities at another, it is natural to weight integrals by appropriate factors of r to make sure that they are scale invariant: our proofs will make use of the quantities[2]

$$\frac{1}{r} \sup_{-r^2 < t < 0} \int_{B_r} |u(t)|^2, \quad \frac{1}{r} \int_{Q_r} |\nabla u|^2, \quad \frac{1}{r^2} \int_{Q_r} |u|^3, \quad \text{and} \quad \frac{1}{r^2} \int_{Q_r} |p|^{3/2}, \quad (15.5)$$

all of which are scale invariant.

15.2 Outline of the proof

Before giving the proof proper we give a simplified version in which we neglect the pressure (which is a significant complicating factor). We consider a suitable weak solution on $Q_1(0, 0)$, the unit cylinder 'centred' at $(0, 0)$, and show that there is a constant $\varepsilon_0 > 0$ such that if [3]

$$\int_{Q_1(0,0)} |u|^3 \le \varepsilon_0 \qquad (15.6)$$

then $u \in L^\infty(Q_{1/2}(0, 0))$. In order to do this we use an inductive argument to show that for every $(a, s) \in Q_{1/2}(0, 0)$

$$\frac{1}{r_n^5} \int_{Q_{r_n}(a,s)} |u|^3 \le \varepsilon_0^{2/3}, \qquad n \in \mathbb{N}, \qquad (15.7)$$

where $r_n = 2^{-n}$. We then use the Lebesgue Differentiation Theorem (Theorem A.2) to show that $u \in L^\infty(Q_{1/2}(0, 0))$.

The argument proceeds by iterating back and forth between two inequalities. The first is the local energy inequality (15.3),

$$\int_{B_1} |u(s)|^2 \phi(s) + 2 \int_{-1}^{s} \int_{B_1} |\nabla u|^2 \phi$$
$$\le \int_{-1}^{s} \int_{B_1} |u|^2 (\partial_t \phi + \Delta \phi) + \int_{-1}^{s} \int_{B_1} |u|^2 u \cdot \nabla \phi \qquad (15.8)$$

for any $s \in (-1, 0)$ (we have neglected the pressure, which should appear in the final term).

If it was possible to take ϕ in (15.8) to be the solution of the backwards[4] heat equation $\partial_t \phi + \Delta \phi = 0$, with $\phi(s) = \delta_x(a)$, then this would yield an estimate

---

[2] It is perhaps tempting to abbreviate these; for example, in CKN the first three quantities in (15.5) are denoted by $A(r)$, $\delta(r)$, and $G(r)$. While this abbreviated notation makes things more concise, the argument seems easier to follow when all the quantities remain explicit.

[3] The space–time L^3 norm is not scale invariant. When we rescale this assumption to a cylinder of radius r it becomes $\frac{1}{r^2} \int_{Q_r(0,0)} |u|^3 < \varepsilon_0$, see Theorem 15.4.

[4] Note that although ϕ would satisfy the backwards heat equation, we would be interested in the values of $\phi(x, t)$ only for $t < s$, which avoids the bad properties of the backwards heat equation (in fact we would be taking $\psi(s - t, x)$ with ψ a solution of the standard heat equation for positive times).

of the point value $u(a, s)$. However, this choice of ϕ is too singular to allow us to bound the second term on the right-hand side of (15.8).

The broadest of outlines of the proof (cf. Kohn, 1982) is that instead of taking this singular ϕ we take a series of test functions ϕ_n that are essentially approximations to ϕ obtained by shifting the singularity on the time axis so that it falls outside the range of integration. With a convenient scaling factor we will choose ϕ_n with $\partial_t \phi_n + \Delta \phi_n \approx 0$,

$$\phi_n \sim r_n^{-1} \quad \text{on} \quad Q_{r_n}(a, s),$$

and

$$|\nabla \phi_n| \leq \begin{cases} C r_n^{-2} & \text{on } Q_{r_n}(a, s) \\ C r_n^2 r_k^{-4} & \text{on } Q_{r_k}(a, s) \setminus Q_{r_{k+1}}(a, s), \quad k = 0, \ldots, n-1, \end{cases}$$

where $r_n = 2^{-n}$.

Supposing that we already have our inductive hypothesis

$$\frac{1}{r_k^2} \int_{Q_{r_k}(a,s)} |u|^3 \leq \varepsilon_0^{2/3} r_k^3$$

for every $k = 1, \ldots, n$ then we can use the local energy inequality to give an estimate like

$$\frac{1}{r_n} \int_{B_{r_n}(a)} |u(s)|^2 + \frac{1}{r_n} \int_{Q_{r_n}(a,s)} |\nabla u|^2 \leq \int_{Q_{1/2}(a,s)} |u|^3 |\nabla \phi_n|$$

$$\leq C r_n^2 \sum_{k=1}^{n} r_k^{-4} \int_{Q_{r_k}(a,s)} |u|^3$$

$$\leq C r_n^2 \sum_{k=1}^{n} r_k^{-4} r_k^5 \varepsilon_0^{2/3}$$

$$= C r_n^2 \left(\sum_{k=1}^{n} r_k \right) \varepsilon_0^{2/3}$$

$$\leq C r_n^2 \varepsilon_0^{2/3},$$

where we dropped the first term on the right-hand side of (15.8) since we have chosen ϕ_n so that $\partial_t \phi_n + \Delta \phi_n \approx 0$.

Noting that this implies that

$$\frac{1}{r_{n+1}} \int_{B_{r_{n+1}}(a)} |u(s)|^2 + \frac{1}{r_{n+1}} \int_{Q_{r_{n+1}}(a,s)} |\nabla u|^2 \leq 8 C r_{n+1}^2 \varepsilon_0^{2/3}$$

we now use the interpolation inequality[5]

$$\frac{1}{r^2} \int_{Q_r(a,s)} |u|^3 \le C_0 \left[\frac{1}{r} \sup_{s-r^2<t<s} \int_{B_r(a)} |u(t)|^2 + \frac{1}{r} \int_{Q_r(a,s)} |\nabla u|^2 \right]^{3/2} \quad (15.9)$$

to obtain

$$\frac{1}{r_{n+1}^2} \int_{Q_{r_{n+1}}(a,s)} |u|^3 \le C_0 (8C)^{3/2} r_{n+1}^3 \varepsilon_0.$$

Using the fact that ε_0 is small we can now absorb the constants and obtain

$$\frac{1}{r_{n+1}^2} \int_{Q_{r_{n+1}}(a,s)} |u|^3 \le \varepsilon_0^{2/3} r_{n+1}^3,$$

which is the next stage of the induction.

Modulo the technicalities (which are considerable), the 'trick' is thus that the right-hand side of the local energy inequality (15.8) is essentially cubic in u, while the left-hand side is only quadratic; this allows the initial smallness assumption to come into play to preserve the inductive hypothesis at the next level. We now give the argument in more detail, neglecting the pressure initially.

We let $r_n = 2^{-n}$, and for each $(a, s) \in Q_{1/2}(0, 0)$ we use induction to move between two sets of bounds,

$$(A_n') : \qquad \frac{1}{r_n^2} \int_{Q_{r_n}(a,s)} |u|^3 \le \varepsilon_0^{2/3} r_n^3 \qquad (15.10)$$

and

$$(B_n) : \qquad \frac{1}{r_n} \sup_{s-r_n^2<t<s} \int_{B_{r_n}(a)} |u(t)|^2 + \frac{1}{r_n} \int_{Q_{r_n}(a,s)} |\nabla u|^2 \le C_B \varepsilon_0^{2/3} r_n^2, \qquad (15.11)$$

where ε_0 and C_B are absolute constants that will be determined during the course of the proof. In order to make sure that these really are *absolute* constants we will keep careful track of numerical factors.

Note that the interpolation inequality in (15.9) says that

$$\text{LHS of (15.10)} \le C_0 \, (\text{LHS of (15.11)})^{3/2},$$

so the implication $(B_n) \Rightarrow (A_n')$ will be easy to prove in this sketch version in which we omit the pressure.

---

[5] This is essentially the simple Lebesgue interpolation $\|u\|_{L^3} \le \|u\|_{L^2}^{1/2} \|u\|_{L^6}^{1/2}$ coupled with the Sobolev embedding theorem $H^1 \subset L^6$ and then integrated in time. A proof of this inequality is given at the end of this chapter in Lemma 15.10.

Step 1. The first inductive step (A_1')

We now consider one fixed $(a, s) \in Q_{1/2}(0, 0)$. To start the induction note that (A_1') follows from (15.6), since

$$4 \int_{Q_{1/2}(a,s)} |u|^3 \leq 4 \int_{Q_1(0,0)} |u|^3 \leq 4\varepsilon_0 \leq \varepsilon_0^{2/3} \left(\frac{1}{2}\right)^3,$$

provided that $\varepsilon_0 \leq 2^{-15}$. For the remainder of this section we drop the (a, s) argument from all cylinders, and the argument (a) from all balls, so that we write $Q_r = Q_r(a, s)$ and $B_r = B_r(a)$ whatever the value of r.

Step 2. Using the local energy inequality, $\{(A_k'), 1 \leq k \leq n\}$ *implies* (B_{n+1})

We use the local energy inequality to show that if (A_k') holds for all $1 \leq k \leq n$ then so does (B_{n+1}). (We never show (B_1).) We choose a test function that is the product of a solution of the backwards heat equation (so $\partial_t \phi + \Delta \phi \approx 0$) multiplied by a cutoff function that localises the support of ϕ: in this way (in Lemma 15.11) we construct a sequence of non-negative test functions $\{\phi_n\}_{n=2}^{\infty}$ such that $\phi_n \in C_c^{\infty}(\mathbb{R}^4)$,

$$(\text{supp } \phi_n) \cap Q_1 \subset Q_{1/3},$$

$$|\partial_t \phi_n + \Delta \phi_n| \leq C_1 r_n^2 \text{ for all } t \leq 0,$$

$$\phi_n \geq C_1^{-1} r_n^{-1} \quad \text{and} \quad |\nabla \phi_n| \leq C_1 r_n^{-2} \quad \text{on} \quad Q_{r_n},$$

and

$$|\nabla \phi_n| \leq C_1 r_n^2 r_k^{-4} \quad \text{on} \quad Q_{r_{k-1}} \setminus Q_{r_k}, \quad 2 \leq k \leq n. \tag{15.12}$$

(It is important that the constant $C_1 \geq 1$ does not depend on n.)

Note that for any function $\psi \geq 0$ whose support satisfies

$$(\text{supp } \psi) \cap Q_1 \subset Q_{1/3}$$

for any $t \in (-1, 0)$ we have

$$\int_{-1}^{t} \int_{B_1} \psi \leq \int_{(\text{supp } \psi) \cap Q_1} \psi \leq \int_{Q_{1/3}} |\psi| \leq \int_{Q_{1/2}} |\psi|.$$

If we use the test function ϕ_n in the local energy inequality (15.8) we obtain, for any $t \in (-1, 0)$,

$$\int_{B_1} |u(t)|^2 \phi_n(t) + 2 \int_{-1}^{t} \int_{B_1} |\nabla u|^2 \phi_n$$

$$\leq \int_{-1}^{t} \int_{B_1} |u|^2 (\partial_t \phi_n + \Delta \phi_n) + \int_{-1}^{t} \int_{B_1} |u|^2 u \cdot \nabla \phi_n. \tag{15.13}$$

If we choose $t \in (s - r_n^2, s)$ then since $\phi_n \geq C_1^{-1} r_n^{-1}$ on Q_{r_n} it follows (dropping the second term on the left-hand side) that

$$C_1^{-1} \frac{1}{r_n} \int_{B_{r_n}} |u(t)|^2 \leq \int_{-1}^{t} \int_{B_1} |u|^2 (\partial_t \phi_n + \Delta \phi_n) + \int_{-1}^{t} \int_{B_1} |u|^2 u \cdot \nabla \phi_n$$

$$\leq C_1 r_n^2 \int_{Q_{1/2}} |u|^2 + \int_{Q_{1/2}} |u|^3 |\nabla \phi_n|,$$

and since this is valid for all such t we obtain

$$C_1^{-1} \frac{1}{r_n} \sup_{s - r_n^2 < t < s} \int_{B_{r_n}} |u(t)|^2 \leq C_1 r_n^2 \int_{Q_{1/2}} |u|^2 + \int_{Q_{1/2}} |u|^3 |\nabla \phi_n|.$$

If we choose $t = s$ in (15.13) then (dropping the first term on the left-hand side)

$$2 \int_{-1}^{s} \int_{B_1} |\nabla u|^2 \phi_n \leq \int_{-1}^{s} \int_{B_1} |u|^2 (\partial_t \phi_n + \Delta \phi_n) + \int_{-1}^{s} \int_{B_1} |u|^2 u \cdot \nabla \phi_n,$$

from which it follows that

$$2 C_1^{-1} \frac{1}{r_n} \int_{-1}^{s} \int_{B_{r_n}} |\nabla u|^2 \leq C_1 r_n^2 \int_{Q_{1/2}} |u|^2 + \int_{Q_{1/2}} |u|^3 |\nabla \phi_n|.$$

Combining these two estimates we obtain

$$C_1^{-1} \frac{1}{r_n} \sup_{s - r_n^2 < t < s} \int_{B_{r_n}} |u(t)|^2 + C_1^{-1} \frac{1}{r_n} \int_{-1}^{s} \int_{B_{r_n}} |\nabla u|^2$$

$$\leq \frac{3}{2} C_1 r_n^2 \int_{Q_{1/2}} |u|^2 + \frac{3}{2} \int_{Q_{1/2}} |u|^3 |\nabla \phi_n|. \qquad (15.14)$$

The first term on the right-hand side can be bounded using Hölder's inequality and our initial smallness assumption (15.6),

$$\frac{3}{2} C_1 r_n^2 \int_{Q_{1/2}} |u|^2 \leq \frac{3}{2} C_1 r_n^2 \left(\int_{Q_{1/2}} 1 \right)^{1/3} \left(\int_{Q_{1/2}} |u|^3 \right)^{2/3}$$

$$= 3 \times 2^{-8/3} (4\pi/3)^{1/3} C_1 \varepsilon_0^{2/3} r_n^2$$

$$< C_1 \varepsilon_0^{2/3} r_n^2, \qquad (15.15)$$

since $3 \times 2^{-8/3} (4\pi/3)^{1/3} \simeq 0.762 < 1$. The second term can be bounded by splitting the domain of integration into 'rings' $Q_{r_{k-1}} \setminus Q_{r_k}$, see Figure 15.2.

We use the upper bounds on $|\nabla \phi_n|$ in each ring from (15.12),

$$|\nabla \phi_n| \leq C_1 r_n^2 r_k^{-4} \quad \text{on} \quad Q_{r_{k-1}} \setminus Q_{r_k}, \quad 2 \leq k \leq n,$$

and

$$|\nabla \phi_n| \leq C_1 r_n^{-2} \quad \text{on} \quad Q_{r_n},$$

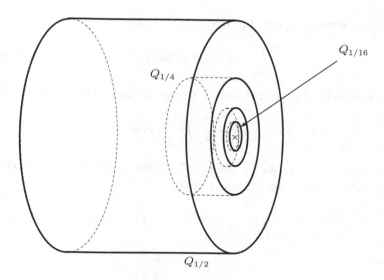

Figure 15.2. Splitting $Q_{1/2}$ into 'rings' to make maximum use of our inductive hypotheses; illustrated is the case $n = 4$.

along with the inductive bound $r_k^{-2} \int_{Q_{r_k}} |u|^3 \le \varepsilon_0^{2/3} r_k^3$ from (A$_k'$):

$$\int_{Q_{1/2}} |u|^3 |\nabla \phi_n| = \left\{ \sum_{k=2}^{n} \int_{Q_{r_{k-1}} \setminus Q_{r_k}} |u|^3 |\nabla \phi_n| \right\} + \int_{Q_{r_n}} |u|^3 |\nabla \phi_n|$$

$$\le \left\{ \sum_{k=2}^{n} \int_{Q_{r_{k-1}} \setminus Q_{r_k}} C_1 r_n^2 r_k^{-4} |u|^3 \right\} + C_1 r_n^{-2} \int_{Q_{r_n}} |u|^3$$

$$\le C_1 r_n^2 \left\{ \sum_{k=2}^{n} r_k^{-4} \int_{Q_{r_{k-1}}} |u|^3 \right\} + C_1 r_n^{-2} \int_{Q_{r_n}} |u|^3.$$

Relabelling the terms in the sum we can write

$$\int_{Q_{1/2}} |u|^3 |\nabla \phi_n| \le C_1 r_n^2 \left\{ \sum_{k=1}^{n-1} 2^4 r_k^{-4} \int_{Q_{r_k}} |u|^3 \right\} + C_1 r_n^{-2} \int_{Q_{r_n}} |u|^3$$

$$\le 2^4 C_1 r_n^2 \sum_{k=1}^{n} r_k^{-2} (\varepsilon_0^{2/3} r_k^3)$$

$$= 2^4 C_1 r_n^2 \varepsilon_0^{2/3} \sum_{k=1}^{n} r_k,$$

and hence, since $\sum_{k=1}^{n} r_k \le 1$ as $r_k = 2^{-k}$,

$$\frac{3}{2} \int_{Q_{1/2}} |u|^3 |\nabla \phi_n| \le 24 C_1 \varepsilon_0^{2/3} r_n^2. \tag{15.16}$$

Combining (15.14), (15.15), and (15.16), for any $(a, s) \in Q_{1/2}(0, 0)$ we have

$$\frac{1}{r_n} \sup_{s-r_n^2 < t < s} \int_{B_{r_n}(a)} |u(t)|^2 + \frac{1}{r_n} \int_{Q_{r_n}(a,s)} |\nabla u|^2 \le 25 C_1^2 \, \varepsilon_0^{2/3} r_n^2.$$

To obtain (B_{n+1}) we use the fact that $r_n = 2r_{n+1}$ and that the integrals over $Q_{r_{n+1}}$ are bounded above by the integrals over Q_{r_n}:

$$\frac{1}{r_{n+1}} \sup_{s-r_{n+1}^2 < t < s} \int_{B_{r_{n+1}}(a)} |u(t)|^2 + \frac{1}{r_{n+1}} \int_{Q_{r_{n+1}}(a,s)} |\nabla u|^2$$
$$\le (8 \times 25) C_1^2 \, \varepsilon_0^{2/3} r_{n+1}^2$$
$$= 200 \, C_1^2 \, \varepsilon_0^{2/3} r_{n+1}^2,$$

which is (B_{n+1}) provided that we take $C_B = 200 C_1^2$.

Step 3. (B_k), $2 \le k \le n$, *implies* (A_n')

In the absence of the pressure it is very simple to use the interpolation inequality (15.9) to show that (B_n) implies (A_n'). Indeed, since (B_n) is exactly a bound on the term in square brackets below, it follows immediately that

$$\frac{1}{r_n^2} \int_{Q_{r_n}} |u|^3 \le C_0 \left[\frac{1}{r_n} \sup_{s-r_n^2 < t < s} \int_{B_{r_n}} |u(t)|^2 + \frac{1}{r_n} \int_{Q_{r_n}} |\nabla u|^2 \right]^{3/2}$$
$$\le C_0 \left[C_B \varepsilon_0^{2/3} r_n^2 \right]^{3/2}$$
$$= C_0 C_B^{3/2} \varepsilon_0 r_n^3$$
$$\le \varepsilon_0^{2/3} r_n^3, \tag{15.17}$$

provided that ε_0 is chosen small enough that $C_0 C_B^{3/2} \varepsilon_0^{1/3} \le 1$. (The pressure will complicate this step considerably.)

Step 4. Use the Lebesgue Differentiation Theorem to finish the proof

This inductive argument proceeds via the chain of implications

$$(A_1') \implies (B_2) \implies \{(A_1'), (A_2')\} \implies \{(B_2), (B_3)\} \implies \cdots$$

to yield, for each $(a, s) \in Q_{1/2}(0, 0)$, the inequalities (A_n') for all $n \ge 1$ and (B_n) for all $n \ge 2$.

In particular, from half of (B_n) we have

$$\sup_{s-r_n^2<t<s} \frac{1}{r_n^3} \int_{B_{r_n}(a)} |u(t)|^2 \leq C_B \, \varepsilon_0^{2/3}$$

for each $(a, s) \in Q_{1/2}(0, 0)$ and every $n \in \mathbb{N}$. It now follows from the Lebesgue Differentiation Theorem (Theorem A.2) that for almost every $(a, s) \in Q_{1/2}(0, 0)$ we have

$$|u(a, s)|^2 \leq C_B \varepsilon_0^{2/3},$$

and hence

$$\|u\|_{L^\infty(Q_{1/2})} \leq C_B^{1/2} \varepsilon_0^{1/3}.$$

15.3 A first local regularity theorem in terms of u and p

We are now in a position to prove the first local regularity result in full. We follow the argument outlined in Section 15.2, but now have to deal with the additional contribution from the pressure term in the local energy inequality. This will require a local pressure estimate, whose proof we postpone until Section 15.5.

Theorem 15.3 *There exist absolute constants $\varepsilon_0^* > 0$ and $c_M > 0$ such that if (u, p) is a suitable weak solution of the Navier–Stokes equations on $Q_1(0, 0)$ and for some $\varepsilon_0 \leq \varepsilon_0^*$*

$$\int_{Q_1(0,0)} |u|^3 + |p|^{3/2} \, dx \, dt \leq \varepsilon_0 \tag{15.18}$$

then $u \in L^\infty(Q_{1/2}(0, 0))$ with $\|u\|_{L^\infty(Q_{1/2}(0,0))} \leq c_M \varepsilon_0^{1/3}$.

Proof Set

$$C_B = 2^{16} C_1^2, \tag{15.19}$$

where $C_1 \geq 1$ is the constant from Lemma 15.11 (construction of cutoff functions), and then choose ε_0^* such that

$$\varepsilon_0^* = (2^{15} C_0^2 C_2^2 C_B^3)^{-3/2},$$

where $C_0 \geq 1$ is the constant from Lemma 15.10 (interpolation inequality) and $C_2 \geq 1$ is the constant from Lemma 15.12 (local pressure estimate). It follows that for any $\varepsilon_0 \leq \varepsilon_0^*$ we have

$$2^{11/2} C_0 C_2 C_B^{3/2} \varepsilon_0 \leq \frac{1}{4} \varepsilon_0^{2/3}. \tag{15.20}$$

The reasons for these choices (which are not optimal) can be seen later, but we fix these constants now to demonstrate clearly that the argument is not circular.

For each $(a, s) \in Q_{1/2}(0, 0)$ we let $r_n = 2^{-n}$ and perform an induction based on the bounds

$$(A_n): \quad \frac{1}{r_n^2} \int_{Q_{r_n}(a,s)} |u|^3 + \frac{1}{r_n^{3/2}} \int_{Q_{r_n}(a,s)} |p - (p)_{r_n}|^{3/2} \le \varepsilon_0^{2/3} r_n^3, \qquad (15.21)$$

where

$$(p)_r := \frac{1}{|B_r(a)|} \int_{B_r(a)} p \, dx$$

denotes the average of p over $B_r(a)$, and

$$(B_n): \quad \frac{1}{r_n} \sup_{s - r_n^2 < t < s} \int_{B_{r_n}(a)} |u(t)|^2 + \frac{1}{r_n} \int_{Q_{r_n}(a,s)} |\nabla u|^2 \le C_B \, \varepsilon_0^{2/3} r_n^2. \qquad (15.22)$$

Note that (A_n) now includes a term involving the pressure, but this is not the scale-invariant quantity

$$\frac{1}{r_n^2} \int_{Q_{r_n}(a,s)} |p|^{3/2}$$

that we might expect. One reason for this is that given bounds on ∇p we can only obtain bounds on p minus its average; the smaller factor of $r_n^{-3/2}$ allows some flexibility in the iterative argument, which turns out to be necessary (see Step 3).

Step 1. The first inductive step (A_1)

Fix $(a, s) \in Q_{1/2}(0, 0)$ and observe that it follows from (15.18) that

$$\int_{Q_{1/2}(a,s)} |u|^3 + |p|^{3/2} \le \varepsilon_0.$$

Since $(\alpha + \beta)^q \le 2^{q-1}(\alpha^q + \beta^q)$ for $\alpha, \beta \ge 0, q \ge 1$ (see Exercise 15.1) we have

$$\int_{Q_{1/2}(a,s)} |p - (p)_{1/2}|^{3/2} \le \sqrt{2} \int_{Q_{1/2}(a,s)} \left(|p|^{3/2} + |(p)_{1/2}|^{3/2} \right);$$

since

$$\int_{Q_r} |(p)_r|^{3/2} \le \int_{Q_r} |p|^{3/2}$$

(see Exercise 15.2) it follows that

$$4 \int_{Q_{1/2}(a,s)} |u|^3 + 2\sqrt{2} \int_{Q_{1/2}(a,s)} |p - (p)_{1/2}|^{3/2}$$

$$\leq 4 \int_{Q_{1/2}(a,s)} |u|^3 + 4 \int_{Q_{1/2}(a,s)} \left(|p|^{3/2} + |(p)_{1/2}|^{3/2}\right)$$

$$= 4 \int_{Q_{1/2}(a,s)} \left(|u|^3 + |p|^{3/2}\right) + 4 \int_{Q_{1/2}(a,s)} |(p)_{1/2}|^{3/2}$$

$$\leq 8\varepsilon_0 \leq \varepsilon_0^{2/3} \left(\frac{1}{2}\right)^3$$

since (15.20) implies that $\varepsilon_0^{1/3} \leq 2^{-6}$, and this is (A$_1$).

For the remainder of the proof we fix (a, s) and drop the (a, s) and (a) arguments as we did before in Section 15.2, so that $Q_r = Q_r(a, s)$ and $B_r = B_r(a)$.

Step 2. Using the local energy inequality, $\{(A_k), 1 \leq k \leq n\}$ implies (B_{n+1})

As in the pressure-free version of the proof we use a sequence of test functions $\phi_n \in C_c^\infty(B_{1/2}(a) \times (s - 1/9, s + r_n^2/2))$, $n \geq 2$, such that

(i) $C_1^{-1} r_n^{-1} \leq \phi_n \leq C_1 r_n^{-1}$ and $|\nabla \phi_n| \leq C_1 r_n^{-2}$ on Q_{r_n};
(ii) $\phi_n \leq C_1 r_n^2 r_k^{-3}$ and $|\nabla \phi_n| \leq C_1 r_n^2 r_k^{-4}$ on $Q_{r_{k-1}} \setminus Q_{r_k}$, $2 \leq k \leq n$;
(iii) (supp ϕ_n) $\cap Q_1 \subset Q_{1/3}$; and
(iv) $|\partial_t \phi_n + \Delta \phi_n| \leq C_1 r_n^2$ on $\mathbb{R}^3 \times (-\infty, 0]$.

The existence of such a family of functions is shown in Lemma 15.11 at the end of the chapter.

Choosing $\phi = \phi_n$ in the local energy inequality

$$\int_{B_1} |u(s)|^2 \phi(s) + 2 \int_{-1}^s \int_{B_1} |\nabla u|^2 \phi$$

$$\leq \int_{-1}^s \int_{B_1} |u|^2 (\partial_t \phi + \Delta \phi) + \int_{-1}^s \int_{B_1} (|u|^2 + 2p) u \cdot \nabla \phi$$

(note that we have restored the pressure term) we obtain, following an argument similar to that employed in the pressure-free version of the proof,

$$C_1^{-1} \frac{1}{r_n} \sup_{s - r_n^2 < t < s} \int_{B_{r_n}} |u(t)|^2 + C_1^{-1} \frac{1}{r_n} \int_{-1}^s \int_{B_{r_n}} |\nabla u|^2$$

$$\leq \frac{3}{2} C_1 r_n^2 \int_{Q_{1/2}} |u|^2 + \frac{3}{2} \int_{Q_{1/2}} |u|^3 |\nabla \phi_n| + 3 \int_{Q_{1/2}} p u \cdot \nabla \phi_n$$

$$=: I_1 + I_2 + I_3. \tag{15.23}$$

We have already estimated I_1 and I_2 in our pressure-free proof (equations (15.15) and (15.16)),

$$I_1 \leq C_1 \varepsilon_0^{2/3} r_n^2 \qquad \text{and} \qquad I_2 \leq 24 \, C_1 \varepsilon_0^{2/3} r_n^2.$$

For the term I_3 involving the pressure a little more care is required; we need to make a decomposition of $Q_{1/2}$ into rings in order to use the fact that our inductive hypothesis (A_k) involves not the pressure but the pressure minus its average. Before we go into the details, note that since u is divergence free, for any constant α and any test function ϕ that is compactly supported within B_r we can integrate by parts to show that

$$\int_{B_r} \alpha u \cdot \nabla \phi = -\int_{B_r} \alpha (\nabla \cdot u)\phi = 0. \tag{15.24}$$

We use this observation to convert I_3 into integrals involving the pressure minus its average, first splitting $Q_{1/2}$ into 'rings' (as we did with I_2 before) so that we can use our inductive hypothesis effectively.

For each $k \geq 1$ we choose a cutoff function $\chi_k \in C_c^\infty(\mathbb{R}^3 \times (-\infty, s])$ with $0 \leq \chi_k \leq 1$,

$$\chi_k \equiv 1 \quad \text{on} \quad \overline{Q_{7r_k/8}}, \qquad \text{supp } \chi_k \cap \overline{Q_1} \subset \overline{Q_{r_k}},$$

and[6] $|\nabla \chi_k| \leq 16 r_k^{-1}$. Noting that $\chi_1 \phi_n = \phi_n$, since the support of ϕ_n is always contained in $Q_{1/3} \subset Q_{1/2 \times 7/8}$, it follows that we can write I_3 as the telescoping sum

$$I_3 = 2\int_{Q_{1/2}} p(u \cdot \nabla)\phi_n$$

$$= 2\sum_{k=1}^{n-1} \int_{Q_{1/2}} p(u \cdot \nabla)[(\chi_k - \chi_{k+1})\phi_n] + 2\int_{Q_{1/2}} p(u \cdot \nabla)(\chi_n \phi_n).$$

Since $(\chi_k - \chi_{k+1})\phi_n$ is compactly supported in $Q_{1/2}$ and u is divergence free, we can use (15.24) to write

$$\int_{Q_{1/2}} p(u \cdot \nabla)[(\chi_k - \chi_{k+1})\phi_n] = \int_{Q_{1/2}} (p - (p)_{r_k})(u \cdot \nabla)[(\chi_k - \chi_{k+1})\phi_n]$$

$$= \int_{Q_{r_k}} (p - (p)_{r_k})(u \cdot \nabla)[(\chi_k - \chi_{k+1})\phi_n],$$

---

[6] To see that this is possible let $\chi \in C_c^\infty(\mathbb{R}^3 \times (-\infty, 0])$ be a cutoff function that is one on $\overline{Q_{7/8}(0,0)}$ and zero off $\overline{Q_1(0,0)}$, with $|\nabla \chi| \leq 16$; then set $\chi_k(x,t) = \chi((x+a)/r_k, (t+s)/r_k^2)$.

since the support of $\chi_k - \chi_{k+1}$ is in fact contained in Q_{r_k}. Now, using the fact that χ_n is supported in Q_{r_n}, a similar argument yields

$$\int_{Q_{1/2}} p(u \cdot \nabla)[\chi_n \phi_n] = \int_{Q_{r_n}} (p - (p)_{r_n})(u \cdot \nabla)[\chi_n \phi_n].$$

Therefore

$$I_3 = 2 \sum_{k=1}^{n-1} \int_{Q_{r_k}} (p - (p)_{r_k})(u \cdot \nabla)[(\chi_k - \chi_{k+1})\phi_n]$$

$$+ 2 \int_{Q_{r_n}} (p - (p)_{r_n})(u \cdot \nabla)[\chi_n \phi_n].$$

Noting that the support of $\chi_k - \chi_{k+1}$ is contained in $Q_{r_k} \setminus Q_{r_{k+2}}$ where $|\nabla \phi_n| \le C_1 r_n^2 r_{k+2}^{-4}$, it follows that

$$\|\nabla[(\chi_k - \chi_{k+1})\phi_n]\|_{L^\infty(Q_{r_k})} \le \|\phi_n\|_{L^\infty(Q_{r_k})}(\|\nabla \chi_k\|_{L^\infty(Q_{r_k})} + \|\nabla \chi_{k+1}\|_{L^\infty(Q_{r_k})})$$

$$+ 2\|\nabla \phi_n\|_{L^\infty(Q_{r_k} \setminus Q_{r_{k+2}})}$$

$$\le 32 C_1 r_n^2 r_{k+1}^{-4} + 2 C_1 r_n^2 r_{k+2}^{-4} = 2^{10} C_1 r_n^2 r_k^{-4}$$

and since $|\nabla \phi_n| \le C_1 r_n^{-2}$ on Q_{r_n}

$$\|\nabla(\chi_n \phi_n)\|_{L^\infty(Q_{r_n})} \le C_1 r_n^{-2} + 16 C_1 r_n^{-2} = 17 C_1 r_n^{-2}.$$

It follows that

$$I_3 \le 3 \times 2^{10} C_1 r_n^2 \sum_{k=1}^{n} \int_{Q_{r_k}} |u||p - (p)_{r_k}| r_k^{-4}$$

$$\le 3 \times 2^{10} C_1 r_n^2 \sum_{k=1}^{n} r_k^{-4} \left(\int_{Q_{r_k}} |u|^3 \right)^{1/3} \left(\int_{Q_{r_k}} |p - (p)_{r_k}|^{3/2} \right)^{2/3}.$$

Including an appropriate factor of r_k so that we can use (A_k), $1 \le k \le n$, we obtain via Young's inequality $(ab \le \frac{1}{3}a^3 + \frac{2}{3}b^{3/2})$

$$I_3 \le 2^{11} C_1 r_n^2 \sum_{k=1}^{n} r_k^{-7/3} \left(\frac{1}{r_k^2} \int_{Q_{r_k}} |u|^3 \right)^{1/3} \left(\frac{1}{r_k^{3/2}} \int_{Q_{r_k}} |p - (p)_{r_k}|^{3/2} \right)^{2/3}$$

$$\le 2^{11} C_1 r_n^2 \sum_{k=1}^{n} r_k^{-7/3} \left(\frac{1}{r_k^2} \int_{Q_{r_k}} |u|^3 + \frac{1}{r_k^{3/2}} \int_{Q_{r_k}} |p - (p)_{r_k}|^{3/2} \right).$$

Recalling (A_k),

$$\frac{1}{r_k^2} \int_{Q_{r_k}} |u|^3 + \frac{1}{r_k^{3/2}} \int_{Q_{r_k}} |p - (p)_{r_k}|^{3/2} \le r_k^3 \varepsilon_0^{2/3},$$

this yields

$$I_3 \leq 2^{11} C_1 r_n^2 \sum_{k=1}^{n} r_k^{2/3} \varepsilon_0^{2/3} \leq 2^{12} C_1 r_n^2 \varepsilon_0^{2/3},$$

since $\sum_{k=1}^{n} r_k^{2/3} < \sum_{k=1}^{\infty} 2^{-2k/3} < 2$.

Returning to (15.23) and combining the estimates for I_1, I_2, and I_3 yields

$$\frac{1}{r_n} \sup_{s-r_n^2 < t < s} \int_{B_{r_n}(a)} |u(t)|^2 + \frac{1}{r_n} \int_{Q_{r_n}(a,s)} |\nabla u|^2 \leq C_1^2 [25 + 2^{12}] \varepsilon_0^{2/3} r_n^2$$

$$< 2^{13} C_1^2 \varepsilon_0^{2/3} r_n^2.$$

As before, if we note that the integral over $Q_{r_{n+1}}$ is no larger than that over Q_{r_n} and that $r_n = 2r_{n+1}$ then

$$\frac{1}{r_{n+1}} \sup_{s-r_{n+1}^2 < t < s} \int_{B_{r_{n+1}}(a)} |u(t)|^2 + \frac{1}{r_{n+1}} \int_{Q_{r_{n+1}}(a,s)} |\nabla u|^2 \leq 2^{16} C_1^2 \varepsilon_0^{2/3} r_{n+1}^2$$

$$= C_B \varepsilon_0^{2/3} r_{n+1}^2,$$

by our choice of C_B in (15.19), and this is (B_{n+1}).

Step 3. (B_k), $2 \leq k \leq n$, *implies* (A_n)

Just as we did in (15.17) we can use the interpolation inequality (15.9) along with (B_n) to obtain the bound we need for u,

$$\frac{1}{r_n^2} \int_{Q_{r_n}} |u|^3 \leq C_0 \left[\frac{1}{r_n} \sup_{s-r_n^2 < t < s} \int_{B_{r_n}(a)} |u(t)|^2 + \frac{1}{r_n} \int_{Q_{r_n}(a,s)} |\nabla u|^2 \right]^{3/2}$$

$$\leq C_0 \left[C_B \varepsilon_0^{2/3} r_n^2 \right]^{3/2}$$

$$= C_0 C_B^{3/2} \varepsilon_0 r_n^3 \leq \frac{1}{4} \varepsilon_0^{2/3} r_n^3, \tag{15.25}$$

by our choice of ε_0 in (15.20).

For the pressure term in (A_n) we need to show that

$$\frac{1}{r_n^{3/2}} \int_{Q_{r_n}} |p - (p)_{r_n}|^{3/2} \leq \frac{3}{4} \varepsilon_0^{2/3} r_n^3. \tag{15.26}$$

Note, as remarked above, that we allow ourselves a little slack here, by replacing the factor of r_n^{-2} that would yield a scale-invariant quantity with only $r_n^{-3/2}$.

The proof of (15.26) uses a local pressure estimate, valid for $0 < r \le \rho/2$,

$$\frac{1}{r^{3/2}}\int_{Q_r}|p-(p)_r|^{3/2} \le C_2\frac{1}{r^{3/2}}\int_{Q_{2r}}|u|^3$$
$$+C_2r^5\left\{\sup_{s-r^2<t<s}\int_{2r<|y|<\rho}\frac{|u(t)|^2}{|y|^4}\,\mathrm{d}y\right\}^{3/2}+C_2\frac{r^3}{\rho^{9/2}}\int_{Q_\rho}|u|^3+|p|^{3/2}, \tag{15.27}$$

where C_2 is an absolute constant. We prove this result in Lemma 15.12 at the end of the chapter.

Using this estimate with $\rho = 1/2$ and $r = r_n$ (so valid for $n \ge 2$), recalling that $2r_n = r_{n-1}$, yields

$$\frac{1}{r_n^{3/2}}\int_{Q_{r_n}}|p-(p)_{r_n}|^{3/2} \le 2^{3/2}C_2\frac{1}{r_{n-1}^{3/2}}\int_{Q_{r_{n-1}}}|u|^3$$
$$+C_2r_n^5\left\{\sup_{s-r_n^2<t<s}\int_{r_{n-1}<|y|<1/2}\frac{|u(t)|^2}{|y|^4}\,\mathrm{d}y\right\}^{3/2}$$
$$+C_2 2^{9/2}r_n^3\int_{Q_{1/2}}|u|^3+|p|^{3/2}.$$

The different contributions to this bound are illustrated in Figure 15.3.

Our induction hypothesis (B_{n-1}) provides a bound on the first term, using the interpolation inequality (15.9) just as in (15.25),

$$2^{3/2}C_2\frac{1}{r_{n-1}^{3/2}}\int_{Q_{r_{n-1}}}|u|^3 \le 2^{3/2}C_2r_{n-1}^{1/2}[C_0C_{\mathrm{B}}^{3/2}\varepsilon_0r_{n-1}^3]$$
$$=2^5C_0C_2C_{\mathrm{B}}^{3/2}\varepsilon_0r_n^{7/2} \le \frac{1}{4}\varepsilon_0^{2/3}r_n^3,$$

by our choice of ε_0 in (15.20) and since $r_n < 1$. The final term is also simple to bound, using our initial smallness assumption (15.18),

$$C_2 2^{9/2}r_n^3\int_{Q_{1/2}}|u|^3+|p|^{3/2} \le C_2 2^{9/2}r_n^3\int_{Q_1}|u|^3+|p|^{3/2}$$
$$\le C_2 2^{9/2}r_n^3\varepsilon_0 \le \frac{1}{4}\varepsilon_0^{2/3}r_n^3,$$

again by our choice of ε_0 in (15.20).

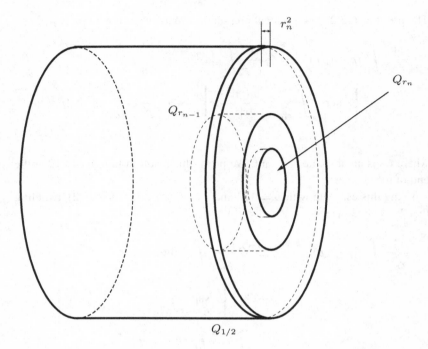

Figure 15.3. Bounding the pressure. The integral over Q_{r_n} can be bounded using three contributions: an integral of $|u|^3$ over $Q_{r_{n-1}}$ (bounded by our inductive hypotheses), the supremum over a short time interval of the integral of $|u|^2/|y|^4$ over $B_{1/2} \setminus B_{r_{n-1}}$ (bounded after a little work by our inductive hypotheses), and the integral of $|u|^3 + |p|^{3/2}$ over $Q_{1/2}$ (bounded by our initial assumption).

The middle term requires a little work. As we have done twice previously, we decompose the domain of integration into a series of 'rings' within which we can use our inductive hypothesis $(\mathbf{B}_k)$:

$$\sup_{s-r_n^2 < t < s} \int_{r_{n-1} < |y| < 1/2} \frac{|u(t)|^2}{|y|^4} \, dy = \sup_{s-r_n^2 < t < s} \sum_{k=2}^{n-1} \int_{r_k \leq |y| \leq r_{k-1}} \frac{|u(t)|^2}{|y|^4} \, dy.$$

We can take the supremum over a longer time interval (depending on k), and then use the inequality

$$\frac{1}{r_k} \sup_{s-r_k^2 < t < s} \int_{B_{r_k}(a)} |u(t)|^2 \leq C_B \, \varepsilon_0^{2/3} r_k^2,$$

which follows from (B_k), to obtain

$$\sup_{s-r_n^2<t<s}\int_{r_{n-1}<|y|<1/2}\frac{|u(t)|^2}{|y|^4}\,dy \le \sum_{k=2}^{n-1}\sup_{s-r_{k-1}^2<t<s}\int_{r_k\le|y|\le r_{k-1}}\frac{|u(t)|^2}{|y|^4}\,dy$$

$$\le \sum_{k=2}^{n-1}r_k^{-4}\left\{\sup_{s-r_{k-1}^2<t<s}\int_{B_{r_{k-1}}}|u(t)|^2\,dy\right\}$$

$$\le 2^3\sum_{k=2}^{n-1}C_B\varepsilon_0^{2/3}r_k^{-1}$$

$$\le 2^3C_B\varepsilon_0^{2/3}r_n^{-1},$$

since $\sum_{k=2}^{n-1}r_k^{-1}=\sum_{k=2}^{n-1}2^k\le 2^n=r_n^{-1}$. Therefore for the middle term we have

$$C_2r_n^5\left\{\sup_{s-r_n^2<t<s}\int_{r_{n-1}<|y|<1/2}\frac{|u(t)|^2}{|y|^4}\,dy\right\}^{3/2} \le C_2r_n^5\left\{2^3C_B\varepsilon_0^{2/3}r_n^{-1}\right\}^{3/2}$$

$$\le C_22^{9/2}C_B^{3/2}\varepsilon_0 r_n^{7/2}$$

$$\le \frac{1}{4}\varepsilon_0^{2/3}r_n^3,$$

once again using the fact that ε_0 has been chosen sufficiently small in (15.20) and[7] that $r_n < 1$.

Combining these three estimates we obtain (15.26), which together with (15.25) yields

$$\frac{1}{r_n^2}\int_{Q_{r_n}}|u|^3 + \frac{1}{r_n^{3/2}}\int_{Q_{r_n}}|p-p_n|^{3/2} \le \varepsilon_0^{2/3}r_n^3,$$

which is $(A)_n$, as required.

Step 4. Finishing the proof

Running the inductive argument above we have

$$(A_1) \Rightarrow (B_2) \Rightarrow \{(A)_1, (A_2)\} \Rightarrow \{(B_2), (B_3)\} \Rightarrow \ldots,$$

and hence we have shown that (A_n) holds for all $n = 1, 2, \ldots$ and that (B_n) holds for all $n = 2, 3, \ldots$

---

[7] Note that the additional factor of $r_n^{1/2}$ in the pressure part of $(A)_n$ was only required to obtain the correct dependence on r_n in the final of the three terms: for the first two we would have obtained a bound of the order of r_n^3 without this additional factor.

In particular, we have shown that for each $(a, s) \in Q_{1/2}(0, 0)$ and for all $n \geq 1$,

$$\sup_{s-r_n^2 < t < s} \frac{1}{r_n^3} \int_{B_{r_n}(a)} |u(t)|^2 \leq C_B \varepsilon_0^{2/3}.$$

The result of the theorem now follows using the Lebesgue Differentiation Theorem, as in the pressure-free case, since

$$C_B^{1/2} \varepsilon_0^{1/3} \leq [2^{16} C_1^2]^{1/2} \varepsilon_0^{1/3} \leq 2^8 C_1 \varepsilon_0^{1/3},$$

recalling from (15.19) that $C_B = 2^{16} C_1^2$. For the result as stated take $c_M = 2^8 C_1$.

$\square$

We now use rescaling to obtain a local regularity theorem whose initial smallness assumption is only required on some parabolic cylinder $Q_r(0, 0)$.

Theorem 15.4 (First Local Regularity Theorem) *There exist absolute constants $\varepsilon_0^* > 0$ and $c_M > 0$ such that if (u, p) is a suitable weak solution of the Navier–Stokes equations on $Q_r(0, 0)$ and for some $\varepsilon_0 \leq \varepsilon_0^*$*

$$\frac{1}{r^2} \int_{Q_r(0,0)} |u|^3 + |p|^{3/2} < \varepsilon_0 \tag{15.28}$$

then $\|u\|_{L^\infty(Q_{r/2})} \leq c_M \varepsilon_0^{1/3} r^{-1}$.

Proof The rescaled function $u_r(x, t) = r u(rx, r^2 t)$ solves the Navier–Stokes equations on Q_1 (see Section 15.1). Using Theorem 15.3, if

$$\int_{Q_1(0,0)} |u_r|^3 + |p_r|^{3/2} < \varepsilon_0 \tag{15.29}$$

then $\|u_r\|_{L^\infty(Q_{1/2})} \leq c_M \varepsilon_0^{1/3}$. In terms of the original functions u and p, the condition (15.29) is

$$\int_{Q_1(0,0)} r^3 |u(rx, r^2 t)|^3 + r^3 |p(rx, r^2 t)|^{3/2} \, dx \, dt$$

$$= \frac{1}{r^2} \int_{Q_r(0,0)} |u(y, s)|^3 + |p(y, s)|^{3/2} \, dy \, ds,$$

by substituting $rx = y$ ($dx = r^{-3} \, dy$) and $r^2 t = s$ ($dt = r^{-2} \, ds$). The conclusion follows since $\|u\|_{L^\infty(Q_{r/2})} = r^{-1} \|u_r\|_{L^\infty(Q_{1/2})}$. $\square$

Applying the local regularity result of Theorem 13.4 and the time regularity result of Proposition 13.10 we immediately obtain the following corollary.

Corollary 15.5 *Under the same assumptions as in Theorem 15.4 it follows that u is a C^∞ function of the space variables on $Q_{r/2}$. If u is the restriction to Q_r*

of a solution of the Navier–Stokes initial-value problem then u is also Hölder continuous in time on $Q_{r/2}$.

Note that it is a peculiar feature of Theorem 15.4 that it does not guarantee that $(0, 0)$ is a regular point, since $u \in L^\infty(Q_{r/2})$ only ensures that u is bounded near $x = 0$ for $t < 0$. It is therefore interesting to give a variant of Theorem 15.4 that deals with this issue. Recall that we denote by $Q_r^*(x, t)$ the parabolic cylinder centred at (x, t),

$$Q_r^*(x, t) = B_r(x) \times (t - r^2/2, t + r^2/2),$$

and so in particular

$$Q_{r/2}^*(x, t) = Q_{r/2}(x, t + r^2/8).$$

Corollary 15.6 *There exist absolute constants $\varepsilon_0^* > 0$ and $c_M > 0$ such that if (u, p) is a suitable weak solution of the Navier–Stokes equations on $Q_r(x, t)$ and for some $\varepsilon_0 \le \varepsilon_0^*$*

$$\frac{1}{r^2} \int_{Q_r(x, t+r^2/8)} |u|^3 + |p|^{3/2} < \varepsilon_0$$

then $\|u\|_{L^\infty(Q_{r/2}^*(x,t))} \le c_M \varepsilon_0^{1/3} r^{-1}$ and in particular (x, t) is a regular point.

For a weaker result involving only centred cylinders, note that the domain of integration could be replaced by $Q_{2r}^*(x, t)$, since $Q_r(x, t + r^2/8) \subset Q_{2r}^*(x, t)$.

15.4 Partial regularity I: $\dim_B(S) \le 5/3$

We now show that the set S of singular points in space–time of any suitable weak solution cannot be too large. Recall that a point (x, t) is 'regular' if u is bounded in a neighbourhood of (x, t), and 'singular' otherwise.

An immediate consequence of the first local regularity theorem is that the set of singular points must be bounded in space–time.

Proposition 15.7 *Let u be a suitable Leray–Hopf weak solution of the Navier–Stokes equations. Then the set S of space–time singularities is a bounded subset of $\Omega \times (0, \infty)$.*

Proof Theorem 8.1 guarantees that any Leray–Hopf solution is regular for $t \ge T(u_0)$, so the theorem follows immediately on a bounded domain; we only need consider the case $\Omega = \mathbb{R}^3$.

We show that there exists an $R > 0$ such that for any parabolic cylinder $Q_1(x, t)$ that does not intersect $B(0, R) \times (0, T)$ we have must have

$$\int_{Q_1(x,t)} |u|^3 + |p|^{3/2} \, dy \, ds < \varepsilon_0,$$

and hence there can be no singular points with $|x| \ge R$. If this claim is false then there are an infinite number of disjoint parabolic cylinders on which either

$$\int_{Q_1(x,t)} |u|^3 \, dy \, ds \ge \frac{1}{2}\varepsilon_0 \qquad \text{or} \qquad \int_{Q_1(x,t)} |p|^{3/2} \, dy \, ds \ge \frac{1}{2}\varepsilon_0.$$

But the lower bound on $\int_{Q_1} |u|^3$ cannot occur infinitely often, since we know that $u \in L^{10/3}(\mathbb{R}^3 \times (0, T))$ and $u \in L^2(\mathbb{R}^3 \times (0, T))$, from which by simple Lebesgue interpolation we have $u \in L^3(\mathbb{R}^3 \times (0, T))$.

The argument is essentially the same when the lower bound on $\int |p|^{3/2}$ occurs infinitely often. $\square$

We can also deduce an upper bound on the dimension of the singular set from the first local regularity theorem: we will show (more or less) that $\dim_B(S) \le 5/3$, where $\dim_B$ is the (upper) box-counting dimension defined in Chapter 8.

First recall that the upper box-counting dimension of a set X can be defined as

$$\dim_B(X) = \limsup_{\varepsilon \to 0} \frac{\log M(X, \varepsilon)}{-\log \varepsilon},$$

where $M(X, \varepsilon)$ is the maximum number of disjoint open balls of radius ε with centres in X (Lemma 8.6). This form of the definition is well suited to proving bounds on the size of the singular set, and the bound $\dim_B(S) \le 5/3$ follows easily from this definition using the first local regularity theorem and the facts that for any weak solution u we have $u \in L^{10/3}(\Omega \times (0, T))$ (see (5.12)) and $p \in L^{5/3}(\Omega \times (0, T))$ (see (5.14)).

Theorem 15.8 (Partial regularity I) *If S denotes the singular set of a suitable weak solution then $\dim_B(S \cap K) \le 5/3$ for every compact subset K of $\Omega \times (0, T)$.*

(In fact a slightly more refined bound is possible, using a definition of the 'box-counting dimension' based on a covering by parabolic cylinders rather than balls, see Exercises 15.3 and 15.4.)

Proof   Using Corollary 15.6 it follows that if $(x, t) \in S$ then

$$\frac{1}{r^2} \int_{Q_r(x, t+r^2/8)} |u|^3 + |p|^{3/2} \ge \varepsilon_0$$

for every $r > 0$, for otherwise (x, t) would be a regular point. Since Hölder's inequality implies that

$$\int_{Q_r} |u|^3 \leq \left(\int_{Q_r} |u|^{10/3} \right)^{9/10} \left(\int_{Q_r} 1 \right)^{1/10}$$

$$= (4\pi/3)^{1/10} r^{1/2} \left(\int_{Q_r} |u|^{10/3} \right)^{9/10}$$

and similarly

$$\int_{Q_r} |p|^{3/2} \leq (4\pi/3)^{1/10} r^{1/2} \left(\int_{Q_r} |p|^{5/3} \right)^{9/10},$$

it follows that

$$\int_{Q_r(x,t+r^2/8)} |u|^{10/3} + |p|^{5/3} \geq C \varepsilon_0^{10/9} r^{5/3}$$

for some absolute constant $C > 0$.

Now fix a compact subset K of $\Omega \times (0, T)$. Since K is compact there exists an $r_0 > 0$ such that $Q_r(x, t + r^2/8) \subset \Omega \times (0, T)$ for all radii $r < r_0$ and every point $(x, t) \in S \cap K$.

Suppose that $\dim_B(S \cap K) > 5/3$; fix d with $5/3 < d < \dim_B(S \cap K)$. It follows from the definition of the box-counting dimension that there exists a decreasing sequence $\varepsilon_j \to 0$ such that $M_j := M(S \cap K, \varepsilon_j) \geq \varepsilon_j^{-d}$. Take a disjoint collection $\{B_i, \ i = 1, \ldots, M_j\}$ of open ε_j-balls with centres in $S \cap K$. Since for any $0 < \varepsilon < 1$

$$B_\varepsilon(x, t) \supset Q^*_{\varepsilon/2}(x, t) = Q_{\varepsilon/2}(x, t + \varepsilon^2/8) \tag{15.30}$$

it follows that for every j large enough that $\varepsilon_j < r_0$,

$$\int_{\Omega \times (0,T)} |u|^{10/3} + |p|^{5/3} \geq \sum_{i=1}^{M_j} \int_{B_{\varepsilon_j}(x_i, t_i)} |u|^{10/3} + |p|^{5/3}$$

$$\geq \sum_{i=1}^{M_j} \int_{Q_{\varepsilon_j/2}(x_i, t_i + \varepsilon_j^2/8)} |u|^{10/3} + |p|^{5/3}$$

$$\geq \varepsilon_j^{-d} C \varepsilon_0^{10/9} \varepsilon_j^{5/3}$$

$$= C \varepsilon_0^{10/9} \varepsilon_j^{5/3-d}.$$

The left-hand side is finite, but the right-hand side tends to infinity as $j \to \infty$ since $d > 5/3$, and we obtain a contradiction. Thus $\dim_B(S \cap K) \leq 5/3$ as claimed. $\qquad\square$

In Chapter 17 we will use this restriction on the size of the possible singular set to prove the almost-everywhere uniqueness of particle trajectories for suitable weak solutions. However, the following simple corollary will be useful in the proof of the second partial regularity theorem (Theorem 16.2) in the next chapter.

Corollary 15.9 *The space–time singular set S of a suitable weak solution has Lebesgue measure zero.*

Proof   Take a sequence of compact subsets K_j of $\mathbb{R}^4$ such that

$$\Omega \times (0, \infty) = \bigcup_{j=1}^{\infty} K_j.$$

Then by Theorem 15.8 $\dim_B(K_j \cap S) \le 5/3$ for each j, and it follows from Corollary 8.12 that $\mu(K_j \cap S) = 0$. Since $S = \bigcup_{j=1}^{\infty}(K_j \cap S)$ it follows that $\mu(S) = 0$. $\qquad\square$

15.5 Lemmas for the first partial regularity theorem

We complete this chapter with the proofs of three auxiliary results that we have delayed to simplify the presentation of the main argument.

First we prove the interpolation inequality (15.9).

Lemma 15.10 (Interpolation inequality) *There exists a constant C_0, which does not depend on r, such that if*

$$u \in L^{\infty}((s - r^2, s); L^2(B_r(a, s))) \quad and \quad \nabla u \in L^2(B_r(a, s) \times (s - r^2, s))$$

then $u \in L^3(Q_r(a, s))$ and

$$\frac{1}{r^2} \int_{Q_r(a,s)} |u|^3 \le C_0 \left[\frac{1}{r} \sup_{s-r^2 < t < s} \int_{B_r(a)} |u(t)|^2 + \frac{1}{r} \int_{Q_r(a,s)} |\nabla u|^2 \right]^{3/2}. \quad (15.31)$$

Proof   We prove the inequality for $r = 1$; since

$$\frac{1}{r^2} \int_{Q_r(a,s)} |u|^3, \quad \frac{1}{r} \sup_{s-r^2 < t < s} \int_{B_r(a)} |u(t)|^2, \quad and \quad \frac{1}{r} \int_{Q_r(a,s)} |\nabla u|^2$$

are all scale-invariant quantities (15.31) then follows.

We begin with the Lebesgue interpolation inequality

$$\|u\|_{L^3(B_1)} \le \|u\|_{L^6(B_1)}^{1/2} \|u\|_{L^2(B_1)}^{1/2}$$

and then use the Sobolev embedding $H^1(B_1) \subset L^6(B_1)$,

$$\|u\|_{L^6(B_1)} \le c_0 \left(\|u\|_{L^2(B_1)}^2 + \|\nabla u\|_{L^2(B_1)}^2 \right)^{1/2}$$
$$\le c_0 \|u\|_{L^2(B_1)} + c_0 \|\nabla u\|_{L^2(B_1)}$$

for some constant c_0. This yields

$$\|u\|_{L^3(B_1)} \le c_0^{1/2} \|\nabla u\|_{L^2(B_1)}^{1/2} \|u\|_{L^2(B_1)}^{1/2} + c_0^{1/2} \|u\|_{L^2(B_1)}$$

or, raising both sides to the power of 3,

$$\int_{B_1} |u|^3 \le 4c_0^{3/2} \left(\int_{B_1} |\nabla u|^2 \right)^{3/4} \left(\int_{B_1} |u|^2 \right)^{3/4} + 4c_0^{3/2} \left(\int_{B_1} |u|^2 \right)^{3/2},$$

using the inequality $(\alpha + \beta)^q \le 2^{q-1}(\alpha^q + \beta^q)$ from Exercise 15.1.

We now integrate in time from $s-1$ to s, using Hölder's inequality on the first term on the right-hand side, and then Young's inequality

$$\int_{Q_1} |u|^3 \le 4c_0^{3/2} \left(\int_{Q_1} |\nabla u|^2 \right)^{3/4} \left[\int_{s-1}^{s} \left(\int_{B_1} |u(t)|^2 \right)^3 \right]^{1/4}$$
$$+ 4c_0^{3/2} \int_{s-1}^{s} \left(\int_{B_1} |u(t)|^2 \right)^{3/2}$$
$$\le 4c_0^{3/2} \left(\int_{Q_1} |\nabla u|^2 \right)^{3/4} \left[\left(\sup_{s-1<t<s} \int_{B_1} |u(t)|^2 \right)^{3/2} \right]^{1/2}$$
$$+ 4c_0^{3/2} \left(\sup_{s-1<t<s} \int_{B_1} |u(t)|^2 \right)^{3/2}$$
$$\le 2c_0^{3/2} \left(\int_{Q_1} |\nabla u|^2 \right)^{3/2} + 6c_0^{3/2} \left(\sup_{s-1<t<s} \int_{B_1} |u(t)|^2 \right)^{3/2},$$

which, setting $C_0 = 6c_0^{3/2}$, yields

$$\int_{Q_1(a,s)} |u|^3 \le C_0 \left[\sup_{s-1<t<s} \int_{B_1(a)} |u(t)|^2 + \int_{Q_1(a,s)} |\nabla u|^2 \right]^{3/2};$$

the inequality (15.31) follows by rescaling. $\square$

Next we show that we can construct a family of smooth cutoff functions that are very close to solutions of the backwards heat equation, i.e. which will make the term $\partial_t \phi + \Delta \phi$ in the local energy inequality as small as we wish.

Lemma 15.11 *For any fixed* $(a, s) \in \mathbb{R}^4$ *write* $Q_r = Q_r(a, s)$, *and set* $r_n = 2^{-n}$. *There exists a constant* $C_1 \geq 1$ *and a sequence of non-negative functions* $\{\phi_n\}_{n=2}^{\infty}$ *with* $\phi_n \in C_c^{\infty}(B_{1/2}(a) \times (s - 1/9, s + r_n^2/2))$ *such that for all* $n \geq 2$

(i) $C_1^{-1} r_n^{-1} \leq \phi_n \leq C_1 r_n^{-1}$ *and* $|\nabla \phi_n| \leq C_1 r_n^{-2}$ *on* Q_{r_n};
(ii) $\phi_n \leq C_1 r_n^2 r_k^{-3}$ *and* $|\nabla \phi_n| \leq C_1 r_n^2 r_k^{-4}$ *on* $Q_{r_{k-1}} \setminus Q_{r_k}$, $2 \leq k \leq n$;
(iii) $(\text{supp } \phi_n) \cap Q_1 \subset Q_{1/3}$; *and*
(iv) $|\partial_t \phi_n + \Delta \phi_n| \leq C_1 r_n^2$ *on* $\mathbb{R}^3 \times (-\infty, 0]$.

Proof   Without loss of generality we take $(a, s) = (0, 0)$. We set

$$\phi_n(x, t) = r_n^2 \theta_n(x, t),$$

where

$$\theta_n(x, t) = \chi_n(x, t) \psi_n(x, t).$$

Here χ_n is a cutoff function that will be defined below and ψ_n is the fundamental solution of the backwards heat equation

$$\psi_t + \Delta \psi = 0 \qquad \text{with} \qquad \psi(x, r_n^2) = \delta(x),$$

i.e. for all $t < r_n^2$ we take

$$\psi_n(x, t) = \frac{1}{(r_n^2 - t)^{3/2}} \exp\left(\frac{-|x|^2}{4(r_n^2 - t)}\right).$$

We define $\chi_n \in C_c^{\infty}(\mathbb{R}^4)$ by setting

$$\chi_n(x, t) = X(x) T_n(t),$$

where $X \in C_c^{\infty}(\mathbb{R}^3)$, $0 \leq X \leq 1$, with

$$X(x) = \begin{cases} 1 & x \in B_{1/4}, \\ 0 & x \notin B_{1/3}, \end{cases}$$

and for each $n = 1, 2, \ldots, T_n \in C_c^{\infty}(\mathbb{R})$, $0 \leq T_n \leq 1$, and

$$T_n(t) = \begin{cases} 1 & -1/16 \leq t \leq 0, \\ 0 & t < -1/9 \text{ and } t > r_n^2/2. \end{cases}$$

(We need this somewhat awkward definition since it is not possible to define a function $\chi \in C_c^\infty(\mathbb{R}^4)$ such that $\chi \equiv 1$ on $Q_{1/4}$ and $\chi \equiv 0$ off $Q_{1/3}$, since these two cylinders share the same centre-right point.) With this definition it is clear that $|\nabla \chi_n|, |\Delta \chi_n| \le c$ for some absolute constant c independent of n, and $|\partial_t \chi_n| \le c$ for all $t \le 0$. Note that for every n we have $\chi_n \equiv 1$ on $Q_{1/4}$, and that $(\text{supp } \chi_n) \cap Q_1 \subseteq Q_{1/3}$.

Given the identities

$$\nabla \theta_n = \nabla \chi_n(x,t)\, \psi_n(x,t) - \frac{2x}{4(r_n^2 - t)} \chi_n(x,t)\, \psi_n(x,t),$$

$$\partial_t \theta_n = (\partial_t \chi_n)(x,t)\, \psi_n(x,t)$$
$$+ \chi_n(x,t) \left\{ \frac{3}{2(r_n^2 - t)} - \frac{|x|^2}{4(r_n^2 - t)^2} \right\} \psi_n(x,t),$$

$$\Delta \theta_n = \psi_n \Delta \chi_n + 2\nabla \chi_n \cdot \nabla \psi_n - \chi_n \left\{ \frac{3}{2(r_n^2 - t)} - \frac{|x|^2}{4(r_n^2 - t)^2} \right\} \psi_n,$$

and noting that $r_n^2 \le (r_n^2 - t) \le r_n^2 + r_k^2$ and $0 \le |x|^2 \le r_k^2$ for $(x,t) \in Q_{r_k}$, properties (i)–(iv) are verified as follows.

(i) If $(x,t) \in Q_{r_n}$ then $r_n^2 \le (r_n^2 - t) \le 2r_n^2$ and $0 \le |x|^2 \le r_n^2$.
(ii) First note that within $Q_{r_{k-1}} \setminus Q_{r_k}$ we have $|\phi_n(x,t)| \le r_n^2 |\psi_n(x,t)|$. We now divide $Q_{r_{k-1}} \setminus Q_{r_k}$ into two parts. In the region

$$\{(x,t) : |x|^2 \le r_{k-1}^2, \ -r_{k-1}^2 < t \le -r_k^2\}$$

clearly

$$|\psi_n(x,t)| \le \frac{1}{(r_n^2 + r_k^2)^{3/2}} \le r_k^{-3}.$$

The remaining portion of $Q_{r_{k-1}} \setminus Q_{r_k}$ is

$$\{(x,t) : r_k^2 \le |x|^2 \le r_{k-1}^2, \ -r_k^2 < t < 0\}$$

and here we have

$$|\psi_n(x,t)| \le \frac{1}{(r_n^2 - t)^{3/2}} \exp\left(-\frac{r_k^2}{4(r_n^2 - t)}\right) = 8r_k^{-3} \alpha^3 e^{-\alpha^2},$$

setting $\alpha = r_k / 2\sqrt{r_n^2 - t}$. Since $\alpha^3 e^{-\alpha^2}$ is bounded on $[0, \infty)$ it follows that here too we have $|\psi_n| \le C r_k^{-3}$ as required.

The argument for $|\nabla \phi_n|$ is similar.

(iii) is immediate from the definition of χ_n.

(iv) follows from

$$\partial_t \theta_n + \Delta \theta_n = (\partial_t \chi_n + \Delta \chi_n)\psi_n + 2\nabla\chi_n \cdot \nabla\psi_n,$$

noting that derivatives of χ_n are identically zero for $(x, t) \in Q_{1/4}(0, 0)$, and therefore ψ_n and its gradient are only evaluated for (x, t) outside $Q_{1/4}(0, 0)$, where they are bounded independently of n. □

We end the chapter with a proof of the local pressure estimate (15.27), which follows from judicious use of the equation (15.32) that defines the pressure in terms of the velocity. The proof is adapted from that in CKN, but is somewhat simpler since we are taking $p \in L^{3/2}(Q)$.

Lemma 15.12 *There is an absolute constant C_2 such that whenever* $p \in L^{3/2}(Q_\rho)$ *and*

$$-\Delta p = \partial_i \partial_j (u_i u_j) \tag{15.32}$$

on Q_ρ then for any $0 < r \le \rho/2$

$$r^{-3/2} \int_{Q_r} |p - (p)_r|^{3/2} \le C_2 r^{-3/2} \int_{Q_{2r}} |u|^3$$

$$+ C_2 r^5 \left\{ \sup_{-r^2 < t < 0} \int_{2r < |y| < \rho} \frac{|u(y, t)|^2}{|y|^4} \, dy \right\}^{3/2} + C_2 \frac{r^3}{\rho^{9/2}} \int_{Q_\rho} |u|^3 + |p|^{3/2}. \tag{15.33}$$

Proof We choose a function $\phi \in C_c^\infty(\mathbb{R}^3)$ such that $0 \le \phi \le 1$,

$$\phi(y) = \begin{cases} 1 & |y| \le 3\rho/4, \\ 0 & |y| \ge \rho, \end{cases}$$

and

$$|\nabla\phi| \le c\rho^{-1}, \qquad |\partial_i \partial_j \phi| \le c\rho^{-2},$$

where c is an absolute constant that does not depend on ρ.

Then we have

$$(\phi p)(x) = (-\Delta)^{-1}(-\Delta)(\phi p) = -\frac{1}{4\pi} \int_{\mathbb{R}^3} \frac{1}{|x - y|} \Delta(\phi p) \, dy,$$

using the explicit form (C.3) for the inverse Laplacian on $\mathbb{R}^3$ given in Appendix C. Note that this representation is valid, since $p \in L^{3/2}(\mathbb{R}^3)$ implies that $\phi p \in L^q$ for all $1 < q \le 3/2$. Thus

$$
(\phi p)(x) = -\frac{1}{4\pi} \int_{\mathbb{R}^3} \frac{1}{|x-y|} [\phi \, \Delta p + 2(\nabla \phi \cdot \nabla p) + p \, \Delta \phi] \, dy
$$

$$
= \frac{1}{4\pi} \int_{\mathbb{R}^3} \frac{1}{|x-y|} [\phi \, \partial_i \partial_j (u_i u_j) - 2(\nabla \phi \cdot \nabla p) - p \, \Delta \phi] \, dy,
$$

substituting $-\partial_i \partial_j (u_i u_j)$ for Δp (where we sum over repeated indices).

We now integrate by parts to remove any derivatives from p and u to give

$$
(\phi p)(x) = \frac{1}{4\pi} \int \left[\partial_i \partial_j \frac{1}{|x-y|} \right] \phi \, u_i u_j \, dy
$$

$$
- \frac{1}{2\pi} \int \frac{x_j - y_j}{|x-y|^3} (\partial_i \phi) u_i u_j \, dy + \frac{1}{4\pi} \int \frac{1}{|x-y|} (\partial_i \partial_j \phi) u_i u_j \, dy
$$

$$
- \frac{1}{2\pi} \int \frac{x_i - y_i}{|x-y|^3} p \, \partial_i \phi \, dy + \frac{1}{4\pi} \int \frac{1}{|x-y|} p \, \Delta \phi \, dy,
$$

which we write as $\phi p = p_1 + p_{2,1} + p_{2,2} + p_{3,1} + p_{3,2}$.

Noting that $\phi p = p$ on $Q_{3\rho/4}$ and that (neglecting the double subscripts, i.e. briefly writing p_2 rather than both $p_{2,1}$ and $p_{2,2}$)

$$
p - (p)_r = \sum_{i=1}^{3} p_i - (p_i)_r
$$

and hence

$$
|p - (p)_r| \le \sum_{i=1}^{3} |p_i - (p_i)_r|,
$$

it suffices to estimate $\| p_i - (p_i)_r \|_{L^{3/2}(Q_r)}$ $(i = 1, 2, 3)$ since $Q_r \subset Q_{3\rho/4}$.

We decompose p_1 into two parts,

$$
p_1 = \frac{1}{4\pi} \int_{|y|<2r} \left[\partial_i \partial_j \frac{1}{|x-y|} \right] \phi \, u_i u_j \, dy
$$

$$
+ \frac{1}{4\pi} \int_{|y|>2r} \left[\partial_i \partial_j \frac{1}{|x-y|} \right] \phi \, u_i u_j \, dy
$$

$$
=: p_{1,1} + p_{1,2}.
$$

To bound $p_{1,1}$ we use the Calderón–Zygmund Theorem (see Appendix B), which guarantees that for each i, j the operator defined by

$$
T_{ij}\psi = \left[\partial_i \partial_j \frac{1}{4\pi |x|} \right] * \psi
$$

(we have often written this as $T_{ij} = \partial_i \partial_j (-\Delta)^{-1}$) is a bounded linear operator on $L^q(\mathbb{R}^3)$ for any $1 < q < \infty$, so that there exists a constant $C_{3/2}$ such that

$$\|T_{ij}\psi\|_{L^{3/2}(\mathbb{R}^3)} \leq C_{3/2}\|\psi\|_{L^{3/2}(\mathbb{R}^3)}.$$

It follows, taking $\psi = \phi \, u_i u_j|_{B_{2r}}$ (extended by zero) and $q = 3/2$, that

$$\|p_{1,1}\|_{L^{3/2}(B_r)} \leq \|p_{1,1}\|_{L^{3/2}(\mathbb{R}^3)} \leq C_{3/2}\||u|^2\|_{L^{3/2}(B_{2r})} \leq C_{3/2}\|u\|^2_{L^3(B_{2r})}.$$

Therefore (recalling that $\|(p)_r\|_{L^q} \leq \|p\|_{L^q}$, see Exercise 15.2)

$$\|p_{1,1} - (p_{1,1})_r\|_{L^{3/2}(B_r)} \leq 2C_{3/2}\|u\|^2_{L^3(B_{2r})}. \tag{15.34}$$

We bound the remaining terms by estimating their gradients and using the Mean Value Theorem. First, for $p_{1,2}$ we have

$$|\nabla p_{1,2}(x)| \leq c \int_{|y|>2r} \frac{1}{|x-y|^4} \phi \, |u|^2 \, dy,$$

and since $|x - y| > |y|/2$ for $|x| < r$ and $|y| > 2r$, we can bound

$$\|\nabla p_{1,2}\|_{L^\infty(B_r)} \leq 16c \int_{2r<|y|<\rho} \frac{|u|^2}{|y|^4} \, dy,$$

where we have used the fact that ϕ is supported within B_ρ and $|\phi| \leq 1$. Since

$$\|p_{1,2} - (p_{1,2})_r\|_{L^\infty(B_r)} \leq 2r\|\nabla p_{1,2}\|_{L^\infty(B_r)}$$

it follows that

$$\|p_{1,2} - (p_{1,2})_r\|_{L^{3/2}(B_r)} \leq 2(4\pi/3)^{2/3} r^3 \|\nabla p_{1,2}\|_{L^\infty(B_r)}$$

$$\leq cr^3 \int_{2r<|y|<\rho} \frac{|u|^2}{|y|^4} \, dy. \tag{15.35}$$

We estimate $p_{2,1}$ and $p_{2,2}$ similarly. Dealing with

$$p_{2,1} = -\frac{1}{2\pi} \int \frac{x_j - y_j}{|x-y|^3} (\partial_i \phi) u_i u_j \, dy$$

we have

$$|\nabla p_{2,1}(x)| \leq c \sum_{j=1}^3 \int \frac{1}{|x-y|^3} |(\partial_i \phi) u_i u_j| \, dy,$$

and, noting that $\partial_i \phi \equiv 0$ for all $|y| \leq 3\rho/4$ and all $|y| \geq \rho$,

$$|\nabla p_{2,1}(x)| \leq c \sum_{j=1}^3 \int_{3\rho/4 \leq |y| \leq \rho} \frac{1}{|x-y|^3} |\partial_i \phi| \, |u_i u_j| \, dy.$$

When $|x| \leq r \leq \rho/2$ and $|y| \geq 3\rho/4$ it follows that $|x - y| \geq \rho/4$, and so

$$\|\nabla p_{2,1}\|_{L^\infty(B_r)} \leq c4^3 \rho^{-4} \int_{B_\rho} |u|^2 \, dy,$$

from which

$$
\begin{aligned}
\|p_{2,1} - (p_{2,1})_r\|_{L^{3/2}(B_r)} &\leq cr^2 \|p_{2,1} - (p_{2,1})_r\|_{L^\infty(B_r)} \\
&\leq cr^3 \|\nabla p_{2,1}\|_{L^\infty(B_r)} \\
&\leq c \frac{r^3}{\rho^4} \int_{B_\rho} |u|^2 \, dy \\
&\leq c \frac{r^3}{\rho^3} \left(\int_{B_\rho} |u|^3 \, dy \right)^{2/3},
\end{aligned}
\tag{15.36}
$$

using Hölder's inequality. We obtain a bound of the same order, using essentially the same argument, for $p_{2,2}$.

A very similar argument in which we simply replace $|u|^2$ by $|p|$ shows that

$$\|p_{3,i} - (p_{3,i})_r\|_{L^{3/2}(B_r)} \leq c \frac{r^3}{\rho^3} \left(\int_{B_\rho} |p|^{3/2} \, dy \right)^{2/3}, \qquad i = 1, 2. \tag{15.37}$$

If we now combine all these estimates ((15.34), (15.35), (15.36), and (15.37)) then we obtain

$$
\begin{aligned}
\|p - (p)_r\|_{L^{3/2}(B_r)} &\leq c \left(\int_{B_{2r}} |u|^3 \right)^{2/3} + cr^3 \int_{2r<|y|<\rho} \frac{|u|^2}{|y|^4} \, dy \\
&\quad + c \frac{r^3}{\rho^3} \left(\int_{B_\rho} |u|^3 + |p|^{3/2} \, dy \right)^{2/3}.
\end{aligned}
$$

Raising both sides to the power of $3/2$ we obtain

$$
\begin{aligned}
\int_{B_r} |p - (p)_r|^{3/2} &\leq c \int_{B_{2r}} |u|^3 + cr^{9/2} \left\{ \int_{2r<|y|<\rho} \frac{|u|^2}{|y|^4} \, dy \right\}^{3/2} \\
&\quad + c \frac{r^{9/2}}{\rho^{9/2}} \int_{B_\rho} |u|^3 + |p|^{3/2} \, dy.
\end{aligned}
$$

Integrating both sides with respect to t between $-r^2$ and 0 and dividing by $r^{3/2}$ yields (15.33), where we use the fact that $\rho \geq 2r$ to enlarge the cylinder in the final integral term on the right-hand side. $\qquad \square$

Note that if we choose $\alpha \geq 2$ and set $\rho = \alpha r$ in (15.33) then we obtain the dimensionless estimate

$$\frac{1}{r^2} \int_{Q_r} |p - (p)_r|^{3/2} \leq \frac{C_2}{r^2} \int_{Q_{2r}} |u|^3 + C_2 r^{9/2} \left\{ \sup_{-r^2 < t < 0} \int_{2r < |y| < \alpha r} \frac{|u(t)|^2}{|y|^4} \, dy \right\}^{3/2}$$

$$+ C_2 \alpha^{-5/2} \frac{1}{(\alpha r)^2} \int_{Q_{\alpha r}} |u|^3 + |p|^{3/2}.$$

Notes

The first partial regularity results for the Navier–Stokes equations were due to Scheffer (1977, 1980). He showed that the 5/3-dimensional Hausdorff measure of the set of space–time singularities is finite (the argument of Theorem 16.2) and (on a bounded domain) that the one-dimensional measure of the spatial singularities at each time slice is finite.

In this chapter we have followed closely the argument due to CKN, with some simplifications due to our assumption that $p \in L^{3/2}$ rather than only $p \in L^{5/4}$; CKN made this weaker assumption since at the time this was the best known regularity for the pressure in bounded domains (as discussed in the Notes for Chapter 5, Sohr and von Wahl proved that $p \in L^{5/3}(\Omega)$ in 1986, some time after the publication of the CKN paper). A relatively informal but instructive summary of the CKN argument is given in Kohn (1982).

Lin (1998) gave an alternative proof of Theorem 15.4, using rescaling and a contradiction argument. A similar approach was also adopted by Ladyzhenskaya & Seregin (1999), whose argument we briefly sketch here, neglecting the pressure for simplicity. We let $z = (x, t)$ be a point in space–time and define

$$(u)_{z,r} := \frac{1}{r^5} \int_{Q_r(z)} u \, dz \quad \text{and} \quad Y(z, r) := \frac{1}{r^5} \int_{Q_r(z)} |u - (u)_{z,r}|^3 \, dz.$$

The key result is that for each $0 < \theta < 1/2$ there exist ε, R, and c such that whenever

$$r|(u)_{z,r}| \leq 1 \quad \text{and} \quad Y(z, r) < \varepsilon$$

for some $0 < r < R$ then

$$Y(z, \theta r) \leq c\theta^2 Y(z, r). \tag{15.38}$$

If this is not true then (this is not immediate) it is possible to find a sequence u_n of translated and rescaled versions of u with

$$\int_{Q_1} |u_n|^3 = 1 \qquad \text{and} \qquad \int_{Q_\theta} |u_n|^3 \geq c\theta^7 \qquad (15.39)$$

that satisfy on Q_1 the equation

$$\partial_t u_n - \Delta u_n + r_n((u_n)_{0,1} \cdot \nabla)u_n + \varepsilon_n r_n(u_n \cdot \nabla)u_n = 0, \qquad (15.40)$$

where $r_n \to 0$, $|r_n(u_n)_{0,1}| \leq 1$, and $\varepsilon_n \to 0$. From (15.39) u_n converges weakly in $L^3(Q_1)$ to some function $u \in L^3(Q_1)$. In fact this convergence is strong on Q_ρ for any $\rho < 1$: we can find bounds on $\partial_t u_n$ from (15.40) that allow us to use the Aubin–Lions compactness theorem to obtain strong convergence in $L^2(Q_\rho)$, and this yields strong convergence in $L^3(Q_\rho)$ by Lebesgue interpolation since $u \in L^{10/3}(Q_\rho)$ (we used a similar argument in the proof of Theorem 14.2).

This strong convergence means that the limiting solution must satisfy

$$\int_{Q_\theta} |u|^3 \geq c\theta^7;$$

but letting n tend to infinity in (15.40) means that u satisfies the linear equation

$$\partial_t u - \Delta u + (a \cdot \nabla)u = 0 \qquad \text{in } Q_1,$$

for some a with $|a| \leq 1$, and standard parabolic results guarantee that for this solution

$$\int_{Q_\theta} |u|^3 \leq c_1\theta^7.$$

With an appropriate choice of c in (15.38) this leads to a contradiction. Iteration of the 'decay estimate' in (15.38) along with the Campanato Lemma then yields local Hölder regularity of u. Of course, the pressure complicates things considerably.

While this method does ultimately rely on regularity results for the Stokes problem (we neglected the pressure in the above sketch), it appears to be widely applicable: for example, related ideas were used by Escauriaza, Seregin, & Šverák (2003) to show that solutions in $L^\infty(0, T; L^3)$ are in fact regular (see Section 16.4).

Vasseur (2007) developed an alternative iteration, based on considering the functions $v_k := (|u| - (1 - 2^{-k}))_+$ and a sequence of decreasing cylinders, an idea used previously by De Giorgi (1957) in the context of elliptic equations.

Kukavica (2009a) uses another iteration scheme that is closer in spirit to the original argument of CKN; he makes use of the fact that a stronger assumption than (15.28) follows from the gradient bound (16.2) that we will use in the next chapter. His argument is based on regularity results for solutions of the heat equation with data in Morrey spaces as developed in an earlier paper (Kukavica, 2008; see also O'Leary, 2003).

Improvements on the bound for the box-counting dimension of the space–time singular set have been obtained by Kukavica (2009b), who showed that $\dim_B(S) \leq 135/82 \simeq 1.646$; subsequently Kukavica & Pei (2012) lowered this to $\dim_B(S) \leq 45/29 \simeq 1.552$. It is still an open problem to show that $\dim_B(S) \leq 1$ for any suitable weak solution.

Exercises

15.1 Suppose that $q \geq 1$ and $a, b \geq 0$. Show that $(a + b)^q \leq 2^{q-1}(a^q + b^q)$.

15.2 Show that for any $1 \leq q < \infty$

$$\int_U |\bar{p}|^q \, dx \leq \int_U |p|^q \, dx,$$

where $\bar{p} = \frac{1}{\mu(U)} \int_U p \, dx$.

15.3 Denote by $N_Q(X, \varepsilon)$ the minimum number of (centred) parabolic cylinders $Q_r^*(x, t) = B_r(x) \times (t - r^2/2, t + r^2/2)$ required to cover a compact set $X \subset \mathbb{R}^4$ and define

$$\dim_P(X) = \limsup_{\varepsilon \to 0} \frac{\log N_Q(X, \varepsilon)}{-\log \varepsilon}.$$

Show that $\dim_B(X) \leq \dim_P(X)$ and find an example of a set Y such that $\dim_B(Y) < \dim_P(Y)$.

15.4 Adapt the proof of Lemma 8.6 to show that if $M_Q(X, \varepsilon)$ denotes the maximal number of disjoint closed cylinders $Q_\varepsilon^*(z)$ with centres $z \in Q$ then

$$\dim_P(X) = \limsup_{\varepsilon \to 0} \frac{\log M_Q(X, \varepsilon)}{-\log \varepsilon}.$$

[Given this result one can immediately adapt the proof of Theorem 15.8 to show that $\dim_P(S \cap K) \leq 5/3$ for every compact $K \subset \Omega \times (0, T)$. The proof is slightly simpler in this case, since we no longer need to use the observation (15.30).]

16

Partial regularity II: $\dim_H(S) \leq 1$

In this chapter we give a complete proof of the second local regularity theorem of Caffarelli, Kohn, & Nirenberg (1982), and the corresponding bound on the Hausdorff dimension of the singular set, $\dim_H(S) \leq 1$ (in fact the result is a little stronger than this, as we will see). For this second regularity theorem we follow the proofs of Lin (1988) and Kukavica (2009a), whose arguments are significantly simpler than those of CKN due in part to the use of the assumption that $p \in L^{3/2}$.

The statement of this second local regularity theorem is qualitatively different from the first. In the previous chapter we showed that if

$$\frac{1}{r^2} \int_{Q_r(a,s)} |u|^3 + |p|^{3/2} \leq \varepsilon_0 \tag{16.1}$$

for some $r > 0$ then u is regular on $Q_{r/2}(a, s)$. The second local regularity theorem, which we prove in this chapter as Theorem 16.1, replaces this with the limiting condition

$$\limsup_{r \to 0} \frac{1}{r} \int_{Q_r(a,s)} |\nabla u|^2 \leq \varepsilon_1, \tag{16.2}$$

where again ε_1 is an absolute constant.[1] The argument proceeds by showing that (16.2) implies that (16.1) must hold for all r sufficiently small, and hence that u is regular on $Q_\rho(a, s)$ for some $\rho > 0$.

It turns out that (16.2) cannot be used to obtain an improved bound on the box-counting dimension of the singular set, since unlike (16.1) such a limiting

---

[1] As we will see later, a very similar proof yields the same conclusion under the alternative assumption that

$$\limsup_{r \to 0} \sup_{t_0 - r^2 \leq t \leq t_0} \frac{1}{r} \int_{B_r(x_0)} |u(x, t)|^2 \, dx \leq \varepsilon_1.$$

315

condition cannot be used to deduce a uniform lower bound

$$\frac{1}{\varrho} \int_{Q_\varrho(a,s)} |\nabla u|^2 > \varepsilon_1 \qquad \text{for all} \quad (a, s) \in S$$

for one fixed $\varrho > 0$. However, this condition is sufficient to bound the Hausdorff dimension (which allows coverings by sets of different diameters) as we show in Theorem 16.2.

16.1 Outline of the proof

As in the previous chapter we will give a sketch of the argument in which we neglect the pressure. Once again this will yield estimates that we will use in the proof of the full result.

Without loss of generality we take $(a, s) = (0, 0)$ and, as in the previous chapter, write B_r for $B_r(0)$ and Q_r for $Q_r(0, 0)$.

We aim to show that there exists an $\varepsilon_1' > 0$ such that if

$$\limsup_{r \to 0} \frac{1}{r} \int_{Q_r} |\nabla u|^2 \leq \varepsilon_1' \tag{16.3}$$

then there exists an r_1 such that

$$\frac{1}{r^2} \int_{Q_r} |u|^3 \leq \varepsilon_0 \qquad \text{for all} \quad r \leq r_1, \tag{16.4}$$

where ε_0 is the constant from (16.1).

The primary ingredients are the same local energy inequality (15.8) and the interpolation inequality (15.9) that we used before, although we now take a different test function in the local energy inequality and make the estimates in a slightly different way. (The full proof also requires another local pressure estimate.) We will bound

$$\tilde{E}(r) := \frac{1}{r} \sup_{-r^2 < t < 0} \int_{B_r} |u|^2$$

in terms of the same quantity with a larger value of r: for some carefully chosen $0 < \theta < 1$ we will show that

$$\tilde{E}(\theta r) \leq \tfrac{1}{2}\varepsilon_1' + \tfrac{1}{2}\tilde{E}(r).$$

We iterate this inequality to show that $\tilde{E}(r) < 2\varepsilon_1'$ once r is sufficiently small and then combine this with (16.3) to ensure that

$$\frac{1}{r} \sup_{-r^2 < t < 0} \int_{B_r} |u(t)|^2 + \frac{1}{r} \int_{Q_r} |\nabla u|^2 < 4\varepsilon_1'$$

for r sufficiently small. This in turn enables us to use the interpolation inequality (15.9) to show that

$$\frac{1}{r^2} \int_{Q_r} |u|^3 \le C_0 \left[\frac{1}{r} \sup_{-r^2 < t < 0} \int_{B_r} |u(t)|^2 + \frac{1}{r} \int_{Q_r} |\nabla u|^2 \right]^{3/2} \qquad (16.5)$$

$$\le C_0 (4\varepsilon_1')^{3/2},$$

and now the right-hand side is bounded by ε_0 if we choose ε_1' appropriately. This means that (16.4) holds, and then it follows from our previous result that u is bounded on $Q_{r/2}$.

In more detail we begin with the local energy inequality (15.8) (from which we have omitted the pressure) in the form

$$\int_{B_1} |u(t)|^2 \phi(t) + 2 \int_{-1}^t \int_{B_1} |\nabla u|^2 \phi$$

$$\le \int_{-1}^t \int_{B_1} |u|^2 (\partial_t \phi + \Delta \phi) + \int_{-1}^t \int_{B_1} \left\{ |u|^2 - (|u|^2)_r \right\} (u \cdot \nabla) \phi.$$

(Recall that $(f)_r = \frac{1}{|B_r|} \int_{B_r} f$ denotes the average of f over B_r.) Note that we have used the observation that for any constant α and any test function ϕ that is compactly supported in B_1 we have

$$\int_{B_1} \alpha(u \cdot \nabla)\phi = 0$$

(see (15.24)). Previously we used this to deal with the pressure term, but now we are using the same trick with the term that is cubic in u.

We show in Lemma 16.6 that for any fixed $r > 0$ and $0 < \theta \le 1/2$ we can find a function $\phi = \phi_{r,\theta}$ with $\phi \in C_c^\infty(\mathbb{R}^4)$, $(\text{supp}\,\phi) \cap Q_1 \subseteq Q_r$,

$$\phi \ge C_3^{-1}(r\theta)^{-1} \qquad \text{for} \quad (x,t) \in Q_{r\theta},$$

and

$$|\nabla \phi| \le C_3(r\theta)^{-2} \qquad \text{and} \qquad |\partial_t \phi + \Delta \phi| \le C_3 \theta^2 r^{-3} \qquad \text{for} \quad (x,t) \in Q_r,$$

where $C_3 \ge 1$ is an absolute constant. Note that now we really do localise the energy inequality, cutting off all contributions outside Q_r.

Using this test function in the local energy inequality we obtain for every $t \in (-(\theta r)^2, 0)$

$$\frac{1}{\theta r} \int_{B_{\theta r}} |u(t)|^2 + \frac{1}{\theta r} \int_{-(\theta r)^2}^{t} \int_{B_{\theta r}} |\nabla u|^2$$

$$\le C_3^2 \theta^2 r^{-3} \int_{Q_r} |u|^2 + C_3^2 \theta^{-2} r^{-2} \int_{Q_r} \left\{ |u|^2 - (|u|^2)_r \right\} |u|$$

$$\le 2C_3^2 \theta^2 \left(\frac{1}{r^2} \int_{Q_r} |u|^3 \right)^{2/3} + C_3^2 \theta^{-2} r^{-2} \int_{Q_r} \left\{ |u|^2 - (|u|^2)_r \right\} |u|, \qquad (16.6)$$

after using Hölder's inequality and the fact that $(4\pi/3)^{1/3} \le 2$. We will now bound the second term on the right-hand side.

The key observation is that by subtracting the average of $|u|^2$ over B_r we are able to use the 'Sobolev–Poincaré inequality', i.e. a version of the Sobolev inequality involving only the highest derivative. For a function f with zero average on B_r the embedding $W^{1,1} \subset L^{3/2}$ gives $\|f\|_{L^{3/2}(B_r)} \le C_4 \|\nabla f\|_{L^1(B_r)}$, where the constant C_4 does not depend on r (see Exercise 16.1). Using the inequality $|\nabla(|u|^2)| \le 2|u||\nabla u|$ we can therefore estimate

$$\int_{B_r} \left\{ |u|^2 - (|u|^2)_r \right\} |u| \le \left[\int_{B_r} \left\{ |u|^2 - (|u|^2)_r \right\}^{3/2} \right]^{2/3} \left(\int_{B_r} |u|^3 \right)^{1/3}$$

$$\le C_4 \left(\int_{B_r} |\nabla(|u|^2)| \right) \left(\int_{B_r} |u|^3 \right)^{1/3}$$

$$\le 2C_4 \left(\int_{B_r} |u||\nabla u| \right) \left(\int_{B_r} |u|^3 \right)^{1/3}$$

$$\le 2C_4 \left(\int_{B_r} |u|^3 \right)^{1/3} \left(\int_{B_r} |u|^2 \right)^{1/2} \left(\int_{B_r} |\nabla u|^2 \right)^{1/2}.$$

We integrate in time from $-r^2$ to 0 and use Hölder's inequality to obtain

$$\int_{Q_r} \left\{ |u|^2 - (|u|^2)_r \right\} |u|$$

$$\le 2C_4 \left(\int_{Q_r} |u|^3 \right)^{1/3} \left(\int_{Q_r} |\nabla u|^2 \right)^{1/2} \left\{ \int_{-r^2}^{0} \left(\int_{B_r} |u(t)|^2 \right)^3 \right\}^{1/6}$$

$$\le 2C_4 r^{1/3} \left(\int_{Q_r} |u|^3 \right)^{1/3} \left(\int_{Q_r} |\nabla u|^2 \right)^{1/2} \left(\sup_{-r^2 < t < 0} \int_{B_r} |u(t)|^2 \right)^{1/2}$$

$$= 2C_4 r^2 \left(\frac{1}{r^2} \int_{Q_r} |u|^3 \right)^{1/3} \left(\frac{1}{r} \int_{Q_r} |\nabla u|^2 \right)^{1/2} \left(\frac{1}{r} \sup_{-r^2 < t < 0} \int_{B_r} |u(t)|^2 \right)^{1/2}.$$

Using this in (16.6) we obtain

$$\frac{1}{\theta r} \int_{B_{\theta r}} |u(t)|^2 + \frac{1}{\theta r} \int_{-(\theta r)^2}^t \int_{B_{\theta r}} |\nabla u|^2 \leq 2C_3^2 \theta^2 \left(\frac{1}{r^2} \int_{Q_r} |u|^3 \right)^{2/3}$$

$$+ 2C_4 C_3^2 \theta^{-2} \left(\frac{1}{r^2} \int_{Q_r} |u|^3 \right)^{1/3} \left(\frac{1}{r} \int_{Q_r} |\nabla u|^2 \right)^{1/2} \left(\frac{1}{r} \sup_{-r^2 < t < 0} \int_{B_r} |u(t)|^2 \right)^{1/2}.$$

Splitting up the second term with Young's inequality after an appropriate distribution of powers of θ yields

$$\frac{1}{\theta r} \int_{B_{\theta r}} |u(t)|^2 + \frac{1}{\theta r} \int_{-(\theta r)^2}^t \int_{B_{\theta r}} |\nabla u|^2 \leq 3C_3^2 \theta^2 \left(\frac{1}{r^2} \int_{Q_r} |u|^3 \right)^{2/3}$$

$$+ 4C_4^2 C_3^2 \theta^{-6} \left(\frac{1}{r} \int_{Q_r} |\nabla u|^2 \right) \left(\frac{1}{r} \sup_{-r^2 < t < 0} \int_{B_r} |u(t)|^2 \right), \qquad (16.7)$$

and then after using the interpolation inequality (16.5) we have

$$\frac{1}{\theta r} \int_{B_{\theta r}} |u(t)|^2 + \frac{1}{\theta r} \int_{-(\theta r)^2}^t \int_{B_{\theta r}} |\nabla u|^2$$

$$\leq 3C_3^2 C_0^{2/3} \theta^2 \left[\frac{1}{r} \sup_{-r^2 < t < 0} \int_{B_r} |u(t)|^2 + \frac{1}{r} \int_{Q_r} |\nabla u|^2 \right]$$

$$+ 4C_4^2 C_3^2 \theta^{-6} \left(\frac{1}{r} \int_{Q_r} |\nabla u|^2 \right) \left(\frac{1}{r} \sup_{-r^2 < t < 0} \int_{B_r} |u(t)|^2 \right).$$

If we drop the first term on the left-hand side and take $t = 0$, and then drop the second term and take the supremum over $t \in (-(\theta r)^2, 0)$ we finally obtain

$$\max \left(\frac{1}{\theta r} \sup_{-(\theta r)^2 < t < 0} \int_{B_{\theta r}} |u(t)|^2, \ \frac{1}{\theta r} \int_{Q_{\theta r}} |\nabla u|^2 \right)$$

$$\leq 3C_3^2 C_0^{2/3} \theta^2 \left[\frac{1}{r} \sup_{-r^2 < t < 0} \int_{B_r} |u(t)|^2 + \frac{1}{r} \int_{Q_r} |\nabla u|^2 \right]$$

$$+ 4C_4^2 C_3^2 \theta^{-6} \left(\frac{1}{r} \int_{Q_r} |\nabla u|^2 \right) \left(\frac{1}{r} \sup_{-r^2 < t < 0} \int_{B_r} |u(t)|^2 \right).$$

Now, note that the above expression is symmetric in the quantities

$$\frac{1}{r} \sup_{-r^2 < t < 0} \int_{B_r} |u(t)|^2 \qquad \text{and} \qquad \frac{1}{r} \int_{Q_r} |\nabla u|^2 \qquad (16.8)$$

and the corresponding expressions involving θr rather than r. We are therefore able to proceed in two distinct ways, choosing to assume the smallness of one of the two quantities in (16.8) and then deducing the smallness of the other via an iterative argument. In what follows we make a smallness assumption on $\int_{Q_r} |\nabla u|^2$, following CKN, but we will state our results giving both alternatives.

For r sufficiently small we can use our initial assumption (16.3) to bound the gradient terms on the right-hand side,

$$\frac{1}{r} \int_{Q_r} |\nabla u|^2 \leq 2\varepsilon_1', \tag{16.9}$$

and obtain

$$\frac{1}{\theta r} \sup_{-(\theta r)^2 < t < 0} \int_{B_{\theta r}} |u(t)|^2 \leq 3C_3^2 C_0^{2/3} \theta^2 \left[2\varepsilon_1' + \frac{1}{r} \sup_{s - r^2 < t < 0} \int_{B(r)} |u(t)|^2 \right]$$
$$+ 8C_4^2 C_3^2 \varepsilon_1' \theta^{-6} \left(\frac{1}{r} \sup_{-r^2 < t < 0} \int_{B_r} |u(t)|^2 \right).$$

Now we let

$$\tilde{E}(r) = \frac{1}{r} \sup_{-r^2 < t < 0} \int_{B_r} |u(t)|^2$$

and rewrite this inequality as

$$\tilde{E}(\theta r) \leq (6C_3^2 C_0^{2/3} \theta^2) \varepsilon_1' + [3C_3^2 C_0^{2/3} \theta^2 + 8C_4^2 C_3^2 \varepsilon_1' \theta^{-6}] \tilde{E}(r). \tag{16.10}$$

Now choose $\theta \leq 1/2$ so that

$$6C_3^2 C_0^{2/3} \theta^2 \leq \tfrac{1}{2}$$

(this implies that $3C_3^2 C_0^{2/3} \theta^2 \leq \tfrac{1}{4}$) and then ε_1' such that

$$8C_4^2 C_3^2 \varepsilon_1' \theta^{-6} \leq \tfrac{1}{4}$$

(we will decrease ε_1' further before the proof is over). With these choices (16.10) becomes

$$\tilde{E}(\theta r) \leq \tfrac{1}{2}\varepsilon_1' + \tfrac{1}{2}\tilde{E}(r).$$

Iterating this inequality (see Exercise 16.2) we obtain

$$\tilde{E}(\theta^k r) \leq \varepsilon_1' + 2^{-k}\tilde{E}(r),$$

and so $\tilde{E}(\theta^k r) \leq 2\varepsilon_1'$ for k sufficiently large.

Combining this with (16.9) for r sufficiently small we have

$$\frac{1}{r} \sup_{-r^2 < t < 0} \int_{B_r} |u(t)|^2 + \frac{1}{r} \int_{Q_r} |\nabla u|^2 < 4\varepsilon_1'$$

from which we can use the interpolation inequality as in (16.5) to deduce that

$$\frac{1}{r^2} \int_{Q_r(a,s)} |u|^3 \le C_0 (4\varepsilon_1')^{3/2}.$$

Finally, choosing ε_1' smaller if necessary so that $C_0 (4\varepsilon_1')^{3/2} < \varepsilon_0$ we obtain

$$\frac{1}{r^2} \int_{Q_r(a,s)} |u|^3 < \varepsilon_0,$$

which is what we set out to achieve.

16.2 A second local regularity theorem

We now combine the above ingredients with estimates on the pressure to prove the second local regularity theorem in full.

Theorem 16.1 (Second Local Regularity Theorem) *There exists an absolute constant $\varepsilon_1 > 0$ such that if (u, p) is a suitable weak solution of the Navier–Stokes equations on $Q_R(a, s)$ for some $R > 0$ and either*

$$\limsup_{r \to 0} \frac{1}{r} \int_{Q_r(a,s)} |\nabla u|^2 \le \varepsilon_1 \tag{16.11}$$

or

$$\limsup_{r \to 0} \frac{1}{r} \sup_{s-r^2 < t < s} \int_{B_r(a)} |u(t)|^2 \le \varepsilon_1 \tag{16.12}$$

then $u \in L^\infty(Q_\rho(a, s))$ for some ρ with $0 < \rho < R$.

Proof   As in the pressure-free proof, we aim to show that the condition (15.28) of the first local regularity theorem (Theorem 15.4) is satisfied, i.e. that for some $\rho < R$,

$$\frac{1}{\rho^2} \int_{Q_\rho(a,s)} |u|^3 + |p|^{3/2} \le \varepsilon_0.$$

We consider the full version of the local energy inequality (15.8), now including the pressure term,

$$\int_{B_1} |u(s)|^2 \phi(s) + 2 \int_{-1}^{s} \int_{B_1} |\nabla u|^2 \phi \le \int_{-1}^{s} \int_{B_1} |u|^2 (\phi_t + \Delta\phi)$$
$$+ \int_{-1}^{s} \int_{B_1} [|u|^2 - (|u|^2)_r](u \cdot \nabla)\phi + \int_{-1}^{s} \int_{B_1} p(u \cdot \nabla)\phi.$$

As in our pressure-free proof we take advantage of (15.24) to subtract a term from the contribution that is cubic in u. We have separated the cubic term in u (which we have already estimated) and the pressure term (which we have not).

We use the same cutoff function that we used in the pressure-free proof, namely a function $\phi \in C_c^\infty(\mathbb{R}^4)$ such that $(\operatorname{supp}\phi) \cap Q_1 \subset Q_r$ and satisfies the bounds $\phi \ge C_3^{-1}(\theta r)^{-1}$ for $(x, t) \in Q_{\theta r}$, and

$$|\nabla\phi| \le C_3(\theta r)^{-2} \quad \text{and} \quad |\partial_t\phi + \Delta\phi| \le C_3\theta^2 r^{-3}, \qquad \text{for} \qquad (x, t) \in Q_r,$$

where $C_3 \ge 1$ is an absolute constant. Such a function is constructed in Lemma 16.6.

With this test function in the local energy inequality we obtain (16.6) with an additional pressure term,

$$\frac{1}{\theta r}\int_{B_{\theta r}} |u(t)|^2 + \frac{1}{\theta r}\int_{-(\theta r)^2}^{t}\int_{B_{\theta r}} |\nabla u|^2 \le C_3^2\theta^2 r^{-3}\int_{Q_r} |u|^2$$
$$+ C_3^2(\theta r)^{-2}\int_{Q_r}\left\{|u|^2 - (|u|^2)_r\right\}|u| + C_3^2(\theta r)^{-2}\int_{Q_r}|p||u|.$$

We have already estimated the first two terms on the right-hand side in our pressure-free proof, and for the final term we can use Hölder's inequality and then Young's inequality to give

$$C_3^2(\theta r)^{-2}\int_{Q_r}|p||u| \le C_3^2\left\{\theta\left(\frac{1}{r^2}\int_{Q_r}|u|^3\right)^{1/3}\right\}\left\{\theta^{-3}\left(\frac{1}{r^2}\int_{Q_r}|p|^{3/2}\right)^{2/3}\right\}$$
$$\le C_3^2\theta^2\left(\frac{1}{r^2}\int_{Q_r}|u|^3\right)^{2/3} + C_3^2\theta^{-6}\left(\frac{1}{r^2}\int_{Q_r}|p|^{3/2}\right)^{4/3}.$$

If we add this additional term to our previous pressure-free estimate in (16.7) then we obtain

$$\frac{1}{\theta r}\int_{B_{\theta r}} |u(t)|^2 + \frac{1}{\theta r}\int_{-(\theta r)^2}^{t}\int_{B_{\theta r}} |\nabla u|^2 \le 4C_3^2\theta^2\left(\frac{1}{r^2}\int_{Q_r}|u|^3\right)^{2/3}$$
$$+ 4C_4^2 C_3^2\theta^{-6}\left(\frac{1}{r}\int_{Q_r}|\nabla u|^2\right)\left(\frac{1}{r}\sup_{-r^2 < t < 0}\int_{B_r}|u(t)|^2\right)$$
$$+ C_3^2\theta^{-6}\left(\frac{1}{r^2}\int_{Q_r}|p|^{3/2}\right)^{4/3}.$$

As before we use the interpolation inequality (15.9) on the first term on the right-hand side. We then bound each term on the left-hand side separately, taking the supremum over $t \in (-(\theta r)^2, 0)$ in the first and letting $t = 0$ in the second to give

$$\max \left(\frac{1}{\theta r} \sup_{-(\theta r)^2 < t < 0} \int_{B_{\theta r}} |u(t)|^2 , \ \frac{1}{\theta r} \int_{-(\theta r)^2}^{0} \int_{B_{\theta r}} |\nabla u|^2 \right)$$

$$\leq 4 C_3^2 C_0^{2/3} \theta^2 \left[\frac{1}{r} \sup_{-r^2 < t < 0} \int_{B_r)} |u(t)|^2 + \frac{1}{r} \int_{Q_r} |\nabla u|^2 \right]$$

$$+ 4 C_4^2 C_3^2 \theta^{-6} \left(\frac{1}{r} \int_{Q_r} |\nabla u|^2 \right) \left(\frac{1}{r} \sup_{-r^2 < t < 0} \int_{B_r} |u(t)|^2 \right)$$

$$+ C_3^2 \theta^{-6} \left(\frac{1}{r^2} \int_{Q_r} |p|^{3/2} \right)^{4/3}. \tag{16.13}$$

We now want to set up an iterative scheme as we did without the pressure. To do this we will require a local estimate on the pressure, giving a bound on the integral over $Q_{\theta r}$ in terms of an integral over Q_r. We then combine the estimate for u with the estimate for p in a way that allows for an iteration. The estimate we need for the pressure is

$$\left[\frac{1}{(r\theta)^2} \int_{Q_{r\theta}} |p|^{3/2} \right]^{4/3} \leq 2 C_5^{4/3} \theta^{-2} \left(\frac{1}{r} \sup_{-r^2 < t < 0} \int_{B_r} |u(t)|^2 \right) \left(\frac{1}{r} \int_{Q_r} |\nabla u|^2 \right)$$

$$+ 2 C_5^{4/3} \theta^{4/3} \left[\frac{1}{r^2} \int_{Q_r} |p|^{3/2} \right]^{4/3}, \tag{16.14}$$

which is given in Lemma 16.7 which we prove at the end of the chapter. Note that both (16.13) and (16.14) involve the quantities

$$\frac{1}{r} \sup_{-r^2 < t < 0} \int_{B_r} |u(t)|^2 \quad \text{and} \quad \frac{1}{r} \int_{Q_r} |\nabla u|^2$$

(and their rescaled versions) in a symmetric way. So, just as before, we can choose to assume that either one of these is small to begin our iteration, which accounts for the two alternative conditions (16.11) and (16.12) in the statement of Theorem 16.1.

We will use assumption (16.11), but the argument is identical if we use (16.12). Taking r sufficiently small we have from (16.11) that $\frac{1}{r} \int_{Q_r} |\nabla u|^2 \leq 2\varepsilon_1$.

For such r from (16.13) we obtain

$$
\frac{1}{\theta r} \sup_{-(\theta r)^2 < t < 0} \int_{B_{\theta r}} |u(t)|^2 \leq 4C_3^2 C_0^{2/3} \theta^2 \left[2\varepsilon_1 + \frac{1}{r} \sup_{-r^2 < t < 0} \int_{B_r} |u(t)|^2 \right]
$$
$$
+ 8C_4^2 C_3^2 \varepsilon_1 \theta^{-6} \left(\frac{1}{r} \sup_{-r^2 < t < 0} \int_{B_r} |u(t)|^2 \right) + C_3^2 \theta^{-6} \left(\frac{1}{r^2} \int_{Q_r} |p|^{3/2} \right)^{4/3}
$$
$$\tag{16.15}$$

and the pressure estimate in (16.14) becomes

$$
\left[\frac{1}{(r\theta)^2} \int_{Q_{r\theta}} |p|^{3/2} \right]^{4/3} \leq 4C_5^{4/3} \varepsilon_1 \theta^{-2} \left(\frac{1}{r} \sup_{-r^2 < t < 0} \int_{B_r} |u(t)|^2 \right)
$$
$$
+ 2C_5^{4/3} \theta^{4/3} \left[\frac{1}{r^2} \int_{Q_r} |p|^{3/2} \right]^{4/3}. \tag{16.16}
$$

Now, observe that (16.15) gives a bound for u within $Q_{\theta r}$ in terms of u and p within Q_r, and that (16.16) gives a bound for p within $Q_{\theta r}$ in terms of u and p within Q_r. We combine the two estimates in a way that allows for an iteration: we add the two expressions, multiplying that for the pressure by θ^{-7}. (This somewhat mysterious factor is precisely what is required to make the iteration work.) This yields

$$
\frac{1}{\theta r} \sup_{-(\theta r)^2 < t < 0} \int_{B_{\theta r}} |u(t)|^2 + \theta^{-7} \left[\frac{1}{(r\theta)^2} \int_{Q_{r\theta}} |p|^{3/2} \right]^{4/3}
$$
$$
\leq 4C_3^2 C_0^{2/3} \theta^2 \left[2\varepsilon_1 + \frac{1}{r} \sup_{-r^2 < t < 0} \int_{B_r} |u(t)|^2 \right]
$$
$$
+ 8C_4^2 C_3^2 \varepsilon_1 \theta^{-6} \left(\frac{1}{r} \sup_{-r^2 < t < 0} \int_{B_r} |u(t)|^2 \right)
$$
$$
+ C_3^2 \theta^{-6} \left(\frac{1}{r^2} \int_{Q_r} |p|^{3/2} \right)^{4/3} + 4C_5^{4/3} \varepsilon_1 \theta^{-9} \left(\frac{1}{r} \sup_{-r^2 < t < 0} \int_{B_r} |u(t)|^2 \right)
$$
$$
+ 2C_5^{4/3} \theta^{4/3} \theta^{-7} \left[\frac{1}{r^2} \int_{Q_r} |p|^{3/2} \right]^{4/3}. \tag{16.17}
$$

While this looks ungainly, it can be rewritten more concisely if we define

$$
E(r) = \frac{1}{r} \sup_{-r^2 < t < 0} \int_{B_r} |u(t)|^2 + \theta^{-7} \left[\frac{1}{r^2} \int_{Q_r} |p|^{3/2} \right]^{4/3},
$$

which is the same as the $\tilde{E}(r)$ of the pressure-free argument, along with a pressure term that includes the numerical factor of θ^{-7}. With this definition of $E(r)$,

(16.17) implies that

$$E(\theta r) \leq 8C_3^2 C_0^{2/3} \theta^2 \varepsilon_1$$

$$+ [4C_3^2 C_0^{2/3} \theta^2 + 8C_4^2 C_3^2 \varepsilon_1 \theta^{-6} + C_3^2 \theta + 4C_5^{4/3} \varepsilon_1 \theta^{-9} + 2C_5^{4/3} \theta^{4/3}] E(r).$$

$$(16.18)$$

We now want to choose θ and ε_1 to obtain from (16.18) a relation for $E(\theta r)$ in terms of $E(r)$ that we can iterate in order to make $E(\theta^k r)$ small. In order to do this, we first choose $\theta < 1$ such that

$$8C_3^2 C_0^{2/3} \theta^2 \leq 1/5 \qquad \text{and} \qquad C_3^2 \theta + 2C_5^{4/3} \theta^{4/3} \leq 1/5$$

and then ε_1 such that

$$8C_4^2 C_3^2 \varepsilon_1 \theta^{-6} + 4C_5^{4/3} \varepsilon_1 \theta^{-9} \leq 1/5, \qquad 2(\theta^7 \varepsilon_1)^{3/4} \leq \varepsilon_0,$$

and

$$2C_0(2\varepsilon_1)^{3/2} \leq \varepsilon_0.$$

Then it follows from (16.18) that

$$E(\theta r) \leq \frac{1}{5}\varepsilon_1 + \frac{1}{2}E(r);$$

using the result of Exercise 16.2, for $\rho = \theta^k r$ with k sufficiently large

$$E(\rho) = \frac{1}{\rho} \sup_{-\rho^2 < t < 0} \int_{B_\rho} |u(t)|^2 + \theta^{-7} \left[\frac{1}{\rho^2} \int_{Q_\rho} |p|^{3/2} \right]^{4/3} \leq \varepsilon_1. \qquad (16.19)$$

Because of our choice of ε_1 this implies in particular that

$$\frac{1}{\rho^2} \int_{Q_\rho} |p|^{3/2} \leq (\theta^7 \varepsilon_1)^{3/4} \leq \frac{1}{2}\varepsilon_0. \qquad (16.20)$$

Recalling the interpolation result of (15.31),

$$\frac{1}{r^2} \int_{Q_r} |u|^3 \leq C_0 \left[\frac{1}{r} \sup_{s-r^2 < t < 0} \int_{B(r)} |u(t)|^2 + \frac{1}{r} \int_{Q_r} |\nabla u|^2 \right]^{3/2},$$

it follows from (16.19) and our initial assumption (16.11) that for ρ sufficiently small

$$\frac{1}{\rho^2} \int_{Q_\rho} |u|^3 \leq C_0(2\varepsilon_1)^{3/2} \leq \frac{1}{2}\varepsilon_0$$

(by our choice of ε_1), and combining this with (16.20) yields

$$\frac{1}{\rho^2} \int_{Q_\rho(a,s)} |u|^3 + |p|^{3/2} \leq \varepsilon_0.$$

Theorem 15.4 now guarantees that $u \in L^\infty(Q_{\rho/2}(a, s))$. $\qquad \square$

16.3 Partial regularity II: $\mathcal{H}^1(S) = 0$

Using this new local regularity result we can prove another partial regularity result, limiting a certain 'one-dimensional measure' of the singular set. We use a version of the Hausdorff measure, which we introduced in Section 8.2, now adapted for use in the space–time domain by using parabolic cylinders rather than arbitrary sets of a maximum specified diameter.

Given a set $X \subset \mathbb{R}^3 \times \mathbb{R}$, for any $s \ge 0$ we define the *s-dimensional parabolic Hausdorff measure* by setting

$$\mathcal{P}^s(X) = \lim_{\delta \to 0^+} \mathcal{P}^s_\delta(X),$$

where $\mathcal{P}^s_\delta$ is a δ-approximation to the same quantity, obtained by setting

$$\mathcal{P}^s_\delta(X) = \inf \left\{ \sum_{i=1}^\infty r_i^s : X \subset \bigcup_i Q_{r_i}^* : r_i < \delta \right\},$$

where $Q_r^*(x, t)$ is the centred parabolic cylinder $B_r(x) \times (t - r^2/2, t + r^2/2)$ (see Definition 13.3). In particular $\mathcal{P}^s(X) = 0$ if and only if for all $\varepsilon > 0$ the set X can be covered by a collection $\{Q_{r_i}^*\}$ such that $\sum_i r_i^s < \varepsilon$ (see Exercise 16.3). (Compare this with the definition of the Hausdorff measure, Definition 8.8.)

Note that $\mathcal{P}^s(X) \ge 2^{s/2} \mathcal{H}^s(X)$, since for any $r < 1$ the cylinder $Q_r^*(x, t)$ is a subset of the four-dimensional ball of radius $\sqrt{2}\, r$ centred at (x, t). It follows that if $\mathcal{P}^s(X) = 0$ then $\mathcal{H}^s(X) = 0$ and so $\dim_H(X) \le s$ (see Definition 8.10).

Theorem 16.2 (Partial Regularity II) *Let S denote the singular set of a suitable weak solution of the Navier–Stokes equations. Then $\mathcal{P}^1(S) = 0$, and in particular* $\dim_H(S) \le 1$.

We will need the following version of the Vitali Covering Lemma, proved in Lemma 16.8 at the end of the chapter: given a family of parabolic cylinders $Q_r^*(x, t)$ contained in a bounded subset of $\mathbb{R}^3 \times \mathbb{R}$, there exists a finite or countable disjoint subfamily $\{Q_{r_i}^*(x_i, t_i)\}$ such that for any cylinder $Q_r^*(x, t)$ in the original family there exists an i such that $Q_r^*(x, t) \subset Q_{5r_i}^*(x_i, t_i)$.

Proof We have already shown in Corollary 15.9 that $\mu(S) = 0$, and we know that $\nabla u \in L^2(\Omega \times (0, T))$. So given any $\delta > 0$, we can choose a neighbourhood U of S such that

$$\frac{1}{\varepsilon_1} \int_U |\nabla u|^2 < \delta,$$

where ε_1 is the constant from Theorem 16.1. Now for each $(x, t) \in S$, choose a cylinder $Q_r^*(x, t)$ such that $Q_r^*(x, t) \subset U$, $r < \delta$, and

$$\frac{1}{r} \int_{Q_r^*(x,t)} |\nabla u|^2 > \varepsilon_1.$$

This must be possible, for otherwise the point (x, t) would be regular.

Since Proposition 15.7 guarantees that the set S is bounded, we can use the Vitali Covering Lemma (Lemma 16.8) to find a disjoint subcollection of these cylinders $\{Q_{r_i}^*(x_i, t_i)\}$ such that the singular set is still covered by $\{Q_{5r_i}^*(x_i, t_i)\}$. Since these cylinders are disjoint,

$$\int_U |\nabla u|^2 \geq \sum_i \int_{Q_{r_i}^*(x_i,t_i)} |\nabla u|^2 \geq \varepsilon_1 \sum_i r_i. \tag{16.21}$$

This construction furnishes a cover with $\sum_i r_i$ arbitrarily small, and so $\mathcal{P}^1(S) = 0$. $\qquad\qquad\square$

Note that one consequence of this result is that there can be no one-dimensional subsets of S, i.e. any singular point in space–time cannot 'evolve' in time in a continuous way, nor can the singular set contain any curves at any fixed time. Such an observation can be used to exclude the presence of any singular points other than on the symmetry axis for axisymmetric flows (see Chen et al., 2008, for example).

We proved in Theorem 8.14 that the set $\mathcal{T}$ of singular times, defined as the set

$$\{t \geq 0 : \|\nabla u\| \text{ is not essentially bounded in any neighbourhood of } t\},$$

satisfies $\mathcal{H}^{1/2}(\mathcal{T}) = 0$. The relationship between the space–time singular set and the set of singular times, which we prove in the following lemma for the equations on $\mathbb{T}^3$, shows that this result can also be obtained as a consequence of the fact that $\mathcal{H}^1(S) = 0$. We will denote by $\Pi_t S$ the projection of S onto the time axis, i.e.

$$\Pi_t S = \{t \geq 0 : (x, t) \in S \text{ for some } x \in \Omega\},$$

and show that the projection $\Pi_t S$ is essentially the same as the set $\mathcal{T}$. (The direct proof of Theorem 8.14 is, of course, much simpler; the main interest of Lemma 16.3 is the information it provides about the relationship between the singular sets.)

Lemma 16.3 *For the equations on $\mathbb{T}^3$, $\Pi_t S = \mathcal{T} \cap (0, \infty)$.*

Proof We use an argument from Robinson (2006). Recall that the complement of $\mathcal{T}$ consists of a countable collection of 'epochs of regularity' J_q, open intervals on which $\|\nabla u(t)\| < \infty$ and hence on which u is smooth in space (see Section 8.3). In particular, if t_0 is contained in one of the J_q then $u(x, t)$ is uniformly bounded for all $x \in \mathbb{T}^3$ and all $t \in (t_0 - \varepsilon, t_0 + \varepsilon)$ for some $\varepsilon > 0$. It follows that $\Pi_t S \cap J_q = \emptyset$ for every q, and so $\Pi_t S \subseteq \mathcal{T} \cap (0, \infty)$.

Now suppose that $t_0 \notin \Pi_t S$. Then (x, t_0) is a regular point for every $x \in \mathbb{T}^3$, which means that for each $x \in \mathbb{T}^3$ we know that $u(x, t)$ is uniformly bounded on $B_{r(x)}(x) \times (t_0 - r(x)^2, t_0 + r(x)^2)$ for some $r(x) > 0$. In particular, since $\mathbb{T}^3$ is compact, we can find a finite collection $\{x_j\}$ such that

$$\mathbb{T}^3 \subset \bigcup_j B_{r(x_j)}(x_j),$$

and hence u is uniformly bounded on $\mathbb{T}^3 \times (t_0 - r^2, t_0 + r^2)$, where we choose $r = \min_j \{r(x_j)\}$.

Since $u \in L^2(0, T; H^1)$ we can find a $t_1 \in (t_0 - r^2, t_0 - r^2/2)$ such that $\|\nabla u(t_1)\| < \infty$. We know that solutions are smooth for $t > t_1$ while $\|\nabla u(t)\|$ remains finite, so that for such times we can take the inner product of the Navier–Stokes equations with $-\Delta u$,

$$\frac{1}{2} \frac{d}{dt} \|\nabla u\|^2 + \|\Delta u\|^2 - \int_{\mathbb{T}^3} [(u \cdot \nabla)u] \cdot \Delta u \, dx = 0,$$

and use the fact that u is uniformly bounded for $t \in (t_0 - r^2, t_0 + r^2)$ to estimate

$$\frac{1}{2} \frac{d}{dt} \|\nabla u\|^2 + \|\Delta u\|^2 \leq \|u\|_{L^\infty} \|\nabla u\| \|\Delta u\|$$

$$\leq \frac{1}{2} \|u\|_{L^\infty}^2 \|\nabla u\|^2 + \frac{1}{2} \|\Delta u\|^2.$$

Therefore

$$\frac{d}{dt} \|\nabla u\|^2 \leq \|u\|_{L^\infty}^2 \|\nabla u\|^2,$$

which implies that $\|\nabla u(t)\|^2$ remains finite on $(t_0 - r^2/2, t_0 + r^2)$, and hence that $t_0 \in J_q$ for some q. It follows that $\mathcal{T} \cap (0, \infty) \subseteq \Pi_t S$. $\qquad \square$

It follows from this lemma that any cover of S by parabolic cylinders with radii $\{r_i\}$ produces a cover of $\mathcal{T}$ by intervals of length $\{s_i\}$ where $s_i = r_i^2$. Since $\mathcal{P}^1(S) = 0$, as these covers are refined the quantity $\sum r_i$ can be made as small as we like, and so $\sum s_i^{1/2}$ can be made arbitrarily small: hence we recover the result that $\mathcal{H}^{1/2}(\mathcal{T}) = 0$.

16.4 The Serrin condition revisited: $u \in L_t^\infty L_x^3$

In this section we briefly study the regularity of Leray–Hopf weak solutions under the assumption that $u \in L^\infty(0, T; L^3)$, revisiting Theorem 13.5 which we stated in Chapter 13 without proof. We will discuss both the global case, extending Theorem 8.17, as well as the local result, extending Theorem 13.4. We only present a sketch of the results, since the proof is very involved. Full details can be found in the original paper of Escauriaza, Seregin, & Šverák (2003).

While these results treat one of the limiting cases in the local regularity theory presented in Chapter 13, they require partial regularity results along the lines of those from the last two chapters, which is why we have delayed this discussion until now.

Theorem 16.4 *Let u be a Leray–Hopf weak solution of the Navier–Stokes equations on $\mathbb{R}^3 \times [0, T)$ that satisfies $u \in L^\infty(0, T; L^3(\mathbb{R}^3))$. Then u is actually smooth (and hence unique).*

Theorem 16.5 *Let the pair (u, p) be a distributional solution of the Navier–Stokes equations in some parabolic cylinder $Q_R(x_0, t_0)$. Assume that*

$$u \in L_t^\infty L_x^2(Q_R), \quad \nabla u \in L_t^2 L_x^2(Q_R), \quad and \quad p \in L_t^{3/2} L_x^{3/2}(Q_R).$$

If $u \in L_t^\infty L_x^3(Q_R)$ then u is Hölder continuous in $Q_{R/2}(x_0, t_0)$.

We note that Theorem 16.5 implies the spatial smoothness of the function u. This is due to the fact that if u is Hölder continuous in $Q_{R/2}(x_0, t_0)$, then $u \in L_t^5 L_x^5$, for example, and hence Theorem 13.4 implies the spatial smoothness of the velocity.

First we show that the global result of Theorem 16.4 is a consequence of the local result in Theorem 16.5.

Sketch of the proof Theorem 16.4 assuming Theorem 16.5

Step 1. Show that $u(t)$ is bounded in L^3 for every $t \in [0, T)$

Notice that the assumption $u \in L^\infty(0, T; L^3(\mathbb{R}^3))$ only guarantees a priori that $u \in L^3(\mathbb{R}^3)$ for almost every $t \in [0, T)$. However, since u is a weak solution it is weakly continuous, i.e. the map

$$t \mapsto \int_{\mathbb{R}^3} u(x, t)\phi(x)\,\mathrm{d}x$$

is continuous in t for every $\phi \in L^2(\mathbb{R}^3)$, $t \in [0, T)$. However, since we have $u \in L^\infty(0, T; L^3(\mathbb{R}^3))$ we can extend the pairing to all $\phi \in L^{3/2}(\mathbb{R}^3)$, hence obtaining the result. Note that in particular $u(0) \in L^3$, which will be useful later.

Step 2. The associated pressure

Since $u \in L^\infty(0, T; L^3(\mathbb{R}^3))$ Theorem 5.1 guarantees that

$$p \in L^\infty(0, T; L^{3/2}(\mathbb{R}^3))$$

and so (u, p) has sufficient regularity to be a suitable weak solution.

Then, interpolating between $u \in L_t^\infty L_x^3$ and $u \in L_t^2 L_x^6$ we obtain $u \in L_t^4 L_x^4$, a condition that guarantees that the solution satisfies the energy inequality; also, since

$$(u \cdot \nabla)u \in L^{4/3}(\mathbb{R}^3 \times (0, T))$$

we can use results for the Stokes operator (see Notes in Chapter 5) to conclude that

$$u \in L^4(\mathbb{R}^3 \times (0, T)) \quad \text{and} \quad \partial_t u, \partial_x^2 u, \nabla p \in L^{4/3}(\mathbb{R}^3 \times (\delta, T)), \quad \delta > 0.$$

As a consequence of this regularity, it is possible to show that (u, p) satisfies the local energy inequality, and hence the pair (u, p) is a suitable weak solution of the Navier–Stokes equations (see Definition 15.2).

Step 3. Use Theorem 16.5 to obtain spatial regularity

Since (u, p) is a suitable weak solution, we can use Theorem 16.5 to show that for any $(x_0, t_0) \in \mathbb{R}^3 \times (0, T)$ there exists a neighbourhood of (x_0, t_0) in which u is Hölder continuous. As remarked before, this is enough to obtain the spatial smoothness of the solution, as we can apply Theorem 13.4.

Step 4. Additional regularity: use CKN

To obtain the additional time regularity of the solution (we already know Hölder regularity for $t > 0$) we will show that $u \in L^5(\mathbb{R}^3 \times (\delta, T))$, which implies the smoothness of the solution for $t > \delta$ using Theorem 8.17; we then couple this with the fact that we have a smooth solution for a short time (up to time δ) since the initial condition is in L^3. Notice that the estimates we have on the pressure and the velocity imply that,

$$\lim_{|x_0| \to \infty} \int_{Q_R(x_0, t_0)} |u|^3 + |p|^{3/2} = 0.$$

This means that for sufficiently large x_0 we can obtain uniform estimates for $\int_{Q_R(x_0, t_0)} |u|^3 + |p|^{3/2}$, which imply (since the result of Theorem 15.4 shows that

$$\|u\|_{L^\infty(Q_{R/2}(x_0, t_0))} \leq \frac{C}{R}\left(\frac{1}{R^2}\int_{Q_R(x_0, t_0)} |u|^3 + |p|^{3/2}\right)^{1/3}$$

whenever the bracketed expression on the right-hand side is sufficiently small) a uniform bound for the velocity for x_0 large. For x_0 small, we can just use the Hölder estimate for the velocity to conclude that

$$\max_{x \in \mathbb{R}^3, t \in [\delta, T]} |u(x,t)| < C(\delta) < \infty.$$

It follows that $u \in L^\infty(\delta, T; L^\infty(\mathbb{R}^3))$, and so since $|u|^3 \le |u|^2 \|u\|_{L^\infty}$ and $|\nabla |u|^{3/2}|^2 \le |u||\nabla u|^2 \le \|u\|_{L^\infty} |\nabla u|^2$ it follows that

$$|u|^{3/2} \in L^\infty(\delta, T; L^2(\mathbb{R}^3)) \quad \text{and} \quad \nabla(|u|^{3/2}) \in L^2(\delta, T; L^2(\mathbb{R}^3)). \tag{16.22}$$

The interpolation inequality $\|f(\cdot, t)\|_{L^{10/3}} \le \|f(\cdot, t)\|_{L^2}^{2/5} \|\nabla f(\cdot, t)\|_{L^2}^{3/5}$, together with (16.22) implies that $u \in L^5(\mathbb{R}^3 \times (\delta, T))$, which in turn implies the smoothness of the solution (see Theorem 8.17) for times $t \in (\delta, T)$. Now, for small times, since the initial condition is in L^3 (see Step 1), the local existence result from Chapter 11 implies the smoothness for short time, hence completing the proof. □

Sketch of the Proof of Theorem 16.5   The pair (u, p) satisfying the hypothesis of the theorem is a suitable weak solution in the cylinder Q (we take $R = 1$ and $(x_0, t_0) = (0, 0)$).

Step 1. Show that $\|u\|_{L^3(B_{3/4})}$ is bounded for every time

We proceed as in the previous theorem, showing that the map

$$t \mapsto \int_{B_{3/4}(0)} u(x, t)\phi(x) \, dx$$

is continuous for $t \in [-(3/4)^2, 0]$, for any $\phi \in L^{3/2}(B_{3/4}(0))$. The justification of this result requires the use of the local energy inequality and results for the Stokes operator with suitably chosen cutoff functions. The reason for this is that we are dealing with a distributional solution, rather than a Leray–Hopf solution, and so we do not, a priori, know the continuity in time of the pairing $\int u(x, t)\phi(x) \, dx$.

Step 2. Pressure decomposition and estimates

We decompose the pressure p as a sum of two terms, p_1 and p_2, where p_1 vanishes at the boundary of the unit ball, and p_2 is harmonic, that is, p_1 solves (weakly) $-\Delta p_1 = \partial_i \partial_j (u_i u_j)$ in the unit ball with Dirichlet boundary conditions. Meanwhile p_2 acts as a corrector, to take care of the boundary values for p, and given the equation satisfied by p_1 we need to have $\Delta p_2 = 0$. We have the following estimates for p_1 and p_2

$$\|p_1\|_{L_t^\infty L_x^{3/2}(Q)} \leq C\|u\|_{L_t^\infty L_x^3(Q)}$$

and

$$\|p_2\|_{L^\infty(-1,0;L^{3/2}(B_{3/4}))} \leq C\|p\|_{L^\infty(-1,0;L^{3/2}(B))} + C\|u\|_{L^\infty(-1,0;L^3(B))}.$$

Step 3. Necessary condition for blowup

Assume that (x_0, t_0) is a singular point. Following Seregin & Šverák (2002) there exist R_k, with $R_k \to 0$ such that

$$A(R_k) := \sup_{t_0 - R_k^2 \leq t \leq t_0} \frac{1}{R_k} \int_{B_{R_k}(x_0)} |u(x,t)|^2 \, dx > \varepsilon_*,$$

for some absolute constant ε_* (cf. Theorem 16.12).

We denote by $\tilde{u}$ and $\tilde{p}$ the extension of u and p by zero to $\mathbb{R}^4$ respectively; we construct a family of functions rescaled around the singular point

$$u^{R_k}(x,t) := R_k \tilde{u}(x_0 + R_k x, t_0 + R_k^2 t),$$
$$p^{R_k}(x,t) := R_k^2 \tilde{p}(x_0 + R_k x, t_0 + R_k^2 t),$$
$$p_j^{R_k}(x,t) := R_k^2 \tilde{p}_j(x_0 + R_k x, t_0 + R_k^2 t), \qquad j = 1, 2.$$

Notice that the rescalings have been chosen in such a way that

$$\|u^{R_k}(\cdot, t)\|_{L^3(\mathbb{R}^3)} = \|\tilde{u}(\cdot, t_0 + R_k^2 t)\|_{L^3(\mathbb{R}^3)},$$
$$\|p_1^{R_k}(\cdot, t)\|_{L^{3/2}(\mathbb{R}^3)} = \|\tilde{p}_1(\cdot, t_0 + R_k^2 t)\|_{L^{3/2}(\mathbb{R}^3)}, \text{ and}$$
$$\|p_2^{R_k}(\cdot, t)\|_{L^{3/2}(\mathbb{R}; L^\infty(\Omega))} = R_k^{2/3} \|\tilde{p}_2(x_0 + R_k \cdot, \cdot)\|_{L^{3/2}(\mathbb{R}; L^\infty(\Omega))},$$

for any bounded Ω. Hence we can find functions v and q such that (for a subsequence, that we relabel)

$$u^{R_k} \overset{*}{\rightharpoonup} v \text{ in } L^\infty(\mathbb{R}; L^3(\mathbb{R}^3)), \qquad p_1^{R_k} \overset{*}{\rightharpoonup} q \text{ in } L^\infty(\mathbb{R}; L^{3/2}(\mathbb{R}^3)), \qquad \text{and}$$

$$p_2^{R_k} \rightharpoonup 0 \text{ in } L^{3/2}(\mathbb{R}; L^\infty(\Omega)).$$

Step 4. Strong convergence

Using the local energy inequality with a suitable family of test functions we obtain the uniform bounds

$$\|p^{R_k}\|_{L^{3/2}(a,b;L^{3/2}(\Omega))} + \|\nabla u^{R_k}\|_{L^2(a,b;L^2(\Omega))} \leq c(a, b, \Omega) \qquad (16.23)$$

$$\|u^{R_k}\|_{L^4(Q)} + \|\partial_t u^{R_k}\|_{L^{4/3}(Q)} + \|\partial^2 u^{R_k}\|_{L^{4/3}(Q)} + \|\nabla p^{R_k}\|_{L^{4/3}(Q)} \leq c(Q), \qquad (16.24)$$

with $(a, b) \subset\subset (-1, 0)$ and $\Omega \subset\subset B_1(0)$. From these we have $u^{R_k} \to v$ in $C([a, b]; L^2(\Omega))$; this result follows by interpolation, having deduced first from the above inequalities the strong convergence in $C([a, b]; L^3(\Omega))$ and $C([a, b]; L^{4/3}(\Omega))$ via the Aubin–Lions Lemma.

Moreover, since the estimates (16.23) and (16.24) are inherited by v, the pair (v, q) is a suitable weak solution of the Navier–Stokes equations on $\Omega \times [a, b]$ for any $\Omega \subset\subset \mathbb{R}^3$. Notice that due to the strong convergence in $C([a, b]; L^2(\Omega))$ we have

$$\sup_{-1 \le t \le 0} \int_{B_1(0)} |v(x, t)|^2 \, \mathrm{d}x > \varepsilon_*. \tag{16.25}$$

Step 5. Show that curl $v = 0$ *using parabolic theory*

The proof is completed by showing that $v = 0$, obtaining a contradiction with (16.25), required for the existence of singular points. We already know that v is harmonic, so showing that $\omega = $ curl $v = 0$ we will obtain $v = 0$.

The vorticity satisfies a differential inequality of the form

$$|\partial_t \omega - \Delta \omega| \le C(|\omega| + |\nabla \omega|).$$

General results about backwards uniqueness of solutions to the heat equation (or more generally functions satisfying the above differential inequality) yield $\omega = 0$ in the exterior of a large cylinder. That $\omega = 0$ in the interior is obtained via a general 'unique continuation' result for functions satisfying the same type of differential inequality.

Both the backwards uniqueness and unique continuation results are of significant interest, and we refer the reader to Escauriaza, Seregin, & Šverák (2003) and references therein for more details. $\qquad \square$

16.5 Lemmas for the second partial regularity theorem

We now give the proofs that we have delayed from earlier in the chapter. First we show that we can find a family of functions to use in the local energy inequality that localise the estimate to within a cylinder Q_r.

Lemma 16.6 *For any fixed $r > 0$ and $0 < \theta \le 1/2$ there exists a function $\phi \in C_c^\infty(\mathbb{R}^4)$ such that* $(\operatorname{supp}\phi) \cap Q_1 \subseteq Q_r$,

$$\phi \ge C_3^{-1}(\theta r)^{-1} \qquad for \qquad (x, t) \in Q_{\theta r},$$

and for any $(x, t) \in Q_r$

$$0 \le \phi \le C_3(\theta r)^{-1}, \quad |\nabla \phi| \le C_3(\theta r)^{-2}, \quad and \quad |\partial_t \phi + \Delta \phi| \le C_3 \theta^2 r^{-3},$$

where $C_3 \ge 1$ is an absolute constant.

Proof   As in our previous construction (Lemma 15.11) we take ϕ of the form $\phi(x, t) = c\psi(x, t)\vartheta(x, t)$, where c depends on θr, ψ is a solution of the backwards heat equation, and $\vartheta(x, t) = \chi(x, t)\eta(t)$ is a cutoff function. We set

$$\psi(x, t) = G(x, (\theta r)^2 - t),$$

where

$$G(x, t) = \begin{cases} (4\pi t)^{-3/2} \exp(-|x|^2/4t) & t > 0 \\ 0 & t \leq 0. \end{cases}$$

Let $\chi \in C_c^\infty(B_1 \times (-1, 1))$ be such that $\chi \equiv 1$ on $B_{1/2} \times (-1/4, 1/4)$ and set

$$\phi(x, t) = (\theta r)^2 \psi(x, t)\chi(x/r, t/r^2)\eta(t),$$

where $\eta \in C^\infty(\mathbb{R}, [0, 1])$ is 1 on $[-\infty, r^2/4)$ and zero on $[r^2/2, \infty)$.

Verification of the properties stated follows from the arguments used in the proof of Lemma 15.11. $\square$

We now prove the local estimate (16.14) on the pressure, following Kukavica (2009a).

Lemma 16.7   *There is an absolute constant $C_5 \geq 1$ such that whenever $p \in L^{3/2}(Q_r)$ and*

$$-\Delta p = \partial_i \partial_j (u_i u_j) \tag{16.26}$$

in Q_r then for $0 < \theta \leq 1/2$

$$\frac{1}{(r\theta)^2} \int_{Q_{r\theta}} |p|^{3/2} \leq C_5 \theta^{-3/2} \left(\frac{1}{r} \sup_{-r^2 < t < 0} \int_{B_r} |u(t)|^2 \right)^{3/4} \left(\frac{1}{r} \int_{Q_r} |\nabla u|^2 \right)^{3/4}$$
$$+ C_5 \theta \frac{1}{r^2} \int_{Q_r} |p|^{3/2}.$$

Proof   It suffices to prove the estimate with $r = 1$, the result for general r following by rescaling. Let $\eta : \mathbb{R}^3 \to [0, 1]$ be a smooth non-negative function such that $\eta \equiv 1$ on $B_{3/5}$ and $\eta = 0$ outside $B_{4/5}$. Defining

$$U_{ij} = u_i(u_j - (u_j)_1)$$

(recall that $(f)_r = \frac{1}{|B_r|} \int_{B_r} f \, dx$ denotes the average of f over the ball of radius r, here centred at zero) equation (16.26) can be rewritten as

$$-\Delta p = \partial_i \partial_j U_{ij},$$

since $(u_j)_1$ is a constant. Now an elementary but tedious algebraic calculation shows that

$$-\Delta(\eta p) = \partial_i\partial_j(\eta U_{ij}) + (\partial_i\partial_j\eta)U_{ij} - \partial_j(U_{ij}\partial_i\eta)$$
$$- \partial_i(U_{ij}\partial_j\eta) + p\Delta\eta - 2\partial_j(p\partial_j\eta),$$

and using the integral representation for $(-\Delta)^{-1}$ from Appendix C we can write

$$\eta p(x) = \frac{1}{4\pi}\int_{\mathbb{R}^3}\frac{1}{|x-y|}\Big[\partial_i\partial_j(\eta U_{ij}) + (\partial_i\partial_j\eta)U_{ij} - \partial_j(U_{ij}\partial_i\eta)$$

$$- \partial_i(U_{ij}\partial_j\eta) + p\,\Delta\eta - 2\partial_j(p\partial_j\eta)\Big]\mathrm{d}y$$

$$= \frac{1}{4\pi}\int_{\mathbb{R}^3}\Big[\partial_i\partial_j\frac{1}{|x-y|}\Big]\eta U_{ij}\,\mathrm{d}y + \frac{1}{4\pi}\int_{\mathbb{R}^3}\frac{1}{|x-y|}U_{ij}\partial_i\partial_j\eta\,\mathrm{d}y$$

$$+ \frac{1}{4\pi}\int_{\mathbb{R}^3}\frac{x_j-y_j}{|x-y|^3}U_{ij}\partial_i\eta\,\mathrm{d}y + \frac{1}{4\pi}\int_{\mathbb{R}^3}\frac{x_i-y_i}{|x-y|^3}U_{ij}\partial_j\eta\,\mathrm{d}y$$

$$+ \frac{1}{4\pi}\int_{\mathbb{R}^3}\frac{1}{|x-y|}p\,\Delta\eta\,\mathrm{d}y + \frac{1}{2\pi}\int_{\mathbb{R}^3}\frac{x_j-y_j}{|x-y|^3}p\partial_j\eta\,\mathrm{d}y$$

$$=: p_1 + p_2 + p_{3,1} + p_{3,2} + p_4 + p_5,$$

after integrating by parts. We now estimate each term on the right-hand side of the above equality in $L^{3/2}(B_\theta)$, using similar techniques to those employed previously in the proof of Lemma 15.12.

We can bound the first term using the Calderón–Zygmund Theorem (see Appendix B)

$$\|p_1\|_{L^{3/2}(B_\theta)} = \|T_{ij}(\eta U_{ij})\|_{L^{3/2}(B_\theta)} \le \|T_{ij}(\eta U_{ij})\|_{L^{3/2}(\mathbb{R}^3)}$$

$$\le C\sum_{i,j}\|\eta U_{ij}\|_{L^{3/2}(\mathbb{R}^3)}$$

$$\le C\sum_{i,j}\|u_i(u_j - (u_j)_1)\|_{L^{3/2}(B_1)}$$

$$\le C\|u\|_{L^2(B_1)}\|u - (u)_1\|_{L^6(B_1)}$$

$$\le C\|u\|_{L^2(B_1)}\|\nabla u\|_{L^2(B_1)}.$$

For p_2, note that since $\partial_i\partial_j\eta(y)$ is only non-zero for $3/5 < |y| < 4/5$ and we only integrate p_2 (with respect to x) over $B_\theta \subseteq B_{1/2}$, the potential singularity in the kernel $1/|x - y|$ does not appear in the range of integration in the convolution integral,[2] i.e. for $x \in B_\theta$ we have

$$p_2(x) = \frac{1}{4\pi} \int_{3/5 < |y| < 4/5} \frac{1}{|x - y|} U_{ij}(y) \partial_i \partial_j \eta(y) \, dy.$$

Therefore using Young's inequality for convolutions (Theorem A.10) we obtain

$$\|p_2\|_{L^{3/2}(B_\theta)} \leq C \|U_{ij} \partial_i \partial_j \eta\|_{L^{3/2}(B_1)}$$
$$\leq C \sum_{i,j} \|U_{ij}\|_{L^{3/2}(B_1)}$$
$$\leq C \|u\|_{L^2(B_1)} \|\nabla u\|_{L^2(B_1)},$$

estimating $\|U_{ij}\|_{L^{3/2}(B_1)}$ as before. The estimates for the next two terms ($p_{3,i}$) are similar. For the final two terms we use the same argument, except we first move to L^∞ before using Young's inequality; for p_4 the details are

$$\|p_4\|_{L^{3/2}(B_\theta)} \leq c\theta^2 \|p_4\|_{L^\infty(B_\theta)} \leq c\theta^2 \|p\Delta\eta\|_{L^1(B_1)} \leq c\theta^2 \|p\|_{L^{3/2}(B_1)}.$$

Combining these estimates yields

$$\|p\|_{L^{3/2}(B_\theta)} = \|\eta p\|_{L^{3/2}(B_\theta)} \leq C \|u\|_{L^2(B_1)} \|\nabla u\|_{L^2(B_1)} + c\theta^2 \|p\|_{L^{3/2}(B_1)}.$$

If we now raise both sides to the power of $3/2$ and integrate in time we obtain

$$\|p\|_{L^{3/2}(Q_\theta)}^{3/2} \leq C \int_{-\theta^2}^0 \|u(t)\|_{L^2(B_1)}^{3/2} \|\nabla u(t)\|_{L^2(B_1)}^{3/2} \, dt + c\theta^3 \|p\|_{L^{3/2}(Q_1)}^{3/2}$$
$$\leq C \left(\int_{-\theta^2}^0 \|u(t)\|_{L^2(B_1)}^6 \right)^{1/4} \|\nabla u\|_{L^2(Q_1)}^{3/2} \, dt + c\theta^3 \|p\|_{L^{3/2}(Q_1)}^{3/2}$$
$$\leq C\theta^{1/2} \left(\sup_{-\theta^2 < t < 0} \int_{B_1} |u(t)|^2 \, dt \right)^{3/4} \|\nabla u\|_{L^2(Q_1)}^{3/2} + c\theta^3 \|p\|_{L^{3/2}(Q_1)}^{3/2},$$

and so

$$\frac{1}{\theta^2} \int_{Q_\theta} |p|^{3/2} \leq C\theta^{-3/2} \left(\sup_{-\theta^2 < t < 0} \int_{B_1} |u(t)|^2 \right)^{3/4} \left(\int_{Q_1} |\nabla u|^2 \right)^{3/4}$$
$$+ c\theta \int_{Q_1} |p|^{3/2}.$$

The inequality in the statement follows by rescaling. $\qquad \square$

Finally, following CKN, we give a proof of the Vitali Covering Lemma for (centred) parabolic cylinders that we used for Theorem 16.2.

---

[2] Since for $x \in B_\theta \subseteq B_{1/2}$ and $3/5 < |y| < 4/5$ we have $|x - y| > \frac{3}{5} - \frac{1}{2} = \frac{1}{10}$.

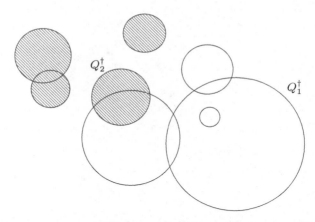

Figure 16.1. If there were a finite number of cylinders, then given $Q_n^\dagger$ we would choose $Q_{n+1}^\dagger$ to be the cylinder of largest radius in $\mathfrak{Q}_n$. Since there are (potentially) an infinite number of cylinders we choose $Q_{n+1}^\dagger$ so that its radius is comparable with the supremum of radii of elements of $\mathfrak{Q}_n$. Here, having chosen $Q_1^\dagger$ the set $\mathfrak{Q}_1$ is shown shaded.

Lemma 16.8 (Vitali Covering Lemma for parabolic cylinders) *Given any family $\mathfrak{Q}$ of centred parabolic cylinders contained in a bounded subset of $\mathbb{R}^3 \times \mathbb{R}$, there exists a finite or countable disjoint subfamily*

$$\mathfrak{Q}^\dagger = \{Q_i^\dagger = Q_{r_i}^*(x_i, t_i)\}$$

such that for any cylinder $Q_r^*(x, t) \in \mathfrak{Q}$ we have $Q_r^*(x, t) \subset Q_{5r_i}^*(x_i, t_i)$ for some i.

Note that we use balls in our pictures rather than cylinders for simplicity.

Proof We use an inductive construction, choosing the first element of $\mathfrak{Q}^\dagger$ to be $Q_1^\dagger = Q_{r_1}^*(x_1, t_1) \in \mathfrak{Q}$ such that $r \leq 2r_1$ for all $Q_r^*(x, t) \in \mathfrak{Q}$.

Now if we have already chosen $Q_i^\dagger$, $1 \leq i \leq n$, then set

$$\mathfrak{Q}_n = \{Q \in \mathfrak{Q} : Q \cap Q_i^\dagger = \varnothing, \ 1 \leq i \leq n\}.$$

Note that $\mathfrak{Q}_{n+1} \subseteq \mathfrak{Q}_n$ for $n \geq 0$. For each $n \geq 0$ such that $\mathfrak{Q}_n$ is non-empty we choose $Q_{n+1}^\dagger = Q_{r_{n+1}}^*(x_{n+1}, t_{n+1}) \in \mathfrak{Q}_n$ such that

$$Q_r^*(x, t) \in \mathfrak{Q}_n \qquad \Rightarrow \qquad r \leq 2r_{n+1}, \tag{16.27}$$

which is possible since all cylinders in $\mathfrak{Q}$ are contained in a bounded subset of $\mathbb{R}^3 \times \mathbb{R}$ and hence

$$\sup\{r : Q_r^*(x, t) \in \mathfrak{Q}_n\} < \infty,$$

see Figure 16.1.

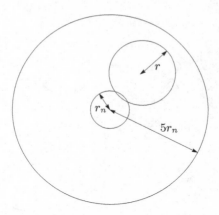

Figure 16.2. If $Q_r^* \cap Q_{r_n}^*(x_n, t_n) \neq \varnothing$ and $r \leq 2r_n$ then $Q_r^* \subset Q_{5r_n}^*(x_n, t_n)$. The figure shows a two-dimensional version of the spatial cross-section of intersecting cylinders.

If $\mathfrak{Q}_n$ is empty for some n then the process terminates, and then $\mathfrak{Q}^\dagger = \cup_{j=1}^{n-1} Q_j^\dagger$ is finite. Otherwise we take $\mathfrak{Q}^\dagger = \cup_{j=1}^{\infty} Q_j^\dagger$ and note that $r_n \to 0$ as $n \to \infty$, for if not we would have obtained a countably infinite collection of disjoint cylinders with a lower bound on their radii, all of which lie within a bounded subset of $\mathbb{R}^3 \times \mathbb{R}$, which is impossible.

It is clear that the sets $\{Q_i^\dagger\}$ constructed in this way are disjoint. Now suppose that $Q = Q_r^*(x, t) \in \mathfrak{Q} \setminus \mathfrak{Q}^\dagger$. Then since $r_n \to 0$ as $n \to \infty$, it follows from (16.27) that there exists an $n \geq 1$ such that $Q \in \mathfrak{Q}_{n-1}$ but $Q \notin \mathfrak{Q}_n$. As $Q \in \mathfrak{Q}_{n-1}$ we have $r \leq 2r_n$, and since $Q \notin \mathfrak{Q}_n$ we have $Q \cap Q_n^\dagger \neq \varnothing$. It follows that $Q_r^*(x, t) \subset Q_{5r_n}^*(x_n, t_n)$, see Figure 16.2. $\qquad\square$

Notes

The argument in this chapter used to obtain Theorem 16.1 is based on the iterative procedure adopted by Lin (1998) and Kukavica (2009a), which greatly simplifies that of CKN. The proof of Theorem 16.2 is essentially that of CKN, but shortened since we have already shown in the previous chapter that the set S has measure zero.

Exercises

16.1 Use the embedding $W^{1,1} \subset L^{3/2}$ and a contradiction argument to show that there exists a constant C_4 that does not depend on r such that for any

$f \in W^{1,1}(B_r)$ with zero average on B_r

$$\|f\|_{L^{3/2}(B_r)} \le C_4 \|\nabla f\|_{L^1(B_r)}. \tag{16.28}$$

(Recall that the embedding $W^{1,1}(B_1) \subset L^1(B_1)$ is compact, and first prove the inequality on B_1; then obtain (16.28) by considering an appropriate rescaling.)

16.2 Suppose that there exist $r_0, \alpha > 0$, and $0 < \theta, \beta < 1$ such that

$$E(\theta r) \le \alpha + \beta E(r) \qquad \text{for all} \quad 0 < r < r_0.$$

Show that $E(\theta^k r) \le 2\alpha/(1 - \beta)$ for all k sufficiently large.

16.3 Show that $\mathcal{P}^s(X) = 0$ if and only if for every $\varepsilon > 0$ the set X can be covered by a collection $\{Q^*_{r_i}\}$ such that $\sum_i r_i^s < \varepsilon$.

17

Lagrangian trajectories

The solution $u(x, t)$ of the Navier–Stokes equations represents the evolving velocity field of a fluid flow. If we imagine the fluid as being composed of an uncountable number of individual 'fluid particles', then we can try to work out the trajectories followed by these particles. In other words, we can try to solve the ordinary differential equation

$$\frac{dX}{dt} = u(X(t), t) \quad \text{with} \quad X(0) = a \tag{17.1}$$

for every a in the domain occupied by the fluid.

Throughout this chapter we treat the simplest case, that of the three-dimensional torus, although a similar analysis is possible on the whole space and in a smooth bounded domain (with Dirichlet boundary conditions). We can understand the motion of a particle on the torus as a trajectory governed by a periodic vector field defined on the whole of $\mathbb{R}^3$; then if we wish to we can think of the 'particle trajectory' confined to one period $[0, 2\pi)^3$, so that a solution of (17.1), that 'leaves' $[0, 2\pi)^3$ reenters as appropriate on the opposite face.

We investigate in what sense particle trajectories can be understood for four different classes of initial data (and the corresponding types of solutions).

(i) u is a classical solution, $u \in C^1([0, T]; C^2(\Omega))$.

In this case u is sufficiently smooth that we can use the classical theory of ODEs to obtain a unique solution of (17.1) for every $a \in \mathbb{T}^3$. An elementary argument shows that since $\nabla \cdot u = 0$ the resulting flow assembled from these individual trajectories is volume preserving. This property of classical divergence-free flows will be useful when we come to analyse what happens in the case that u is only a weak solution.

(ii) $u_0 \in H \cap \dot{H}^{1/2}$ and $u \in L^\infty(0, T; \dot{H}^{1/2}) \cap L^2(0, T; \dot{H}^{3/2})$.

Corollary 10.2 ensures that these solutions are smooth for $t \in (0, T]$, so existence and uniqueness of solutions of (17.1) for positive times is not an issue. Rather the problem is existence and uniqueness 'from $t = 0$'. In order to obtain the existence of solutions, in Section 17.2 we recast the problem as the integral equation

$$X(t) = a + \int_0^t u(X(s), s)\, ds; \tag{17.2}$$

by approximating u by a sequence of smooth functions we can obtain the existence of a solution of (17.2). We then show (following Dashti & Robinson, 2009) that when $u_0 \in H \cap \dot{H}^{1/2}$ we have $\sqrt{t}u \in L^2(0, T; \dot{H}^{5/2})$, which is enough to prove that solutions are unique.

(iii) $u_0 \in H$ and $u \in L^\infty(0, T; L^2) \cap L^2(0, T; H^1)$ is a Leray–Hopf weak solution.

The same approximation argument used in case (ii) can be adapted to show that there is still at least one solution of (17.2) for each $a \in \mathbb{T}^3$, even when u has limited regularity. In Section 17.3 we use a construction due to Foias, Guillopé, & Temam (1985) to show that it is possible to choose one solution for each $a \in \mathbb{T}^3$ in such a way as to define a 'solution mapping' $\Phi \colon \mathbb{T}^3 \times [0, T] \to \mathbb{T}^3$ that maintains the volume-preserving properties one would expect from the fact that u is divergence-free.

(iv) $u_0 \in H \cap \dot{H}^{1/2}$ and (u, p) is a suitable weak solution on $(0, \infty)$.

On some time interval $[0, T]$ there will be a unique solution u that satisfies the conditions of case (ii), so the existence and uniqueness of trajectories is assured for t small. Once the solution u loses regularity it can be continued as a suitable weak solution, although there may be many such possibilities. If we choose one particular continuation then we know that for $t > 0$ the set of singularities is limited by the partial regularity results of Chapter 15. This will allow us to show in Section 17.4 that the solution of (17.1) is unique and C^1 in time for almost every choice of initial point $a \in \mathbb{T}^3$. This implies that for such initial conditions the equation (17.1) is satisfied in a classical sense, and that the solution mapping Φ from case (iii) is unique.

The last of these four results, due to Robinson & Sadowski (2009), involves almost all the properties of (suitable) weak solutions of the Navier–Stokes equations that we have covered in this book:

(i) existence of Leray–Hopf weak solutions for all $t \geq 0$;
(ii) epochs of regularity for weak solutions and $u \in L^{2/3}(0, T; H^2)$;

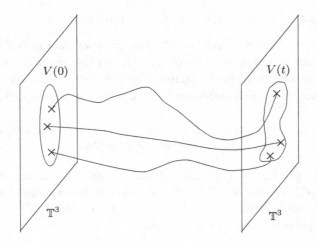

Figure 17.1. The initial volume $V(0)$ evolves to $V(t)$ under the flow; three representative trajectories are shown.

(iii) small-time existence of smooth solutions for $u_0 \in H \cap \dot{H}^{1/2}$;
(iv) the local smoothness of u as a consequence of Serrin's condition;
 (v) Hölder continuity in time via global estimates on the pressure; and
(vi) the partial regularity result $\dim_B(S) \le 5/3$.

17.1 Lagrangian trajectories for classical solutions and the volume-preserving property

If $u \in C^1([0, T]; C^2(\mathbb{T}^3))$ then the equation

$$\dot{X} = u(X, t), \qquad X(0) = a,$$

has a unique solution $X(t; a)$ for every $a \in \mathbb{T}^3$. The classical theory of ODEs (see Hale, 1980, or Hartman, 1973, for example) requires that u is jointly continuous in X and t for existence; the Lipschitz condition

$$|u(X, t) - u(Y, t)| \le L(t)|X - Y|$$

with $L \in L^1(0, T)$ is sufficient to guarantee the uniqueness of the solution on $[0, T]$.

More importantly for what follows, since we have $\nabla \cdot u = 0$ it is easy to show that when u is smooth (as here) the resulting flow preserves volumes. To show this, given an initial volume $V(0)$ define

$$V(t) = \{X(t; a) : a \in V(0)\},$$

see Figure 17.1.

We will write χ_A for the characteristic function of the a set $A \subset \mathbb{T}^3$; recall that we use $\mu(A)$ to denote the Lebesgue measure of the set A. We have

$$\mu(V(t)) = \int_{V(t)} 1 \, dx = \int_{\mathbb{T}^3} \chi_{V(t)}(x) \, dx$$

$$= \int_{\mathbb{T}^3} \chi_{V(t)}(X(t; y)) \, |\det(\nabla X(t; y))| \, dy$$

$$= \int_{\mathbb{T}^3} \chi_{V(0)}(y) \, |\det(\nabla X(t; y))| \, dy,$$

where we have substituted $x = X(t; y)$. In order to evaluate $\det(\nabla X)$, observe that since

$$\frac{d}{dt} X(t; a) = u(X(t; a), t)$$

the gradient ∇X satisfies the linear equation

$$\frac{d}{dt} \nabla X(t; a) = \nabla u(X(t; a), t) \nabla X(t; a),$$

where ∇u is the matrix with components $[\nabla u]_{ij} = \partial_j u_i$.

By Liouville's formula (see Theorem 1.2 in Chapter IV of Hartman, 1973, for example, or Exercise 17.1)

$$\det \nabla X(t) = \exp\left(\int_0^t \mathrm{tr}(\nabla u(s)) \, ds\right) \det \nabla X(0) = \exp\left(\int_0^t \nabla \cdot u(s) \, ds\right),$$

since $\nabla X(0) = I$. As $\nabla \cdot u(t) = 0$ for every t it follows that $|\det(\nabla X)| = 1$ and hence

$$\mu(V(t)) = \int_{V(0)} 1 \, dy = \mu(V(0))$$

and the flow is volume preserving.

17.2 Lagrangian uniqueness for $u_0 \in H \cap \dot{H}^{1/2}$

We know from Corollary 10.2 that when $u_0 \in H \cap \dot{H}^{1/2}$, there is some time interval during which $u \in L^\infty(0, T; \dot{H}^{1/2}) \cap L^2(0, T; \dot{H}^{3/2})$, and indeed u is smooth on $(0, T]$. It is therefore reasonable to expect that the ODE

$$\dot{X} = u(X, t) \qquad \text{with} \qquad X(0) = a \qquad (17.3)$$

should have a unique solution on $[0, T]$. However, there is potentially a problem in proving the existence of solutions of the ODE as we move away from

$t = 0$, and similarly the uniqueness of such solutions at the initial time is not immediate.

In order to find solutions we reformulate (17.3) as the integral equation

$$X(t) = a + \int_0^t u(X(s), s) \, ds \qquad (17.4)$$

and solve a sequence of approximate problems in which u is replaced by smooth functions u_n. Although we will see in Section 17.3 that we can prove the existence of a solution even when u is only a weak solution using this method, the proof is simpler in this case, since we can take account of some additional regularity of u that will also be important in proving uniqueness.

17.2.1 Additional bounds on solutions with $u_0 \in H \cap \dot{H}^{1/2}$

We now obtain some extra information on the regularity of u in $\dot{H}^{5/2}$; by introducing a factor of $\sqrt{t}$ we are able to obtain a bound that is valid on the whole time interval $(0, T)$. Since we are concerned with potential difficulties of existence and uniqueness at $t = 0$ this will turn out to be particularly useful.

Lemma 17.1 *If $u_0 \in H \cap \dot{H}^{1/2}(\mathbb{T}^3)$ gives rise to a solution $u \in L^2(0, T; \dot{H}^{3/2})$ for some $T > 0$ then $\sqrt{t}\, u \in L^2(0, T; \dot{H}^{5/2})$.*

Proof We present the formal calculations that could be made rigorous by arguing with the Galerkin approximation and then taking limits. Recalling that $\|u\|_{\dot{H}^s} = \|\Lambda^s u\|$ (see (10.3)) we apply $\Lambda^{3/2}$ to the equation

$$\partial_t u - \Delta u + (u \cdot \nabla)u + \nabla p = 0$$

and take the inner product with $t \Lambda^{3/2} u$ (equivalently we take the inner product of the equation with $t \Lambda^3 u$). Then, noting that $-\Delta = \Lambda^2$,

$$\langle \Lambda^{3/2} \partial_t u, t \Lambda^{3/2} u \rangle + \langle \Lambda^{7/2} u, t \Lambda^{3/2} u \rangle = \langle \Lambda^{3/2}[(u \cdot \nabla)u], t \Lambda^{3/2} u \rangle,$$

and since

$$\frac{d}{dt}(t\|u\|_{\dot{H}^{3/2}}^2) = \|u\|_{\dot{H}^{3/2}}^2 + 2t\langle \Lambda^{3/2} u, \Lambda^{3/2} \partial_t u \rangle$$

we obtain

$$\frac{1}{2}\frac{d}{dt}(t\|u\|_{\dot{H}^{3/2}}^2) - \frac{1}{2}\|u\|_{\dot{H}^{3/2}}^2 + t\|u\|_{\dot{H}^{5/2}}^2 \le t\langle \Lambda^{3/2}[(u \cdot \nabla)u], \Lambda^{3/2} u \rangle,$$

where we have used the fact that $\langle \Lambda^{7/2} u, \Lambda^{3/2} u \rangle = \|\Lambda^{5/2} u\|^2$, which can easily be seen by considering the expression in the Fourier series representation.

We now use the estimate

$$|\langle \Lambda^{3/2}[(u \cdot \nabla)u], \Lambda^{3/2}u \rangle \le c\|u\|_{\dot{H}^{3/2}}^2 \|u\|_{\dot{H}^{5/2}}, \tag{17.5}$$

whose proof can be found at the end of the chapter in Lemma 17.15, to obtain the inequality

$$\frac{1}{2}\frac{d}{dt}(t\|u\|_{\dot{H}^{3/2}}^2) - \frac{1}{2}\|u\|_{\dot{H}^{3/2}}^2 + t\|u\|_{\dot{H}^{5/2}}^2 \le ct\|u\|_{\dot{H}^{3/2}}^2\|u\|_{\dot{H}^{5/2}}.$$

We apply Young's inequality and then absorb the term $\frac{1}{2}t\|u\|_{\dot{H}^{5/2}}^2$ on the left-hand side to obtain

$$\frac{d}{dt}(t\|u\|_{\dot{H}^{3/2}}^2) + t\|u\|_{\dot{H}^{5/2}}^2 \le ct\|u\|_{\dot{H}^{3/2}}^4 + \|u\|_{\dot{H}^{3/2}}^2.$$

Setting $E(t) = \exp(-c\int_0^t \|u(s)\|_{\dot{H}^{3/2}}^2 \, ds)$ we can rewrite this as

$$\frac{d}{dt}(E(t)t\|u\|_{\dot{H}^{3/2}}^2) + tE(t)\|u\|_{\dot{H}^{5/2}}^2 \le \|u\|_{\dot{H}^{3/2}}^2 E(t),$$

and therefore

$$E(t)t\|u(t)\|_{\dot{H}^{3/2}}^2 + \int_0^t sE(s)\|u(s)\|_{\dot{H}^{5/2}}^2 \, ds \le \int_0^t \|u(s)\|_{\dot{H}^{3/2}}^2 E(s) \, ds.$$

Dropping the first term and using the facts that $0 < E(s) \le 1$ and $E(s)$ is decreasing, we can deduce that

$$\int_0^t s\|u(s)\|_{\dot{H}^{5/2}}^2 \, ds \le \frac{1}{E(t)}\int_0^t \|u(s)\|_{\dot{H}^{3/2}}^2 \, ds$$

$$= \exp\left(c\int_0^t \|u(s)\|_{\dot{H}^{3/2}}^2 \, ds\right)\int_0^t \|u(s)\|_{\dot{H}^{3/2}}^2 \, ds.$$

Since by assumption $u \in L^2(0, T; \dot{H}^{3/2})$, it follows that the right-hand side is finite and therefore $\sqrt{t}\, u \in L^2(0, T; \dot{H}^{5/2})$. □

It is a simple consequence of this result that[1] $u \in L^r(0, T; \dot{H}^{5/2})$ for any $r < 1$ (see Exercise 17.4). Even for solutions of the heat equation with initial data $u_0 \in \dot{H}^{1/2}$, it may be that $u \notin L^1(0, T; \dot{H}^{5/2})$, see McCormick et al. (2016b).

Corollary 17.2 *If*

$$u \in L^\infty(0, T; \dot{H}^{1/2}) \qquad and \qquad \sqrt{t}\, u \in L^2(0, T; \dot{H}^{5/2})$$

then $\|u(t)\|_{L^\infty} \le \theta(t)$, *where* $\theta \in L^1(0, T)$ *and for* $0 \le s < t \le T$

$$\int_s^t \theta(\tau) \, d\tau \le C(t-s)^{1/2}. \tag{17.6}$$

---

[1] We use the notation $u \in L^r(0, T; X)$ to signify that $\int_0^T \|u(t)\|_X^r \, dt < \infty$, even when $0 < r < 1$.

In particular, $u \in L^1(0, T; L^\infty)$ *and* $\int_s^t \|u(\tau)\|_{L^\infty} \, d\tau \le C(t - s)^{1/2}$.

Proof We start with Agmon's inequality (see Theorem 1.20) in the form

$$\|u(t)\|_{L^\infty} \le \theta(t) := c\|u\|_{\dot{H}^{1/2}}^{1/2} \|u\|_{\dot{H}^{5/2}}^{1/2} \tag{17.7}$$

(see Exercise 1.10) and then

$$
\begin{aligned}
\int_s^t \theta(\tau) \, d\tau &= c \int_s^t \|u(\tau)\|_{\dot{H}^{1/2}}^{1/2} \|u(\tau)\|_{\dot{H}^{5/2}}^{1/2} \, d\tau \\
&= c \int_s^t \tau^{-1/4} \|u(\tau)\|_{\dot{H}^{1/2}}^{1/2} \tau^{1/4} \|u(\tau)\|_{\dot{H}^{5/2}}^{1/2} \, d\tau \\
&\le c\|u\|_{L^\infty(0,T;\dot{H}^{1/2})}^{1/2} \left(\int_s^t \tau^{-1/3} \, d\tau \right)^{3/4} \left(\int_s^t \tau \|u(\tau)\|_{\dot{H}^{5/2}}^2 \, d\tau \right)^{1/4} \\
&\le c\|u\|_{L^\infty(0,T;\dot{H}^{1/2})}^{1/2} (t^{2/3} - s^{2/3})^{3/4} \|\sqrt{t}\, u\|_{L^2(0,T;\dot{H}^{5/2})}^{1/2}.
\end{aligned}
$$

The inequality (17.6) now follows, noting that for $0 < \alpha \le 1$ and $t, s \ge 0$ we have $|t^\alpha - s^\alpha| \le |t - s|^\alpha$ (see Exercise 17.5). $\qquad\square$

17.2.2 Existence of trajectories when $u_0 \in H \cap \dot{H}^{1/2}$

Our first task is to show that the equation

$$\dot{X} = u(X, t), \qquad X(0) = a \in \mathbb{T}^3,$$

has at least one solution. To do this we reformulate the problem as the integral equation

$$X(t) = a + \int_0^t u(X(s), s) \, ds. \tag{17.8}$$

It is easy to see that when $u : \Omega \times [0, T] \to \Omega$ and $X : [0, T] \to \Omega$ are continuous the ODE and the integral equation are equivalent.

Lemma 17.3 *If u is divergence free with*

$$u \in L^\infty(0, T; \dot{H}^{1/2}) \qquad \text{and} \qquad \sqrt{t}\, u \in L^2(0, T; \dot{H}^{5/2})$$

then for every $a \in \mathbb{T}^3$ the integral equation (17.8) has at least one solution $X \in C^0([0, T]; \mathbb{T}^3)$.

Proof We approximate u by u_n, where for each $s \in [0, T]$ we set $u_n(s) = P_n u(s)$, with P_n the projection onto the first n eigenfunctions of the Stokes operator (see Section 4.1). Note that u_n is divergence free, since the eigenfunctions of the Stokes operator are divergence free.

Since for any $s \in \mathbb{R}$

$$\|P_n u\|_{\dot{H}^s} \le \|u\|_{\dot{H}^s}$$

it follows that

$$\|u_n\|_{L^\infty} \le C\|u_n\|_{\dot{H}^{1/2}}^{1/2}\|u_n\|_{\dot{H}^{5/2}}^{1/2} \le C\|u\|_{\dot{H}^{1/2}}^{1/2}\|u\|_{\dot{H}^{5/2}}^{1/2} = \theta(t),$$

where θ is defined in (17.7).

Furthermore, since

$$\|P_n u(t) - u(t)\|_{\dot{H}^s} \to 0 \text{ as } n \to \infty$$

for each t (see Exercise 4.1), the Dominated Convergence Theorem implies that $\sqrt{t}\, u_n \to \sqrt{t}\, u$ in $L^2(0, T; \dot{H}^{5/2})$; the analysis in the proof of Corollary 17.2 thus guarantees that

$$
\begin{aligned}
\|u - u_n\|_{L^1(0,T;L^\infty)} & \\
&\le CT^{1/2}\|u - u_n\|_{L^\infty(0,T;\dot{H}^{1/2})}^{1/2}\|\sqrt{t}(u - u_n)\|_{L^2(0,T;\dot{H}^{5/2})}^{1/2} \\
&\le 2CT^{1/2}\|u\|_{L^\infty(0,T;\dot{H}^{1/2})}^{1/2}\|\sqrt{t}(u - u_n)\|_{L^2(0,T;\dot{H}^{5/2})}^{1/2}
\end{aligned}
\tag{17.9}
$$

and so $u \to u_n$ in $L^1(0, T; L^\infty)$.

Now for each $n \in \mathbb{N}$ we consider the integral equation

$$X_n(t) = a + \int_0^t u_n(X_n(s), s)\, ds.$$

Since u_n is smooth, this is equivalent to the ordinary differential equation

$$\dot{X}_n(t) = u_n(X_n(t), t), \quad \text{with} \quad X_n(0) = a,$$

which has a unique solution on $[0, T]$ for any $a \in \mathbb{T}^3$.

For each fixed $a \in \mathbb{T}^3$, $X_n(t)$ is bounded uniformly in n and for all $t \in [0, T]$, since

$$
\begin{aligned}
|X_n(t)| &\le |a| + \int_0^t |u_n(X_n(s), s)|\, ds \\
&\le |a| + \int_0^t \|u_n(s)\|_{L^\infty}\, ds \\
&\le |a| + \int_0^t \theta(s)\, ds;
\end{aligned}
$$

in particular solutions of the ODE (17.11) cannot 'blow up'. (This short calculation is useful if we consider the trajectory X_n as a subset of $\mathbb{R}^3$ rather than the bounded set $\mathbb{T}^3$.)

Note also that for $t' > t$ we have

$$
\begin{aligned}
|X_n(t') - X_n(t)| &= \left| \int_t^{t'} u_n(X_n(s), s) \, ds \right| \\
&\le \int_t^{t'} |u_n(X_n(s), s)| \, ds \\
&\le \int_t^{t'} \|u_n(s)\|_{L^\infty} \, ds \\
&\le \int_t^{t'} \theta(s) \, ds \\
&\le C(t' - t)^{1/2},
\end{aligned} \tag{17.10}
$$

independent of n. So the (X_n) form a bounded equicontinuous family of functions. It follows using the Arzelà–Ascoli Theorem (Theorem 1.1) that there is a subsequence (which we relabel) such that $X_n \to X$ uniformly on $[0, T]$, where $X \in C^0([0, T]; \mathbb{T}^3)$.

To show that $X(\cdot)$ is a solution of the integral equation (17.8) we want to pass to the limit in

$$
X_n(t) = a + \int_0^t u_n(X_n(s), s) \, ds.
$$

To do this it suffices to show that $u_n(X_n(\cdot), \cdot)$ converges to $u(X(\cdot), \cdot)$ in $L^1(0, T; \mathbb{R}^3)$. We know from Corollary 10.2 that $u(t)$ is a continuous function of x for all $t \in (0, T)$, and so for every $t > 0$ we have

$$
\begin{aligned}
|u_n&(X_n(t), t) - u(X(t), t)| \\
&\le |u_n(X_n(t), t) - u(X_n(t), t)| + |u(X_n(t), t) - u(X(t), t)| \\
&\le \|u_n(t) - u(t)\|_{L^\infty} + |u(X_n(t), t) - u(X(t), t)|.
\end{aligned}
$$

We already know from (17.9) that $u_n \to u$ in $L^1(0, T; L^\infty)$, which deals with the first term. For the second term it follows from the fact that $u(t) \in C^0(\mathbb{T}^3)$ that $u(X_n(t), t) \to u(X(t), t)$ for almost every t, and since

$$
|u(X_n(t), t) - u(X(t), t)| \le 2\|u(t)\|_{L^\infty} \le 2\theta(t)
$$

and $\theta \in L^1(0, T)$, the required convergence follows from the Dominated Convergence Theorem.

It follows that for every $a \in \mathbb{T}^3$ there exists at least one continuous solution $X(t; a)$ of (17.8) on $[0, T]$ as claimed. $\qquad\square$

In what follows we will often use the bound

$$|X(t') - X(t)| \le \int_t^{t'} \|u(s)\|_{L^\infty} \, ds, \tag{17.11}$$

which is a byproduct of the chain of inequalities in (17.10).

17.2.3 Uniqueness of trajectories when $u_0 \in H \cap \dot{H}^{1/2}$

We now show that the solution found in the previous section is in fact unique. To do this we will require the following 'borderline' Sobolev embedding result due to Zuazua (2002); the embedding is borderline in the sense that although $u \in H^{5/2}$ implies that $u \in C^\alpha$ for any $0 < \alpha < 1$ (see Exercise 17.3), $u \in H^{5/2}$ 'just fails' to be Lipschitz ($\alpha = 1$). We delay the proof until the end of the chapter (see Lemma 17.16).

Lemma 17.4 *There exists a constant $C_b > 0$ such that if $u \in \dot{H}^{5/2}(\mathbb{T}^3)$ then*

$$|u(x) - u(y)| \le C_b \|u\|_{\dot{H}^{5/2}} |x - y| |\log |x - y||^{1/2} \tag{17.12}$$

for all $x, y \in \mathbb{T}^3$ with $|x - y| < 1/2$.

Given this result and the regularity proved in Lemma 17.1, we can now prove the uniqueness of Lagrangian trajectories when $u_0 \in H \cap \dot{H}^{1/2}$. (A very similar analysis guarantees uniqueness of trajectories given a weak solution of the two-dimensional Navier–Stokes equations, based on the regularity of $\sqrt{t}u$ proved in Exercise 17.2.)

Proposition 17.5 *Suppose that $u_0 \in H \cap \dot{H}^{1/2}$ gives rise to a solution u that satisfies $u \in L^\infty(0, T; \dot{H}^{1/2}) \cap L^2(0, T; \dot{H}^{3/2})$ for some $T > 0$. Then for every $a \in \mathbb{T}^3$ there exists a unique solution of the equation*

$$X(t) = a + \int_0^t u(X(s), s) \, ds, \tag{17.13}$$

for $t \in [0, T]$.

Proof Due to Lemma 17.1, solutions with $u_0 \in H \cap \dot{H}^{1/2}$ that belong to $L^2(0, T; \dot{H}^{3/2})$ have $\sqrt{t}\, u \in L^2(0, T; \dot{H}^{5/2})$. The result of Lemma 17.3 then implies that under the assumptions of this proposition there is at least one solution of (17.13).

Now observe that since $u(t)$ is smooth for $t > 0$, if solutions of (17.13) are unique on $[0, t^*]$ for some $t^* > 0$ then they are unique on $[0, T]$. In what follows we find a $t^* > 0$ sufficiently small such that the solutions are indeed unique on $[0, t^*]$.

To this end we suppose that X and Y are two solutions of (17.13) with the same initial condition and we set $W = X - Y$. Since $u(t)$ is smooth for

$t \in (0, T)$, it follows that W satisfies the differential equation

$$\frac{dW}{dt} = u(X(t), t) - u(Y(t), t)$$

and so, assuming that $|W| < 1/2$ and using (17.12),

$$\frac{d}{dt}|W|^2 \le C_b \|u(t)\|_{\dot{H}^{5/2}} |W|^2 (-\log |W|)^{1/2}.$$

Therefore, provided that $W \ne 0$,

$$\frac{2}{(-\log |W|)^{1/2}|W|} \frac{d}{dt}|W| \le C_b \|u(t)\|_{\dot{H}^{5/2}}.$$

Integrating both sides between $s > 0$ and $t > s$ we obtain

$$\left[-(-\log |W|)^{1/2}\right]_s^t \le C_b \int_s^t \|u(r)\|_{\dot{H}^{5/2}} \, dr,$$

yielding

$$-(-\log |W(t)|)^{1/2} \le -(-\log |W(s)|)^{1/2} + C_b \int_s^t \|u(r)\|_{\dot{H}^{5/2}} \, dr. \quad (17.14)$$

If we knew that $u \in L^1(0, T; \dot{H}^{5/2})$ then we could prove uniqueness by letting $s \to 0$ in this inequality, since

$$|W(s)| = \left| \int_0^s u(X(r), r) - u(Y(r), r) \, dr \right| \le 2 \int_0^s \|u(r)\|_{L^\infty} \, dr, \quad (17.15)$$

which tends to zero as $s \to 0$ since $u \in L^1(0, T; L^\infty)$. However, we do not have sufficient regularity of u to argue so directly. Instead we show that

$$\lim_{s \to 0^+} \left[(-\log |W(s)|)^{1/2} - C_b \int_s^t \|u(r)\|_{\dot{H}^{5/2}} \, dr \right] = +\infty.$$

To do this we find upper bounds on $|W(s)|$ and on $\int_s^t \|u(r)\|_{\dot{H}^{5/2}} \, dr$.

For $|W(s)|$ we have the crude (but sufficient) upper bound from (17.15)

$$|W(s)| \le 2 \int_0^s \|u(t)\|_{L^\infty} \, dt \le K s^{1/2},$$

for some constant $K > 0$ (which depends on u) using the result of Corollary 17.2. It follows that

$$(-\log |W(s)|)^{1/2} \ge \left(-\log K - \frac{1}{2} \log s \right)^{1/2} \ge \frac{1}{4}(-\log s)^{1/2}$$

for all s in the range $0 < s \le s_0$ for some suitably small $s_0 < 1$.

To obtain an upper bound on the integral term we use Hölder's inequality and the fact that $\sqrt{t}\, u \in L^2(0, T; \dot{H}^{5/2})$ to write

$$
\int_s^t \|u(r)\|_{\dot{H}^{5/2}}\, dr = \int_s^t r^{-1/2} r^{1/2} \|u(r)\|_{\dot{H}^{5/2}}\, dr
$$

$$
\leq \left(\int_s^t r^{-1}\, dr \right)^{1/2} \left(\int_s^t r \|u(r)\|_{\dot{H}^{5/2}}^2\, dr \right)^{1/2}
$$

$$
= (\log t - \log s)^{1/2} \left(\int_0^t r \|u(r)\|_{\dot{H}^{5/2}}^2\, dr \right)^{1/2}.
$$

Using these estimates in (17.14) it follows that for all $0 < s \leq s_0$

$$
-(-\log |W(t)|)^{1/2}
$$

$$
\leq -\frac{1}{4}(-\log s)^{1/2} + C_b(\log t - \log s)^{1/2} \left(\int_0^t r \|u(r)\|_{\dot{H}^{5/2}}^2\, dr \right)^{1/2}.
$$

Since $r\|u(r)\|_{\dot{H}^{5/2}}^2$ is integrable on $(0, T)$, it is possible to choose $t^* > 0$ with $t^* \leq \min(s_0, 1)$ such that

$$
C_b \left(\int_0^t r \|u(r)\|_{\dot{H}^{5/2}}^2\, dr \right)^{1/2} \leq \frac{1}{8} \qquad \text{for all} \quad t \leq t^*.
$$

It follows that for all $t \leq t^*$

$$
-(-\log |W(t)|)^{1/2} \leq -\frac{1}{4}(-\log s)^{1/2} + \frac{1}{8}(\log t - \log s)^{1/2},
$$

and so for $t \leq t^*$ we have

$$
-\log |W(t)| \geq \left[\frac{1}{4}(-\log s)^{1/2} - \frac{1}{8}(\log t - \log s)^{1/2} \right]^2
$$

$$
\geq \left[\frac{1}{8}(-\log s)^{1/2} \right]^2 = \frac{1}{64}(-\log s),
$$

since $(\log t - \log s)^{1/2} \leq (-\log s)^{1/2}$ for all $0 < s \leq t \leq 1$. Therefore

$$
\log |W(t)| \leq \frac{1}{64} \log s
$$

and so $|W(t)| \leq s^{1/64}$. Now letting $s \to 0^+$ shows that $|W(t)| = 0$, which guarantees the uniqueness of solutions on $[0, t^*]$ and hence on $[0, T]$. $\square$

17.3 Existence of a Lagrangian flow map for weak solutions

If we want to consider the problem of the existence of Lagrangian trajectories for weak solutions then, as in the previous section, we reformulate the problem as the integral equation

$$X(t) = a + \int_0^t u(X(s), s)\, ds. \tag{17.16}$$

The key observation that leads us to hope that we can still find solutions of (17.16) even when u is only a weak solution, is that for such solutions we have $u \in L^1(0, T; L^\infty)$. We proved this at the end of Chapter 8, but since the steps in the proof as well as the end result will be needed in what follows we recall the short argument here.

We proved in Lemma 8.15, using the epochs of regularity, that any weak solution u is an element of $L^{2/3}(0, T; H^2)$. Thus, using Agmon's inequality

$$\|u\|_{L^\infty} \le c\|u\|_{\dot{H}^1}^{1/2}\|u\|_{\dot{H}^2}^{1/2} \tag{17.17}$$

(Theorem 1.20), it follows (via Hölder's inequality) that

$$\int_0^T \|u(s)\|_{L^\infty}\, ds \le c \left(\int_0^T \|u(s)\|_{\dot{H}^1}^2\, ds \right)^{1/4} \left(\int_0^T \|u(s)\|_{\dot{H}^2}^{2/3}\, ds \right)^{3/4},$$

i.e. that $u \in L^1(0, T; L^\infty)$.

The existence of at least one solution of (17.4) for each $a \in \mathbb{T}^3$ can be obtained by a slightly more elaborate version of the approximation argument we have already used in the previous section. However, our aim in this section is more ambitious. Following Foias, Guillopé, & Temam (1985) we will construct a solution mapping (see the statement of the following theorem for the definition) that combines these individual solutions in such a way as to produce a flow that is volume preserving.

In the statement of the theorem and what follows we use the notation $W^{1,1}(0, T; \mathbb{T}^3)$ to denote the collection of all functions $f\colon (0, T) \to \mathbb{T}^3$ such that $f \in L^1(0, T; \mathbb{T}^3)$ and $df/dt \in L^1(0, T; \mathbb{R}^3)$.

Theorem 17.6 *For any Leray–Hopf weak solution u there exists at least one 'solution mapping' $\Phi\colon \mathbb{T}^3 \times [0, \infty) \to \mathbb{T}^3$ such that*

(i) *for each $a \in \mathbb{T}^3$, $X_a(\cdot) := \Phi(a, \cdot)$ is absolutely continuous (i.e. an element of $W^{1,1}(0, T; \mathbb{T}^3)$ for each $T > 0$) and satisfies*

$$X_a(t) = a + \int_0^t u(X_a(s), s)\, ds \tag{17.18}$$

for all $t > 0$;

(ii) *the mapping $a \mapsto \Phi(a, \cdot)$ belongs to $L^\infty(\mathbb{T}^3; C([0, T], \mathbb{T}^3))$ for all $T > 0$; and*

(iii) *$\Phi(\cdot, t)$ is measurable and volume preserving: for any $t > 0$ and any Borel set $B \subset \mathbb{T}^3$*

$$\mu[\Phi(\cdot, t)^{-1}(B)] = \mu(B), \tag{17.19}$$

where μ denotes the three-dimensional Lebesgue measure.

We refer the mapping Φ as 'a solution mapping corresponding to u'.

We prove this theorem in a number of steps. First we show in Proposition 17.7 that for every $a \in \mathbb{T}^3$ there is at least one solution of (17.18); the argument is very similar to that used already to prove Lemma 17.3 (existence of trajectories when $u_0 \in H \cap \dot{H}^{1/2}$).

By choosing one of these possible solutions for each $a \in \mathbb{T}^3$ we would obtain a solution mapping Φ with the properties required by (i) in Theorem 17.6. But this would be a fairly weak result, since the fact that the solution map Φ is 'volume-preserving' is a key property of the corresponding fluid flow.

In order even to consider whether or not the volume-preserving property holds, the flow mapping Φ must be measurable (this is essentially encoded in property (ii) in the statement of the theorem). To achieve this we will appeal to an abstract 'measurable selection theorem' for multi-valued maps: if $\Lambda \colon X \to Y$ is multi valued (i.e. for every $x \in X$, $\Lambda(x)$ is a subset of Y, not necessarily a point) then, under certain measurability and compactness conditions on Λ, it is possible to choose a measurable single-valued map $\Phi \colon X \to Y$ such that $\Phi(x) \in \Lambda(x)$ for every $x \in X$ (the result we will use is Corollary E.3 from Appendix E).

Finally, we show that the resulting measurable solution mapping Φ is volume preserving in the sense of (iii). We do this by using the fact that u has 'epochs of regularity' on which it is smooth. On these epochs of regularity Φ is volume preserving (as in Section 17.1), and using arguments based on absolute continuity this is enough to show that Φ is volume preserving for all $t \in [0, T]$ (Proposition 17.10).

17.3.1 Existence of a solution for each initial condition

First we show that (17.18) has at least one solution for each $a \in \Omega$.

Proposition 17.7 *If u is divergence free and*

$$u \in L^2(0, T; H^1) \cap L^{2/3}(0, T; H^2)$$

then for every $a \in \mathbb{T}^3$ there is at least one solution $X_a \in W^{1,1}(0, T; \mathbb{T}^3)$ of the integral equation

$$X(t) = a + \int_0^t u(X(s), s) \, ds, \qquad (17.20)$$

and there is a non-negative function $\theta \in L^1(0, T)$ such that for every choice of $a \in \mathbb{T}^3$ all such solutions satisfy

$$|\dot{X}_a(t)| \le \theta(t)$$

for almost every $t \in (0, T)$.

Proof The proof of the existence of solutions of (17.20) given a weak solution follows almost exactly the proof in Lemma 17.3, except that instead of (17.7) we use the inequality

$$\|u(t)\|_{L^\infty} \le \theta(t) := c\|u(t)\|_{\dot{H}^1}^{1/2} \|u(t)\|_{\dot{H}^2}^{1/2}, \qquad (17.21)$$

see (17.17). The remainder of the argument is almost identical, except that given (17.10),

$$|X_n(t) - X_n(t')| \le \int_t^{t'} \theta(s) \, ds, \qquad (17.22)$$

we now use the fact that $\theta \in L^1(0, T)$ to deduce that $\int_0^t \theta(s) \, ds$ is absolutely continuous, which in turn implies that the (X_n) are equicontinuous (see Corollary A.6). From (17.22) the limiting function X satisfies

$$|X(t) - X(t')| \le \int_t^{t'} \theta(s) \, ds, \qquad \text{which implies that} \qquad |\dot{X}(t)| \le \theta(t).$$

We now use the fact that u is continuous for almost every t, since we know that for almost every t we have $u(t) \in \dot{H}^2(\mathbb{T}^3) \subset C^0(\mathbb{T}^3)$. □

17.3.2 Construction of a measurable flow map Φ

We now have to show that it is possible to put together a measurable flow $\Phi(\cdot, \cdot)$ by choosing one of the (potentially many) possible solutions of the equation $\dot{X} = u(X, t)$ for each initial condition $a \in \mathbb{T}^3$. In order to do this we need to appeal to a selection theorem due to Castaing (1967), which guarantees that we can make such a choice in a measurable way given certain measurability properties of our initial collection.

First we state the abstract theorem, whose proof can be found in Appendix E, and then we show how it applies in our particular situation.

Theorem 17.8 *Let $\mathcal{X}$ and $\mathcal{Y}$ be two separable metric spaces, where $\mathcal{X}$ is equipped with a σ-algebra of measurable sets that contains all closed sets. Suppose that Λ is a measurable mapping from $\mathcal{X}$ into the set of non-empty compact subsets of $\mathcal{Y}$ whose graph is a closed subset of $\mathcal{X} \times \mathcal{Y}$. Then there exists a measurable single-valued mapping $\Phi \colon \mathcal{X} \to \mathcal{Y}$ such that $\Phi(x) \in \Lambda(x)$ for every $x \in \mathcal{X}$.*

For a particular choice of initial condition, $X(0) = a$, let us denote by $\Lambda(a)$ the collection of all possible solutions of

$$\dot{X} = u(X, t) \qquad \text{with} \qquad X(0) = a$$

that satisfy $|\dot{X}(t)| \le \theta(t)$ for almost every $t \in [0, T]$, where θ is defined in (17.21). Then Λ is a mapping from $\mathcal{X} = \mathbb{T}^3$ into the metric space $\mathcal{Y}$, where

$$\mathcal{Y} = \{X \in W^{1,1}(0, T; \mathbb{T}^3) : |\dot{X}(t)| \le \theta(t) \text{ a.e. } t \in (0, T)\}.$$

Note that $\mathcal{Y}$ is a closed subspace of the Banach space $W^{1,1}(0, T; \mathbb{T}^3)$ (this follows from the Dominated Convergence Theorem); thus $\mathcal{Y}$ is a complete separable metric space when equipped with the (metric derived from the) $W^{1,1}$ norm.

A simpler version of the convergence argument used in the proof of Proposition 17.7 shows that for each $a \in \mathbb{T}^3$ the set $\Lambda(a)$ is a compact subset of $\mathcal{Y}$. Indeed, if $X_n(\cdot) \in \mathcal{Y}$ and

$$X_n(t) = a + \int_0^t u(X_n(s), s) \, \mathrm{d}s$$

then (17.22) shows that

$$|X_n(t) - X_n(t')| \le \int_t^{t'} \theta(s) \, \mathrm{d}s$$

and as $\theta \in L^1(0, T)$ the functions $X_n(\cdot)$ are equicontinuous. Since they are also bounded we can use the Arzelà–Ascoli Theorem (Theorem 1.1) to conclude that there is a subsequence that converges uniformly on the interval $[0, T]$ to some $X \in C^0([0, T]; \mathbb{T}^3)$.

As before, we can now argue that $u(\cdot, t)$ is continuous for almost every t, and so

$$u(X_n(t), t) \to u(X(t), t) \qquad \text{for almost every} \quad t \in [0, T].$$

This is enough to show that

(i) $X_n \to X$ in $W^{1,1}$ with $|\dot{X}(t)| \le \theta(t)$ and that

(ii) $X(t) = a + \int_0^t u(X(s), s) \, \mathrm{d}s$,

from which it is clear that $X \in \Lambda(a)$ (from (ii)) and that the required compactness holds (from (i)).

A very similar argument shows that the graph of Λ,

$$\mathcal{G}(\Lambda) = \{(a, \Lambda(a)) : a \in \mathbb{T}^3\} \subset \mathbb{T}^3 \times \mathcal{Y},$$

is closed. Suppose that $(a_n, X_n(\cdot)) \in \mathcal{G}(\Lambda)$, i.e. that $X_n(\cdot)$ is a solution of

$$X_n(t) = a_n + \int_0^t u(X_n(s), s)\, ds.$$

Suppose also that $a_n \to a$ in $\mathbb{T}^3$ and $X_n \to X$ in $\mathcal{Y}$. Then we can repeat the latter parts of the convergence argument of Proposition 17.7 to show that $X(\cdot)$ satisfies

$$X(t) = a + \int_0^t u(X(s), s)\, ds.$$

This says exactly that $(a, X) \in \mathcal{G}(\Lambda)$, and hence the graph of Λ is closed.

It now follows using Theorem 17.8 that there exists a measurable single-valued map $\Psi : \mathbb{T}^3 \to \mathcal{Y}$ such that for each choice of $a \in \mathbb{T}^3$ the function $X(t) := \Psi(a)(t)$ satisfies

$$X(t) = a + \int_0^t u(X(s), s)\, ds.$$

We now define our flow map $\Phi : \mathbb{T}^3 \times (0, T) \to \mathbb{T}^3$ by setting

$$\Phi(a, t) = \Psi(a)(t);$$

clearly Φ satisfies properties (i) and (ii) in Theorem 17.6.

Finally we need to check property (iii), i.e. that $\Phi(\cdot, t) : \mathbb{T}^3 \to \mathbb{T}^3$ is measurable for each $t \in (0, T)$. To do this, fix $t \in (0, T)$ and let K be a closed subset of $\mathbb{T}^3$; we will show that the subset of $W^{1,1}(0, T; \mathbb{T}^3)$ given by

$$K_t := \{\Phi(a, \cdot) : \Phi(a, t) \in K\}$$

is also closed, and hence Lebesgue measurable. It then follows that

$$\Phi(\cdot, t)^{-1}(K) = \Psi^{-1}(K_t)$$

is measurable, and so $\Phi(\cdot, t)$ is measurable as required.

In order to show that K_t is closed, observe that if $y(\cdot) \in K_t$ then for every $\tau \in (0, T)$ we have

$$y(\tau) = y(t) + \int_t^\tau u(y(s), s)\, ds.$$

In particular, if $y_n \in K_t$ and $y_n \to y$ then we can follow the argument that we used above to show that the graph of Λ is closed; this implies that $y \in K_t$ and hence K_t is closed as claimed.

17.3.3 The solution mapping Φ is volume preserving

It remains only to show that this mapping Φ is volume preserving. Note that if $\Psi : \mathbb{T}^3 \to \mathbb{T}^3$ is smooth and volume preserving then for any $g \in C^\infty(\mathbb{T}^3)$ the equality

$$\int_{\mathbb{T}^3} g(\Psi(a))\, da = \int_{\mathbb{T}^3} g(a)\, da \qquad (17.23)$$

holds for every $t \in [0, T]$ (change variables and then use the fact that $|\det \nabla \Psi| = 1$). When Ψ is not a smooth mapping, we can use (17.23) as an alternative definition of what it means to be volume preserving, as shown in the following lemma.

Lemma 17.9 *If* $\Psi \colon \mathbb{T}^3 \to \mathbb{T}^3$ *is a measurable mapping such that*

$$\int_{\mathbb{T}^3} g(\Psi(x))\, dx = \int_{\mathbb{T}^3} g(a)\, da \qquad for\ all \quad g \in C^\infty(\mathbb{T}^3) \qquad (17.24)$$

then for any Borel set $A \subset \mathbb{T}^3$

$$\mu(\Psi^{-1}(A)) = \mu(A).$$

Proof We can extend (17.24) to every bounded Borel function on $\mathbb{T}^3$ by continuity. Then taking g to be the characteristic function of A we obtain

$$\mu(A) = \int_A 1\, da = \int_{\mathbb{T}^3} \chi_A(a)\, da$$

$$= \int_{\mathbb{T}^3} \chi_A(\Psi(x))\, dx = \int_{\Psi^{-1}(A)} 1\, da = \mu(\Psi^{-1}(A)),$$

as required. $\qquad \square$

Proposition 17.10 *The solution mapping* Φ *obtained via the measurable selection theorem of the previous section is volume preserving, i.e. for any Borel set* $B \subset \mathbb{T}^3$ *we have*

$$\mu[\Phi(\cdot, t)^{-1}(B)] = \mu(B) \qquad for\ any\ t \geq 0,$$

where μ *denotes the Lebesgue measure on* $\mathbb{R}^3$.

Proof Given any $g \in C^\infty(\mathbb{T}^3)$, define

$$J(t) = \int_{\mathbb{T}^3} g(\Phi(a, t))\, da.$$

We will show that J is constant by showing that J is absolutely continuous with time derivative $\dot{J} = 0$ almost everywhere (see Theorem A.7).

To show that J is absolutely continuous, we show that for any collection (t_i, s_i) of disjoint intervals in $[0, T]$ of total measure δ,

$$\sum_i |J(s_i) - J(t_i)| \to 0 \qquad \text{as} \qquad \delta \to 0$$

(see Definition A.3).

First, observe that for $t' > t$ we have

$$|J(t') - J(t)| = \left| \int_{\mathbb{T}^3} g(\Phi(a, t')) - g(\Phi(a, t)) \, da \right|$$

$$\leq \|\nabla g\|_{L^\infty} \int_{\mathbb{T}^3} |\Phi(a, t') - \Phi(a, t)| \, da$$

$$\leq \|\nabla g\|_{L^\infty} \int_{\mathbb{T}^3} \int_t^{t'} |u(\Phi(a, s), s)| \, da \, ds$$

$$\leq \|\nabla g\|_{L^\infty} \int_{\mathbb{T}^3} \int_t^{t'} \|u(s)\|_{L^\infty} \, da \, ds$$

$$\leq |\mathbb{T}^3| \|\nabla g\|_{L^\infty} \int_t^{t'} \theta(s) \, ds.$$

Therefore for any collection (t_i, s_i) of disjoint intervals in $[0, T]$ of total measure δ we have

$$\sum_i |J(s_i) - J(t_i)| \leq c \|\nabla g\|_{L^\infty} \sum_i \int_{t_i}^{s_i} \theta(s) \, ds,$$

which tends to zero as δ tends to zero since $\theta \in L^1(0, T)$ and hence J is absolutely continuous as claimed.

On each of the intervals on which u is regular, J is differentiable and $\dot{J} = 0$. Hence $\dot{J} = 0$ almost everywhere, and then, since J is absolutely continuous, J is constant, which proves (17.23). $\qquad\square$

17.4 Lagrangian a.e. uniqueness for suitable weak solutions

The result of the previous section gives solutions for any $a \in \mathbb{T}^3$, but says nothing about their uniqueness. However, if we require u to be not only a weak solution but a suitable weak solution, i.e. one for which the partial regularity result of Chapter 15 is valid, we can show that the Lagrangian trajectories are unique for almost every $a \in \mathbb{T}^3$. (Note that we prove the a.e.-uniqueness of the trajectories given a particular choice of suitable weak solution; nothing is currently known about the uniqueness (or otherwise) of suitable weak solutions themselves.)

More precisely, we begin with $u_0 \in H \cap \dot{H}^{1/2}$, so that the solution is smooth for a short time interval, during which the (existence and) uniqueness of trajectories is ensured by Proposition 17.5. If the solution loses regularity it can be continued as a suitable weak solution, whose space–time singular set S is sufficiently small that $\dim_B(S) \leq 5/3$, due to the partial regularity result of Chapter 15. The combination of the $L^1(0, T; L^\infty)$ integrability of the vector field u and the bound on the box-counting dimension of the singular set is enough to show that almost all trajectories avoid S, which in turn guarantees their uniqueness.

We use a general result guaranteeing that volume-preserving flows in $\mathbb{R}^n$ that arise from vector fields in $L^1(0, T; L^\infty)$ avoid sets with box-counting dimension less than $n - 1$. This is a consequence of the following simple observation relating the volume of neighbourhoods of a subset of $\mathbb{R}^n$ to its box-counting dimension.[2]

Lemma 17.11 *Let K be a bounded subset of $\mathbb{R}^n$ and denote by $O(K, \varepsilon)$ the ε-neighbourhood of K, i.e.*

$$O(K, \varepsilon) = \{z \in \mathbb{R}^n : z = x + y, \ x \in K, \ |y| < \varepsilon\}.$$

Then for any $d > \dim_B(K)$ and any $\varepsilon_0 > 0$ there exists a constant C_d such that

$$\mu(O(K, \varepsilon)) \leq C_d \varepsilon^{n-d} \qquad \text{for all} \quad 0 < \varepsilon < \varepsilon_0.$$

Proof For any $d > \dim_B(K)$ and $\varepsilon_0 > 0$ there exists a $C_d' > 0$ such that for any $0 < \varepsilon < \varepsilon_0$ the set K can be covered by $N_\varepsilon \leq C_d' \varepsilon^{-d}$ balls of radius ε,

$$K \subset \bigcup_{i=1}^{N_\varepsilon} B_\varepsilon(x_i).$$

It follows that for any such ε,

$$O(K, \varepsilon) \subset \bigcup_{i=1}^{N_\varepsilon} B_{2\varepsilon}(x_i)$$

and so

$$\mu(O(K, \varepsilon)) \leq N_\varepsilon V_n (2\varepsilon)^n \leq 2^n V_n C_d' \varepsilon^{n-d} =: C_d \varepsilon^{n-d}, \qquad (17.25)$$

where V_n is the volume of the unit ball in $\mathbb{R}^n$. $\qquad \square$

---

[2] This can be used to give another definition of the box-counting dimension, the 'Minkowski' definition, see Exercise 17.6.

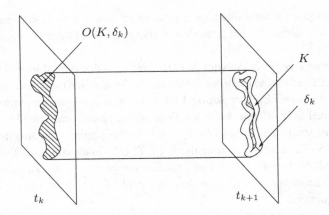

$O(K, \delta_k)$

K

δ_k

t_k t_{k+1}

Figure 17.2. Avoiding the set K: only trajectories that start in $O(K, \delta_k)$ can enter K on the time interval $[t_k, t_{k+1}]$.

Proposition 17.12 *Take $\Omega \subset \mathbb{R}^n$ and let $\Phi \colon \Omega \times [0, \infty) \to \Omega$ be a volume-preserving solution mapping (as in Theorem 17.6) corresponding to a vector field u that satisfies $u \in L^1(0, T; L^\infty(\Omega))$ for every $T > 0$. If K is a compact subset of Ω with $\dim_B(K) < n - 1$ then*

$$\Phi(t, a) \notin K \qquad \text{for all} \qquad t \geq 0$$

for almost every initial condition $a \in \Omega$.

Proof For a more compact notation we write $X_a(t) := \Phi(a, t)$. Fix $T > 0$, choose $N \in \mathbb{N}$, and partition the time interval $[0, T]$ by setting $t_j = jT/N$. We will consider the problem of avoiding K on the time interval $[t_k, t_{k+1}]$ for some $k \in \{0, \ldots, N - 1\}$. Since

$$X_a(t) - X_a(s) = \int_s^t u(X_a(r), r)\,dr,$$

it follows that for all $t \in [t_k, t_{k+1}]$,

$$|X_a(t) - X_a(t_k)| \leq \delta_k := \int_{t_k}^{t_{k+1}} \|u(t)\|_{L^\infty}\,dt. \qquad (17.26)$$

So if $X_a(t) \in K$ for some $t \in [t_k, t_{k+1}]$, we must have

$$X_a(t_k) \in O(K, \delta_k),$$

see Figure 17.2.

It follows from (17.25) and the fact that $\dim_B(K) < n - 1$ that

$$\mu(O(K, \delta_k)) \leq c_n \delta_k^r$$

for some $r > 1$. Since Φ is volume preserving (see (17.19)) the set of initial conditions for which $X_a(t_k) \in O(K, \delta_k)$ has measure bounded by $c_n \delta_k^r$,

$$\mu\{a \in \mathbb{T}^3 : X_a(t_k) \in O(K, \delta_k)\} \leq c_n \delta_k^r.$$

Now let

$$\Omega_K = \{a \in \mathbb{T}^3 : X_a(t) \in K \text{ for some } t \in [0, T]\}.$$

Then for any choice of N

$$\mu(\Omega_K) \leq \sum_{k=0}^{N-1} \mu\{a \in \mathbb{T}^3 : X_a(t) \in K \text{ for some } t \in [t_k, t_{k+1}]\}$$

$$\leq \sum_{k=0}^{N-1} \mu\{a \in \mathbb{T}^3 : X_a(t_k) \in O(K, \delta_k)\}$$

$$\leq c_n \sum_{k=0}^{N-1} \delta_k^r = c_n \sum_{k=0}^{N-1} \left(\int_{t_k}^{t_{k+1}} \|u(s)\|_{L^\infty} \, ds \right)^r, \tag{17.27}$$

recalling (17.26).

Since $u \in L^1(0, T; L^\infty)$ the integral of $\|u(\cdot)\|_{L^\infty}$ is absolutely continuous. Thus (see Theorem A.5) given an $\varepsilon > 0$, there exists an $N \in \mathbb{N}$ such that for each $k \in \{0, \ldots, N - 1\}$,

$$\int_{t_k}^{t_{k+1}} \|u(s)\|_{L^\infty} \, ds < \varepsilon.$$

Using this in (17.27) it follows that

$$\mu(\Omega_K) \leq c_n \sum_{k=0}^{N-1} \left(\int_{t_k}^{t_{k+1}} \|u(s)\|_{L^\infty} \, ds \right)^{r-1} \left(\int_{t_k}^{t_{k+1}} \|u(s)\|_{L^\infty} \, ds \right)$$

$$\leq c_n \varepsilon^{r-1} \sum_{k=0}^{N-1} \int_{t_k}^{t_{k+1}} \|u(s)\|_{L^\infty} \, ds$$

$$= c_n \varepsilon^{r-1} \int_0^T \|u(s)\|_\infty \, ds.$$

Since $r > 1$ and both ε and T are arbitrary this completes the proof. $\qquad\square$

It is now easy to show that almost every Lagrangian trajectory avoids the singular set.

Theorem 17.13 *If u is a Leray–Hopf suitable weak solution and Φ is a corresponding solution mapping then for almost every $a \in \mathbb{T}^3$, $\Phi(a, t) \notin S$ for all $t \geq 0$.*

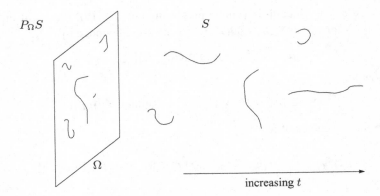

Figure 17.3. The singular set S and its projection onto Ω.

Proof   Fix $T > 0$ and let K_n be a sequence of compact subsets of $\mathbb{T}^3 \times (0, T)$ such that $\mathbb{T}^3 \times (0, T) = \bigcup_{n=1}^{\infty} K_n$, e.g. $K_n = \mathbb{T}^3 \times [\frac{1}{n}, T - \frac{1}{n}]$.

We know from Theorem 15.8 that $\dim_B(S \cap K_n) \leq 5/3$. Denote by P_Ω the projection from $\Omega \times (0, T)$ onto Ω, i.e.

$$P_\Omega(x, t) = x,$$

see Figure 17.3. Clearly any cover of $S \cap K_n$ by N balls in $\mathbb{R}^4$ of radius ε provides a cover of $P_\Omega(S \cap K_n)$ by N balls in $\mathbb{R}^3$ of radius ε, and so

$$\dim_B(P_\Omega(S \cap K_n)) \leq \dim_B(S \cap K_n) \leq 5/3.$$

It now follows from Proposition 17.12 that almost every trajectory avoids $P(S \cap K_n)$ for all $t \geq 0$, and so, since by definition

$$S \cap K_n \subseteq P_\Omega(S \cap K_n) \times (0, T),$$

almost every trajectory avoids the set $S \cap K_n$. Since the countable intersection of sets of full measure still has full measure, almost every trajectory avoids the whole set of singularities S for all $t \geq 0$. $\square$

Uniqueness almost everywhere now follows using the local regularity results from Chapters 13 and 15.

Corollary 17.14   *If u is a Leray–Hopf suitable weak solution arising from the initial condition $u_0 \in H \cap \dot{H}^{1/2}(\mathbb{T}^3)$ then the equation*

$$\dot{X} = u(X, t), \qquad X(0) = a$$

has a unique solution for almost every $a \in \mathbb{T}^3$, i.e. gives rise to a unique particle trajectory $X(t; a)$, which is a C^1 function of time. In particular, the solution mapping Φ corresponding to u is uniquely defined for almost every $a \in \mathbb{T}^3$.

Proof   Since $u_0 \in H \cap \dot{H}^{1/2}$ it follows (using Corollary 10.2) that

$$u \in L^\infty(0, T; \dot{H}^{1/2}) \cap L^2(0, T; \dot{H}^{3/2})$$

for some $T > 0$, and hence Proposition 17.5 guarantees that trajectories are unique on $[0, T]$. So we need only be concerned with non-uniqueness that might arise for some $t > 0$. Let $X_a(t)$ be a trajectory that avoids the singular set for all $t \geq 0$, and suppose that t is the first time at which the trajectory splits, i.e. that there are two distinct trajectories that emanate from the space–time point $(X_a(t), t)$. Since S is compact there is a neighbourhood of $(X_a(t), t)$ that does not intersect S and within which u is Hölder continuous, see Figure 17.4.

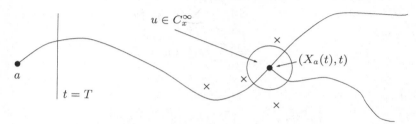

Figure 17.4. A trajectory that avoids the singular set (here marked by crosses) cannot become non-unique at a space–time point $(X_a(t), t)$. Also marked is the first blow-up time $t = T$.

The regularity result of Theorem 13.4 now guarantees that u is uniformly Lipschitz (in fact C^∞) in the spatial variables in some neighbourhood of $(X_a(t), t)$, and hence the solution of $\dot{X} = u(X, t)$ is unique at $(X_a(t), t)$, a contradiction. Since u is also Hölder continuous in t on the complement of S, it follows that X is a C^1 function of time. $\square$

17.5 Proof of the inequality (17.5)

We now prove the inequality for the nonlinear term that we needed for Lemma 17.1. The elegant argument is due to Montero (2015).

Lemma 17.15   *If $u \in \dot{H}^{5/2}(\mathbb{T}^3)$ with $\nabla \cdot u = 0$ then*

$$|\langle \Lambda^{3/2}[(u \cdot \nabla)u], \Lambda^{3/2}u \rangle| \leq C\|u\|_{\dot{H}^{3/2}}^2 \|u\|_{\dot{H}^{5/2}}, \tag{17.28}$$

where powers of Λ are defined in (1.18).

Proof   The first step is to show that

$$|\langle \Lambda^{3/2}(u \cdot \nabla)u, \Lambda^{3/2}u \rangle| \le C \left(\sum_{k \in \mathbb{Z}^3} |k|^{1/2}|\hat{u}_k| \right) \|u\|_{\dot{H}^{3/2}} \|u\|_{\dot{H}^2}, \qquad (17.29)$$

for which we use Montero's variant of the argument in Robinson, Sadowski, & Silva (2012), see also Constantin & Foias (1988).

First we note that

$$\langle \Lambda^{3/2}[(u \cdot \nabla)u], \Lambda^{3/2}u \rangle = \langle \Lambda^{3/2}[(u \cdot \nabla)u] - (u \cdot \nabla)\Lambda^{3/2}u, \Lambda^{3/2}u \rangle, \quad (17.30)$$

as $\langle (u \cdot \nabla)\Lambda^{3/2}u, \Lambda^{3/2}u \rangle = 0$. Since

$$\begin{aligned}
\Lambda^{3/2}[(u \cdot \nabla)u] &= i\Lambda^{3/2} \sum_{j,\ell} (\hat{u}_\ell \cdot j)\hat{u}_j e^{i(j+\ell)\cdot x} \\
&= i\Lambda^{3/2} \sum_{j,k} (\hat{u}_{k-j} \cdot j)\hat{u}_j e^{ik\cdot x} \\
&= i \sum_{j,k} |k|^{3/2}(\hat{u}_{k-j} \cdot j)\hat{u}_j e^{ik\cdot x}
\end{aligned}$$

and

$$(u \cdot \nabla)\Lambda^{3/2}u = i \sum_{j,k} |j|^{3/2}(\hat{u}_{k-j} \cdot j)\hat{u}_j e^{ik\cdot x}$$

it follows from (17.30) that

$$\langle \Lambda^{3/2}[(u \cdot \nabla)u], \Lambda^{3/2}u \rangle = (2\pi)^3 i \sum_{j,k} (|k|^{3/2} - |j|^{3/2})(\hat{u}_{k-j} \cdot j)(\hat{u}_j \cdot \overline{\hat{u}_k})|k|^{3/2}.$$

Therefore

$$\begin{aligned}
|\langle \Lambda^{3/2}[(u \cdot \nabla)u], \Lambda^{3/2}u \rangle| &\le (2\pi)^3 \sum_{j,k} \left| |k|^{3/2} - |j|^{3/2} \right| |\hat{u}_{k-j}||j||\hat{u}_j||\hat{u}_k||k|^{3/2} \\
&\le c \sum_{j,k} |k|^{3/2}|j||\hat{u}_{k-j}||\hat{u}_j||\hat{u}_k||k-j|(|k-j|^{1/2} + |j|^{1/2}) \\
&= c \sum_{j,k} |k|^{3/2}|\hat{u}_{k-j}||\hat{u}_j||\hat{u}_k|(|j||k-j|^{3/2} + |j|^{3/2}|k-j|) \\
&= 2c \sum_{j,k} |k|^{3/2}|j||k-j|^{3/2}|\hat{u}_{k-j}||\hat{u}_j||\hat{u}_k| \\
&= 2c \sum_{j} \left(|j|^{1/2}|\hat{u}_j| \sum_{k} |j|^{1/2}|k|^{3/2}|k-j|^{3/2}|\hat{u}_k||\hat{u}_{k-j}| \right),
\end{aligned}$$

using the result of Exercise 17.7 between the first and second lines. Now observe that since

$$|j|^{1/2} \le (|j - k| + |k|)^{1/2} \le |j - k|^{1/2} + |k|^{1/2}$$

it follows that for every fixed j

$$\sum_k |j|^{1/2}|k|^{3/2}|k - j|^{3/2}|\hat{u}_k||\hat{u}_{k-j}|$$

$$\le \sum_k (|k|^2|k - j|^{3/2} + |k|^{3/2}|k - j|^2)||\hat{u}_k||\hat{u}_{k-j}|$$

$$\le \left(\sum_k |k|^4|\hat{u}_k|^2\right)^{1/2} \left(\sum_k |k - j|^3|\hat{u}_{k-j}|^2\right)^{1/2}$$

$$+ \left(\sum_k |k|^3|\hat{u}_k|^2\right)^{1/2} \left(\sum_k |k - j|^4|\hat{u}_{k-j}|^2\right)^{1/2}$$

$$= 2\|u\|_{\dot{H}^{3/2}}\|u\|_{\dot{H}^2}$$

and so

$$|\langle \Lambda^{3/2}(u \cdot \nabla)u, \Lambda^{3/2}u\rangle| \le 4C\|u\|_{\dot{H}^{3/2}}\|u\|_{\dot{H}^2} \sum_j |j|^{1/2}|\hat{u}_j|,$$

which is (17.29).

In order to obtain (17.28) we estimate the sum from (17.29) in as follows:[3]

$$\sum_k |k|^{1/2}|\hat{u}_k| = \sum_k |k|^{1/2}|\hat{u}_k|\frac{\sqrt{a|k|^2 + b|k|^4}}{\sqrt{a|k|^2 + b|k|^4}}$$

$$\le \left(\sum_k (a|k|^3 + b|k|^5)|\hat{u}_k|^2\right)^{1/2} \left(\sum_{k \ne 0} \frac{1}{a|k|^2 + b|k|^4}\right)^{1/2}$$

$$= \left(\sum_{k \ne 0} \frac{1}{a|k|^2 + b|k|^4}\right)^{1/2} \left(a\|u\|_{\dot{H}^{3/2}}^2 + b\|u\|_{\dot{H}^{5/2}}^2\right)^{1/2}$$

$$\le c\left(\frac{1}{\sqrt{ab}}\int_0^\infty \frac{dy}{1 + y^2}\right)^{1/2} \left(a\|u\|_{\dot{H}^{3/2}}^2 + b\|u\|_{\dot{H}^{5/2}}^2\right)^{1/2}.$$

---

[3] This is Hardy's trick for proving Carlson's inequality (Hardy, 1936). A straightforward approach, writing

$$\sum_j |j|^{1/2}|\hat{u}_j| = \sum_j |j|^{-3/2}|j|^{3/4}|\hat{u}_j|^{1/2}|j|^{5/4}|\hat{u}_j|^{1/2}$$

and using Hölder's inequality with exponents $(2, 4, 4)$ fails, since $\sum_j |j|^{-3}$ diverges.

See Exercise 17.8 for the bound on the sum used in the final step.

Choosing $a = \|u\|^2_{\dot{H}^{5/2}}$ and $b = \|u\|^2_{\dot{H}^{3/2}}$ we obtain

$$\sum_k |k|^{1/2} |\hat{u}_k| \le C \|u\|^{1/2}_{\dot{H}^{3/2}} \|u\|^{1/2}_{\dot{H}^{5/2}},$$

and so

$$|\langle \Lambda^{3/2}(u \cdot \nabla)u, \Lambda^{3/2}u \rangle| \le C \|u\|^{3/2}_{\dot{H}^{3/2}} \|u\|_{\dot{H}^2} \|u\|^{1/2}_{\dot{H}^{5/2}}$$

$$\le C \|u\|^2_{\dot{H}^{3/2}} \|u\|_{\dot{H}^{5/2}},$$

using the Sobolev space interpolation $\|u\|_{\dot{H}^2} \le \|u\|^{1/2}_{\dot{H}^{3/2}} \|u\|^{1/2}_{\dot{H}^{5/2}}$. $\qquad\square$

17.6 Proof of the borderline Sobolev embedding inequality

We now give a proof of the 'almost Lipschitz' Sobolev embedding result used in the proof of Proposition 17.5, due to Zuazua (2002).

Lemma 17.16 *There exists a constant $C > 0$ such that if $u \in \dot{H}^{5/2}(\mathbb{T}^3)$ then $u \in C^0(\mathbb{T}^3)$ and*

$$|u(x) - u(y)| \le C \|u\|_{\dot{H}^{5/2}} |x - y| \, |\log |x - y||^{1/2} \qquad (17.31)$$

for all $x, y \in \mathbb{T}^3$ with $|x - y| < 1/2$.

Proof We expand u as a Fourier series and split the sum into low modes and high modes:

$$\begin{aligned}
u(x) &= \sum_{k \in \dot{\mathbb{Z}}^3} \hat{u}_k e^{ik \cdot x} \\
&= \sum_{1 \le |k| \le N} \hat{u}_k e^{ik \cdot x} + \sum_{|k| > N} \hat{u}_k e^{ik \cdot x} \\
&=: u^-_N(x) + u^+_N(x),
\end{aligned}$$

where $N \in \mathbb{N}$.

Now, for any $x \in \mathbb{T}^3$ we have

$$\begin{aligned}
|\nabla u^-_N(x)| &\le \sum_{1 \le |k| \le N} |\hat{u}_k||k| \\
&\le \left(\sum_{1 \le |k| \le N} |k|^{-3} \right)^{1/2} \left(\sum_{1 \le |k| \le N} |\hat{u}_k|^2 |k|^5 \right)^{1/2} \\
&\le C \left(1 + \int_1^N \frac{1}{\kappa} \, d\kappa \right)^{1/2} \|u\|_{\dot{H}^{5/2}}
\end{aligned}$$

from which it follows that

$$\|\nabla u^-_N\|_{L^\infty} \le C(1 + \log N)^{1/2} \|u\|_{\dot{H}^{5/2}}.$$

For the high modes we have

$$\|u_N^+\|_{L^\infty} \le \sum_{|k|>N} |\hat{u}_k|$$

$$\le \left(\sum_{|k|>N} |k|^{-5}\right)^{1/2} \left(\sum_{|k|>N} |\hat{u}_k|^2 |k|^5\right)^{1/2}$$

$$\le C \left(\int_N^\infty \kappa^{-3}\, d\kappa\right)^{1/2} \|u\|_{\dot{H}^{5/2}}$$

$$= CN^{-1} \|u\|_{\dot{H}^{5/2}}.$$

Combining these two estimates we obtain

$$|u(x) - u(y)| \le |u_N^-(x) - u_N^-(y)| + |u_N^+(x) - u_N^+(y)|$$

$$\le \|\nabla u_N^-\|_{L^\infty} |x - y| + 2\|u_N^+\|_{L^\infty}$$

$$\le C\|u\|_{\dot{H}^{5/2}} \left[(1 + \log N)^{1/2}|x - y| + N^{-1}\right].$$

Now we choose N so that $(1 + \log N)^{1/2}|x - y|$ and N^{-1} are comparable, i.e.

$$N \sim \frac{1}{|x - y|\,|\log|x - y||^{1/2}},$$

which yields (17.31). □

Notes

The proof given here of the uniqueness of Lagrangian trajectories for initial data $u_0 \in H^{1/2}$ is due to Dashti & Robinson (2009); with some changes to the argument the result is also valid on the whole space and on bounded domains. An earlier proof of the same result on the whole space, using Littlewood–Paley theory, is due to Chemin & Lerner (1995); they show that if $u_0 \in H^{1/2}$ then u is in a mixed space–time Besov-type space of functions that have a sufficiently nice modulus of continuity to guarantee the uniqueness of solutions of the ODE $\dot{X} = u(X, t)$.

The result of Theorem 17.6 can be found in Temam (1983), but without a proof of the important volume-preserving property, which followed in the full treatment given in the paper by Foias, Guillopé, & Temam (1985). We stated as part of Corollary 17.14 that the solution mapping Φ must be unique; but the existence and uniqueness of such a Φ also follows from the general theory due to DiPerna & Lions (1989) that relates solutions of $\dot{X} = u(X, t)$ to solutions of the transport equation $\partial_t \phi + (u \cdot \nabla)\phi = 0$, when $u \in L^1(0, T; W^{1,1})$ and is divergence free (in fact their theory is more general). However, the existence of a unique Φ is not the same as the almost everywhere uniqueness that we provide in Corollary 17.14, as shown by Robinson, Sadowski, & Sharples (2013).

One can prove a more general version of the inequality in Lemma 17.1: if $u \in \dot{H}^{3/2}(\mathbb{T}^3) \cap \dot{H}^{s+1}(\mathbb{T}^3)$ with $\nabla \cdot u = 0$ then

$$|\langle \Lambda^s[(u \cdot \nabla)u], \Lambda^s u \rangle| \leq C\|u\|^2_{\dot{H}^{3/2}} \|u\|_{\dot{H}^{s+1}}.$$

The proof (in the periodic case) is implicit in Montero (2015) and given for the whole space case in McCormick et al. (2016a).

Exercises

17.1 Show that if M and A are 3×3 matrices and $\mathrm{d}M/\mathrm{d}t = A(t)M$ then

$$\frac{\mathrm{d}}{\mathrm{d}t}(\det M) = (\mathrm{tr}\, A)(\det M).$$

17.2 By taking the inner product of the Navier–Stokes equations with $-t\Delta u$ show that weak solutions of the 2D Navier–Stokes equations on $\mathbb{T}^2$ satisfy $\sqrt{t}u \in L^2(0, T; H^2)$. (Weak solutions of the 2D Navier–Stokes equations are smooth for $t > 0$, and in the 2D case in the absence of boundaries $\langle (u \cdot \nabla)u, \Delta u \rangle = 0$, which can be proved using integration by parts and repeated use of the divergence-free property of u.)

17.3 Use Morrey's Theorem (Theorem 1.7 part (ii)) and the Sobolev embedding $H^{3/2}(\mathbb{T}^3) \subset L^p(\mathbb{T}^3)$ for every $1 \leq p < \infty$ from Theorem 1.18 part (i) to show that $u \in H^{5/2}(\mathbb{T}^3)$ implies that $u \in C^{0,\alpha}(\mathbb{T}^3)$ for any $0 < \alpha < 1$.

17.4 Show that if $\sqrt{t}\,u \in L^2(0, T; X)$ then $u \in L^r(0, T; X)$ for every $0 < r < 1$.

17.5 Show that for $t \geq s \geq 0$ and $0 < \alpha < 1$

$$t^\alpha - s^\alpha \leq (t - s)^\alpha.$$

17.6 Show that if K is a bounded subset of $\mathbb{R}^n$ then

$$\dim_{\mathrm{B}}(K) = n - \sup\{s : \mu(O(K, \varepsilon)) \leq C\varepsilon^s \text{ for some } C > 0\}.$$

17.7 Use the Mean Value Theorem to show that

$$\left||k|^{3/2} - |j|^{3/2}\right| \leq C|k - j|(|k - j|^{1/2} + |j|^{1/2}).$$

17.8 Use Exercise 1.7 to show that

$$\sum_{k \neq 0} \frac{1}{a|k|^2 + b|k|^4} \leq c\sqrt{ab} \int_0^\infty \frac{\mathrm{d}y}{1 + y^2}.$$

Appendix A

Functional analysis: miscellaneous results

In this appendix we collate several standard results in Analysis that are used throughout the book. For the most part we do not provide proofs, but rather only references to texts that provide detailed expositions.

A.1 L^p spaces

Theorem A.1 (Minkowski's inequality) *Let $f\colon \mathbb{R}^n \times \mathbb{R}^n \to \mathbb{R}$ be a measurable function (with respect to Lebesgue measure). Then (assuming the finiteness of all the integrals involved) we have*

$$\left\| \int_{\mathbb{R}^n} f(\cdot, y)\, dy \right\|_{L^p(\mathbb{R}^n)} \le \int_{\mathbb{R}^n} \| f(\cdot, y) \|_{L^p(\mathbb{R}^n)}\, dy,$$

for $1 \le p \le \infty$.

Theorem A.2 (Lebesgue Differentiation Theorem) *Take $\Omega \subset \mathbb{R}^n$ measurable and let $f\colon \Omega \subset \mathbb{R}^n \to \mathbb{R}$ be a measurable function in $L^1_{loc}(\Omega)$, i.e. $\int_K |f| < \infty$ for every compact set $K \subset \Omega$. Then for any family of balls $\{B_r\}$ with $\cap_{r>0} B_r = \{x\}$ we have*

$$\lim_{r\to 0} \frac{1}{|B_r|} \int_{B_r} f(y)\, dy = f(x), \qquad a.e.\ x \in \Omega$$

or more strongly

$$\lim_{r\to 0} \frac{1}{|B_r|} \int_{B_r} |f(y) - f(x)|\, dy = 0, \qquad a.e.\ x \in \Omega.$$

A proof of this result can be found in Folland (1999), Theorem 3.2. Note that in the theorem above we can replace balls by cubes for example, or in fact by any family of sets that 'shrink nicely' to a point (see Folland for the precise definition).

A.2 Absolute continuity

We briefly recall some basic facts about absolute continuity: details and proofs can be found in Section 3.5 in Folland (1999), for example.

Definition A.3 A function $f: \mathbb{R} \to \mathbb{R}$ is absolutely continuous if for every $\varepsilon > 0$ there exists $\delta > 0$ such that for any finite set of disjoint intervals (a_i, b_i), $i = 1, \ldots, n$,

$$\sum_{i=1}^{n} |b_i - a_i| < \delta \quad \Rightarrow \quad \sum_{i=1}^{n} |f(b_i) - f(a_i)| < \varepsilon.$$

Proposition A.4 (Corollary 3.33 in Folland) *If $f \in L^1(\mathbb{R})$ then the function $F(x) = \int_{-\infty}^{x} f(t)\, dt$ is absolutely continuous and $F' = f$ almost everywhere.*

Two immediate corollaries of this result will be useful in Chapter 17.

Corollary A.5 *If $g \in L^1(\mathbb{R})$ then given $\varepsilon > 0$ there exists $\delta > 0$ such that*

$$|b - a| < \delta \quad \Rightarrow \quad \left| \int_{a}^{b} g(t)\, dt \right| < \varepsilon.$$

Corollary A.6 *Let $\theta \in L^1(0, T)$. If $X_n : [0, T] \to \mathbb{R}$ satisfy*

$$|X_n(t) - X_n(t')| \leq \int_{t}^{t'} \theta(s)\, ds$$

for all n and $t, t' \in [0, T]$, then (X_n) is an equicontinuous family of functions.

Finally, we give a version of the Fundamental Theorem of Calculus.

Theorem A.7 (Theorem 3.35 in Folland) *If $-\infty < a < b < \infty$ and $F: [a, b] \to \mathbb{R}$ then the following are equivalent:*

(i) *F is absolutely continuous on $[a, b]$;*
(ii) *$F(x) - F(a) = \int_{a}^{x} f(t)\, dt$ for some $f \in L^1(a, b)$; and*
(iii) *F is differentiable almost everywhere on $[a, b]$, $F' \in L^1(a, b)$ and*

$$F(x) - F(a) = \int_{a}^{x} F'(t)\, dt.$$

The following simple corollary is also useful in Chapter 17.

Corollary A.8 *If $F\colon [a, b] \to \mathbb{R}$ is absolutely continuous and $F' = 0$ almost everywhere on (a, b) then F is constant on $[a, b]$.*

A.3 Convolution and mollification

We will denote the convolution of two functions $f, g\colon \mathbb{R}^n \to \mathbb{R}$ by

$$(f * g)(x) = \int_{\mathbb{R}^n} f(y)g(x - y)\, dy = \int_{\mathbb{R}^n} f(x - y)g(y)\, dy.$$

We review a few of the results concerning convolutions that feature more prominently in the book. It is a fundamental property of convolutions that

$$\widehat{f * g} = \hat{f}\hat{g}$$

(see Grafakos, 2008, for example).

Theorem A.9 (Approximation to the identity) *Let $\phi\colon \mathbb{R}^n \to \mathbb{R}$ be an integrable function with $\int_{\mathbb{R}^n} \phi = 1$; for every $\varepsilon > 0$ set $\phi_\varepsilon(x) := \varepsilon^{-n}\phi(x/\varepsilon)$. Then*

$$\|(\phi_\varepsilon * f) - f\|_{L^p} \to 0 \quad as \quad \varepsilon \to 0$$

for every $f \in L^p$, $1 \le p < \infty$.

The result above is in general not true for $p = \infty$, but one can obtain uniform convergence if, for example, $f \in C_c^0$, the space of continuous functions with compact support. The family ϕ_ε in the theorem above is usually referred to as an approximation to the identity. One of the fundamental consequences of this result is, as we will see in Theorem D.3, the convergence in the L^p sense of the solution to the heat equation to the initial data (as t goes to zero).

Theorem A.10 (Young's inequality for convolutions) *If $1 \le p, q, r \le \infty$ satisfy*

$$1 + \frac{1}{p} = \frac{1}{q} + \frac{1}{r}$$

then there exists $C = C(p, q, r)$ such that

$$\|f * g\|_{L^p} \le C\|f\|_{L^q}\|g\|_{L^r}.$$

The proof can be found in Grafakos (2008) or McCormick, Robinson, & Rodrigo (2013).

Now suppose that $\rho\colon \mathbb{R}^n \to \mathbb{R}$ is a smooth, non-negative function with compact support (say in the unit ball) such that $\int \rho = 1$. For $\varepsilon > 0$, we define the mollifier $\rho_\varepsilon(x) = \varepsilon^{-n}\rho(x/\varepsilon)$. We will refer to $\rho_\varepsilon * u$ as the mollification of u.

The following theorem contains some of the boundedness and regularisation results for mollifiers that we use throughout the book.

Theorem A.11 (Mollification) *Let ρ be as above. We have the following properties for the mollification of u.*

1. Boundedness. *The operation of mollification $\rho_\varepsilon * u$ is bounded in L^p, $(1 \le p \le \infty)$ and H^s, $s \ge 0$, i.e.*

$$\|\rho_\varepsilon * u\|_{L^p} \le c\|u\|_{L^p} \qquad and \qquad \|\rho_\varepsilon * u\|_{H^s} \le c\|u\|_{H^s}.$$

2. Regularisation. *We have*

$$\|\rho_\varepsilon * u\|_{L^q} \le c\varepsilon^{(1/q - 1/p)n}\|u\|_{L^p}, \qquad q \ge p$$

and

$$\|\rho_\varepsilon * u\|_{H^s} \le c\varepsilon^{-s}\|u\|_{L^2}, \qquad s \ge 0.$$

Proof The proof of these facts follows easily from Young's inequality for convolutions (Theorem A.10) and the fact that derivatives (including fractional derivatives) of the mollification can be moved directly to ρ or to u, that is

$$\Lambda^s(\rho_\varepsilon * u) = \rho_\varepsilon * \Lambda^s u = (\Lambda^s(\rho_\varepsilon)) * u.$$

In order to see this, notice that if we take the Fourier transform of the term on the left-hand side we would obtain the product of $|\xi|^s$ (corresponding to the fractional derivative) with the Fourier transforms of ρ_ε and u. Since multiplication is commutative we can rearrange this product to obtain the other equalities. Returning to our estimate, Young's inequality yields

$$\|\rho_\varepsilon * u\|_{L^p} \le \|\rho_\varepsilon\|_{L^1}\|u\|_{L^p} = \|\rho\|_{L^1}\|u\|_{L^p} \le \|u\|_{L^p},$$

which proves the L^p boundedness. As for H^s, since we have already shown boundedness in L^2, it suffices to prove the bound for $\dot{H}^s$. We have

$$\|\rho_\varepsilon * u\|_{\dot{H}^s} = \|\Lambda^s(\rho_\varepsilon * u)\|_{L^2} = \|\rho_\varepsilon * (\Lambda^s u)\|_{L^2} \le \|\Lambda^s u\|_{L^2} \le \|u\|_{H^s}.$$

Concerning the smoothing properties, similar arguments yield the results:

$$\|\rho_\varepsilon * u\|_{L^q} \le \|\rho_\varepsilon\|_{L^r}\|u\|_{L^p},$$

provided that $1 + \frac{1}{q} = \frac{1}{r} + \frac{1}{p}$. Now $\|\rho_\varepsilon\|_{L^r} = \varepsilon^{(1/r-1)n}\|\rho\|_{L^r}$, which in terms of p and q becomes $c\varepsilon^{(1/q-1/p)n}$, completing the estimate in the L^p case. For the Sobolev case, notice that, as before, it suffices to show the result for $\dot{H}^s$. We have

$$\|\rho_\varepsilon * u\|_{\dot{H}^s} = \|\Lambda^s(\rho_\varepsilon * u)\|_{L^2} = \|(\Lambda^s\rho_\varepsilon) * u\|_{L^2} \le \|\Lambda^s(\rho_\varepsilon)\|_{L^1}\|u\|_{L^2}$$
$$\le \varepsilon^{-s}\|(\Lambda^s\rho)_\varepsilon\|_{L^1}\|u\|_{L^2} \le c\varepsilon^{-s}\|u\|_{L^2},$$

where we have used that $\Lambda^s(\rho_\varepsilon) = \varepsilon^{-s}(\Lambda^s\rho)_\varepsilon$. $\qquad\qquad\square$

A.4 Weak L^p spaces

In this section we present a quick overview of weak L^p spaces. The use of such spaces will be necessary when considering singular integral operators (Appendix B). As we will see there, the image of an L^1 function by a singular integral is not necessarily an L^1 function, but rather it belongs to weak L^1. We do not systematically cover this area, but rather briefly review the results that will be more useful for us.

Definition A.12 Given a measurable function $f: \Omega \to \mathbb{R}$ and $1 \le p < \infty$ we say that f belongs to the space *weak L^p*, denoted by $L^{p,\infty}$ if there exists M such that

$$\mu\{x \in \Omega : |f(x)| > \alpha\} \le \left(\frac{M}{\alpha}\right)^p. \tag{A.1}$$

We define $\|f\|_{L^{p,\infty}}$ to be infimum of all M for which (A.1) is satisfied. This defines a seminorm in the space $L^{p,\infty}$.

We also define *weak L^∞* as L^∞.

It is easy to see that $L^p \subset L^{p,\infty}$ with $\|f\|_{L^{p,\infty}} \le \|f\|_{L^p}$. Indeed, if $f \in L^p$, with $1 \le p < \infty$, then, denoting $\{x \in \Omega : |f(x)| > \alpha\}$ by $\{|f(x)| > \alpha\}$,

$$\|f\|_{L^p}^p = \int |f|^p \ge \int_{\{|f(x)|>\alpha\}} |f|^p$$
$$\ge \int_{\{|f(x)|>\alpha\}} \alpha^p = \alpha^p\mu(\{|f(x)| > \alpha\}),$$

which rearranged yields

$$\mu(\{|f(x)| > \alpha\}) \le \left(\frac{\|f\|_{L^p}}{\alpha}\right)^p. \tag{A.2}$$

This inequality is known as the generalised Chebyshev inequality (with $p = 1$ corresponding to the original Chebyshev inequality). To see that $L^{p,\infty}$ is strictly larger than L^p (for $1 \le p < \infty$) we consider the function $f(x) = |x|^{-n/p}$ in $\mathbb{R}^n$.

This function just fails to be in L^p (near the origin and infinity) but is indeed in $L^{p,\infty}$.

Definition A.13 Let T be an operator defined on L^p that maps into the space of measurable functions. We say that T is weak (p, q), $q < \infty$, if there exists C such that

$$|\{x : |Tf(x)| > \alpha\}| \le \left(\frac{C\|f\|_{L^p}}{\alpha}\right)^q,$$

i.e. $\|Tf\|_{L^{q,\infty}} \le C\|f\|_{L^p}$. We say that T is weak (p, ∞) if it is *strong* (p, ∞), that is if it satisfies $\|Tf\|_{L^\infty} \le c\|f\|_{L^p}$, for some c independent of f.

Definition A.14 We say that T is sublinear if

$$|T(f_0 + f_1)(x)| \le |Tf_0(x)| + |Tf_1(x)| \quad \text{and} \quad |T(\lambda f)| = |\lambda||T(f)|.$$

Theorem A.15 (Marcinkiewicz Interpolation Theorem) *Suppose that* $1 \le p_0 < p_1 \le \infty$ *and let T be a sublinear operator from $L^{p_0} + L^{p_1}$ into a space of measurable functions. Assume that T is weak (p_0, p_0) and weak (p_1, p_1). Then T is a bounded operator from L^p into itself for $p_0 < p < p_1$.*

Proofs of this result can be found in the books by Duoandikoetxea (2001) or Grafakos (2008), among others. We will use this result in the next appendix, when studying boundedness properties for singular integrals.

We will also require a generalisation of the Young inequality for convolutions (Theorem A.10) involving weak L^p spaces.

Theorem A.16 (Weak Young Inequality) *Let $1 < p, q, r < \infty$ satisfy the condition $1 + \frac{1}{p} = \frac{1}{q} + \frac{1}{r}$. Then there exists $C = C(p, q, r)$ such that*

$$\|f * g\|_{L^p} \le C\|f\|_{L^{q,\infty}}\|g\|_{L^r}.$$

The proof, which uses the Marcinkiewicz Interpolation Theorem, can be found in Grafakos (2008); see also McCormick, Robinson, & Rodrigo (2013).

A.5 Weak and weak-∗ convergence and compactness

If X is a Banach space (over $\mathbb{R}$) then its dual space X^* consists of all bounded linear functionals from X into $\mathbb{R}$, and is itself a Banach space when equipped with the norm

$$\|f\|_{X^*} := \sup_{x \in X: \|x\|=1} \langle f, x \rangle_{X^* \times X}.$$

We omit the $X^* \times X$ subscript from the 'duality pairing' in the remainder of this section.

We recall here a number of results concerning dual spaces and weak and weak-∗ convergence. For proofs see Robinson (2001), for example.

A fundamental result is the Hahn–Banach Theorem, which allows for the extension of linear functionals defined on subspaces.

Theorem A.17 (Hahn–Banach Theorem) *Let X be a Banach space and Z a linear subspace of X. If $f \in Z^*$ then there exists an $F \in X^*$ that satisfies $F(x) = f(x)$ for every $x \in Z$ and $\|F\|_{X^*} \leq \|f\|_{Z^*}$.*

Corollary A.18 *Let X be a Banach space and take $x \in X$ with $x \neq 0$. Then there exists $f \in X^*$ such that $\|f\|_{X^*} = 1$ and $\langle f, x \rangle = \|x\|_X$.*

A sequence (x_n) in X converges weakly to $x \in X$, which we write as $x_n \rightharpoonup x$ in X, if

$$\langle f, x_n \rangle \to \langle f, x \rangle \qquad \text{for every} \qquad f \in X^*.$$

The following properties of weak limits are simple to prove, apart from (iii), which requires the Banach–Steinhaus Principle of Uniform Boundedness.

Proposition A.19 *Let X be a Banach space and $(x_n), x \in X$.*

(i) *Weak limits are unique;*
(ii) *if $x_n \to x$ then $x_n \rightharpoonup x$;*
(iii) *weakly convergent sequences are bounded; and*
(iv) *if $x_n \rightharpoonup x$ then*

$$\|x\|_X \leq \liminf_{n \to \infty} \|x_n\|_X,$$

so in particular if $\|x_n\|_X \leq M$ for every n it follows that $\|x\|_X \leq M$.

The following result is useful in the Hilbert space case.

Lemma A.20 *If H is a Hilbert space and $(x_n) \in H$ then $x_n \to x$ if and only if $x_n \rightharpoonup x$ and $\|x_n\| \to \|x\|$.*

Proof The 'only if' part is immediate. If $x_n \rightharpoonup x$ and $\|x_n\| \to \|x\|$ then consider

$$\|x_n - x\|^2 = (x_n - x, x_n - x) = \|x_n\|^2 - 2(x, x_n) + \|x\|^2.$$

Since x is fixed and $x_n \rightharpoonup x$ it follows that $(x, x_n) \to (x, x) = \|x\|^2$, and since $\|x_n\|^2 \to \|x\|^2$ we have $\|x_n - x\|^2 \to 0$ and $x_n \to x$. $\qquad\qquad \square$

A sequence (f_n) in X^* converges weakly-$*$ to $f \in X^*$, which we write as $f_n \overset{*}{\rightharpoonup} f$ in X^*, if

$$\langle f_n, x \rangle \to \langle f, x \rangle \qquad \text{for each} \qquad x \in X$$

(despite its name this is in some sense the natural definition[1] of convergence in X^*). Weak-$*$ convergence satisfies the following properties.

Proposition A.21 *Let X be a Banach space and (f_n), $f \in X^*$.*

  (i) *Weak-$*$ limits are unique;*
 (ii) *weakly-$*$ convergent sequences are bounded; and*
(iii) *if $f_n \overset{*}{\rightharpoonup} f$ then*

$$\|f\|_{X^*} \le \liminf_{n \to \infty} \|f_n\|_{X^*},$$

so in particular if $\|f_n\|_X^* \le M$ for every n it follows that $\|f\|_{X^*} \le M$.

The following two results provide weak and weak-$*$ compactness and are central to the arguments that construct solutions as limits of approximations. The following result can be proved using a relatively simple diagonal argument.

Theorem A.22 (Banach–Alaoglu Theorem) *If X is a separable Banach space then any bounded sequence in X^* has a weakly-$*$ convergent subsequence.*

Without separability, the closed unit ball in X^* is compact in the weak-$*$ topology (the proof uses Tychonoff's Theorem), but this does not imply sequential compactness in general.

The following weak-compactness result requires X to be reflexive (and this condition is necessary, see James (1964) or Yosida (1980), for example).

Corollary A.23 (Weak compactness) *If X is a reflexive Banach space then any bounded sequence in X has a weakly convergent subsequence.*

Since L^p is reflexive for $1 < p < \infty$ with $(L^p)^* \simeq L^q$ (p and q conjugate), if we have a bounded sequence (f_n) in such a space we can use the weak compactness provided by Corollary A.23 to find a weakly convergent subsequence (f_{n_j}), i.e. there exists $f \in L^p$ such that

$$\int f_{n_j} g \to \int f g \qquad \text{for every} \qquad g \in L^q.$$

---

[1] The notion of weak convergence could be applied to a sequence in X^*: $f_n \rightharpoonup f$ in X^* if

$$\langle F, f_n \rangle_{X^{**} \times X^*} \to \langle F, f \rangle_{X^{**} \times X^*} \qquad \text{for each} \qquad F \in X^{**}$$

such weak convergence implies weak-$*$ convergence, with the two notions being equivalent if X is reflexive.

Due to (iv) of Proposition A.19 bounds on $\|f_n\|_{L^p}$ are inherited by f.

Although L^∞ is not reflexive we can still use Theorem A.22 and appeal to weak-∗ compactness in L^∞, since we can characterise L^∞ as a dual space, $(L^\infty) \simeq (L^1)^*$, and L^1 is separable. It follows that any bounded sequence (f_n) in L^∞ has a weakly-∗ convergent subsequence (f_{n_j}), i.e. there exists $f \in L^\infty$ such that

$$\int f_{n_j} g \to \int fg \qquad \text{for every} \qquad g \in L^1.$$

Again, bounds on $\|f_n\|_{L^\infty}$ are inherited by f (now due to part (iii) of Proposition A.21).

A similar trick is not available for sequences bounded in L^1.

A.6 Gronwall's Lemma

Lemma A.24 *Let $\eta \colon [0, T] \to [0, \infty)$ be an absolutely continuous function that satisfies the differential inequality*

$$\eta'(t) \le \phi(t)\eta(t) + \psi(t),$$

where ϕ and ψ are non-negative integrable functions. Then

$$\eta(t) \le e^{\int_0^t \phi(s)\,ds} \left[\eta(0) + \int_0^t \psi(s)\,ds \right] \qquad \text{for all } t \in [0, T].$$

The proof of Lemma A.24 can be found in Evans (1998); one simply uses the integrating factor $\exp\left(\int_0^t \phi(s)\,ds\right)$. We will also need the following integral version of the Gronwall Lemma, proved by setting $y(t) = \int_0^t \phi(s)\eta(s)\,ds$.

Lemma A.25 *Let $\eta \colon [0, T] \to [0, \infty)$ be a continuous function satisfying*

$$\eta(t) \le a(t) + \int_0^t \phi(s)\eta(s)\,ds,$$

where $a, \phi \colon [0, T] \to [0, \infty)$ are integrable functions and a is increasing. Then

$$\eta(t) \le a(t)\exp\left(\int_0^t \phi(s)\,ds\right) \qquad \text{for all } t \in [0, T].$$

Appendix B

Calderón–Zygmund Theory

In this appendix we present a review of boundedness properties for singular integral operators, focusing on the Calderón–Zygmund theory. In a nutshell we will study sufficient conditions on a function K, the kernel, so that the convolution operator $K * f$ is bounded in L^p (for $1 < p < \infty$), and maps L^1 into $L^{1,\infty}$. The main result we will require is contained in Theorem B.3. In the second part of the appendix we also include some results for Riesz transforms and applications to partial differential equations.

B.1 Calderón–Zygmund decompositions

In this section we present the Calderón–Zygmund decomposition for non-negative functions in $\mathbb{R}^n$, one of the fundamental ingredients in the proof of the Calderón–Zygmund Theorem B.3. Loosely speaking the result can be described as follows. Given a non-negative function $f\colon \mathbb{R} \to [0, \infty)$ and a positive α, it is possible to decompose $\mathbb{R}^n$ into two sets, one where the function is essentially smaller than the given α, and its complement, which can be decomposed into essentially disjoint cubes, where we control the average of f. The family of cubes in the decomposition is structured using dyadic cubes, which are defined as follows.

Definition B.1 Let $Q = [0, 1)^n$. We denote by $\mathcal{Q}_0$ the collection of all cubes that are congruent with Q and whose vertices lie on the lattice $\mathbb{Z}^n$. For $k \in \mathbb{Z}$ we denote by $\mathcal{Q}_k$ the family of cubes congruent to $2^{-k}Q$ and with vertices on the lattice $(2^{-k}\mathbb{Z})^n$.

The cubes in $\bigcup_{k\in\mathbb{Z}} \mathcal{Q}_k$ are called dyadic cubes.

Theorem B.2 *Let $f\colon \mathbb{R}^n \to \mathbb{R}$ be non-negative and integrable, and let α be a given positive number. Then there exists a collection of disjoint dyadic cubes $\{Q_j\}$ such that the following properties are satisfied:*

(i) *$f(x) \le \alpha$ almost everywhere in $\left(\bigcup Q_j\right)^c$;*

(ii) *$\left|\bigcup Q_j\right| \le \dfrac{1}{\alpha}\|f\|_{L^1}$; and*

(iii) *$\alpha < \dfrac{1}{|Q_j|}\displaystyle\int_{Q_j} f \le 2^n\alpha$ for every j.*

Proof We will construct the family of dyadic cubes in the theorem inductively. Notice that since $f \in L^1$, we can find k large enough that

$$\frac{1}{2^{kn}}\int_{\mathbb{R}^n} f(x)\,dx \le \alpha,$$

which means that for any dyadic cube Q of side 2^k we have

$$\frac{1}{|Q|}\int_Q f(x)\,dx < \alpha.$$

For any of these cubes, we consider the 2^n cubes obtained by bisecting every side of Q. For any such cube, $\tilde{Q}$ say, we consider the quantity

$$\frac{1}{|\tilde{Q}|}\int_{\tilde{Q}} f(x)\,dx.$$

If this quantity is greater than α, then we keep $\tilde{Q}$ for our collection of dyadic cubes. Notice that in this case we have

$$\alpha < \frac{1}{|\tilde{Q}|}\int_{\tilde{Q}} f(x)\,dx \le \frac{2^n}{2^n|\tilde{Q}|}\int_Q f(x)\,dx \le 2^n\frac{1}{|Q|}\int_Q f(x)\,dx \le 2^n\alpha,$$

with the last inequality being satisfied by our hypothesis on Q. This shows that the cube selected for our collection satisfies property (iii) above.

For every cube that has not been selected above, we repeat the bisecting procedure iteratively, until (if ever) the corresponding cube is selected. It is clear that all cubes selected will satisfy property (iii) above, and that moreover all the selected cubes are disjoint, from which property (ii) follows. Indeed by the first inequality from property (iii), we have

$$|Q_j| \le \frac{1}{\alpha}\int_{Q_j} f(x)\,dx.$$

Taking the union in j, and using the fact that the cubes are disjoint we obtain property (ii).

As for property (i), notice that if $x \in (\bigcup Q_j)^c$, then for every dyadic cube of every generation that we have considered we have always had

$$\frac{1}{|Q|} \int_Q f(x)\,dx < \alpha,$$

regardless of how small the cube Q is. Property (i) now follows immediately from the Lebesgue Differentiation Theorem (see Theorem A.2), as the family of cubes Q shrinks nicely to the point x. $\qquad\qquad\square$

B.2 The Calderón–Zygmund Theorem

In this section we present the main Calderón–Zygmund Theorem that we will use in the text to obtain boundedness properties for singular integral operators. The version presented relies on an L^∞ bound on the Fourier transform of the kernel (B.1) as well as on the 'Hörmander condition' (B.2). Corollary B.4 contains an alternative version where (B.2) is replaced by a pointwise constraint on the gradient of the kernel.

Theorem B.3 *Let K be a tempered distribution in $\mathbb{R}^n$ that coincides with a locally integrable function in $\mathbb{R}^n\setminus\{0\}$ and satisfies*

$$\|\hat{K}\|_{L^\infty} \le A \tag{B.1}$$

and

$$\int_{\{|x|>2|y|\}} |K(x-y) - K(x)|\,dx \le B \quad \text{for all} \quad y \in \mathbb{R}^n. \tag{B.2}$$

Then

$$|\{x:\ |K*f(x)| > \alpha\}| \le \frac{C}{\alpha}\|f\|_{L^1} \tag{B.3}$$

and

$$\|K*f\|_{L^p} \le c_p\|f\|_{L^p}, \qquad 1 < p < \infty. \tag{B.4}$$

Proof The proof is reduced to showing (B.3) together with (B.4) for the special case $p = 2$, together with some standard interpolation and duality arguments. Indeed, given the two endpoint results we can use the Marcinkiewicz Interpolation Theorem (Theorem A.15) to obtain the result for $1 < p < 2$. The remaining part of the range, $2 < p < \infty$ is obtained by what is known as a duality argument, which we outline now. Notice that if p satisfies $2 < p < \infty$, its dual exponent q satisfies $1 < q < 2$. Now, define $\tilde{K}(x) = K(-x)$, and notice that $\tilde{K}$ satisfies the two hypotheses (B.1) and (B.2) of the theorem and so is bounded in L^q for $1 < q < 2$. We have

$$
\begin{aligned}
\|K * f\|_{L^p} &= \sup_{g:\|g\|_{L^q}=1} \int K * f(x) g(x)\, dx \\
&= \sup_{g:\|g\|_{L^q}=1} \int \int K(x-y) f(y)\, dy\, g(x)\, dx \\
&= \sup_{g:\|g\|_{L^q}=1} \int \tilde{K} * g(y) f(y)\, dy \\
&\leq \sup_{g:\|g\|_{L^q}=1} \|f\|_{L^p} \|\tilde{K} * g\|_{L^q} \\
&\leq \sup_{g:\|g\|_{L^q}=1} c_q \|f\|_{L^p} \|g\|_{L^q} = c_q \|f\|_{L^p},
\end{aligned}
$$

which proves the boundedness result for the remaining range $2 < p < \infty$.

The case $p = 2$ follows trivially from Plancherel's formula. Indeed

$$
\|K * f\|_{L^2} = \|\widehat{K * f}\|_{L^2} = \|\hat{K}\hat{f}\|_{L^2} \leq \|\hat{K}\|_{L^\infty} \|\hat{f}\|_{L^2} \leq A\|f\|_{L^2}.
$$

So all that remains is to prove estimate (B.3).

By applying the result to the positive and the negative part of f we can assume, without loss of generality, that f is non-negative. Then, given $\alpha > 0$ we can apply Theorem B.2 to obtain a Calderón–Zygmund decomposition. With the aid of this decomposition we are going to split $f = g + b$, with g usually referred to as the good part, and b as the bad part. We define

$$
g(x) =
\begin{cases}
f(x) & x \in \left(\bigcup Q_j \right)^c, \\
\dfrac{1}{|Q|_j} \displaystyle\int_{Q_j} f(y)\, dy & x \in Q_j.
\end{cases}
$$

This implicitly defines b as $f - g$, with support in $\bigcup Q_j$. It will be convenient for us to write

$$
b = \sum_j b_j \quad \text{with} \quad b_j = \left(f - \frac{1}{|Q|_j} \int_{Q_j} f(y)\, dy \right) \chi_{Q_j}.
$$

The following properties of g and b follow directly from the definitions, coupled with the fact that $\{Q_j\}$ is a Calderón–Zygmund decomposition:

$$
g \leq 2^n \alpha, \qquad \int_{Q_j} g = \int_{Q_j} f, \qquad g \in L^2 \ (\text{since } g \subset L^1 \cap L^\infty),
$$

$$
\int_{Q_j} b_j = \int_{Q_j} b = \int_{\mathbb{R}^n} b = 0, \quad \text{and} \quad \int_{Q_j} |b_j(x)|\, dx \leq 2 \int_{Q_j} f(y)\, dy.
$$

We now return to the proof of estimate (B.3). Notice that

$$|\{x : \ |K * f(x)| > \alpha\}| \leq \left|\left\{x : \ |K * g(x)| > \frac{\alpha}{2}\right\}\right|$$
$$+ \left|\left\{x : \ |K * b(x)| > \frac{\alpha}{2}\right\}\right|,$$

allowing us to consider separate estimates for g and b. Using the generalised Chebyshev inequality

$$|\{x : \ f(x) > \alpha\}| \leq \frac{1}{\alpha^2} \int f^2$$

(see (A.2)) we obtain

$$\left|\left\{x : \ |K * g(x)| > \frac{\alpha}{2}\right\}\right| \leq \frac{4}{\alpha^2}\|K * g\|_{L^2}^2 \leq \frac{4c}{\alpha^2}\|g\|_{L^2}^2 \leq \frac{4c}{\alpha^2}\|g\|_{L^1}\|g\|_{L^\infty}$$
$$\leq \frac{2^{n+2}c}{\alpha}\|g\|_{L^1} \leq \frac{2^{n+2}c}{\alpha}\|f\|_{L^1}, \tag{B.5}$$

completing the estimate for g.

For b things are more delicate. First we define the cubes Q_j^* to be cubes of side $2\sqrt{n}$ times the side of Q_j and with the same centre. Then

$$\left|\left\{x : |K * b(x)| > \frac{\alpha}{2}\right\}\right| \leq \left|\bigcup Q_j^*\right| + \left|\left\{x \in \left(\bigcup Q_j^*\right)^c : |K * b(x)| > \frac{\alpha}{2}\right\}\right|.$$

Notice that

$$\left|\bigcup Q_j^*\right| \leq c\left|\bigcup Q_j\right| \leq \frac{c}{\alpha} \int f \tag{B.6}$$

and so the first term satisfies the correct estimate. As for the second, Chebyshev's inequality yields

$$\left|\left\{x \in \left(\bigcup Q_j^*\right)^c : |K * b(x)| > \frac{\alpha}{2}\right\}\right| \leq \frac{2}{\alpha}\int_{(\bigcup Q_j^*)^c} |K * b(x)|\, dx,$$

reducing the problem to showing that

$$\int_{(\bigcup Q_j^*)^c} |K * b(x)|\, dx \leq c \int f.$$

Now,

$$\int_{\left(\bigcup Q_j^*\right)^c} |K * b(x)| \, dx \le \sum_j \int_{\left(\bigcup Q_j^*\right)^c} |K * b_j(x)| \, dx$$

$$= \sum_j \int_{\left(\bigcup Q_j^*\right)^c} \left| \int K(x-y) b_j(y) \, dy \right| dx$$

$$\le \sum_j \int_{\left(Q_j^*\right)^c} \left| \int_{Q_j} K(x-y) b_j(y) \, dy \right| dx$$

$$\le \sum_j \int_{\left(Q_j^*\right)^c} \left| \int_{Q_j} [K(x-y) - K(x-c_j)] b_j(y) \, dy \right| dx,$$

using the fact that $\int_{Q_j} b_j = 0$ to subtract the counterterm $K(x - c_j)$, with c_j the centre of the cube Q_j. Moving the absolute value inside the integral and applying Fubini's theorem we obtain

$$\int_{\left(\bigcup Q_j^*\right)^c} |K * b(x)| \, dx \le \sum_j \int_{Q_j} \int_{\left(Q_j^*\right)^c} |K(x-y) - K(x-c_j)| \, dx \, |b_j(y)| \, dy$$

$$\le \sum_j \int_{Q_j} \int_{|x-c_j| > 2|y-c_j|} |K(x-y) - K(x-c_j)| \, dx \, |b_j(y)| \, dy,$$

since $(Q_j^*)^c \subset \{|x - c_j| > 2|y - c_j|\}$. Then by hypothesis (B.2) we obtain

$$\le B \sum_j \int_{Q_j} |b_j(y)| \, dy \le B \sum_j 2 \int_{Q_j} f(x) \, dx \le 2B \|f\|_{L^1},$$

which shows that

$$\left| \left\{ x \in \left(\bigcup Q_j^*\right)^c : |K * b(x)| > \frac{\alpha}{2} \right\} \right| \le 2B \|f\|_{L^1}.$$

This estimate, together with (B.5) and (B.6) yields the desired result. $\qquad\square$

We conclude this section with a corollary containing a weaker result than Theorem B.3, but with a condition on the kernel that is easier to check in particular examples.

Corollary B.4 *Let K be a tempered distribution in $\mathbb{R}^n$ that coincides with a locally integrable function in $\mathbb{R}^n \setminus \{0\}$ and satisfies*

$$\|\hat{K}(\cdot)\|_{L^\infty} \le A$$

and

$$|\nabla K(x)| \leq \frac{C}{|x|^{n+1}}, \qquad x \neq 0. \tag{B.7}$$

Then

$$|\{x: |K * f(x)| > \alpha\}| \leq \frac{C}{\alpha}\|f\|_{L^1}$$

and

$$\|K * f\|_{L^p} \leq c_p\|f\|_{L^p}, \qquad 1 < p < \infty.$$

Notice that condition (B.2) in the theorem above is satisfied by every K that satisfies (B.7), and so the result follows immediately from Theorem B.3.

B.3 Riesz transforms

In this section we discuss a prominent example of a convolution-type operator to which we will apply the theory developed in the previous section. Namely, we consider the family of Riesz transforms, which we define in two possible ways:

$$\widehat{R_j(f)}(\xi) = i\frac{\xi_j}{|\xi|}\hat{f}(\xi), \tag{B.8}$$

or

$$R_j f(x) = c_n \, \text{p.v.} \int_{\mathbb{R}^n} \frac{y_j}{|y|^{n+1}} f(x-y)\,dy, \qquad c_n = \frac{\Gamma((n+1)/2)}{\pi^{(n+1)/2}}, \tag{B.9}$$

where p.v. above stands for principal value, meaning that we consider the integral as $\lim_{\varepsilon \to 0} \int_{|y| \geq \varepsilon}$. It is easy to see (by considering the Fourier Transform of homogeneous distributions) that the two operators are the same.

Notice that (B.8) guarantees (B.1) while (B.9) guarantees (B.2) (see Corollary B.4). Hence as a result of Theorem B.3 we obtain that all Riesz transforms are bounded in L^p for $1 < p < \infty$.

This result has immediate applications to PDEs.

Theorem B.5 *Given $1 < p < \infty$ there exists a constant c_p such that for all $f \in C_c^\infty(\mathbb{R}^n)$ we have the estimate*

$$\|\partial_j\partial_k f\|_{L^p} \leq c_p\|\Delta f\|_{L^p}. \tag{B.10}$$

Proof The result is an immediate consequence of the boundedness of the Riesz transforms once we notice that $\partial_j\partial_k f = -R_jR_k\Delta f$, as can be seen by taking the Fourier transform of both sides and using (B.8). $\qquad\square$

It clear that once the estimate (B.10) has been established for smooth functions with compact support it is possible to extend it to less regular functions using density.

It is customary to write the relationship $\partial_j \partial_k f = -R_j R_k \Delta f$ in the form $T_{jk} f := R_j R_k f = \partial_j \partial_k (-\Delta)^{-1} f$.

Theorem B.6 *The linear operator T_{jk} defined by*

$$T_{jk} f = \partial_j \partial_k (-\Delta)^{-1} f$$

is a bounded linear operator from $L^q(\mathbb{R}^3)$ into $L^q(\mathbb{R}^3)$ for all $1 < q < \infty$.

We conclude with an application of the results to the Poisson equation in $\mathbb{R}^3$ and $\mathbb{T}^3$.

Theorem B.7 *Let Ω be either $\mathbb{R}^3$ or $\mathbb{T}^3$. Assume that $-\Delta u = f$ in Ω and that f belongs to $L^p(\Omega)$ for some $p \in (1, \infty)$. Then all second derivatives of u can be bounded in L^p in terms of f namely*

$$\|\partial_j \partial_k u\|_{L^p} \le c_p \|f\|_{L^p}.$$

Proof We present the proof in the case of $\mathbb{R}^3$ and sketch the required modifications for $\mathbb{T}^3$.

We start by writing $\partial_j \partial_k u = \partial_j \partial_k (-\Delta)^{-1} \Delta u$. Then Theorem B.6 yields the result

$$\|\partial_j \partial_k u\|_{L^p} = \|\partial_j \partial_k (-\Delta)^{-1} \Delta u\|_{L^p} \le c_p \|\Delta u\|_{L^p} = c_p \|f\|_{L^p}.$$

The proof for $\mathbb{T}^n$ relies on the same strategy, writing

$$\partial_j \partial_k u = \partial_j \partial_k (-\Delta)^{-1} \Delta u$$

and using the boundedness (in L^p) of the operator $\partial_j \partial_k (-\Delta)^{-1}$.

In this appendix we have only discussed the boundedness of these operators in $\mathbb{R}^n$; however, the results can be translated to the periodic case, as we now show. In Appendix C we show that the solution of $-\Delta u = f$ in the periodic case is given (see Theorem C.5) by

$$u(x) = \int_{\mathbb{T}^n} K(x, y) f(y) \, dy,$$

where the kernel K can decomposed as

$$\frac{1}{|x - y|} \phi(x - y) + S(x, y),$$

where $\phi(z)$ is smooth and satisfies $\phi(z) = 1$ for $|z| < 1/10$ and $\phi(z) = 0$ for $|z| > 1/4$, and S is smooth. Hence

$$u(x) = \int_{\mathbb{T}^3} \frac{1}{|x-y|}\phi(x-y)f(y)\,dy + \int_{\mathbb{T}^3} S(x,y)f(y)\,dy =: u_m + u_s.$$

It is clear that the term involving S poses no problem, as we can differentiate S directly and use Young's inequality to obtain the required L^p estimate. As for the first term, notice that due to the support properties of ϕ we have,

$$u_m = \int_{\mathbb{T}^3} \frac{1}{|x-y|}\phi(x-y)f(y)\,dy = \int_{\mathbb{R}^3} \frac{1}{|x-y|}\phi(x-y)f(y)\,dy,$$

and so if $Q = [-\pi, \pi)^3$ we obtain, for $x \in Q$

$$\partial_j\partial_k u_m(x) = \partial_j\partial_k \int_{\mathbb{R}^3} \frac{1}{|x-y|}\phi(x-y)f(y)\chi_Q(y)\,dy.$$

Therefore, we can use the result for $\mathbb{R}^3$, as the kernel $\frac{1}{|x|}\phi(x)$ satisfies the same properties as $\frac{1}{|x|}$ to obtain

$$\|u_m\|_{L^p(Q)} \le c_p\|f\chi_Q\|_{L^p(\mathbb{R}^n)} = c_p\|f\|_{L^p(Q)},$$

completing the proof. $\qquad\qquad\qquad\qquad\qquad\qquad\qquad\qquad\qquad\qquad$ $\square$

Appendix C

Elliptic equations

In this appendix we collect some standard results on properties of solutions of the Laplace and Poisson problems.

C.1 Harmonic and weakly harmonic functions

Let Ω be an open subset of $\mathbb{R}^n$ (which need not be bounded). We say that u is harmonic in Ω if $u \in C^2(\Omega)$ and $\Delta u = 0$ in Ω.

We begin with the mean value representations of harmonic functions (Theorem 2.6 in Gilbarg & Trudinger, 1983).

Lemma C.1 (Harmonic mean-value theorems) *If $u \in C^2(\Omega)$ and $\Delta u = 0$ in Ω then for any ball $B = B_r(x) \subset \Omega$*

$$u(x) = \frac{1}{n\omega_n r^{n-1}} \int_{\partial B} u \, dS \tag{C.1}$$

and

$$u(x) = \frac{1}{\omega_n r^n} \int_B u \, dy, \tag{C.2}$$

where ω_n is the volume of the unit ball in $\mathbb{R}^n$.

A useful consequence of these representation formulae are the following interior estimates (Theorem 2.10 in Gilbarg & Trudinger, 1983).

Theorem C.2 (Interior estimates on derivatives) *Let u be a harmonic function in an open set Ω, and let Ω' be any compact subset of Ω. Then for any* *multi-index α we have*

$$\sup_{\Omega'}|D^\alpha u| \le \frac{n|\alpha|^{|\alpha|}}{\text{dist}(\Omega, \Omega')}\sup_{\Omega}|u|.$$

Remarkably, functions that satisfy Laplace's equation weakly are in fact smooth and therefore 'harmonic' in the classical sense. This is the content of Weyl's Lemma; for details of the proof see Theorem 4.7 in Dacorogna (2004), for example.

Theorem C.3 (Weyl's Lemma) *If $u \in L^1_{\text{loc}}(\Omega)$ is weakly harmonic in Ω, i.e.*

$$\int_\Omega u \, \Delta\phi \, dx = 0 \qquad \text{for all } \phi \in C_c^\infty(\Omega),$$

then $u \in C^\infty(\Omega)$ and $\Delta u = 0$ in Ω.

C.2 The Laplacian

C.2.1 Fundamental solution on $\mathbb{R}^3$

For completeness we state a basic result concerning the fundamental solution for the Laplace operator. Define

$$K_3(x) = \frac{1}{4\pi}\frac{1}{|x|}.$$

Theorem C.4 *If f is an integrable function on $\mathbb{R}^3$ then $u = K_3 * f$, i.e.*

$$u(x) = \frac{1}{4\pi}\int_{\mathbb{R}^3}\frac{1}{|x-y|}f(y)\,dy, \tag{C.3}$$

solves the equation $-\Delta u = f$.

We refer the reader to Evans (1998) for more details.

C.2.2 Fundamental solution on $\mathbb{T}^3$

Theorem C.5 *Given a function $f: \mathbb{R}^3 \to \mathbb{R}$, periodic of period 2π, with*

$$\int_{[0,2\pi)^3} f(x)\,dx = 0,$$

there exists a unique solution of the equation

$$-\Delta u = f,$$

which is given by

$$u(x) = \int_{\mathbb{T}^3} K(x, y) f(y) \, dy,$$

with $K(x, y)$ doubly periodic and smooth off the diagonal, that is smooth unless $x = y$ mod 2π. The singularity along the diagonal is of the form $|x - y|^{-1}$, assuming that $|x - y| < \pi$. Here $\mathbb{T}^3$ denotes $\mathbb{R}^3 / (2\pi \mathbb{Z})^3$.

The idea of the proof is to consider f in the unit cube, and extend it to $\mathbb{R}^3$ by zero. We denote this function by f_T. Then we solve $-\Delta v = f_T$ in $\mathbb{R}^3$, obtaining

$$v(x) = \frac{1}{4\pi} \int_{\mathbb{R}^3} \frac{1}{|x - y|} f_T(y) \, dy.$$

We would hope to find u by periodising v, that is by considering

$$\sum_{k \in \mathbb{Z}^3} v(x + 2\pi k).$$

Unfortunately the direct periodisation of the kernel $|x|^{-1}$ is not possible, since

$$\sum_{k \in \mathbb{Z}^3} \frac{1}{|x + 2\pi k|} = \infty.$$

The fact that $\int f_T(y) \, dy = 0$ given the original assumption on f allows us to subtract a counterterm in the kernel, but still that is not enough to ensure that the series converges.

However, the approach sketched above would work if the first few moments of f vanished, as we could subtract additional counterterms. We will now find a set of correction functions that will allow us to adjust the function f_T so that we can apply the periodisation procedure and find a solution to the original problem.

• We need to find periodic functions ϕ_i, $i = 1, 2, 3$ such that

(i) $\int_{\mathbb{T}^3} \Delta \phi_i \, dy = 0$, for every i;
(ii) $\int_{\mathbb{T}^3} y_l \Delta \phi_i \, dy = \delta_{il}$, for every i and l; and
(iii) $\int_{\mathbb{T}^3} y_l y_k \Delta \phi_i \, dy = 0$, for every i, l, and k.

One such example is $\phi_i(y) = c_i \sin(y_i)$, with c_i chosen to satisfy condition (ii).

• Additionally, for every multi-index $\alpha = (\alpha_1, \alpha_2, \alpha_3)$ with $|\alpha| = 2$ we need to find functions ψ_α such that

(i) $\int_{\mathbb{T}^3} \Delta \psi_\alpha \, dy = 0$, for every α;
(ii) $\int_{\mathbb{T}^3} y_l \Delta \psi_\alpha \, dy = 0$, for every α and $l = 1, 2, 3$;
(iii) $\int_{\mathbb{T}^3} y^\beta \Delta \psi_\alpha \, dy = \delta_{\alpha\beta}$, for every $|\beta| = 2$.

One such example is $\psi_\alpha(y) = c_\alpha(\sin(y_i))^2$, in the case in which $\alpha_i = 2$ for some i, and $\psi_\alpha(y) = c_\alpha \sin(y_i) \sin(y_j)$ when $\alpha_i = \alpha_j = 1$ for some $i \neq j$. Again, the c_α are chosen to satisfy condition (iii).

Proof (Theorem C.5) We want to solve $-\Delta u = f$ in $\mathbb{T}^3$, for a periodic function f. It is easy to see that a necessary requirement is that $\int_{\mathbb{T}^3} f = 0$. Coupled with an integrability condition this will also be sufficient.

We consider the function

$$v := u + \left(\int y_j f(y) \, dy \right) \phi_j + \sum_{|\alpha|=2} \left(\int y^\alpha f(y) \, dy \right) \psi_\alpha, \qquad (C.4)$$

where the functions ϕ_i and ψ_α are as above. We notice that

$$-\Delta v = f - \left(\int y_j f(y) \, dy \right) \Delta \phi_j - \sum_{|\alpha|=2} \left(\int y^\alpha f(y) \, dy \right) \Delta \psi_\alpha =: g.$$

It is easy to check that $\int g = 0$, $\int y_i g = 0$ for every $i = 1, 2, 3$, and $\int y^\alpha g = 0$ for every $|\alpha| = 2$.

We will construct below a solution to the equation $-\Delta v = g$, under the above assumptions on the vanishing of the moments of g, and then use (C.4) to find an expression for u. We start by defining the function g_T in the whole of $\mathbb{R}^3$, by $g_T(x) = g(x)\chi_Q$, with $Q = [-\pi, \pi]^3$. The solution of $-\Delta w = g_T$ on $\mathbb{R}^3$ is given by

$$w(x) = \frac{1}{4\pi} \int_{\mathbb{R}^3} \frac{1}{|x-y|} g_T(y) \, dy.$$

We would like to find v by periodising this expression for w, but the kernel is too singular. It is at this point that we use the fact that we have additional information about the integrals of g. Notice that

$$\int \frac{1}{|x-y|} g_T(y) \, dy$$

$$= \int \left[\frac{1}{|x-y|} - \frac{1}{|x|} - \sum_i \frac{x_i}{|x|^3} y_i - \sum_{|\alpha|=2} \left(\frac{\sum_j \delta_{\alpha_j=2}}{|x|^3} + 3\frac{x^\alpha}{|x|^5} \right) y^\alpha \right] g_T(y) \, dy.$$

Given that $g_T(y)$ is supported in $[-\pi, \pi]^3$ we can periodise the expression above to obtain

$$\int_{\mathbb{T}^3} K_1(x, y) g(y) \, dy, \tag{C.5}$$

with $K_1(x, y)$ the periodisation in y, for fixed $x \in \mathbb{R}$ of

$$\left[\frac{1}{|x - y|} - \frac{1}{|x|} - \sum_i \frac{x_i}{|x|^3} y_i - \sum_{|\alpha|=2} \left(\frac{\sum_j \delta_{\alpha_j=2}}{|x|^3} + 3 \frac{x^\alpha}{|x|^5} \right) y^\alpha \right] \chi_Q(y).$$

Notice that, for any fixed $x \in \mathbb{R}$ and $y \in (x - \pi, x + \pi)$, the kernel $K_1(x, y)$ has a sole singularity at $y = x$, where the function locally looks like $|x - y|^{-1}$. We also observe that as x tends to infinity, the function $K_1(x, y)$ decays like $|x|^{-4}$, as a consequence of Taylor's Theorem. Hence it is possible to periodise the expression in x, since

$$\sum_{k \in \mathbb{Z}^3} \frac{1}{|x + 2\pi k|^4}$$

is finite for all $x \neq 0$.

When we periodise expression (C.5), by considering

$$\sum_{k \in \mathbb{Z}^3} \int_{\mathbb{T}^3} K_1(x + 2\pi k, y) g(y) \, dy, \tag{C.6}$$

we obtain

$$\int_{\mathbb{T}^3} K_2(x, y) g(y) \, dy,$$

where $K_2(x, y)$ is doubly periodic, and singular along the diagonal. That is, for a given x, $K(x, y)$ is singular for $y = x \bmod 2\pi$. Assuming that, for a given x, we always consider $y \in (x - \pi, x + \pi)$, we only have the singularity $x = y$, where the function looks like $|x - y|^{-1}$.

Going back to the expression for u, using (C.4), we find

$$u = v - \left(\int y_j f(y) \, dy \right) \phi_j + \sum_{|\alpha|=2} \left(\int y^\alpha f(y) \, dy \right) \psi_\alpha$$

$$= \int_{\mathbb{T}^3} K_2(x, y) f(y) \, dy + \int_Q \left[y_j \phi_j(x) + \sum_{|\alpha|=2} y^\alpha \psi_\alpha(x) \right] f(y) \, dy$$

$$= \int_{\mathbb{T}^3} K(x, y) f(y) \, dy,$$

where

$$K(x, y) = K_2(x, y) + \mathcal{P}\left[y_j\phi_j(x) + \sum_{|\alpha|=2} y^\alpha \psi_\alpha(x)\right]$$

satisfies all the stated properties, where $\mathcal{P}w(x, y)$ denotes the periodisation of w in both x and y – see (C.6) for periodisation in one variable. $\quad\square$

Appendix D
Estimates for the heat equation

A fundamental part of our analysis of the Navier–Stokes equations relies on estimates for weak solutions of heat equations, primarily of the form

$$\partial_t \omega - \Delta \omega = f(x, t), \qquad \omega(x, 0) = 0, \tag{D.1}$$

with right-hand sides having various degrees of regularity.

We will not give a systematic treatment for parabolic equations like (D.1), but rather we will focus on providing complete proofs for the $L_t^q L_x^q$ estimates that we require. For completeness we state some relevant existence and uniqueness results in the following section. There are many books where parabolic theory is treated systematically: we refer the reader to Ladyzhenskaya, Solonnikov, & Uraltseva (1967), Lieberman (1996), and Krylov (1996, 2008), for example.

D.1 Existence, uniqueness, and regularity

We review the well-known explicit formula for the heat kernel below, which, together with the following uniqueness theorem, will suffice for all the results covered in this appendix. Note that since equation (D.1) is linear, for the general case it suffices to consider zero right-hand side to guarantee uniqueness. It is well known that a growth condition on the solution needs to be imposed to obtain uniqueness (see Tychonoff, 1935). We refer the reader to Chung (1999) for a nice review, including sharper results than that presented here.

Theorem D.1 (Shapiro, 1966) *Let ω be a weak solution of the equation*

$$\partial_t \omega - \Delta \omega = 0, \qquad in \ \mathbb{R}^n \times [0, T),$$

that satisfies

393

(i) $\|\omega(x,t)\|_{L^\infty} = o(1/t)$ as $t \to 0$;

(ii) $\lim_{t \to 0} \omega(x,t) = 0$, *except possibly in a countable set E; and*

(iii) $\liminf_{t \to 0} t^{1/2}\omega(x,t) = 0$ *for every x in E.*

Then $\omega(x,t) = 0$ in $\mathbb{R}^n \times [0, T)$.

It is well known that the solution of equation (D.1) with $\omega(x, 0) = 0$ can be written as

$$\omega(x,t) = \int_0^t e^{(t-s)\Delta} f(x, s) \, ds, \qquad (D.2)$$

where

$$e^{t\Delta} f(x) := \int_{\mathbb{R}^n} K(x - y, t) f(y) \, dy$$

and

$$K(x,t) = \frac{1}{(4\pi t)^{n/2}} e^{-|x|^2/4t}. \qquad (D.3)$$

For convenience sometimes we will denote the right-hand side of equation (D.2) by $(\partial_t - \Delta)^{-1} f$.

For the sake of completeness we also state an existence theorem for the heat equation with lower-order terms. Notice that significant improvements can be made on the regularity of the coefficients, but the result in this form suffices for our purposes. We refer the reader to Ladyzhenskaya, Solonnikov, & Uraltseva (1967) for additional results.

Theorem D.2 *Let $a_j(x,t)$, $b(x,t)$ be smooth functions with compact support and $f(x,t) \in L_t^{r'} L_x^r$ for some $1 \le r, r' \le \infty$. Then there exists a unique solution of the equation*

$$\partial_t \omega - \Delta \omega = a_j \partial_j \omega + b\omega + f, \qquad \omega(x, 0) = 0,$$

that belongs to $L_t^{q'} L_x^q$ for some $q, q' \ge 1$.

D.2   Estimates for $e^{t\Delta}\omega_0$

In several instances in Chapter 11 we require results and estimates on the solution of the following initial value problem for the heat equation:

$$\partial_t \omega - \Delta \omega = 0, \qquad \omega(x, 0) = \omega_0. \qquad (D.4)$$

The results above (under some mild growth assumptions) guarantee that the solution is unique. It is well known that the solution ω is given by

$$\omega = e^{t\Delta}\omega_0 = \int K(x - y, t)\omega_0(y)\, dy, \tag{D.5}$$

where K is defined in (D.3).

The following theorem provides a way to justify that the initial condition is attained, given that we are dealing with functions in L^p only.

Theorem D.3 *Let* $\omega_0 \in L^p(\mathbb{R}^n)$, *for some* $1 \le p < \infty$. *Let* ω *be the unique solution of equation* (D.4) *given by* (D.5). *Then*

$$\|\omega(\cdot, t) - \omega_0\|_{L^p} \xrightarrow[t \to 0^+]{} 0.$$

Proof The proof follows almost immediately from Theorem A.9 about approximations to the identity. Indeed, notice that if we define

$$\phi(x) = \frac{1}{(4\pi)^{n/2}} e^{-|x|^2/4t},$$

then ϕ is integrable and $\int \phi = 1$. Notice that if we set $\varepsilon = \sqrt{t}$ then we find that $\phi_\varepsilon(x) = K(x, t)$, where, just as in Theorem A.9 we let $\phi_\varepsilon(x) = \varepsilon^{-n}\phi(x/\varepsilon)$. The result now follows directly from Theorem A.9. $\qquad\square$

We will start by reviewing some integrability properties of $e^{t\Delta}f$ in L_x^p, for fixed $t > 0$.

Theorem D.4 (Basic estimates for $e^{t\Delta}$) *For any* $1 \le l \le r \le \infty$ *we have, for* $t > 0$,

$$\|e^{t\Delta}f\|_{L_x^r} \le C \frac{1}{t^{\frac{n}{2}(\frac{1}{l}-\frac{1}{r})}} \|f\|_{L_x^l} \quad and \tag{D.6}$$

$$\|e^{t\Delta}\nabla f\|_{L_x^r} \le C \frac{1}{t^{\frac{1}{2}+\frac{n}{2}(\frac{1}{l}-\frac{1}{r})}} \|f\|_{L_x^l}. \tag{D.7}$$

Proof Both inequalities are just a simple application of Young's inequality for convolutions (Theorem A.10): if we define a by $1 + \frac{1}{r} = \frac{1}{a} + \frac{1}{l}$ we have

$$\|e^{t\Delta}f\|_{L_x^r} \le C \left\| \frac{1}{t^{n/2}} e^{-\frac{|x|^2}{4t}} \right\|_{L_x^a} \|f\|_{L_x^l}$$

$$= C \frac{1}{t^{n/2}} t^{\frac{n}{2}(1-\frac{1}{l}+\frac{1}{r})} \|f\|_{L_x^l}$$

$$= C \frac{1}{t^{\frac{n}{2}(\frac{1}{l}-\frac{1}{r})}} \|f\|_{L_x^l}.$$

The proof of the second inequality is very similar. For the case $l < \infty$ we use a density argument. We start by assuming that f is actually a Schwartz function ($f \in \mathscr{S}$) or C_c^∞ so that we can justify the integration by parts formula that moves the derivative from f to the kernel K. Then we use the same approach as before using Young's inequality, estimating the time derivatives of K in L^α. This proves the required estimate for Schwartz functions (or C_c^∞). A standard density argument completes the proof, as both classes are dense in L^l, since $l < \infty$. In the case $l = r = \infty$, where the density argument would not be possible, we interpret $e^{t\Delta}\nabla f$ as $\nabla e^{t\Delta} f$, for which we can obtain the desired result using Young's inequality, noting that the derivative of K can be estimated in L^1 (notice $\alpha = 1$ when $l = r = \infty$); we have

$$\left\| \frac{x - y}{t^{n/2+1}} e^{\frac{-|x-y|^2}{4t}} \right\|_{L^1_x} = \frac{C}{t^{1/2}},$$

for some constant C depending only on the dimension. □

Theorem D.4 provides precise estimates for L^r norms of the solution (and the gradient) of equation (D.4), just based on L^r norms of ω_0, showing that the equation regularises the initial data. In fact, the following result shows that for any positive time the solution is actually smooth.

Theorem D.5 *Let $\omega_0 \in L^l(\mathbb{R}^n)$, for some $1 \le l \le \infty$. Then for $t > 0$ the solution $\omega = e^{t\Delta}\omega_0$ is a smooth function.*

Proof The proof follows directly from the fact that for positive times t the function K describing $e^{t\Delta}$ is a Schwartz function, and so all derivatives can be applied to K (that is $\partial^\alpha e^{t\Delta}\omega_0 = \partial^\alpha K * \omega_0$, with the resulting integral still making sense). In fact, repeating the argument used to prove estimate (D.7) we find that the estimate deteriorates by a factor of $t^{-1/2}$ for every space derivative that we take, just as happens between estimates (D.6) and (D.7), and so for any $1 \le l \le r \le \infty$ we have

$$\|\partial^\alpha e^{t\Delta} f\|_{L^r_x} \le C \frac{1}{t^{\frac{|\alpha|}{2} + \frac{n}{2}(\frac{1}{l} - \frac{1}{r})}} \|f\|_{L^l_x}. \qquad \Box$$

D.3 Estimates for $(\partial_t - \Delta)^{-1} f$

With the estimates from Theorem D.4 we move on to $(\partial_t - \Delta)^{-1} f$. For convenience of presentation we split the results for $(\partial_t - \Delta)^{-1} f$ and $(\partial_t - \Delta)^{-1}\nabla f$ in the following two theorems.

Theorem D.6 *Assume that* l, l', r, r' *satisfy either*

(case $r' < \infty$)
$$\begin{cases} 1 \le l \le r \le \infty, & 1 < l' \le r' < \infty, \\ \dfrac{n}{l} + \dfrac{2}{l'} \le \dfrac{n}{r} + \dfrac{2}{r'} + 2 \end{cases} \tag{D.8}$$

or

(case $r' = \infty$)
$$\begin{cases} 1 \le l \le r \le \infty, & 1 < l' \le r' = \infty, \\ \dfrac{n}{l} + \dfrac{2}{l'} < \dfrac{n}{r} + \dfrac{2}{r'} + 2. \end{cases} \tag{D.9}$$

Then there exists $c = c(n, l, l', r, r')$ *such that*

$$\|(\partial_t - \Delta)^{-1} f\|_{L_t^{r'} L_x^r} \le c \|f\|_{L_t^{l'} L_x^l}.$$

We do not make explicit reference to the domain $\mathbb{R}^n \times [0, T)$ in the norms below, but we notice that it will be important for us that we are dealing with a finite time interval.

Observe that the only difference between the two cases treated in the theorem is that when $r' = \infty$ we require a strict inequality in the expression relating all the indices, namely $\frac{n}{l} + \frac{2}{l'} < \frac{n}{r} + \frac{2}{r'} + 2$.

Proof The complete proof is actually rather long and requires several advanced technical tools (see Fabes, Lewis, & Rivière, 1977). On the other hand the cases we use in this book all have simple proofs, which we present now.

Case 1. $l' < r' < \infty$

We consider just the L^r norm in space and use Minkowski's inequality (see Theorem A.1) to obtain

$$\|(\partial_t - \Delta)^{-1} f\|_{L_x^r} \le \int_0^t \left\| \int_{\mathbb{R}^n} \frac{1}{(4\pi(t-s))^{n/2}} e^{\frac{-|\cdot - y|^2}{4(t-s)}} f(y, s) \, dy \right\|_{L_x^r} ds.$$

We now use (D.6) from Theorem D.4 to obtain

$$\|(\partial_t - \Delta)^{-1} f\|_{L_x^r} \le C \int_0^t \frac{1}{(t-s)^{\frac{n}{2}\left(\frac{1}{l} - \frac{1}{r}\right)}} \|f(\cdot, s)\|_{L_x^l} \, ds.$$

Now we use the following weak Young inequality (Theorem A.16)

$$\|g * h\|_{L_t^{r'}} \le C \|g\|_{L_t^{a,\infty}} \|h\|_{L_t^{l'}},$$

where $1 + \frac{1}{r'} = \frac{1}{a} + \frac{1}{l'}$ and $1 < l', r', a < \infty$; we obtain

$$\|(\partial_t - \Delta)^{-1} f\|_{L_t^{r'} L_x^r} \le C \left\| \chi_{t>0}(t) \frac{1}{t^{\frac{n}{2}\left(\frac{1}{l} - \frac{1}{r}\right)}} \right\|_{L_t^{a,\infty}} \|f\|_{L_t^{l'} L_x^l}.$$

Notice that the conditions $1 < l' < r' < \infty$, guarantee that we can use the inequality as all indices are different from 1 and ∞. In order to complete the proof we need to check that

$$a\frac{n}{2}\left(\frac{1}{l} - \frac{1}{r}\right) \le 1,$$

i.e.

$$\frac{n}{2}\left(\frac{1}{l} - \frac{1}{r}\right) \le \frac{1}{a} = \frac{1}{l'} - \frac{1}{r'} + 1,$$

which is just a rearrangement of the hypothesis (D.8) of the theorem.

Case 2. $l' = r' < \infty$ *and* $\frac{n}{l} < \frac{n}{r} + 2$

This case allows for $l' = r'$, but demands a strict inequality for condition (D.8), similar to condition (D.9) even though r' is not necessarily infinity. In this case we apply Young's inequality for convolutions, obtaining

$$\|(\partial_t - \Delta)^{-1}f\|_{L_t^{r'} L_x^r} \le C \left\|\frac{1}{t^{n/2}} e^{\frac{-|x|^2}{4t}}\right\|_{L_t^a L_x^a} \|f\|_{L_t^{l'} L_x^l},$$

provided $1 + \frac{1}{r} = \frac{1}{a} + \frac{1}{l}$ and $1 + \frac{1}{r'} = \frac{1}{a'} + \frac{1}{l'}$, which in our case implies $a' = 1$. Notice that

$$\left\|\frac{1}{t^{n/2}} e^{\frac{-|x|^2}{4t}}\right\|_{L_x^a} = C \frac{1}{t^{n/2}} \frac{1}{t^{-n/2a}},$$

which upon using the constraint for a becomes

$$\frac{1}{t^{\frac{n}{2}\left(\frac{1}{l} - \frac{1}{r}\right)}},$$

which is an element of L_t^1 since the exponent in the denominator is strictly less than 1 by the assumption $\frac{1}{l} < \frac{1}{r} + \frac{2}{n}$.

Case 3. $r' = \infty$

In this case (D.9) reduces to $r' = \infty$ and

$$\frac{n}{l} + \frac{2}{l'} < \frac{n}{r} + 2, \qquad 1 \le l < r \le \infty, \qquad 1 < l' \le \infty,$$

observing that $l = r$ is impossible.

The proof proceeds as before, using Minkowski's inequality to obtain

$$\|(\partial_t - \Delta)^{-1}f\|_{L_x^r} \le \int_0^t \left\|\int_{\mathbb{R}^n} \frac{1}{(4\pi(t-s))^{n/2}} e^{\frac{-|\cdot - y|^2}{4(t-s)}} f(y,s)\, dy\right\|_{L_x^r} ds.$$

We now use (D.6) from Theorem D.4 to obtain

$$\|(\partial_t - \Delta)^{-1}f\|_{L^r_x} \leq C \int_0^t \frac{1}{(t-s)^{\frac{n}{2}(\frac{1}{l}-\frac{1}{r})}} \|f(\cdot, s)\|_{L^l_x}\, ds.$$

We need to control the L^∞_t norm of the above expression. For this we use Hölder's inequality, with exponents l' and $\frac{l'}{l'-1}$. We need

$$\frac{n}{2}\left(\frac{1}{l} - \frac{1}{r}\right)\frac{l'}{l'-1} < 1$$

but we notice that this is equivalent (after rearranging terms) to $\frac{n}{l} + \frac{2}{l'} < \frac{n}{r} + 2$. This also covers the case $l' = \infty$, where we understand $\frac{l'}{l'-1}$ as 1. $\qquad\square$

We now present a similar result to Theorem D.6 but involving derivatives on the left-hand side.

Theorem D.7 *Assume that m, m', r, r' satisfy either*

$$(\text{case } r' < \infty) \qquad \begin{cases} 1 \leq m \leq r \leq \infty, \qquad 1 < m' \leq r' < \infty, \\[2mm] \dfrac{n}{m} + \dfrac{2}{m'} \leq \dfrac{n}{r} + \dfrac{2}{r'} + 1 \end{cases}$$

or

$$(\text{case } r' = \infty) \qquad \begin{cases} 1 \leq m \leq r \leq \infty, \qquad 1 < m' \leq r' = \infty, \\[2mm] \dfrac{n}{m} + \dfrac{2}{m'} < \dfrac{n}{r} + \dfrac{2}{r'} + 1. \end{cases}$$

Then there exists $c = c(n, m, m', r, r')$ such that on $\mathbb{R}^n \times [0, T)$ we have

$$\|(\partial_t - \Delta)^{-1}\nabla f\|_{L^{r'}_t L^r_x} \leq c\|f\|_{L^{m'}_t L^m_x}.$$

We remark that, as in Theorem D.6, the only difference in the case $r' = \infty$ is that we require a strict inequality in $\frac{n}{m} + \frac{2}{m'} < \frac{n}{r} + \frac{2}{r'} + 1$.

Proof The proof of the full result can be found in Fabes, Lewis, & Rivière (1977). We will only require certain particular cases, and we only discuss these.

For the case $m' < r'$ the proof of this theorem follows exactly the same argument as in Theorem D.6, replacing every use of estimate (D.6) by (D.7). This produces a different power of t in the estimate, which is turn is responsible for the different conditions of the two theorems. From there on, every line can be repeated, replacing l, l' by m, m'.

For the case $m' = r'$ and $\frac{n}{m} < \frac{n}{r} + 1$ the proof is the same as in Theorem D.6 except that the estimates are performed on the derivative of the heat kernel (i.e. attaching the derivative to the kernel, rather than f). Young's inequality for weak L^p spaces completes the proof as before.

Finally for the case $r' = \infty$, the same applications of Young's inequality yield the result, again having attached the derivative to the kernel rather than the function. □

D.4 Higher regularity – Hölder estimates

The goal of this section is to prove a higher regularity result. We focus exclusively on covering the applications required for the Navier–Stokes equations and make no attempt to optimise the theorem.

Theorem D.8 *Let W be a solution of the heat equation*

$$W_t - \Delta W = \partial f$$

in the parabolic cylinder $Q_R^*(x_0, t_0)$. Assume that f vanishes outside $Q_{\rho_s R}^*$, for some $\rho_s < 1$ and that $W(x, t_0 - R^2/2) = 0$, i.e. W vanishes on the left lid of the cylinder. Then, for any $\rho_i < \rho_s$

  (i) $f \in L_t^\infty L_x^\infty(Q_R^*) \Rightarrow W \in C_x^{0,\alpha}(Q_{\rho_i R}^*(x_0, t_0))$, *for any $0 < \alpha < 1$;*
 (ii) $f \in L_t^\infty C_x^k \Rightarrow W \in C_x^{k,\alpha}(Q_{\rho_i R}^*(x_0, t_0))$, *for any $0 < \alpha < 1$;*
(iii) $f \in L_t^\infty C_x^{0,\alpha}$ *for some $0 < \alpha < 1 \Rightarrow \nabla W \in L_x^\infty(Q_{\rho_i R}^*(x_0, t_0))$, with the derivatives understood in the classical sense; and*
 (iv) $f \in L_t^\infty C_x^{k,\alpha}$ *for some $0 < \alpha < 1 \Rightarrow \partial^\beta W \in L_x^\infty(Q_{\rho_i R}^*(x_0, t_0))$ for every $|\beta| \leq k + 1$, with the derivatives understood in the classical sense.*

Proof Notice that statements (ii) and (iv) follow directly from (i) and (iii) (respectively) given the linearity of the equation.

(i) *From L^∞ to Hölder*

We may assume, via a translation and a scaling argument, that $R = 1$ and $(x_0, t_0) = (0, 0)$. We want to show that there exists C such that

$$|W(x) - W(z)| \leq C|x - z|^\alpha \qquad \text{for all} \qquad x, z \in Q_{\rho_i}^*$$

and some α. Given that the only assumption on f is boundedness, the additional regularity must arise from the exponential kernel. Notice that the assumptions on the support and the initial data allow us to extend the problem to $\mathbb{R}^3$. We write

$$W(x, t) - W(z, t) = c \int_{-1/2}^{t} \int_{\mathbb{R}^3} \left[\frac{x_j - y_j}{(t-s)^{n/2+1}} e^{-\frac{|x-y|^2}{4(t-s)}} - \frac{z_j - y_j}{(t-s)^{n/2+1}} e^{-\frac{|z-y|^2}{4(t-s)}} \right]$$
$$\times f(y, s) \, dy \, ds.$$

Introducing as new coordinates

$$\xi_j = \frac{x_j}{(t-s)^{1/2}}, \qquad \eta_j = \frac{y_j}{(t-s)^{1/2}}, \qquad \text{and} \qquad \rho_j = \frac{z_j}{(t-s)^{1/2}}$$

the right-hand side becomes

$$c \int_{-1/2}^{t} \int \frac{1}{(t-s)^{1/2}} \left[(\xi_j - \eta_j) e^{-|\xi-\eta|^2} - (\rho_j - \eta_j) e^{-|\rho-\eta|^2} \right]$$
$$\times f((t-s)^{1/2}\eta, s) \, d\eta \, ds. \tag{D.10}$$

Hence

$$\frac{W(x) - W(z)}{|x - z|^\alpha} = c \int_{-1/2}^{t} \int \frac{1}{(t-s)^{1/2+\beta/2}} \frac{[(\xi_j - \eta_j) e^{-|\xi-\eta|^2} - (\rho_j - \eta_j) e^{-|\rho-\eta|^2}]}{|\xi - \rho|^\alpha}$$
$$\times f((t-s)^{1/2}\eta, s) \, d\eta \, ds.$$

We consider two cases, $|\xi - \rho| < 1$ and $|\xi - \rho| \geq 1$. In the second case, we can simply remove the denominator and estimate f in L^∞. This leaves the integral

$$\int_{-1/2}^{t} \frac{1}{(t-s)^{1/2+\beta/2}} \int [|\xi_y - \eta_j| e^{-|\xi-\eta|^2} + |\rho_j - \eta_j| e^{-|\rho-\eta|^2}] \, d\eta \, ds$$
$$\leq C \int_{-1/2}^{t} \frac{1}{(t-s)^{1/2+\beta/2}} \, 2 \int [|\eta| e^{-|\eta|^2} \, d\eta \, ds \leq C(T).$$

In the case $|\xi - \rho| < 1$ we replace $|\xi - \rho|^\beta$ by $|\xi - \rho|$ and apply the Mean Value Theorem to get

$$c \int_{-1/2}^{t} \int \frac{1}{(t-s)^{1/2+\beta/2}} \frac{[(\xi_y - \eta_j) e^{-|\xi-\eta|^2} - (\rho_j - \eta_j) e^{-|\rho-\eta|^2}]}{|\xi - \rho|}$$
$$\times f((t-s)^{1/2}\eta, s) \, d\eta \, ds$$
$$\leq C \|f\|_{L_t^\infty L_x^\infty} \int_{-1/2}^{t} \frac{1}{(t-s)^{1/2+\beta/2}} \int e^{-|c-\eta|^2} + 2|c - \eta|^2 e^{-|c-\eta|^2} \, d\eta \, ds$$
$$\leq C \|f\|_{L_t^\infty L_x^\infty} \int_{-1/2}^{t} \frac{1}{(t-s)^{1/2+\beta/2}} \int e^{-|\eta|^2} + 2|\eta|^2 e^{-|\eta|^2} \, d\eta \, ds \leq C(T),$$

completing the proof of statement (i).

(iii) *From Hölder to C^1*

We want to show that if f is $C^{0,\alpha}$ in the space variables then ∇W is bounded, where

$$W(x, t) = c \int_{-1/2}^{t} \int_{\mathbb{R}^3} \frac{x_j - y_j}{(t - s)^{n/2+1}} e^{-\frac{|x-y|^2}{4(t-s)}} f(y, s) \, dy \, ds.$$

First we observe that

$$\int_{\mathbb{R}^3} \frac{x_j - y_j}{(t - s)^{n/2+1}} e^{-\frac{|x-y|^2}{4(t-s)}} \, dy = \int_{\mathbb{R}^3} \frac{y_j}{(t - s)^{n/2+1}} e^{-\frac{|y|^2}{4(t-s)}} \, dy = 0$$

and as a result, differentiating with respect to x_k we obtain

$$\int_{\mathbb{R}^3} \frac{\delta_{jk}}{(t - s)^{n/2+1}} e^{-\frac{|x-y|^2}{4(t-s)}} - 2\frac{(x_j - y_j)(x_k - y_k)}{(t - s)^{n/2+2}} e^{-\frac{|x-y|^2}{4(t-s)}} \, dy = 0.$$

For $\varepsilon > 0$ we define

$$W_\varepsilon(x, t) = c \int_{-1/2}^{t-\varepsilon} \int_{\mathbb{R}^3} \frac{x_j - y_j}{(t - s)^{n/2+1}} e^{-\frac{|x-y|^2}{4(t-s)}} f(y, s) \, dy \, ds$$

so that we can safely differentiate with respect to x_k. We obtain

$$\partial_k W_\varepsilon(x, t) = \int_{-1/2}^{t-\varepsilon} \int_{\mathbb{R}^3} \left[\frac{\delta_{jk}}{(t - s)^{n/2+1}} e^{-\frac{|x-y|^2}{4(t-s)}} - 2\frac{(x_j - y_j)(x_k - y_k)}{(t - s)^{n/2+2}} e^{-\frac{|x-y|^2}{4(t-s)}} \right]$$
$$\times f(y, s) \, dy \, ds$$
$$= \int_{-1/2}^{t-\varepsilon} \int_{\mathbb{R}^3} \left[\frac{\delta_{jk}}{(t - s)^{n/2+1}} e^{-\frac{|x-y|^2}{4(t-s)}} - 2\frac{(x_j - y_j)(x_k - y_k)}{(t - s)^{n/2+2}} e^{-\frac{|x-y|^2}{4(t-s)}} \right]$$
$$\times [f(y, s) - f(x, s)] \, dy \, ds$$
$$= \int_{-1/2}^{t-\varepsilon} \int_{\mathbb{R}^3} \left[\frac{\delta_{jk}}{(t - s)^{n/2+1}} e^{-\frac{|x-y|^2}{4(t-s)}} - 2\frac{(x_j - y_j)(x_k - y_k)}{(t - s)^{n/2+2}} e^{-\frac{|x-y|^2}{4(t-s)}} \right]$$
$$\times |x - y|^\alpha \frac{[f(y, s) - f(x, s)]}{|x - y|^\alpha} \, dy \, ds$$
$$\leq \|f\|_{C^{0,\alpha}} \int_{-1/2}^{t-\varepsilon} \int_{\mathbb{R}^3} \frac{(t - s)^{\alpha/2}}{t - s} |\xi - \eta|^\alpha$$
$$\times \left[\delta_{jk} e^{-|\xi-\eta|^2} + 2|\xi_j - \eta_j| \, |\xi_k - \eta_k| e^{-|\xi-\eta|^2} \right] d\eta \, ds$$
$$\leq C(t) \|f\|_{C^{0,\alpha}},$$

since as before we can carry out the space integration by a translation. Here we have used the same change of variables as in the proof of statement (i) (see (D.10)). Since the estimate above is independent of ε we obtain the desired result. $\square$

A simpler result, which can be obtained as a corollary (by considering the equation for the derivative of the solution) when the right-hand side does not contain a derivative is presented below. We do not attempt to find an optimal result.

Theorem D.9 *Let W be a solution of the heat equation*

$$W_t - \Delta W = g$$

in the parabolic cylinder $Q_R^*(x_0, t_0)$. Assume that g vanishes outside $Q_{\rho_s R}^*$, for some $\rho_s < 1$ and that $W(x, t_0 - R^2/2) = 0$, i.e. W vanishes on the left lid of the cylinder. Then, for any $\rho_i < \rho_s$ we have

(i) $g \in L_t^\infty L_x^\infty(Q_R^*) \Rightarrow W \in C_x^{1,\alpha}(Q_{\rho_i R}^*(x_0, t_0))$, for any $0 < \alpha < 1$;

(ii) $g \in L_t^\infty C_x^k \Rightarrow W \in C_x^{k+1,\alpha}(Q_{\rho_i R}^*(x_0, t_0))$, for any $0 < \alpha < 1$;

(iii) $g \in L_t^\infty C_x^{0,\alpha}$ for some $0 < \alpha < 1 \Rightarrow \partial^\beta W \in L_x^\infty(Q_{\rho_i R}^*(x_0, t_0))$ for any $|\beta| \leq 2$, with the derivatives understood in the classical sense; and

(iv) $g \in L_t^\infty C_x^{k,\alpha}$ for some $0 < \alpha < 1 \Rightarrow \partial^\beta W \in L_x^\infty(Q_{\rho_i R}^*(x_0, t_0))$ for every $|\beta| \leq k + 2$, with the derivatives understood in the classical sense.

The result above is stated for centred cylinders, but we remark that the proof above is carried out in $\mathbb{R}^3$, as the conditions on the support of f and initial data of W allow for an extension to zero of the solution. The result is still valid for cylinders $Q_R(x_0, t_0)$, as in Chapter 15 (see Figure 15.1), where (x_0, t_0) corresponds to the centre of the right lid of the parabolic cylinder, as the same extension of the solutions is still valid.

D.5 Maximal regularity for the heat equation

We now prove 'maximal regularity' for solutions of the heat equation on $\mathbb{R}^n$: we show that if u satisfies

$$\partial_t u - \Delta u = f, \qquad u(0) = 0,$$

then there is a constant $C_{p,q}$ such that

$$\|\partial_t u\|_{L^p(0,T;L^q(\mathbb{R}^n))} + \|\Delta u\|_{L^p(0,T;L^q(\mathbb{R}^n))} \leq C_{p,q} \|f\|_{L^p(0,T;L^q(\mathbb{R}^n))},$$

for any $1 < p, q < \infty$. We follow the instructive sketch of the argument given by Krylov (2001).

We will require two results from harmonic analysis: a multiplier theorem (Theorem D.10) due to Mihlin (1957) and Hörmander (1960), and a result on Banach-space-valued convolution operators that is a consequence of a generalisation of the Calderón–Zygmund Theorem (Theorem B.3) due to Benedek, Calderón, & Panzone (1962).

We begin with the multiplier theorem, which can be found (along with a proof) as Theorem 3 in Chapter IV, Section 3, of Stein (1970).

Theorem D.10 (Mihlin, Hörmander) *Suppose that m is C^k on $\mathbb{R}^n \setminus \{0\}$ for some $k > n/2$. If*

$$|\partial^\alpha m(x)| \leq C|x|^{-|\alpha|} \qquad \text{for every} \quad |\alpha| \leq k$$

then the map T_m defined by

$$\widehat{T_m f}(\xi) = m(\xi)\hat{f}(\xi)$$

is bounded from $L^2(\mathbb{R}^n) \cap L^p(\mathbb{R}^n)$ into $L^p(\mathbb{R}^n)$ for every $1 < p < \infty$.

The following result can be found as Theorem 2 in Benedek, Calderón, & Panzone (1962). We give a much less general version of their result; we follow the formulation in Krylov (2001), but tailor it to the application in this chapter (we take $B_1 = B_2 = X$ and an operator K that depends on only one variable). In the statement of the theorem, $L_0^\infty(\mathbb{R}; X)$ denotes the collection of all strongly measurable, essentially bounded mappings from $\mathbb{R}$ into X that have compact support.

Theorem D.11 (Benedek, Calderón, and Panzone) *Let X be a Banach space and suppose that for $t \in \mathbb{R}$, $t \neq 0$, $K(t): X \to X$ is a bounded operator such that*

 (i) *K is strongly measurable;*
 (ii) *$\|K(t)\|_{\mathscr{L}(X,X)}$ is bounded outside any neighbourhood of $t = 0$; and*
(iii) *$K(t)$ is weakly differentiable in t and*

$$|\partial_t K(s)| \leq M|s|^{-2}.$$

For $f \in L_0^\infty(\mathbb{R}; X)$ define

$$Tf(t) = \int_{\mathbb{R}^d} K(t - s)f(s)\,\mathrm{d}s$$

and suppose that for some p with $1 < p < \infty$ we have

$$\|Tf\|_{L^p(\mathbb{R};X)} \leq c\|f\|_{L^p(\mathbb{R};X)}.$$

Then

$$\|Tf\|_{L^q(\mathbb{R};X)} \leq c\|f\|_{L^q(\mathbb{R};X)}$$

for every $1 < q < \infty$, and consequently T has a unique extension to a bounded operator from $L^q(\mathbb{R}; X)$ into $L^q(\mathbb{R}; X)$ for any $1 < q < \infty$.

Theorem D.12 (Maximal regularity for the heat equation) *If u is the solution of*

$$\partial_t u - \Delta u = f, \qquad x \in \mathbb{R}^n, \qquad u(0) = 0, \tag{D.11}$$

with $f \in L^q(0, T; L^p(\mathbb{R}^n))$ *given by (D.2) then*

$$\|\partial_t u\|_{L^q(0,T;L^p(\mathbb{R}^n))} + \|\Delta u\|_{L^q(0,T;L^p(\mathbb{R}^n))} \le C \|f\|_{L^q(0,T;L^p(\mathbb{R}^n))}. \tag{D.12}$$

Proof We concentrate on proving the second derivative estimates in (D.12); the time derivative estimate follows identical reasoning.

We use the fact that we have an explicit form for the solution of (D.11) in terms of the heat kernel, namely

$$u(x, t) = \int_0^t \int_{\mathbb{R}^n} G(x - y, t - s) f(y, s) \, dy \, ds,$$

where $G(x, t) = (4\pi t)^{-n/2} e^{-|x|^2/4t}$. We therefore have

$$\Delta u(x, t) = Tf := \int_0^t \int_{\mathbb{R}^n} [\Delta G(x - y, t - s)] f(y, s) \, dy \, ds,$$

which we can write as

$$\int_{\mathbb{R} \times \mathbb{R}^n} \Gamma(x - y, t - s) F(y, s) \, dy \, ds \tag{D.13}$$

if we define $\Gamma \colon \mathbb{R}^{n+1} \to \mathbb{R}$ and $F \colon \mathbb{R}^{n+1} \to \mathbb{R}$ by setting

$$\Gamma(x, t) = \begin{cases} \Delta G(x, t) & t > 0, \\ 0 & t < 0, \end{cases} \quad \text{and} \quad F(x, t) = \begin{cases} f(x, t) & 0 < t < T, \\ 0 & \text{otherwise.} \end{cases}$$

In order to apply Theorem D.11 we first have to show maximal regularity in the case $p = q$. To do this we follow the argument in Ladyzhenskaya, Solonnikov, & Uraltseva (1967), Chapter IV, Section 3, pages 289–290 where they prove their inequality (3.1), which is our (D.12).

We only have to check that the operator T corresponds to a Fourier multiplier that satisfies the conditions of the Mihlin–Hörmander Theorem (Theorem D.10). This is simple, since taking the Fourier transform of Γ in all variables (x and t) we obtain

$$\hat{\Gamma}(\xi, \xi_0) = -\frac{|\xi|^2}{i\xi_0 + |\xi|^2},$$

which does indeed satisfy the conditions of Theorem D.10. We immediately obtain

$$\|\Delta u\|_{L^p((0,T)\times\mathbb{R}^3)} \le \|\Gamma * F\|_{L^p(\mathbb{R}^4)} \le C \|F\|_{L^p(\mathbb{R}^4)} = C \|f\|_{L^p(0,T;L^p(\mathbb{R}^3))}.$$

We now apply Theorem D.11 with $X = L^p(\mathbb{R}^n)$. We denote the heat semi-group on $\mathbb{R}^n$ by $e^{\Delta t}$; then by definition

$$\partial_t[e^{\Delta t} f] = \Delta[e^{\Delta t} f]$$

and by a simple application of Young's inequality for convolutions (see the proof of Theorem D.5) we have

$$\|\partial^\alpha e^{\Delta t} f\|_{L^p} \le N_k t^{-k/2} \|f\|_{L^p}, \qquad \text{where} \quad k = |\alpha|. \tag{D.14}$$

For $t \le 0$ we set $K(t) = 0$, and for $t > 0$ we define

$$K(t) = \Delta e^{\Delta t};$$

by (D.14) we have $K(t) \colon L^p(\mathbb{R}^n) \to L^p(\mathbb{R}^n)$ with

$$\|K(t) f\|_{L^p} \le N t^{-1} \|f\|_{L^p}.$$

For $t \le 0$ we set $K(t) = 0$. The expression in (D.13) for $\Delta u(t)$ can be written in this notation as

$$\int_{\mathbb{R}} K(t - s) f(s) \, ds =: (T f)(t).$$

The only assumption on K required for Theorem D.11 that is not immediate is (iii), which is satisfied since for $t > 0$

$$\|\partial_t K(s) f\|_{L^p} = \|\partial_t \Delta e^{\Delta s} f\|_{L^p} = \|\Delta^2 e^{\Delta s} f\|_{L^p} \le N |s|^{-2} \|f\|_{L^p}$$

by (D.14) and the result follows. □

Appendix E

A measurable-selection theorem

In this appendix we prove a measurable-selection theorem that we use in Chapter 17. We start with some basic definitions. Throughout this chapter we assume that X and Y are metric spaces and that X is equipped with a σ-algebra of measurable sets.

A multi-function $\Lambda \colon X \to Y$ is a map from X into the subsets of Y. We say that Λ is compact valued if $\Lambda(x)$ is a compact subset of Y for every $x \in X$, and that Λ is measurable if $\Lambda^{-1}(A)$ is a measurable subset of X for any closed subset A of Y. A single-valued mapping $f \colon X \to Y$ is measurable if $f^{-1}(A)$ is measurable for each closed subset A of Y.

In the case of a single-valued mapping into a subset Y of $\mathbb{R}^n$, the requirement that $f^{-1}(A)$ is measurable for each closed subset A of Y is equivalent to the same property holding for open subsets, and both are equivalent to this holding for any Borel subset of Y. For general metric spaces Y these two notions are equivalent when Y is metrisable; if not the notion defined here is stronger than requiring the same condition for only open sets A – see Robertson (1974) for details.

The following theorem, due to Castaing (1967), guarantees the existence of a 'measurable selection' provided that Λ is measurable and compact-valued.

Theorem E.1 *If X and Y are separable metric spaces and $\Lambda \colon X \to Y$ is a measurable compact-valued multi-function then there exists a measurable (single-valued) function $f \colon X \to Y$ such that $f(x) \in \Lambda(x)$ for every $x \in X$ (see Figure E.1).*

We prove this theorem following Robertson (1974). We first make two simple observations that allow us to construct new measurable multi-functions. First, suppose that $\Lambda \colon X \to Y$ is a measurable multi-function, and B a closed subset

407

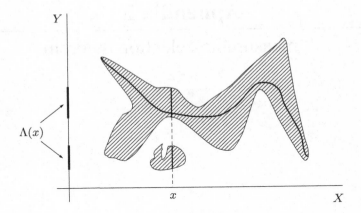

Figure E.1. Schematic graph of a multi-function $\Lambda\colon X \to Y$ (the shaded area) along with a measurable selection $f\colon X \to Y$ (in bold). For each $x \in X$, $\Lambda(x)$ is a subset of Y.

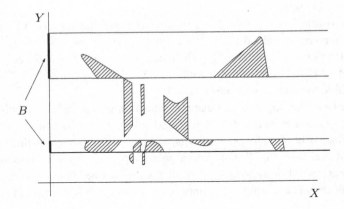

Figure E.2. The graph of the multi-function $[\Lambda \cap B]$, for the multi-function Λ in Figure E.1 and the choice of $B \subset Y$ as indicated.

of Y. Then the multi-function $[\Lambda \cap B]\colon X \to Y$, defined by setting

$$[\Lambda \cap B](x) = \begin{cases} \Lambda(x) \cap B & x \in \Lambda^{-1}(B), \\ \Lambda(x) & \text{otherwise,} \end{cases} \tag{E.1}$$

(see Figure E.2) is still measurable, since for any closed set $A \subset Y$ the set

$$[\Lambda \cap B]^{-1}(A) = (\Lambda^{-1}(A \cap B)) \cup (\Lambda^{-1}(A) \setminus \Lambda^{-1}(B))$$

is measurable. We make the trivial observation (which will be useful later) that

$$[\Lambda \cap B](x) \subseteq \Lambda(x) \qquad \text{for all} \quad x \in X. \tag{E.2}$$

Now suppose that Λ_n is a decreasing sequence of compact-valued measurable multi-functions, i.e. $\{\Lambda_n(x)\}$ is a decreasing sequence of compact sets for every $x \in X$. Then for each $x \in X$ the set

$$\Lambda_\infty(x) := \bigcap_{n=1}^{\infty} \Lambda_n(x)$$

is non-empty and hence Λ_∞ is a (compact-valued) multi-function. Now, observe that if A is closed then

$$\Lambda_\infty^{-1}(A) = \bigcap_{n=1}^{\infty} \Lambda_n^{-1}(A), \tag{E.3}$$

from which it follows that $\Lambda_\infty^{-1}(A)$ is measurable and so Λ_∞ is a measurable multi-function.

To check that (E.3) holds, observe that if $x \in \Lambda_\infty^{-1}(A)$ then $\Lambda_\infty(x) \cap A \neq \varnothing$, and so $\Lambda_n(x) \cap A \neq \varnothing$ for every n, i.e. $x \in \cap_{n=1}^{\infty} \Lambda_n^{-1}(A)$; on the other hand, if $\Lambda_n(x) \cap A \neq \varnothing$ for every n then $\{\Lambda_n(x) \cap A\}$ is a decreasing sequence of non-empty compact sets, and so

$$\Lambda_\infty(x) \cap A = \bigcap_{n=1}^{\infty} \{\Lambda_n(x) \cap A\} \neq \varnothing,$$

i.e. $x \in \Lambda_\infty^{-1}(A)$ and we have equality in (E.3).

We can now prove Theorem E.1.

Proof of Theorem E.1 Since Y is a separable metric space it has a countable base $\{U_n\}$ of open sets. We set $\Lambda_0 = \Lambda$ and define a sequence of compact-valued multi-functions Λ_n by induction, setting

$$\Lambda_{n+1} = [\Lambda_n \cap \overline{U_{n+1}}].$$

Because of (E.2) this is a decreasing sequence, and so it follows from the discussion above that the multi-function $\Lambda_\infty : X \to Y$ defined by

$$\Lambda_\infty(x) := \bigcap_{n=1}^{\infty} \Lambda_n(x)$$

is measurable.

We complete the proof by showing that in fact $\Lambda_\infty(x)$ consists of a single point for each $x \in X$. Fix some $x \in X$, choose $y \in \Lambda_\infty(x)$, and take $z \in Y$ with

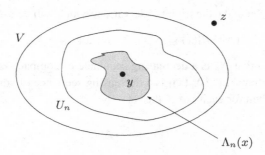

Figure E.3. If $y \in \Lambda_\infty(x)$ and $z \neq y$ then $z \notin \Lambda_\infty(x)$, i.e. $\Lambda_\infty \colon X \to Y$ is single valued. Unlike the other figures in this appendix, this figure shows only points and sets in the space Y.

$z \neq y$. Then there is a closed neighbourhood V of y in Y that does not contain z. Since $\{U_n\}$ is a base of open sets for Y, there exists an n such that $U_n \subseteq V$, and so also $\overline{U_n} \subseteq V$, see Figure E.3.

Now, since $y \in \Lambda_\infty(x)$, in particular $y \in \Lambda_{n-1}(x)$, and by our choice of n we know that $y \in \overline{U_n}$. So by the definition of $[\Lambda_{n-1} \cap \overline{U_n}]$ in (E.1) it follows that

$$\Lambda_n(x) = \Lambda_{n-1}(x) \cap \overline{U_n}.$$

In particular, $\Lambda_\infty(x) \subseteq \Lambda_n(x) \subseteq \overline{U_n}$ and so $z \notin \Lambda_\infty(x)$. It follows that $\Lambda_\infty \colon X \to Y$ is a single-valued measurable function, as required. $\qquad \square$

In fact the measurability condition on the original multi-function Λ can be easily checked if every closed set in X is measurable and we make an assumption on the graph of Λ. Recall that the graph of $\Lambda \colon X \to Y$ is the subset of $X \times Y$ given by

$$\mathcal{G}(\Lambda) := \{(x, y) \colon y \in \Lambda(x), \, x \in X\}.$$

Lemma E.2 *Suppose that every closed subset of X is measurable. Let $\Lambda \colon X \to Y$ be a closed-valued multi-function whose graph is a closed subset of $X \times Y$. Then $\Lambda \colon X \to Y$ is measurable.*

Proof Denote the graph of Λ by $\mathcal{G}$, and let A be a closed subset of Y. Then

$$\Lambda^{-1}(A) = P_X((X \times A) \cap \mathcal{G}),$$

where P_X denotes the projection onto X, $P_X(x, y) = x$, see Figure E.4.

Indeed, if $x \in \Lambda^{-1}(A)$ then there exists a $z \in Y$ such that $z \in \Lambda(x) \cap A$, and so $(x, z) \subset \mathcal{G}$ and $(x, z) \subset X \times A$, yielding one inclusion. In the opposite direction, if $(x, z) \in X \times A$ and $(x, z) \in \mathcal{G}$ then $z \in \Lambda(x)$ and $z \in A$, which implies that $\Lambda(x) \cap A \neq \varnothing$ and so $x \in \Lambda^{-1}(A)$.

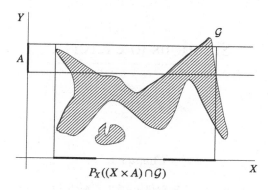

Figure E.4. $\Lambda^{-1}(A) = P_X((X \times A) \cap \mathcal{G})$ is shown as a darker subset of X.

Since X and A are closed so is $X \times A$. Since $\mathcal{G}$ is closed so is $(X \times A) \cap \mathcal{G}$, and so $P_X((X \times A) \cap \mathcal{G})$ is closed. It follows that $\Lambda^{-1}(A)$ is closed, and therefore measurable. $\qquad \square$

We apply Theorem E.1 in a variant that takes into account this result.

Corollary E.3 *Let X and Y be separable metric spaces and suppose that $\Lambda : X \to Y$ is a compact-valued multi-function whose graph is closed. Then if every closed subset of X is measurable there is a measurable function $f : X \to Y$ such that $f(x) \in \Lambda(x)$ for every $x \in X$.*

Solutions to exercises

Chapter 1

1.1 Write $ab = \exp(\log(ab)) = \exp(\log a + \log b)$, and then

$$ab = \exp\left(\frac{1}{p}\log a^p + \frac{1}{q}\log b^q\right)$$

$$\leq \frac{1}{p}\exp(\log a^p) + \frac{1}{q}\exp(\log b^q)$$

$$= \frac{a^p}{p} + \frac{b^q}{q},$$

where we can use the convexity of the exponential function as $1/p + 1/q = 1$.

1.2 Let $u = (1 + |x|)^{-3/q}$ then

$$\int_{\mathbb{R}^3} |u(x)|^q \, dx = \int_{\mathbb{R}^3} \frac{1}{(1 + |x|)^3} \, dx = \infty,$$

but clearly $u \in L^\infty(\mathbb{R}^3)$ and

$$\int_{\mathbb{R}^3} |u(x)|^p \, dx = \int_{\mathbb{R}^3} \frac{1}{(1 + |x|)^{3p/q}} \, dx \leq 4\pi \int_0^\infty \frac{1}{(1 + r)^{3p/q - 2}} \, dr < \infty.$$

1.3 Take (p, q, r) such that $p^{-1} + q^{-1} + r^{-1} = 1$. First apply Hölder's inequality with exponents (p, p'), where (p, p') are conjugate exponents, to yield

$$\int_\Omega |uvw| \leq \left(\int_\Omega |u|^p\right)^{1/p} \left(\int_\Omega |vw|^{p'}\right)^{1/p'}.$$

Now note that since

$$\frac{p'}{q} + \frac{p'}{r} + \frac{p'}{p} = p'$$

it follows that

$$\frac{p'}{q} + \frac{p'}{r} = p'\left(1 - \frac{1}{p}\right) = p'\left(\frac{1}{p'}\right) = 1,$$

i.e. the exponents $(q/p', r/p')$ are conjugate, and so applying Hölder's inequality again we obtain

$$\int_\Omega |uvw| \le \|u\|_{L^p} \|v\|_{L^q} \|w\|_{L^r}$$

as required.

1.4 Write

$$\int_\Omega |u|^r \, dx = \int_\Omega |u|^{\alpha r} |u|^{(1-\alpha)r} \, dx.$$

We want to use Hölder's inequality with exponents $(p/\alpha r, q/(1-\alpha)r)$. To ensure that these are conjugate we need

$$\frac{\alpha r}{p} + \frac{(1-\alpha)r}{q} = 1;$$

dividing by r yields the relationship between the exponents in (1.7).

1.5 Writing $I(u, v, w) = [(u \cdot \nabla)v] \cdot w$ in terms of components yields

$$I = \sum_{i,j} u_j (\partial_j v_i) w_i.$$

Then using the Cauchy–Schwarz inequality twice, first in j and then in i,

$$|I| \le \sum_{i,j} |u_j| |\partial_j v_i| |w_i|$$

$$\le \sum_i \left(\sum_j |u_j|^2 \right)^{1/2} \left(\sum_j |\partial_j v_i|^2 \right)^{1/2} |w_i|$$

$$\le |u| \left(\sum_{i,j} |\partial_j v_i|^2 \right)^{1/2} \left(\sum_i |w_i|^2 \right)^{1/2}$$

$$= |u| |\nabla v| |w|.$$

1.6 We simply use Hölder's inequality with exponents $1/\theta$ and $1/(1-\theta)$,

$$\|u\|_{\dot H^s}^2 = (2\pi)^3 \sum_{k \in \mathbb{Z}^3} |k|^{2s} |\hat u_k|^2$$

$$= (2\pi)^3 \sum_{k \in \mathbb{Z}^3} |k|^{2\theta s_1} |\hat u_k|^{2\theta} |k|^{2(1-\theta)s_2} |\hat u_k|^{2(1-\theta)}$$

$$\le \left((2\pi)^3 \sum_{k \in \mathbb{Z}^3} |k|^{2s_1} |\hat u_k|^2 \right)^\theta \left((2\pi)^3 \sum_{k \in \mathbb{Z}^3} |k|^{2s_2} |\hat u_k|^2 \right)^{1-\theta}$$

$$= \|u\|_{\dot H^{s_1}}^{2\theta} \|u\|_{\dot H^{s_2}}^{2(1-\theta)}.$$

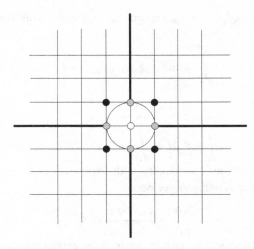

Figure S.1. Decomposing $\sum_{k\in\dot{\mathbb{Z}}^2} f(|k|)$ (the origin (open circle) is not included in the sum). Four one-dimensional sums (bold lines) can be removed and treated separately: the grey circles each contribute $f(1)$ and the remaining one-dimensional sums can be bounded by $\int_1^\infty f(r)\,dr$. The remaining two-dimensional sum can be bounded by treating the four black circles separately ($f(\sqrt{2}) \le f(1)$) and then bounding the remaining sum with an integral, yielding $4f(1) + \int_1^\infty (2\pi r)f(r)\,dr$ and overall $8f(1) + \int_1^\infty (1 + 2\pi r)f(r)\,dr \le 8f(1) + (2\pi + 1)\int_1^\infty rf(r)\,dr.$

1.7 First, notice that we can partition elements $k = (k_1, k_2, k_3) \in \dot{\mathbb{Z}}^3$ according to how many of the k_j are non-zero, and then we can write

$$\sum_{k\in\dot{\mathbb{Z}}^3} f(|k|) = 6\sum_{k_1=1}^\infty f(k_1) + 3\sum_{k'=(k_1,k_2):\, k_1k_2\neq 0} f(|k'|) + \sum_{k:\, k_1k_2k_3\neq 0} f(|k|);$$

the idea is illustrated in Figure S.1.

We can now bound each term above and below by suitable integrals. For an upper bound we observe that $f(|k|) \le f(x)$ on the unit cube that has k as a vertex and within which $|x| \le |k|$. In this way, excluding from the integration the finite number of k for which this includes a cube with an element of modulus < 1, we can bound the sum by

$$26f(1) + 3\int_{x_1\in\mathbb{R}:\, |x_1|\ge 1} f(|x_1|)\,dx_1 + 3\int_{x'\in\mathbb{R}^2:\, |x'|\ge 1} f(|x'|)\,dx' + \int_{x\in\mathbb{R}^3:\, |x|\ge 1} f(|x|)\,dx$$

$$\le C\left(f(1) + \int_1^\infty (1 + r + r^2)f(r)\,dr\right) \le 3C\left(f(1) + \int_1^\infty r^2 f(r)\,dr\right),$$

where we have used the fact that $f(\sqrt{3}) \le f(\sqrt{2}) \le f(1)$ to deal with the excluded points (see Figure S.1, again). The lower bound follows similarly.

1.8 We have

$$\int_{\mathbb{R}^d} \frac{1}{1 + |x|^{2s}} \, dx = c_d \int_0^\infty \frac{r^{d-1}}{1 + r^{2s}} \, dr$$

and hence

$$\frac{c_d}{2} \int_1^\infty r^{d-1-2s} \, dr \le \int_{\mathbb{R}^d} \frac{1}{1 + |x|^{2s}} \, dx \le c_d \left(1 + \int_1^\infty r^{d-1-2s} \, dr\right)$$

and so the integral is finite if and only if $s > d/2$.

For the sum we use Exercise 1.7; upper and lower bounds are determined by

$$\int_1^\infty \frac{r^2}{1 + r^{2s}} \, dr,$$

and the argument is identical to that above.

1.9 We show that u is continuous by showing that the partial Fourier sums (which are continuous) converge uniformly on $\mathbb{T}^3$. If $u_n(x) = \sum_{|k| \le n} \hat{u}_k e^{ik \cdot x}$ it follows that

$$|u_n(x)| \le \sum_{|k| \le n} |\hat{u}_k| = \sum_{|k| \le n} (1 + |k|^{2s})^{-1/2} (1 + |k|^{2s})^{1/2} |\hat{u}_k|$$

$$\le \left(\sum_{|k| \le n} \frac{1}{1 + |k|^{2s}}\right)^{1/2} \left(\sum_{|k| \le n} (1 + |k|^{2s}) |\hat{u}_k|^2\right)^{1/2}.$$

Noting that the second term is bounded by $\|u\|_{H^s}$ and that $\sum_k \frac{1}{1+|k|^{2s}}$ is finite provided that $2s > 3$ (see Exercise 1.7) it follows that u_n converges uniformly to u as $n \to \infty$. The bound (1.32) follows, along with the continuity of u.

1.10 Since $u \in H^s(\mathbb{T}^3)$, $s > 3/2$, we know that $u \in C^0(\mathbb{T}^3)$ and that the Fourier expansion converges uniformly to u on $\mathbb{T}^3$. So we can write

$$|u(x)| \le \sum_k |\hat{u}_k|$$

$$= \sum_{|k| \le N} |\hat{u}_k| + \sum_{|k| > N} |\hat{u}_k|$$

$$\leq \left(\sum_{|k| \leq N} \frac{1}{1 + |k|^{2r}} \right)^{1/2} \left(\sum_{|k| \leq N} (1 + |k|^{2r}) |\hat{u}_k|^2 \right)^{1/2}$$

$$+ \left(\sum_{|k| > N} \frac{1}{1 + |k|^{2s}} \right)^{1/2} \left(\sum_{|k| > N} (1 + |k|^{2s}) |\hat{u}_k|^2 \right)^{1/2}$$

$$\leq cN^{\frac{3}{2} - r} \|u\|_{H^r} + cN^{-(s - \frac{3}{2})} \|u\|_{H^s};$$

now choose N so that both terms are of the same order of magnitude, that is $N^{s-r} \sim \|u\|_{H^s} / \|u\|_{H^r}$, which yields (1.33).

1.11 Consider the map $F \colon (0, 1) \to L^\infty(0, 1)$ defined by setting

$$F(t)(x) = f(x, t).$$

Note that for any $t \in (0, 1)$, $F(s) \notin B_{1/2}(F(t))$ for all $s \neq t$, since whenever $s \neq t$ we have

$$\|F(t) - F(s)\|_{L^\infty} = \|f(\cdot, t) - f(\cdot, s)\|_{L^\infty} = 1.$$

It follows that for any $t \in (0, 1)$

$$F^{-1}(B_{1/2}(F(t))) = \{s \in (0, 1) : \ F(s) \in B_{1/2}(F(t))\} = \{t\}. \qquad \text{(S.1)}$$

Now let E be a non-measurable subset of $(0, 1)$ and consider the open subset of $L^\infty(0, 1)$ defined by

$$U = \bigcup_{t \in E} B_{1/2}(F(t)).$$

It follows from (S.1) that $F^{-1}(U) = E$; since U is open but $F^{-1}(U) = E$ is not measurable, the map F is not measurable.

1.12 From the interpolation inequality we have

$$\|u\|_{H^{k_1}} \leq \|u\|_{H^{l_1}}^\alpha \|u\|_{H^{l_2}}^{1-\alpha}$$

$$\|u\|_{H^{k_2}} \leq \|u\|_{H^{l_1}}^\beta \|u\|_{H^{l_2}}^{1-\beta},$$

where

$$k_1 = \alpha l_1 + (1 - \alpha) l_2 \quad \text{and} \quad k_2 = \beta l_1 + (1 - \beta) l_2.$$

Since $k_1 + k_2 = l_1 + l_2$ it follows that $\alpha + \beta = 1$ and we obtain (1.34).

1.13 Let $k \in \mathbb{N}$ be such that $0 \le k \le s$. From the interpolation inequality (1.22) we have

$$
\begin{aligned}
\|u\|_{\dot{H}^k}^2 &\le \|u\|^{2\alpha}\|u\|_{\dot{H}^s}^{2(1-\alpha)} \\
&\le \alpha\|u\|^{2\alpha\frac{1}{\alpha}} + (1-\alpha)\|u\|_{\dot{H}^s}^{2(1-\alpha)\frac{1}{(1-\alpha)}} \\
&\le \alpha\|u\|^2 + (1-\alpha)\|u\|_{\dot{H}^s}^2,
\end{aligned}
$$

which immediately proves the result.

Chapter 2

2.1 Let $I = \int_{\mathbb{R}^3} h(\nabla \times g)$. From integration by parts

$$
\begin{aligned}
I_i &= \int_{\mathbb{R}^3} h\epsilon_{ijk}\partial_j g_k \\
&= -\int_{\mathbb{R}^3} (\partial_j h)\epsilon_{ijk}g_k = \int_{\mathbb{R}^3} (\partial_j h)\epsilon_{ikj}g_k \\
&= \int_{\mathbb{R}^3} (g \times \nabla h)_i.
\end{aligned}
$$

2.2 We have $(\nabla \times A)_k = \epsilon_{kmn}\partial_m A_n$ and so

$$
\begin{aligned}
(\nabla \times (\nabla \times A))_i &= \epsilon_{ijk}\partial_j(\epsilon_{kmn}\partial_m A_n) = \epsilon_{ijk}\epsilon_{kmn}\partial_j\partial_m A_n \\
&= [\delta_{im}\delta_{jn} - \delta_{in}\delta_{jm}]\partial_j\partial_m A_n = \partial_j\partial_i A_j - \partial_j\partial_j A_i \\
&= (\nabla(\nabla \cdot A))_i - (\Delta A)_i,
\end{aligned}
$$

completing the proof.

2.3 From Theorem C.4 we have

$$
w(x) = \frac{1}{4\pi} \int_{\mathbb{R}^3} \frac{-\Delta w(y)}{|x-y|}\, dy,
$$

so from Exercises 2.1 and 2.2, and integration by parts:

$$
\begin{aligned}
w(x) &= \frac{1}{4\pi} \int_{\mathbb{R}^3} \frac{-\nabla \mathrm{div}\, w(y) + \nabla \times (\nabla \times w(y))}{|x-y|}\, dy \\
&= \frac{1}{4\pi} \int_{\mathbb{R}^3} \mathrm{div}\, w(y)\nabla_y\left(\frac{1}{|x-y|}\right) dy \\
&\quad + \frac{1}{4\pi} \int_{\mathbb{R}^3} (\nabla \times w(y)) \times \nabla_y\left(\frac{1}{|x-y|}\right) dy
\end{aligned}
$$

$$= -\frac{1}{4\pi} \int_{\mathbb{R}^3} \text{div } w(y) \nabla_x \left(\frac{1}{|x-y|} \right) \, dy$$

$$- \frac{1}{4\pi} \int_{\mathbb{R}^3} (\nabla \times w(y)) \times \nabla_x \left(\frac{1}{|x-y|} \right) \, dy$$

$$= -\nabla \left(\frac{1}{4\pi} \int_{\mathbb{R}^3} \frac{\text{div } w(y)}{|x-y|} \, dy \right) + \nabla \times \left(\frac{1}{4\pi} \int_{\mathbb{R}^3} \frac{\nabla \times w(y)}{|x-y|} \, dy \right).$$

2.4 Notice that in L^2 we do have $\check{\hat{u}} = u$ and so we will describe the desired decomposition in terms of the Fourier transform.

$$\hat{u}(k) = \frac{\hat{u}(k) \cdot k}{|k|^2} k + \left[\hat{u}(k) - \frac{\hat{u}(k) \cdot k}{|k|^2} k \right]$$

$$= 2\pi i k \frac{\hat{u}(k) \cdot k}{2\pi i |k|^2} + \left[\hat{u}(k) - \frac{\hat{u}(k) \cdot k}{|k|^2} k \right].$$

As a result we have $u = \nabla g + h$, where

$$\hat{g} = \frac{\hat{u}(k) \cdot k}{2\pi i |k|^2} \quad \text{and} \quad \hat{h} = \hat{u}(k) - \frac{\hat{u}(k) \cdot k}{|k|^2} k.$$

Since $\hat{h}(k)$ is orthogonal to k the function h is divergence free, and $h \in L^2(\mathbb{R}^3)$ since $|\hat{h}(k)| \leq |\hat{u}(k)|$. Furthermore,

$$\hat{g} = \frac{\hat{u}(k) \cdot k}{2\pi i |k|^2} \in L^1_{\text{loc}}(\mathbb{R}^3)$$

and

$$\int_{\mathbb{R}^3} |\hat{g}(k)|^2 |k|^2 \, dk \leq \|u\|_{L^2(\mathbb{R}^3)}^2 < \infty,$$

so $g \in \dot{H}^1(\mathbb{R}^3)$.

2.5 If $f = \nabla g = c$ then we have

$$(2\pi)^3 |c|^2 = \|f\|^2 = \|\nabla g\|^2 = \langle \nabla g, \nabla g \rangle = \langle \nabla g, c \rangle = \int_{\mathbb{T}^3} \nabla g = 0.$$

2.6 It suffices to show that $\widehat{\mathbb{P} \partial_j u} = \widehat{\partial_j \mathbb{P} u}$. Recall that $\widehat{\mathbb{P} f} = \hat{f}(k) - \frac{\hat{f}(k) \cdot k}{|k|^2} k$. Therefore

$$\widehat{\mathbb{P} \partial_j u} = \widehat{\partial_j u}(k) - \frac{\widehat{\partial_j u}(k) \cdot k}{|k|^2} k = 2\pi k_j \left(\hat{u}(k) - \frac{\widehat{u}(k) \cdot k}{|k|^2} k \right) = \widehat{\partial_j \mathbb{P} u}.$$

2.7 For example

$$u = \nabla(x_1 x_2 + x_3^2) - (0, x_1, 0)$$
$$= \nabla(2x_1 x_2 + x_3^2) - (x_2, 2x_1, 0).$$

This decomposition is not unique and cannot be regarded as the Helmholtz–Weyl decomposition because there is no requirement on the boundary.

2.8 The function u is divergence free:

$$\text{div } u = \partial_1(x_2^3) + \partial_2(-x_1^3) + \partial_3(0) = 0.$$

It remains to check that its normal component vanishes on the boundary. Consider first

$$\Gamma_1 := \{(x_1, x_2, x_3) : x_1^4 + x_2^4 = 1; -1 < x_3 < 1\}.$$

The vector $n_{\Gamma_1} = (4x_1^3, 4x_2^3, 0)$ is perpendicular to both Γ_1 and u. So the normal component of u vanishes on Γ_1. Obviously the normal component of u vanishes also on

$$\Gamma_2 = \{(x_1, x_2, x_3) : x_1^4 + x_2^4 \le 1; |x_3| = 1\}.$$

It follows that $u \in H(\Omega)$. Hence we have $\mathbb{P}u = u$, so

$$-\Delta\mathbb{P}u = -\Delta u = (6x_2, -6x_1, 0).$$

Hence $-\Delta\mathbb{P}u$ is divergence free but does not belong to $H(\Omega)$ because its normal component does not vanish on $\partial\Omega$. Therefore $-\Delta\mathbb{P}u \ne -\mathbb{P}\Delta u \in H$.

2.9 We define a functional $h: V \times V \to \mathbb{R}$ by

$$h(u, v) = \langle \nabla u, \nabla v \rangle.$$

The functional h is clearly bilinear, coercive since

$$h(u, u) = \|\nabla u\|^2 \ge c_1 \|u\|_{H^1}^2,$$

and continuous on $V \times V$ as

$$|h(u, v)| = |\langle \nabla u, \nabla v \rangle| \le c_2 \|u\|_{H^1} \|v\|_{H^1}.$$

If $f \in V^*$ then $\langle f, v \rangle$ is a continuous functional on V. Hence from the Lax–Milgram Theorem for each $f \in V^*$ there is exactly one $u \in V$ such that

$$\langle \nabla u, \nabla v \rangle = \langle f, v \rangle \quad \text{for all} \quad v \in V.$$

2.10 If

$$u = \sum_{k=1}^{\infty} \tilde{u}_k a_k \text{ and } v = \sum_{k=1}^{\infty} \tilde{v}_k a_k$$

then we have

$$\langle A^{s_1} u, A^{s_2} v \rangle = \sum_{k=1}^{\infty} \lambda_k^{s_1} \tilde{u}_k \lambda_k^{s_2} \tilde{v}_k = \sum_{k=1}^{\infty} \lambda_k^{s} \tilde{u}_k \tilde{v}_k = \langle u, A^s v \rangle.$$

2.11 We have

$$\|\nabla u\|^2 = \langle A^{1/2} u, A^{1/2} u \rangle = \langle u, Au \rangle \leq \|u\| \|Au\| \leq c \|\nabla u\| \|Au\|,$$

where we have used the Poincaré inequality. The result follows when we divide both sides by $\|\nabla u\|$ (if $\|\nabla u\| = 0$ then there is nothing to prove).

2.12 If $m = 2n$ then

$$\begin{aligned}
\|a_k\|_{H^{2n}} &\leq c \|Aa_k\|_{H^{2n-2}} = c\lambda_k \|a_k\|_{H^{2n-2}} \\
&\leq c^2 \lambda_k \|Aa_k\|_{H^{2n-4}} = c^2 \lambda_k^2 \|a_k\|_{H^{2(n-2)}} \\
&\leq \cdots \\
&\leq C\lambda_k^n \|a_k\| = C\lambda_k^{m/2}.
\end{aligned}$$

If $m = 2n + 1$ then in a similar way we have

$$\begin{aligned}
\|a_k\|_{H^{2n+1}} &\leq c \|Aa_k\|_{H^{2n-1}} = c\lambda_k \|a_k\|_{H^{2n-1}} \\
&\leq \cdots \\
&\leq c^n \lambda_k^n \|a_k\|_{H^1} \leq C\lambda_k^n \lambda_k^{1/2} = C\lambda_k^{m/2},
\end{aligned}$$

because

$$\|\nabla a_k\|^2 = \langle \nabla a_k, \nabla a_k \rangle = \langle Aa_k, a_k \rangle = \lambda_k.$$

2.13 (i) If $m = 2n$ then

$$\|u\|_{H^{2n}} \leq c \|Au\|_{H^{2n-2}} \leq \cdots \leq c^n \|A^n u\| = C \|A^{m/2}\|.$$

If $m = 2n + 1$ then

$$\|u\|_{H^{2n+1}} \leq c \|Au\|_{H^{2n-1}} \leq \cdots \leq c^n \|A^n u\|_{H^1}.$$

We now notice that

$$\|A^n u\|_{H^1} = \|A^{1/2}(A^n u)\| = \|A^{m/2} u\|,$$

where we have used the fact that $V = D(A^{1/2})$.

We prove parts (ii) and (iii) for v of the form $v = \sum_{j=1}^{n} \tilde{u}_j a_j$, $a_j \in \mathcal{N}$; the general case of $u \in D(A^s)$ then follows from the density of such elements in $D(A^s)$.

(ii) To show that $A^m v = \mathbb{P}(-\Delta)^m v$ we notice that applying $-\Delta$ to both sides of

$$-\Delta a_k = -\nabla p_k + \lambda_k a_k$$

we obtain

$$(-\Delta)^2 a_k = -\nabla(-\Delta p_k) + \lambda_k(-\Delta a_k) = \lambda_k^2 a_k - \nabla p_k$$

since $\Delta p_k = 0$. Hence

$$\mathbb{P}(-\Delta)^2 a_k = A^2 a_k.$$

Continuing in this vein and using linearity of A we obtain $\mathbb{P}(-\Delta)^m v = A^m v$ for any $m \in \mathbb{N}$.

(iii) To show that $\|A^m v\| \le c\|v\|_{H^{2m}}$ we only need to notice that

$$\|A^m v\| = \|\mathbb{P}(-\Delta)^m v\| \le c\|(-\Delta)^m v\| \le c\|v\|_{H^{2m}}.$$

2.14 Suppose that $u \in H_q \cap G_q$. Since $u \in G_q$ it follows that $u = \nabla g$, so for each $k \in \mathbb{Z}^3$:

$$\hat{u}_k = (2\pi)^{-3} \int_{\mathbb{T}^3} e^{-ik\cdot x} \nabla g(x)\,dx = -(2\pi)^{-3} \int_{\mathbb{T}^3} g(x)\nabla e^{-ik\cdot x}\,dx = \alpha_k k,$$

for some $\alpha_k \in \mathbb{C}$. However, if $u \in H_q$ then

$$0 = \int_{\mathbb{T}^3} u(x) \cdot \nabla e^{-ik\cdot x}\,dx = -ik \cdot \int_{\mathbb{T}^3} u(x)e^{-ik\cdot x} = -(2\pi)^3 ik \cdot \hat{u}_k.$$

It follows that for each k $\hat{u}_k$ is both parallel and perpendicular to k, so $\hat{u}_k = 0$ for every $k \in \mathbb{Z}^3$.

2.15 We have

$$\begin{aligned}
(-\Delta)^{-1} \operatorname{div} S_N(u) &= (-\Delta)^{-1} \left(\sum_{k \in \mathbb{Z}^3 \cap Q_N} \hat{u}_k \cdot k e^{ik\cdot x} \right) \\
&= \sum_{k \in \mathbb{Z}^3 \cap Q_N} \frac{\hat{u}_k \cdot k}{|k|^2} e^{ik\cdot x} \\
&= \operatorname{div} \left(\sum_{k \in \mathbb{Z}^3 \cap Q_N} \frac{\hat{u}_k}{|k|^2} e^{ik\cdot x} \right) \\
&= \operatorname{div}(-\Delta)^{-1} S_N(u).
\end{aligned}$$

Chapter 3

3.1 See Example 13.2.

3.2 We use the Hölder inequality, interpolation between L^2 and L^6, and the Sobolev embedding $H^1 \subset L^6$ to obtain

$$|\langle (u \cdot \nabla)v, w \rangle| \leq \|u\|_{L^4} \|\nabla v\| \|w\|_{L^4}$$
$$\leq c\|u\|^{1/4} \|u\|_{L^6}^{3/4} \|\nabla v\| \|w\|^{1/4} \|w\|_{L^6}^{3/4}$$
$$\leq c\|u\|_{H^1} \|v\|_{H^1} \|w\|_{H^1}.$$

3.3 If u is a weak solution of the Navier–Stokes equations then $u(t) \in V$ for a.e. $t \in (0, T)$, so from Lemma 3.2 it follows that $\langle (u(t) \cdot \nabla)u(t), u(t) \rangle = 0$ for a.e. $t \in (0, T)$. Hence $\int_0^T \langle (u \cdot \nabla)u, u \rangle = 0$.

Now let us choose $\tilde{u}, \tilde{v} \in V(\mathbb{R}^3)$ such that $\langle (\tilde{u} \cdot \nabla)\tilde{u}, \tilde{v} \rangle > 0$. We define

$$u(x, t) = t^{-2/3} \tilde{u}(t^{-1/2}x) \quad \text{and} \quad v(x, t) = t^{-2/3} \tilde{v}(t^{-1/2}x).$$

Then

$$\|u(t)\|^2 = t^{1/6} \|\tilde{u}\|^2 \in L^\infty(0, T)$$
$$\|\nabla u(t)\|^2 = t^{-5/6} \|\nabla \tilde{u}\|^2 \in L^1(0, T).$$

Obviously, also $v \in L^\infty(0, T; H) \cap L^2(0, T; V)$. Furthermore

$$\int_0^T \langle (u \cdot \nabla)u, v \rangle = \int_0^T t^{-1} \langle (\tilde{u} \cdot \nabla \tilde{u}), \tilde{v} \rangle = \infty.$$

3.4 It is easy to see that if u satisfies (3.4) then also for all $a_k \in \mathcal{N}$ we have

$$\int_{t_1}^{t_2} \langle \nabla u, \nabla a_k \rangle + \int_{t_1}^{t_2} \langle (u \cdot \nabla)u, a_k \rangle = \langle u(t_1), a_k \rangle - \langle u(t_2), a_k \rangle.$$

(We note that $a_k \in V$ so a_k can be approximated by smooth functions with compact support in the norm of H^1, which is what is needed to pass to the limit in the integrals on the left-hand side, see Lemma 3.4.) It follows that for any a_k the function $f(t) = \langle u(t), a_k \rangle$ is absolutely continuous with its derivative given by

$$f' = -\langle \nabla u, \nabla a_k \rangle - \langle (u \cdot \nabla)u, a_k \rangle \in L^1(0, T),$$

where the integrability of f' follows easily from Lemma 3.4. Let $\varphi \in \tilde{\mathcal{D}}_\sigma$ be given by $\varphi = \alpha(t)a_k(x)$, where $\alpha \in C_c^1[0, \infty)$. The product

$$f(t)\alpha(t) = \langle u(t), \alpha(t)a_k \rangle$$

is absolutely continuous, with derivative given by $f'\alpha + \alpha'f$. Therefore

$$f(t)\alpha(t) - f(0)\alpha(0) = \int_0^t (f'\alpha + \alpha'f)(s)\,\mathrm{d}s.$$

So we have

$$\langle u(t), \varphi(t)\rangle - \langle u(s), \varphi(s)\rangle = -\int_s^t \langle \nabla u, \nabla\varphi\rangle - \int_s^t \langle (u \cdot \nabla)u, \varphi\rangle + \int_s^t \langle u, \partial_t\varphi\rangle,$$

which is (3.3).

3.5 Let $I = \int_s^{s+1/n} \langle (u \cdot \nabla u), \varphi_n\rangle$. From Hölder's inequality and Lemma 3.4 we have

$$|I| \le \left(\int_s^{s+1/n} \|(u \cdot \nabla)u\|_{L^{6/5}}^{4/3}\right)^{3/4} \left(\int_s^{s+1/n} \|\varphi_n\|_{L^6}^4\right)^{3/4}$$

$$\le \|(u \cdot \nabla)u\|_{L^{4/3}(0,T;L^{6/5})} |\alpha(s)|^3 \|a_k\|_{L^6}^3 n^{-3/4} \to 0.$$

3.6 If $u \in V$ then we have from the Sobolev embedding $H^{1/2} \subset L^4$ in 2D:

$$\|u\|_{L^4} \le c\|u\|_{H^{1/2}} \le c\|u\|^{1/2}\|u\|_{\dot{H}^1}^{1/2} = c\|u\|^{1/2}\|\nabla u\|^{1/2}.$$

In 3D for $u \in V$ we first interpolate between L^2 and L^6, and then use the 3D Sobolev embedding $H^1 \subset L^6$:

$$\|u\|_{L^4} \le \|u\|^{1/4}\|u\|_{L^6}^{3/4} \le c\|u\|^{1/4}\|\nabla u\|^{3/4}.$$

3.7 As in the proof of Lemma 3.7 we obtain

$$\frac{\mathrm{d}}{\mathrm{d}t}\langle u(t), \varphi\rangle = -\langle \nabla u(t), \nabla\varphi\rangle - \langle (u(t) \cdot \nabla)u(t), \varphi\rangle$$

$$= -\langle \nabla u(t), \nabla\varphi\rangle + \langle (u(t) \cdot \nabla)\varphi, u(t)\rangle.$$

Consider the functional on the right-hand side. We use the Ladyzhenskaya inequality in dimension 2

$$\|u\|_{L^4}^2 \le c\|u\|\|\nabla u\|,$$

to obtain

$$|\langle (u \cdot \nabla)\varphi, u\rangle| \le \|u\|_{L^4}^2\|\nabla\varphi\| \le c\|u\|\|\nabla u\|\|\nabla\varphi\|.$$

Hence

$$\|\partial_t u\|_{V^*} \le \|\nabla u\| + c\|u\|\|\nabla u\|$$

and the result follows since for a weak solution u

$$\int_0^T \|u\|^2\|\nabla u\|^2 \le \left(\operatorname*{ess\,sup}_{t\in[0,T]} \|u(t)\|^2\right) \int_0^T \|\nabla u\|^2 < \infty.$$

3.8 We integrate the equation

$$\partial_t u - \Delta u + (u \cdot \nabla)u + \nabla p = 0$$

over $\mathbb{T}^3$ to obtain

$$\frac{\mathrm{d}}{\mathrm{d}t} \int_{\mathbb{T}^3} u = \int_{\mathbb{T}^3} \Delta u - \int_{\mathbb{T}^3} (u \cdot \nabla)u - \int_{\mathbb{T}^3} \nabla p.$$

From integration by parts (here $\cdot$ means only multiplication, not a dot product)

$$\int_{\mathbb{T}^3} \Delta u_i = \int_{\mathbb{T}^3} (\Delta u_i) \cdot 1 = -\int_{\mathbb{T}^3} \nabla u_i \nabla 1 = 0,$$

$$\int_{\mathbb{T}^3} \nabla p = \int_{\mathbb{T}^3} \nabla p \cdot 1 = -\int_{\mathbb{T}^3} p \nabla 1 = 0.$$

Moreover (using Einstein's summation convention)

$$\int_{\mathbb{T}^3} u_j \partial_j u_i = -\int_{\mathbb{T}^3} \partial_j u_j u_i = 0,$$

so $\frac{\mathrm{d}}{\mathrm{d}t} \int_{\mathbb{T}^3} u = 0$.

Chapter 4

4.1 Let $u \in H \cap H^s$ be given by

$$u = \sum_{j=1}^{\infty} u_j a_j,$$

where $a_j \in \mathcal{N}$ corresponds to the eigenvalue λ_j. Then

$$\|u\|_{\dot{H}^s}^2 = \|A^{s/2} u\|^2 = \sum_{j=1}^{\infty} \lambda_j^s u_j^2.$$

Since this sum is finite it follows that

$$\|P_n u - u\|_{\dot{H}^s}^2 = \left\| \sum_{j=n+1}^{\infty} u_j a_j \right\|_{\dot{H}^s}^2 = \sum_{j=n+1}^{\infty} \lambda_j^s u_j^2 \to 0,$$

when $n \to 0$. We also have

$$\|P_n u\|_{\dot{H}^s}^2 = \|A^{s/2} P_n u\|^2 \le \|A^{s/2} u\|^2 = \|u\|_{\dot{H}^s}^2.$$

4.2 Notice that

$$\langle \partial_t u_n, a_k \rangle = \frac{\mathrm{d}}{\mathrm{d}t} c_k^n$$

so for $t \geq s$

$$
\begin{aligned}
|c_k^n(t) - c_k^n(s)| &\leq \int_s^t |\langle \partial_t u_n, a_k \rangle| \\
&\leq \int_s^t |\langle Au_n, a_k \rangle| + \int_s^t |\langle (u_n \cdot \nabla)u_n, a_k \rangle| \\
&\leq \int_s^t \|\nabla u_n\|\|\nabla a_k\| + \int_s^t \|u_n\|\|\nabla u_n\|\|a_k\|_\infty \\
&\leq \|\nabla a_k\| \int_s^t \|\nabla u_n\| + \|a_k\|_\infty \sup_{0 \leq t \leq T} \|u_n(t)\| \int_s^t \|\nabla u_n\| \\
&\leq C_k \sqrt{t - s},
\end{aligned}
$$

where we used the estimate

$$
\int_s^t \|\nabla u_n\| \leq \left(\int_0^T \|\nabla u_n\|^2 \right)^{1/2} (t - s)^{1/2} \leq \|u_0\|(t - s)^{1/2}.
$$

4.3 From the uniform boundedness

$$
\sup_{0 \leq t \leq T} |c_k^n(t)| \leq \|u_0\|
$$

and equicontinuity (Exercise 4.2) of c_k^n for each fixed k it follows that we can choose (using a standard diagonal method) a subsequence of the Galerkin approximations such that for each fixed k

$$
c_k^n \to c_k \quad \text{in} \quad C([0, T]).
$$

We will now prove that $u_n(t) \rightharpoonup u(t)$ weakly in L^2 uniformly in t, where

$$
u(t) = \sum_{j=1}^\infty c_j(t)a_j.
$$

Fix $\varepsilon > 0$ and $\varphi \in L^2(\Omega)$. We can choose N so large that

$$
\mathbb{P}\varphi = P_N\varphi + \tilde{\varphi},
$$

where

$$
P_N\varphi(x) = \sum_{j=1}^N b_j a_j(x), \quad b_j \in \mathbb{R}, \ a_j \in \mathcal{N},
$$

and

$$
\|\tilde{\varphi}\| \leq \frac{\varepsilon}{4\|u_0\|T}.
$$

Therefore for $n \geq N$ we have

$$\int_0^T |\langle u_n - u, \varphi \rangle| = \int_0^T |\langle u_n - u, P_N \varphi \rangle| + \int_0^T |\langle u_n - u, \tilde{\varphi} \rangle|$$

$$\leq \int_0^T \sum_{j=1}^N |\langle (c_j^n(t) - c_j(t))a_j, b_j a_j \rangle| + \int_0^T (\|u_n\| + \|u\|)\|\tilde{\varphi}\|$$

$$\leq \sum_{j=1}^N |b_j| \int_0^T |c_j^n(t) - c_j(t)| + \varepsilon/2$$

and the result follows from the uniform convergence of c_j^n to c_j on $[0, T]$ for $j = 1, 2, \ldots, N$.

4.4 We choose a subsequence of u_n denoted by u_n^1 such that u_n^1 is weakly convergent in $L^2(0, 1; L^2)$. From this subsequence we choose a further subsequence that is weakly convergent on $L^2(0, 2; L^2)$. Continuing in this vein we obtain a subsequence of (u_n) such that u_n^j is weakly convergent in $L^2(0, j; L^2)$. Choosing $v_n = u_n^n$ we obtain the desired subsequence.

4.5 Since $u_n \overset{*}{\rightharpoonup} v_1$ in $L^\infty(0, T; L^2)$ it follows that

$$\langle u_n, \varphi \rangle \to \langle v_1, \varphi \rangle$$

for every $\varphi \in L^1(0, T; L^2)$ and in particular for every $\varphi \in L^2(0, T; L^2)$. Hence v_1 is a weak limit of (u_n) in $L^2(0, T; L^2)$. Since v_2 is the strong limit of (u_n) in $L^2(0, T; L^2)$ it is also a weak limit in this space. Since weak limits are unique it follows that $v_1 = v_2$.

To prove that $\nabla v_2 = g$ we denote by g_i the ith component of g and notice that passing to the limit in

$$\int_\Omega (\partial_i u_n)\varphi = -\int_\Omega u_n \partial_i \varphi,$$

where $\varphi \in C_c^\infty(\Omega)$ we obtain

$$\int_\Omega g_i \varphi = -\int_\Omega v_2 \partial_i \varphi$$

for every $\varphi \in C_c^\infty(\Omega)$. Hence $\nabla v_2 = g$.

4.6 Fix any $a_k \in \mathcal{N}$. We have

$$\langle u_n(t_2), a_k \rangle - \langle u_n(t_1), a_k \rangle = \int_{t_1}^{t_2} \langle \partial_t u_n, a_k \rangle.$$

Passing to the limit, we get for almost every pair (t_1, t_2)

$$\langle u(t_2), a_k \rangle - \langle u(t_1), a_k \rangle = \int_{t_1}^{t_2} \langle g, a_k \rangle.$$

Since this holds for all $a_k \in \mathcal{N}$ and $\mathcal{N}$ is a basis of V, the assertion follows from Corollary 1.32.

4.7 It is clear that it is enough to prove the inequality for zero-mean functions. Let $Q_1 = [0, 1]^3$ be the unit cube. From Theorem 1.9 we know that there is a constant C such that

$$\|v\| \leq C\|\nabla v\|$$

for every zero-mean function $v \in H^1(Q_1)$. Now, let

$$Q = [p, p + a] \times [q, q + a] \times [r, r + a]$$

and let $u \in H^1(Q)$ have mean zero. Consider a rescaled function

$$v(x, y, z) = u(ax + p, ay + q, az + r).$$

Then $v \in H^1(Q_1)$, $\int_{Q_1} v = 0$, so

$$\|v\|_{L^2(Q_1)} \leq C\|\nabla v\|_{L^2(Q_1)}.$$

Since $\|v\|_{L^2(Q_1)} = \frac{1}{a^{3/2}} \|u\|_{L^2(Q)}$ and $\|\nabla v\|_{L^2(Q_1)} = \frac{a}{a^{3/2}} \|\nabla u\|_{L^2(Q)}$, we obtain immediately

$$\|u\|_{L^2(Q)} \leq Ca\|\nabla u\|_{L^2(Q)}.$$

4.8 We use Exercise 4.7 and the identity

$$\int_{Q_i} u = \langle u, \chi_{Q_i} \rangle$$

to obtain (using an idea from Galdi, 2011)

$$\|u\|_{L^2(Q)}^2 = \sum_{i=1}^{M^3} \|u\|_{L^2(Q_i)}^2$$

$$= \sum_{i=1}^{M^3} \left\| u - \frac{M^3}{a^3} \int_{Q_i} u + \frac{M^3}{a^3} \int_{Q_i} u \right\|_L^2$$

$$\leq \sum_{i=1}^{M^3} 2 \left\| u - \frac{M^3}{a^3} \int_{Q_i} u \right\|_{L^2(Q_i)}^2 + 2 \sum_{i=1}^{M^3} \left\| \frac{M^3}{a^3} \int_{Q_i} u \right\|_{L^2(Q_i)}^2$$

$$= \sum_{i=1}^{M^3} 2 \left\| u - \frac{M^3}{a^3} \int_{Q_i} u \right\|^2_{L^2(Q_i)} + 2 \frac{M^6}{a^6} \sum_{i=1}^{M^3} \int_{Q_i} \langle u, \chi_{Q_i} \rangle^2$$

$$\leq \sum_{i=1}^{M^3} C \frac{a^2}{M^2} \|\nabla u\|^2_{L^2(Q_i)} + 2 \frac{M^3}{a^3} \sum_{i=1}^{M^3} \langle u, \chi_{Q_i} \rangle^2$$

$$\leq C \frac{a^2}{M^2} \|\nabla u\|^2_{L^2(Q)} + 2 \frac{M^3}{a^3} \sum_{i=1}^{M^3} \langle u, \chi_{Q_i} \rangle^2.$$

4.9 Let Q be a cube with side of length a such that $K \subset Q$. We will prove that u_n is a Cauchy sequence in $L^2(0, T; L^2(Q))$.

Choose any $\varepsilon > 0$. Since $u_n(t) - u_m(t) \in H^1(Q)$ for almost all t, we can integrate Friedrich's inequality in time to obtain, for all $M \in \mathbb{N}$:

$$\int_0^T \|u_n - u_m\|^2_{L^2(Q)}$$

$$\leq C \frac{a^2}{M^2} \int_0^T \|\nabla u_n - \nabla u_m\|^2_{L^2(Q)} + 2 \frac{M^3}{a^3} \sum_{i=1}^{M^3} \int_0^T \langle \chi_{Q_i}, u_n - u_m \rangle^2$$

$$\leq C \frac{a^2}{M^2} C_1 + 2 \frac{M^3}{a^3} \sum_{i=1}^{M^3} \int_0^T \langle \chi_{Q_i}, u_n - u_m \rangle^2.$$

We choose a natural number M so large that

$$C \frac{a^2}{M^2} C_1 \leq \frac{\varepsilon}{2}.$$

Then from the uniform convergence of $\langle u_n, \chi_{Q_i} \rangle$ on $[0, T]$ it follows that for each $i = 1, 2, 3, \ldots, M^3$ we can find N_i such that for $n, m > N_i$

$$\langle u_n(t) - u_m(t), \chi_{Q_i} \rangle^2 \leq \frac{\varepsilon a^3}{4 T M^6}.$$

Choosing $N > N_i$ for all $i = 1, 2, \ldots, M_1^3$ we have for all $n, m > N$

$$\sum_{i=1}^{M^3} \int_0^T \langle \chi_{Q_i}, u_n - u_m \rangle^2 \leq \frac{\varepsilon a^3}{4 M^3}.$$

It follows that for $n, m > N$ we have $\int_0^T \|u_n - u_m\|^2 \leq \varepsilon$, so u_n is strongly convergent in $L^2(0, T; L^2(Q))$ and, hence in $L^2(0, T; L^2(K))$.

4.10 Fix any $T > 0$ and take any $\xi \in C^1([0, T])$. Integrating by parts the first term on the right-hand side of (3.3) with $s = T$ we obtain

$$\int_0^T \xi(t) \left(\langle \partial_t u_n, a_k \rangle + \langle A u_n, a_k \rangle + \langle (u_n \cdot \nabla) u_n, a_k \rangle \right) dt = 0.$$

Since ξ is arbitrary it follows that

$$\langle \partial_t u_n, a_k \rangle + \langle A u_n, a_k \rangle + \langle (u_n \cdot \nabla) u_n, a_k \rangle = 0, \quad k = 1, 2, \ldots, n$$

or in a more compact form:

$$\partial_t u_n + A u_n + P_n (u_n \cdot \nabla) u_n = 0.$$

4.11 It is obvious that $w \in L^2(0, T; H) \cap L^\infty(0, T; V)$. So we only have to prove that for all test functions and all $0 \le s \le T$ the weak formulation of the Navier–Stokes equations (3.3) is satisfied. If $s \in [0, T_1]$ there is nothing to prove. So suppose that $s \in [T_1, T]$. Then we have

$$\int_0^s -\langle w, \partial_t \varphi \rangle + \int_0^s \langle \nabla w, \nabla \varphi \rangle + \int_0^s \langle (w \cdot \nabla) w, \varphi \rangle$$

$$= \int_0^{T_1} -\langle u, \partial_t \varphi \rangle + \int_0^{T_1} \langle \nabla u, \nabla \varphi \rangle + \int_0^{T_1} \langle (u \cdot \nabla) u, \varphi \rangle$$

$$+ \int_{T_1}^s -\langle v, \partial_t \varphi \rangle + \int_{T_1}^s \langle \nabla v, \nabla \varphi \rangle + \int_{T_1}^s \langle (v \cdot \nabla) v, \varphi \rangle$$

$$= \langle u_0, \varphi(0) \rangle - \langle u(T_1), \varphi(T_1) \rangle + \langle v(T_1), \varphi(T_1) \rangle - \langle v(s), \varphi(s) \rangle$$

$$= \langle w_0, \varphi(0) \rangle - \langle w(s), \varphi(s) \rangle.$$

Chapter 5

5.1 Given $u \in L^r(\mathbb{R}^3)$ we want to find q such that $\nabla q \in L^r(\mathbb{R}^3)$ and

$$-\Delta q = \nabla \cdot u, \tag{S.2}$$

since then $\nabla \cdot (u + \nabla q) = 0$ and we can define $\mathbb{P}_r u = u + \nabla q$.

To solve (S.2) we write $\nabla q = (-\Delta)^{-1} \nabla (\nabla \cdot u)$; the Calderón–Zygmund Theorem in the form of Theorem B.6 now guarantees that $\|\nabla q\|_{L^r} \le C \|u\|_{L^r}$, which shows that $\mathbb{P}_r$ is bounded from $L^r(\mathbb{R}^3)$ into itself as required.

5.2 We repeat the argument of the previous exercise, observing now that if $u \in \mathscr{S}$ then $\nabla(\nabla \cdot u) \in \mathscr{S}$. If $f \in \mathscr{S}$ and $-\Delta v = f$ then $v \in \mathscr{S}$: it follows that $\nabla q = (-\Delta)^{-1} \nabla(\nabla \cdot u) \in \mathscr{S}$ and hence $u - \nabla q \in \mathscr{S}$ as well.

5.3 We start with the weak form of the equations

$$\int_0^s -\langle u, \partial_t \psi \rangle + \int_0^s \langle \nabla u, \nabla \psi \rangle + \int_0^s \langle (u \cdot \nabla) u, \psi \rangle = \langle u_0, \psi(0) \rangle - \langle u(s), \psi(s) \rangle,$$

valid for all test functions $\psi \in \mathcal{D}_\sigma$ and almost every $s > 0$.

If $\psi(\cdot, t) = 0$ for all t outside a compact subset of $(0, \infty)$, $[a, b]$, say, then (as in the proof of Proposition 5.3) we observe that by taking $s > b$ both terms on the right-hand side vanish, and we can extend the range of time integration to $(0, \infty)$ to give

$$\int_0^s -\langle u, \partial_t \psi \rangle + \int_0^s \langle \nabla u, \nabla \psi \rangle + \int_0^s \langle (u \cdot \nabla)u, \psi \rangle = 0.$$

Using the fact that ψ is smooth and has compact support we can integrate by parts in space in the second and third terms to obtain

$$\int_0^s \langle u, \partial_t \psi \rangle + \int_0^s \langle u, \Delta \psi \rangle + \int_0^s \langle u \otimes u, \nabla \psi \rangle = 0. \qquad \text{(S.3)}$$

The result now follows since we can approximate the test function φ by a sequence $\psi_n \in \mathscr{D}_\sigma$ such that all the terms in (S.3) converge as $n \to \infty$.

5.4 Given that

$$u(t) = \sum_{k \in \mathbb{Z}^3} \left(\int_0^t e^{-|k|^2(t-s)} \hat{f}_k(t) \, ds \right) e^{ik \cdot x}$$

we have

$$\Lambda^2 u(t) = \sum_{k \in \mathbb{Z}^3} |k|^2 \left(\int_0^t e^{-|k|^2(t-s)} \hat{f}_k(t) \, ds \right) e^{ik \cdot x}.$$

If we set

$$G(t, k) = \begin{cases} |k|^2 e^{-|k|^2 t} & t \geq 0 \\ 0 & t < 0 \end{cases} \quad \text{and} \quad F_k(t) = \begin{cases} \hat{f}_k(t) & 0 < t < T \\ 0 & \text{otherwise} \end{cases}$$

then we can write

$$\Lambda^2 u(t) = \sum_{k \in \mathbb{Z}^3} \left(\int_{\mathbb{R}} G(t - s, k) F_k(s) \, ds \right) e^{ik \cdot x}$$

and so

$$\int_0^T \|\Lambda^2 u(t)\|^2 = \sum_{k \in \mathbb{Z}^3} \left\| \int_{\mathbb{R}} G(t - s, k) F_k(s) \, ds \right\|_{L^2(0,T)}^2.$$

Using Young's inequality for convolutions we have

$$\left\| \int_{\mathbb{R}} G(t - s, k) F_k(s) \, ds \right\|_{L^2(\mathbb{R})} \leq \|G(\cdot, k)\|_{L^1(\mathbb{R})} \|F_k\|_{L^2(\mathbb{R})}$$

$$= \|\hat{f}_k\|_{L^2(0,T)},$$

and so

$$\int_0^T \|\Lambda^2 u(t)\|^2 \, dt \leq \sum_{k \in \mathbb{Z}^3} \|F_k\|_{L^2(0,T)}^2$$

$$= \int_0^T \|f(\cdot, t)\|^2 \, dt,$$

as required.

Chapter 6

6.1 Let $u \in L^2(0, T; D(A))$ with $\partial_t u \in L^2(0, T; H)$ and $u(t) = \sum_{j=1}^{\infty} c_j(t)a_j$, where $a_j \in \mathcal{N}$. Then

$$\|\partial_t u\|^2_{L^2(0,T;L^2)} = \int_0^T \sum_{k=1}^{\infty} |\dot{c}_k(s)|^2$$

$$\|u\|^2_{L^2(0,T;D(A))} = \int_0^T \sum_{k=1}^{\infty} \lambda_{k^2} |c_k(s)|^2.$$

Let $\bar{u}_n = P_n u$, so $\bar{u}_n(t) = \sum_{j=1}^{n} c_j(t)a_j$. We can find n so large that

$$\|\bar{u}_n - u\|^2_{L^2(0,T;D(A))} + \|\partial_t \bar{u}_n - \partial_t u\|^2_{L^2(0,T;L^2)} \le \varepsilon/2.$$

Since $c_k \in H^1(0, T)$, $k = 1, 2, \ldots, n$, we can extend these functions to some $\bar{c}_k \in H^1(\mathbb{R})$ and denoting by $c_{k,\varepsilon}$ the mollification of $\bar{c}_k$ we have

$$\|c_k - c_{k,\varepsilon}\|^2_{L^2(0,T)} \le \frac{\varepsilon}{4n\lambda_{k^2}} \quad \text{and} \quad \|\dot{c}_k - \dot{c}_{k,\varepsilon}\|^2_{L^2(0,T;L^2)} \le \frac{\varepsilon}{4n}.$$

Define $u_n(t) = \sum_{k=1}^{n} c_{k,\varepsilon}(t)a_k$. Then

$$\|\partial_t u_n - \partial_t \bar{u}_n\|^2_{L^2(0,T;L^2)} = \sum_{j=1}^{n} |\dot{c}_k(t) - \dot{c}_{k,\varepsilon}(t)|^2 \le \varepsilon/4 \qquad \text{and}$$

$$\|u_n - \bar{u}_n\|^2_{L^2(0,T;H^2)} = \sum_{j=1}^{n} \lambda_{j^2} |c_k(t) - c_k^{\varepsilon}(t)|^2 \le \varepsilon/4.$$

The result now follows using the triangle inequality.

6.2 Define $Z = Y - X$. Then $Z(0) \ge 0$ and

$$Z' = Y' - X' \ge Y^3 - X^3 = (Y - X)(Y^2 + XY + X^2).$$

Hence $Z'(t) \ge 0$ as long as $Z(t) \ge 0$.

6.3 We have

$$\int_0^T |\langle (u \cdot \nabla)u, v \rangle| \le \int_0^T \|u\| \|\nabla u\| \|v\|_{\infty}$$

$$\le c \int_0^T \|u\| \|\nabla u\| \|v\|_{H^1}^{1/2} \|v\|_{H^2}^{1/2}$$

$$\le c \int_0^T \|\nabla u\|^2 + c \int_0^T \|u\|^2 \|v\|_{H^1} \|v\|_{H^2} < \infty,$$

as $u \in L^{\infty}(0, T; L^2) \cap L^2(0, T; H^1)$ and $v \in L^{\infty}(0, T; H^1) \cap L^2(0, T; H^2)$.

6.4 From Lemma 3.2 we have for almost all time

$$\langle (u \cdot \nabla)u, v \rangle + \langle (v \cdot \nabla)v, u \rangle = \langle (w \cdot \nabla w), u \rangle.$$

Hence we only need to notice that all integrals appearing in the equation are finite when u is strong and v is weak.

6.5 Define $w = (u - v)$. Then taking the L^2 inner product with w we obtain two equations:

$$\langle \partial_t u, w \rangle + \langle Au, w \rangle + \langle (u \cdot \nabla)u, w \rangle = 0,$$
$$\langle \partial_t v, w \rangle + \langle Av, w \rangle + \langle (v \cdot \nabla)v, w \rangle = 0.$$

Taking the difference of these equations we obtain

$$\frac{1}{2}\frac{d}{dt}\|w\|^2 + \|\nabla w\|^2 \le c|\langle (w \cdot \nabla)w, u \rangle|$$
$$\le c\|w\|\|\nabla w\|\|u\|_{L^\infty}$$
$$\le \|\nabla w\|^2 + c\|u\|_{H^2}^2\|w\|^2.$$

Hence

$$\frac{d}{dt}\|w\|^2 \le c\|u\|_{H^2}^2\|w\|^2,$$

and uniqueness follows from the Gronwall Lemma since if u is a strong solution on $[0, T]$ then the function $\|u\|_{H^2}^2$ is an integrable function of time.

6.6 To obtain bounds on the time derivative we take the inner product of the Galerkin equation with $\partial_t u_n$. This yields

$$\|\partial_t u_n\|^2 + \langle Au_n, \partial_t u_n \rangle + \langle (u_n \cdot \nabla)u_n, \partial_t u_n \rangle = 0,$$

and hence

$$\|\partial_t u_n\|^2 \le \|Au_n\|\|\partial_t u_n\| + |\langle (u_n \cdot \nabla)u_n, \partial_t u_n \rangle|.$$

We estimate the right-hand side, using the standard bounds for the nonlinear term:

$$\|Au_n\|\|\partial_t u_n\| \le \frac{1}{4}\|\partial_t u_n\|^2 + \|Au_n\|^2$$
$$|\langle (u_n \cdot \nabla)u_n, \partial_t u_n \rangle| \le c\|\nabla u_n\|^{3/2}\|Au_n\|^{1/2}\|\partial_t u_n\|$$
$$\le c\|\nabla u_n\|^3\|Au_n\| + \frac{1}{4}\|\partial_t u_n\|^2$$
$$\le c\|\nabla u_n\|^6 + \frac{1}{2}\|Au_n\|^2 + \frac{1}{4}\|\partial_t u_n\|^2.$$

It follows that

$$\|\partial_t u_n\|^2 \le c_1 \|\nabla u_n\|^6 + c_2 \|A u_n\|^2.$$

If we integrate this in time between 0 and T and use (6.9) and (6.10) then we obtain

$$\int_0^T \|\partial_t u_n\|^2 \le cT \|\nabla u_0\|^6 + c \|\nabla u_0\|^2 \le C.$$

6.7 We argue by contradiction. Suppose that there exists a sequence $t_n \to 0^+$ for which $\|u(t_n)\|_{H^1} \le C$ for some $C > 0$. Then we can choose a subsequence of (t_n), which we relabel, such that $u(t_n) \rightharpoonup v$ weakly in H^1. It is simple to finish the argument on a bounded domain: since H^1 is compactly embedded in L^2, we can take a further subsequence (which we relabel again) such that $u(t_n) \to u_0$ in L^2 and so we have $u_0 = v \in V$, contradicting our initial assumption on u_0. On the whole space we take a subsequence for which $u(t_n)|_K \to u_0|_K$ in $L^2(K)$ for every compact $K \subset \mathbb{R}^3$, from which it follows once more that $u_0 = v$.

Chapter 7

7.1 We know that there is a constant $c > 0$ such that every $u_0 \in V$ gives rise to a strong solution u obtained as a limit of Galerkin approximations for all $t > 0$ such that

$$t \le c \|\nabla u_0\|^{-4}.$$

Let $a \in \mathbb{R}$ be such that

$$a < c \|u\|_{L^\infty(0,T;H^1)}^{-4}.$$

From the uniqueness of strong solutions we deduce that u coincides on the time interval $[0, a]$ with a strong solution constructed as a limit of Galerkin approximations, so from the reasoning in the proof of Theorem 7.1 we deduce that

$$u \in L^\infty(0, a; H^m) \cap L^2(0, a; H^{m+1}).$$

Now $u(a) \in H^m$. Since u coincides on the time interval $[a, 2a]$ with the solution constructed as a limit of Galerkin approximations we deduce that

$$u \in L^\infty(a, 2a; H^m) \cap L^2(a, 2a; H^{m+1}).$$

Continuing in this vein we deduce that

$$u \in L^\infty(ka, (k+1)a; H^m) \cap L^2(ka, (k+1)a; H^{m+1})$$

for every $k \geq 0$ such that $ka \in [0, T)$. Hence

$$u \in L^\infty(0, T; H^m) \cap L^2(0, T; H^{m+1}).$$

7.2 From the equality

$$\partial_t u = \Delta u - \mathbb{P}(u \cdot \nabla)u$$

and since H^m is a Banach algebra for $m > 3/2$, we deduce that

$$\|\partial_t u\|_{H^{m-1}} \leq \|\Delta u\|_{H^{m-1}} + \|\mathbb{P}(u \cdot \nabla)u\|_{H^{m-1}}$$
$$\leq \|u\|_{H^{m+1}} + c\|u\|_{H^m}\|u\|_{H^{m+1}}$$

and the result easily follows.

7.3 Using the fact that u is bounded in all spaces $\dot{H}^k$ when u is strong, we take the inner product in $\dot{H}^2$ of the equation

$$\partial_t u - \Delta u + \mathbb{P}(u \cdot \nabla)u = 0$$

with u. We obtain

$$\frac{1}{2}\frac{d}{dt}\|u\|_{\dot{H}^2}^2 + \|u\|_{\dot{H}^3}^2 \leq \left|\langle \partial_i \partial_j (u \cdot \nabla)u, \partial_i \partial_j u\rangle\right|.$$

Since

$$\langle (u \cdot \nabla)\partial_i \partial_j u, \partial_i \partial_j u\rangle = 0$$

we only need to estimate two types of terms:

$$\left|\langle \partial_i(u \cdot \nabla)\partial_j u, \partial_i \partial_j u\rangle\right| \leq \|\nabla u\|_\infty \|u\|_{\dot{H}^2}\|u\|_{\dot{H}^2}$$
$$\left|\langle ((\partial_i \partial_j u) \cdot \nabla)u, \partial_i \partial_j u\rangle\right| \leq \|u\|_{\dot{H}^2}\|\nabla u\|_\infty \|u\|_{\dot{H}^2}$$

and the result follows.

7.4 If $u_0 \in V$ then on some time interval u must be strong. If u loses its regularity before time T and $T^* \leq T$ is the first blowup time, then $\|u(t)\|_{\dot{H}^2} \to \infty$ when $t \to T^*$. However, u is smooth on $(0, T^*)$, so for any $t \in (0, T^*)$ we have $u(t) \in H^2$ and on $[t, T^*)$

$$\frac{d}{dt}\|u\|_{\dot{H}^2}^2 + \|u\|_{\dot{H}^3}^2 \leq c\|\nabla u\|_\infty \|u\|_{\dot{H}^2}^2,$$

according to Exercise 7.3. Hence from the Gronwall Lemma and our assumption it follows that

$$\|u(T^*)\|_{\dot{H}^2} < \infty,$$

contradicting our assumption on the loss of regularity of u at time T^*.

7.5 From Exercise 7.3 and Agmon's inequality we deduce that

$$\frac{d}{dt}\|u\|_{\dot{H}^2}^2 + \|u\|_{\dot{H}^3}^2 \le c\|\nabla u\|_{\dot{H}^1}^{1/2}\|\nabla u\|_{\dot{H}^2}^{1/2}\|u\|_{\dot{H}^2}^2.$$

Hence from Young's inequality we obtain

$$\frac{d}{dt}\|u\|_{\dot{H}^2}^2 \le c\|u\|_{\dot{H}^2}^{10/3}.$$

We set $X = \|u\|_{\dot{H}^2}^2$ and compare it with Y, the solution of

$$Y' = cY^{5/3}, \quad Y(0) = X(0).$$

We have

$$Y^{2/3}(t) = \frac{X^{2/3}(0)}{1 - ctX^{2/3}(0)},$$

so if $u(0) \in \dot{H}^2$ then the $\|u(t)\|_{\dot{H}^2}$ stays finite for all $t > 0$ such that

$$t < c^{-1}\|u(0)\|_{\dot{H}^2}^{-4/3}.$$

Hence if u blows up at time T we must have for every $t \in (0, T)$:

$$T - t > c^{-1}\|u(t)\|_{\dot{H}^2}^{-4/3}$$

so

$$\|u(t)\|_{\dot{H}^2} \ge c^{-3/4}(T - t)^{-3/4}.$$

7.6 Taking the L^2 inner product of the Galerkin equation with $A^m u_n$ we get

$$\frac{d}{dt}\|A^{m/2}u_n\|^2 + \|A^{(m+1)/2}u_n\|^2 \le |\langle P_n(u_n \cdot \nabla)u_n, A^m u_n\rangle|.$$

Note that $P_n(u_n \cdot \nabla)u_n \in D(A^m)$.

If $m = 2k$, $k \ge 1$, then we estimate

$$\begin{aligned}
|\langle P_n(u_n \cdot \nabla)u_n, A^m u_n\rangle| &\le |\langle A^{m/2}[P_n(u_n \cdot \nabla)u_n], A^{m/2}u_n\rangle| \\
&\le \|A^k[P_n(u_n \cdot \nabla)u_n]\|\|A^{m/2}u_n\| \\
&\le c\|P_n(u_n \cdot \nabla)u_n\|_{H^{2k}}\|A^{m/2}u_n\| \\
&\le c\|u_n\|_{H^m}\|\nabla u_n\|_{H^m}\|A^{m/2}u_n\| \\
&\le c\|u_n\|_{H^m}\|u_n\|_{H^{(m+1)}}\|A^{m/2}u_n\| \\
&\le c\|u_n\|_{H^m}\|A^{(m+1)/2}u_n\|\|A^{m/2}u_n\| \\
&\le c\|u_n\|_{H^m}^2\|A^{m/2}u_n\|^2 + \frac{1}{2}\|A^{(m+1)/2}u_n\|^2.
\end{aligned}$$

If $m = 2k + 1$, $k \geq 1$, then we estimate

$$
\begin{aligned}
|\langle P_n(u_n \cdot \nabla)u_n, A^m u_n\rangle| &\leq \left|\langle A^{m/2-1/2}[P_n(u_n \cdot \nabla)u_n], A^{m/2+1/2}u_n\rangle\right| \\
&\leq \|A^k[P_n(u_n \cdot \nabla)u_n]\|\,\|A^{(m+1)/2}u_n\| \\
&\leq c\|P_n(u_n \cdot \nabla)u_n\|_{H^{2k}}\|A^{(m+1)/2}u_n\| \\
&\leq c\|u_n\|_{H^{m-1}}\|\nabla u_n\|_{H^{m-1}}\|A^{(m+1)/2}u_n\| \\
&\leq c\|u_n\|_{H^m}\|u_n\|_{H^m}\|A^{(m+1)/2}u_n\| \\
&\leq c\|u_n\|_{H^m}\|A^{m/2}u_n\|\,\|A^{(m+1)/2}u_n\| \\
&\leq c\|u_n\|_{H^m}^2\|A^{m/2}u_n\|^2 + \frac{1}{2}\|A^{(m+1)/2}u_n\|^2.
\end{aligned}
$$

7.7 We will give bounds on the Galerkin approximations. From Theorem 7.1 (valid also on a smooth bounded domain) we know that if $u_0 \in H^2 \cap V$ then we can obtain uniform bounds on a subsequence of Galerkin approximations u_n in $L^\infty(\varepsilon, T; H^2) \cap L^\infty(\varepsilon, T; H^3)$. From the equality

$$
\partial_t u_n = -Au_n - P_n(u_n \cdot \nabla)u_n
$$

and the fact that $\partial_t u_n \in D(A^{1/2})$ for every n we easily deduce that $\partial_t u_n$ is uniformly bounded in $L^2(0, T; V)$.

7.8 We need to show that

$$
\left|\int_{\mathbb{R}^3} \left[\partial_x^\alpha((u \cdot \nabla)u) - (u \cdot \nabla)\partial_x^\alpha u\right] \cdot \partial_x^\alpha u \, \mathrm{d}x\right| \leq c\|\nabla u\|_{L^\infty}\|u\|_{H^k}^2.
$$

Using Hölder's inequality the left-hand side is controlled by

$$
\|\partial_x^\alpha((u \cdot \nabla)u) - (u \cdot \nabla)\partial_x^\alpha u\|_{L^2}\|\partial_x^\alpha u\|_{L^2},
$$

and it remains only to show that

$$
\|\partial_x^\alpha((u \cdot \nabla)u) - (u \cdot \nabla)\partial_x^\alpha u\|_{L^2} \leq c\|\nabla u\|_{L^\infty}\|u\|_{H^k}.
$$

By Leibniz's rule every term on the left-hand side has the form $\partial^a\partial_i u_i \partial^b\partial_j u_j$, where $|a| + |b| = |\alpha| - 1$. Now, using Hölder's inequality to obtain

$$
\|(\partial^a\partial u_i)\partial^{b+1}u_j\|_{L^2} \leq \|\partial^a\partial u\|_{L^{2(|\alpha|-1)/a}}\|\partial^b\partial u\|_{L^{2(|\alpha|-1)/b}}.
$$

For each of the terms on the right-hand side we use the interpolation inequality (with $f = \partial u$ and $s = a, b$, respectively)

$$
\|D^s f\|_{L^{2l/s}} \leq c\|f\|_{L^\infty}^{1-s/l}\|D^l f\|_{L^2}^{s/l}, \qquad 0 \leq s \leq l,
$$

to obtain the upper bound

$$
c\|\partial u\|_{L^\infty}^{1-a/|\alpha|-1}\|\partial^{|\alpha|-1}\partial u\|_{L^2}^{a/|\alpha|-1}\|\partial u\|_{L^\infty}^{1-b/|\alpha|-1}\|\partial^{|\alpha|-1}\partial u\|_{L^2}^{b/|\alpha|-1} \leq \|\partial u\|_{L^\infty}\|\partial^{|\alpha|}u\|_{L^2}.
$$

Summing over all $|\alpha| \leq k$ yields the result.

Chapter 8

8.1 Choose any $\varepsilon > 0$. Consider the difference between two consecutive terms a_n and a_{n-1} and bound it from below:

$$a_{n-1} - a_n = \frac{1}{(n-1)^k} - \frac{1}{n^k} > \frac{1}{n^{k+1}}.$$

If $n \leq \varepsilon^{-\frac{1}{k+1}}$ then the gap between a_n and a_{n-1} is greater than ε. It follows that to cover all elements $a_1, a_2, \ldots, a_N$, where $N \leq \varepsilon^{-\frac{1}{k+1}}$ we need at least $N/2$ balls of radius ε. Hence we obtain a lower bound on the number of balls of radius ε needed to cover S_k,

$$\frac{1}{2}\varepsilon^{-\frac{1}{k+1}} \leq N(S_k, \varepsilon).$$

On the other hand if $n \geq \varepsilon^{-\frac{1}{k+1}}$ then $a_n \in [0, \varepsilon^{\frac{k}{k+1}}]$ and we can cover this interval by no more than

$$(\varepsilon^{\frac{k}{k+1}}/\varepsilon) + 1 = \varepsilon^{-\frac{1}{k+1}} + 1$$

balls of radius ε. Hence we obtain an upper bound of the form

$$N(S_k, \varepsilon) \leq 2\varepsilon^{-\frac{1}{k+1}} + 1$$

and the assertion follows.

8.2 We know that

$$\limsup_{\varepsilon \to 0^+} \left(-\log_\varepsilon M(X, \varepsilon) \right) = d.$$

Since $d' > d$ there is $\varepsilon_0 > 0$ such that

$$-\log_\varepsilon M(X, \varepsilon) < d' \qquad \text{for all} \quad 0 < \varepsilon < \varepsilon_0.$$

Hence, since we can assume that $\varepsilon < 1$, we obtain

$$\varepsilon^{-\log_\varepsilon M(X,\varepsilon)} > \varepsilon^{d'} \qquad \text{for all} \quad 0 < \varepsilon < \varepsilon_0,$$

so $M(X, \varepsilon) < \varepsilon^{-d'}$ for all positive $\varepsilon < \varepsilon_0$.

8.3 There is a constant $c > 0$ (depending only on the domain Ω) such that if $u(s) \in V$ then u is strong on time interval $[s, s + c\|\nabla u(s)\|^{-4}]$. Suppose that

$$\frac{1}{\sqrt{r}} \int_{t-r}^{t} \|\nabla u(s)\|^2 \, ds \leq \sqrt{c/2}.$$

Then there must be $s \in [t - r, t]$ such that $\|\nabla u(s)\|^2 \leq \sqrt{c/2}r^{-1/2}$, so

$$c\|\nabla u(s)\|^{-4} \geq 2r.$$

Hence $u(s) \in V$ and u must be strong on the time interval

$$[s, s + 2r] \supset [t, t + r].$$

Thus we can choose $c_* := \sqrt{c/2}$.

8.4 Changing variables we obtain

$$\|\nabla u_\lambda(s)\|^2 = \int_{\mathbb{R}^3} |\nabla_x \lambda u(\lambda x, \lambda^2 s)|^2 \, dx = \lambda \|\nabla u(\lambda^2 s)\|^2.$$

Therefore

$$\int_0^{T/\lambda^2} \|\nabla u_\lambda(s)\|^4 \, ds = \int_0^{T/\lambda^2} \lambda^2 \|\nabla u(\lambda^2 s)\|^4 \, ds = \int_0^T \|\nabla u(s)\|^4 \, ds$$

and

$$\frac{1}{\sqrt{r/\lambda^2}} \int_{(t-r)/\lambda^2}^{t/\lambda^2} \|\nabla u_\lambda(s)\|^2 \, ds = \frac{\lambda^2}{\sqrt{r}} \int_{(t-r)/\lambda^2}^{t/\lambda^2} \|\nabla u(\lambda^2 s)\|^2 \, ds$$

$$= \frac{1}{\sqrt{r}} \int_{t-r}^t \|\nabla u(s)\|^2 \, ds.$$

8.5 Let $r_0 > 0$ be so large that $\|u_0\|^2 \le 2\sqrt{r_0} c_*$, where c_* is defined in Exercise 8.3. Then for every $t > r_0$ we have

$$\frac{1}{\sqrt{r_0}} \int_{t-r_0}^t \|\nabla u(s)\|^2 \, ds \le \frac{1}{2\sqrt{r_0}} \|u_0\|^2 \le c_*,$$

so u is regular on every interval $[t, t + r_0]$, where $t > r_0$, and hence for $t \ge r_0$.

8.6 Fix $\varepsilon > 0$. We choose $t_i \in \mathcal{T}$, $i = 1, 2, \ldots, n$, such that $(t_i - r/3, t_i + r/3)$ form a maximal family of disjoint intervals of length $(2r)/3$ centred at points of $\mathcal{T}$. Then the family of intervals $(t_i - r, t_i + r)$, $i = 1, 2, \ldots, n$ covers the set $\mathcal{T}$. Since $d_B(\mathcal{T}) < 1/2$, we can choose $\delta > 0$ such that for all $0 < r \le \delta$ the Lebesgue measure of the chosen cover $\mathcal{O}$ is so small that

$$\int_{\mathcal{O}} \|\nabla u(s)\|^2 \, ds \le \varepsilon.$$

Since $t_i \in \mathcal{T}$ it follows from Exercise 8.3 that for all $i = 1, 2, 3, \ldots, n$ we have

$$\int_{t_i - r/3}^t \|\nabla u(s)\|^2 \, ds > c_* (r/3)^{1/2}.$$

Thus from the fact that the intervals $(t_i - r/3, t_i)$, $i = 1, 2, 3, \ldots, n$ are disjoint we have

$$nc_*(r/3)^{1/2} \leq \sum_{i=1}^{n} \int_{t_i-r/3}^{t_i} \|\nabla u(s)\|^2 \, ds$$

$$\leq \int_{\mathcal{O}} \|\nabla u(s)\|^2 \, ds \leq \varepsilon.$$

8.7 Take $\varepsilon > 0$ sufficiently small and let n be such that $4^{-(n+1)} < \varepsilon \leq 4^{-n}$. Since

$$\tilde{\mathcal{T}} \subset \bigcup_{t_j \in B_{n+1}} [t_j, t_{j+1}]$$

and since there are no more than $\tilde{c}2^{n+2}$ points in B_n we can cover the set $\tilde{\mathcal{T}}$ by $N(\varepsilon) = \tilde{c}2^{n+2}$ intervals of length ε. Since $N(\varepsilon) = \tilde{c}2^{n+2} \leq c_1\varepsilon^{-1/2}$ the assertion follows. (This argument is due to Kukavica, 2009b.)

8.8 We follow the proof of Lemma 8.15, recalling that the set of regular times $\mathcal{R}$ is given by

$$(T, \infty) \cup \bigcup_i (a_i, b_i)$$

and is of full measure in $(0, \infty)$.

We take the inner product in H^2, then (since H^3 is an algebra)

$$\frac{1}{2}\frac{d}{dt}\|u\|_{H^2}^2 + \|u\|_{H^3}^2 \leq c\|u\|_{H^2}^2\|u\|_{H^3},$$

which after using Young's inequality gives

$$\frac{d}{dt}\|u\|_{H^2}^2 + \|u\|_{H^3}^2 \leq c\|u\|_{H^2}^4.$$

Now divide by $\|u\|_{H^2}^4$ and integrate from a_i to b_i to yield

$$\int_{a_i}^{b_i} \frac{\|u\|_{H^3}^2}{\|u\|_{H^2}^4} \, ds \leq c(b_i - a_i),$$

from which (8.14) follows.

The bounds in (8.15) now follow by an identical to argument to that used in Lemma 8.15.

Chapter 9

9.1 Let $u_0 \in K$, and let u be the global-in-time strong solution arising from u_0. First we choose a time $T > 0$ such that if an initial condition $v_0 \in V$ satisfies the condition

$$\|\nabla v_0\| \le \|\nabla u_0\| + 1,$$

then a weak solution v arising from v_0 is regular after time T (see Theorem 8.1). For such T we compute $R(u)$ given by (9.1). Choose $R = \min(R(u), 1)$ and let

$$B(u_0, R) = \{v_0 \in V : \|\nabla v_0 - \nabla u_0\| \le R\}.$$

We claim that $B(u_0, R) \subset K$. Theorem 9.1 guarantees that if $v_0 \in B(u_0, R)$ then v_0 gives rise to a strong solution v on $[0, T]$. Since $R \le 1$ our choice of T ensures that v remains strong for all times greater than T. Hence K is open in V.

9.2 (i) First we prove the strict inequality $X(t) < Y(t)$, assuming that $a < b$. Let T be maximal existence time for Y and suppose that $t^* \in (0, T)$ is the first time for which we have $X(t^*) = Y(t^*)$. Then for all $t \in (0, t^*]$

$$\frac{d}{dt}Y(t) = \alpha Y^3(t)$$

$$\ge \alpha X^3(t) \ge \frac{d}{dt}X - f,$$

so integrating this inequality between 0 and t^* we obtain

$$Y(t^*) - Y(0) \ge X(t^*) - X(0) - \int_0^{t^*} f(s)\,ds.$$

Since $X(t^*) = Y(t^*)$ we have

$$X(0) + \int_0^{t^*} f(s)\,ds \ge Y(0),$$

but

$$X(0) + \int_0^{t^*} f(s)\,ds \le a + \int_0^T f(s)\,ds$$

$$< b + \int_0^T f(s)\,ds = Y(0).$$

Contradiction.

(ii) Choose $b_n \to a$, $b_n > a$, and a sequence of corresponding solutions Y_n of

$$dY_n/dt = \alpha Y_n^3, \quad Y_n(0) = b_n + \int_0^T f(s)\,ds.$$

From (i) we have $Y_n(t) > X(t)$, so passing to the limit (and using the continuous dependence of Y_n on initial data) we obtain (ii).

9.3 In the proof of Theorem 9.3 we have already shown that

$$u_n \to u \quad \text{in} \quad C(0, T; H^1),$$

i.e.

$$\sup_{0 \le t \le T} \|\nabla w_n(t)\| \to 0.$$

Now we integrate inequality (9.11) to obtain

$$\int_0^T \|A w_n\|^2 \le \int_0^T \alpha \|\nabla w_n\|^2 + c \int_0^T \|\nabla w_n\|^6 + c \int_0^T \|Q_n \mathbb{P}(u \cdot \nabla) u\|^2 + \|\nabla w_n(0)\|^2.$$

Clearly $\|\nabla w_n(0)\| \to 0$ and we have shown before that

$$\int_0^T \|Q_n \mathbb{P}(u(t) \cdot \nabla) u(t)\|^2 \, dt \to 0.$$

Furthermore

$$\int_0^T \alpha(t) \|\nabla w_n(t)\|^2 \, dt \le \sup_{0 \le t \le T} \|\nabla w_n(t)\|^2 \int_0^T \alpha(t) \, dt \to 0,$$

$$\int_0^T \|\nabla w_n(t)\|^6 \, dt \le T \sup_{0 \le t \le T} \|\nabla w_n(t)\|^6 \to 0.$$

Hence $\int_0^T \|A w_n(s)\|^2 \, ds \to 0$.

9.4 We follow the steps of the proof of Theorem 9.1 obtaining on the right-hand side the additional term $\langle f, Aw \rangle$ that can be estimated as follows:

$$|\langle f, Aw \rangle| \le \|f\|^2 + \frac{1}{4} \|Aw\|^2.$$

Using similar estimates as before, we obtain

$$\frac{d}{dt} X \le \beta X^3 + \|f\|^2 e^{-\int_0^t \alpha(s) \, ds}$$

$$\le \beta X^3 + \|f\|^2,$$

where $X(t) = \|\nabla w\|^2 e^{-\int_0^t \alpha(s) \, ds}$, where α and β are as in the proof of Theorem 9.1. To finish the proof we compare X with the solution of

$$dY/dt = \beta Y^3, \qquad Y(0) = \|\nabla w(0)\|^2 + \int_0^T \|f(s)\|^2 \, ds.$$

It follows that $\|\nabla w\|$ remains bounded as long as

$$Y(0) < (2\beta t)^{-1/2} = (2tc)^{-1/2} \exp\left(-c \int_0^T \alpha(s) \, ds\right).$$

So w remains bounded on $[0, T]$ if

$$\|\nabla u_0 - \nabla v_0\|^2 + \int_0^T \|f(s)\|^2 \, ds < \frac{1}{\sqrt{2cT}} \exp\left(-c \int_0^T \|\nabla u\| \|Au\| + \|\nabla u\|^4 \, ds\right).$$

9.5 Using the triangle inequality, we obtain

$$\|Q_n \mathbb{P}(u_n \cdot \nabla)u_n\|^2 \leq 2\|Q_n \mathbb{P}(u_n \cdot \nabla)u_n - Q_n \mathbb{P}(u \cdot \nabla)u\|^2 + 2\|Q_n \mathbb{P}(u \cdot \nabla)u\|^2.$$

We have already proved that

$$\int_0^T \|Q_n \mathbb{P}(u \cdot \nabla)u]\|^2 \to 0$$

as $n \to \infty$, so it remains to estimate the term

$$I_n := \|Q_n \mathbb{P}[(u_n \cdot \nabla)u_n] - Q_n \mathbb{P}[(u \cdot \nabla)u]\|^2.$$

Notice that

$$
\begin{aligned}
I_n &\leq 2\|Q_n \mathbb{P}[(u_n \cdot \nabla)u_n - (u \cdot \nabla)u]\|^2 \\
&\leq 2\|Q_n \mathbb{P}[((u_n - u) \cdot \nabla)u_n]\|^2 + 2\|Q_n \mathbb{P}[(u \cdot \nabla)(u - u_n)]\|^2 \\
&\leq 2\|((u_n - u) \cdot \nabla)u_n\|^2 + 2\|(u \cdot \nabla)(u - u_n)\|^2 \\
&\leq 2\|u_n - u\|_{L^4}^2 \|\nabla u_n\|_{L^4}^2 + 2\|u\|_{L^4}^2 \|\nabla(u - u_n)\|_{L^4}^2.
\end{aligned}
$$

To conclude the proof we notice that the convergence

$$u_n \to u \text{ in } L^\infty(0, T; H^1) \cap L^2(0, T; H^2),$$

the Sobolev embedding $H^1 \subset L^6$, and the embedding $L^6 \subset L^4$ on $\mathbb{T}^3$ imply that

$$\int_0^T \|u_n - u\|_{L^4}^2 \|\nabla u_n\|_{L^4}^2 \leq \|u_n - u\|_{L^\infty(0,T;L^6)}^2 \|\nabla u_n\|_{L^2(0,T;L^6)}^2 \to 0$$

and

$$\int_0^T \|u\|_{L^4}^2 \|\nabla(u_n - u)\|_{L^4}^2 \leq \|u\|_{L^\infty(0,T;L^6)}^2 \|\nabla(u_n - u)\|_{L^2(0,T;L^6)}^2 \to 0.$$

9.6 It is obvious that $\|u_n(0) - u_0\|^2 \to 0$. From Exercise 9.5 we know that

$$\int_0^T \|f_n(s)\|^2 \, ds \to 0.$$

Finally since $u_n \to u$ in $L^\infty(0, T; H^1) \cap L^2(0, T; H^2)$ it follows that we have a uniform bound

$$I_n := \int_0^T \|Au_n\| \|\nabla u_n\| + \|\nabla u_n\|^4 \leq C,$$

so $e^{-I_n} > \varepsilon_0$ for some $\varepsilon_0 > 0$ for all n. The assertion now follows.

Chapter 10

10.1 Since $\widehat{u_\lambda}(\xi) = \lambda^{-2}\hat{u}(\xi/\lambda)$ it follows that

$$
\|u_\lambda\|^2_{\dot{H}^s(\mathbb{R}^3)} = \int |\xi|^{2s}|\hat{u}_\lambda(\xi)|^2 \, d\xi = \lambda^{-4}\int |\xi|^{2s}|\hat{u}(\xi/\lambda)|^2 \, d\xi
$$

$$
= \lambda^{-4}\int |\lambda\xi|^{2s}|\hat{u}(\xi)|^2 \, \lambda^3 \, d\xi = \lambda^{2s-1}\int |\xi||\hat{u}(\xi)|^2 \, d\xi
$$

$$
= \lambda^{2s-1}\|u\|^2_{\dot{H}^s(\mathbb{R}^3)}.
$$

For the L^p norms simply write

$$
\|u_\lambda\|^p_{L^p} = \int |u_\lambda(x)|^p \, dx = \lambda^p \int |u(\lambda x)|^p \, dx
$$

$$
= \lambda^p \int |u(y)|^p \lambda^{-3} \, dy = \lambda^{p-3}\|u\|^p_{L^p}.
$$

10.2 We have

$$
\int_{\mathbb{T}^3} |f(\lambda x)|^p \, dx = \lambda^{-3} \int_{\lambda\mathbb{T}^3} |f(y)|^p \, dy = \int_{\mathbb{T}^3} |f(y)|^p \, dy,
$$

since y is $\mathbb{T}^3$-periodic.

10.3 We have

$$
\|e^{\Delta s}u_0\|^2_{\dot{H}^{3/2}} = \sum_{k \in \dot{\mathbb{Z}}^3} |k|^3 e^{-2|k|^2 t}|\hat{u}_k|^2,
$$

and so

$$
\int_0^T \|e^{\Delta s}u_0\|^2_{\dot{H}^{3/2}} \, ds = \frac{1}{2}\sum_{k \in \dot{\mathbb{Z}}^3}(1 - e^{-2|k|^2 T})|k||\hat{u}_k|^2. \tag{S.4}
$$

Choosing $\ell_n \in \dot{\mathbb{Z}}^3$ with $|\ell_n| = n$ and taking u_n to be the function with Fourier coefficients

$$
\hat{u}_k = \begin{cases} \frac{1}{\sqrt{2n}} & k = \pm\ell_n, \\ 0 & \text{otherwise} \end{cases}
$$

yields $u_n \in \dot{H}^{1/2}$ with $\|u_n\|_{\dot{H}^{1/2}} = 1$ but for which

$$
\int_0^T \|e^{\Delta s}u_n\|^2_{\dot{H}^{3/2}} \, ds = \frac{1}{2}(1 - e^{-2n^2 T}).
$$

If α is the solution of $1 - e^{-2\alpha} = 2\varepsilon_0$ then $T_n = \alpha/2n^2 \to 0$ as $n \to \infty$.

10.4 The equality (S.4) still holds when $u_0 \in \dot{H}^1(\mathbb{T}^3)$, but now we can write the right-hand side as

$$\frac{1}{2} \sum_{k \in \mathbb{Z}^3} \frac{1 - e^{-2|k|^2 T}}{|k|} |k|^2 |\hat{u}_k|^2 \le \frac{1}{2} c T^{1/2} \sum_{k \in \mathbb{Z}^3} |k|^2 |\hat{u}_k|^2 = \frac{c}{2} T^{1/2} \|u_0\|^2_{\dot{H}^1},$$

since

$$\sup_{\kappa > 0} \frac{1 - e^{-2\kappa^2 T}}{\kappa} = T^{1/2} \sup_{\kappa > 0} \frac{1 - e^{-2\kappa^2 T}}{\kappa T^{1/2}} = c T^{1/2}.$$

The required result now follows easily.

10.5 Consider the Galerkin approximation u_n that satisfies

$$\frac{\mathrm{d}u_n}{\mathrm{d}t} - \Delta u_n + P_n(u_n \cdot \nabla)u_n = 0, \qquad u_n(0) = P_n u_0.$$

Since u_n is smooth we can take the inner product with Λu_n to yield

$$\frac{1}{2} \frac{\mathrm{d}}{\mathrm{d}t} \|u_n\|^2_{\dot{H}^{1/2}} + \|u_n\|^2_{\dot{H}^{3/2}} \le |\langle (u_n \cdot \nabla)u_n, \Lambda u_n \rangle$$

$$\le \|u_n\|_{L^3} \|\nabla u_n\|_{L^3} \|\Lambda u_n\|_{L^3}$$

$$\le C \|u_n\|_{\dot{H}^{1/2}} \|u_n\|^2_{\dot{H}^{3/2}}.$$

We now use Young's inequality on the right-hand side to write

$$\frac{1}{2} \frac{\mathrm{d}}{\mathrm{d}t} \|u_n\|^2_{\dot{H}^{1/2}} + \|u_n\|^2_{\dot{H}^{3/2}} \le \frac{1}{2} \|u_n\|^2_{\dot{H}^{3/2}} + \frac{1}{2} C^2 \|u_n\|^2_{\dot{H}^{3/2}} \|u_n\|^2_{\dot{H}^{1/2}}$$

and therefore

$$\frac{\mathrm{d}}{\mathrm{d}t} \|u_n\|^2_{\dot{H}^{1/2}} \le C^2 \|u_n\|^2_{\dot{H}^{3/2}} \|u_n\|^2_{\dot{H}^{1/2}}.$$

An application of Gronwall's Lemma (Lemma A.24) shows that

$$\|u_n(t)\|^2 \le \|u_n(0)\|^2_{\dot{H}^{1/2}} \exp\left[C^2 \int_0^t \|u_n(s)\|^2_{\dot{H}^{3/2}} \, \mathrm{d}s \right]$$

$$\le \|u_0\|^2_{\dot{H}^{1/2}} \exp\left[C^2 \int_0^t \|u(s)\|^2_{\dot{H}^{3/2}} \, \mathrm{d}s \right],$$

and this bound on u_n, uniform in $L^\infty(0, T; H^{1/2})$, persists for the limiting function u.

10.6 If u and v are two solutions with the stated regularity and the same initial condition $u(0) = v(0) = u_0$ then their difference $w = u - v$ satisfies

$$\partial_t w + Aw + (u \cdot \nabla)w + (w \cdot \nabla)v + \nabla q = 0.$$

Taking the inner product of this equation with Λw yields

$$\frac{1}{2}\frac{d}{dt}\|w\|_{\dot{H}^{1/2}}^2 + \|w\|_{\dot{H}^{3/2}}^2 \le \|u\|_{L^6}\|\nabla w\|_{L^2}\|\Lambda w\|_{L^3} + \|w\|_{L^3}\|\nabla v\|_{L^3}\|\Lambda w\|_{L^3}$$

$$\le c\|u\|_{\dot{H}^1}\|w\|_{\dot{H}^1}\|w\|_{\dot{H}^{3/2}} + c\|w\|_{\dot{H}^{1/2}}\|v\|_{\dot{H}^{3/2}}\|w\|_{\dot{H}^{3/2}}$$

$$\le c\|u\|_{\dot{H}^1}\|w\|_{\dot{H}^{1/2}}^{1/2}\|w\|_{\dot{H}^{3/2}}^{3/2} + c\|w\|_{\dot{H}^{1/2}}\|v\|_{\dot{H}^{3/2}}\|w\|_{\dot{H}^{3/2}}$$

$$\le \|w\|_{\dot{H}^{3/2}}^2 + c\|u\|_{\dot{H}^1}^4\|w\|_{\dot{H}^{1/2}}^2 + c\|v\|_{\dot{H}^{3/2}}^2\|w\|_{\dot{H}^{1/2}}^2,$$

and so

$$\frac{d}{dt}\|w\|_{\dot{H}^{1/2}}^2 \le c\left(\|u\|_{\dot{H}^1}^4 + \|v\|_{\dot{H}^{3/2}}^2\right)\|w\|_{\dot{H}^{1/2}}^2.$$

Since $u \in L^4(0, T; \dot{H}^1)$, $v \in L^2(0, T; \dot{H}^{3/2})$, and $w(0) = 0$, $\|w(t)\|_{\dot{H}^{1/2}} = 0$ for all $t \in (0, T)$ as required.

Chapter 11

11.1 Integrating by parts and using (11.2) we obtain

$$-\int (\partial_i\partial_i u_j)u_j|u|^{\alpha-2} = \int (\partial_i u_j)(\partial_i u_j)|u|^{\alpha-2} + (\alpha-2)\int (\partial_i u_j)u_j u_k(\partial_i u_k)|u|^{\alpha-3}$$

$$= \int |\nabla u|^2|u|^{\alpha-2} + (\alpha-2)\sum_i \int |(\partial_i u_j)u_j|^2|u|^{\alpha-3}$$

$$\ge \int |\nabla u|^2|u|^{\alpha-2}.$$

11.2 To prove the inequality consider first $u \in C_c^\infty(\mathbb{R}^3)$, and note that

$$\|u\|_{L^{3\alpha}}^\alpha = \left(\int_{\mathbb{R}^3} |u|^{3\alpha}\,dx\right)^{1/3} = \||u|^{\alpha/2}\|_{L^6}^2 \le C\|\nabla|u|^{\alpha/2}\|_{L^2}^2.$$

Now, using (11.2) we have $|\nabla|u|^{\alpha/2}| \le \frac{\alpha}{2}|\nabla u||u|^{\frac{\alpha}{2}-1}$, and so

$$\|u\|_{L^{3\alpha}}^\alpha \le C\int_{\mathbb{R}^3} |\nabla u|^2|u|^{\alpha-2}\,dx$$

as claimed. The general case follows by density.

11.3 Using (11.2) and an integration by parts,

$$\int u_j(\partial_j u_i)u_i|u|^\alpha\,dx = -\int (\partial_j u_j)|u|^{\alpha+2}\,dx - \int u_j u_i(\partial_j u_i)|u|^\alpha\,dx$$

$$-\alpha\int u_j|u|^2 u_k(\partial_j u_k)|u|^{\alpha-2}\,dx$$

$$= -(\alpha+1)\int u_j u_k(\partial_j u_k)|u|^\alpha\,dx,$$

from which it follows that $\int [(u \cdot \nabla)u] \cdot u|u|^\alpha = 0$.

11.4 Take the inner product of the equation with $|u|^{\alpha-2}u$ to obtain

$$\frac{1}{\alpha}\frac{d}{dt}\|u\|_{L^\alpha}^\alpha + \int |u|^{\alpha-2}|\nabla u|^2 \le -\int \nabla p \cdot u|u|^{\alpha-2};$$

we have used Exercise 11.1 for the Laplacian term, and the nonlinear term vanishes due to the result of Exercise 11.3. To estimate the pressure term we integrate by parts,

$$\begin{aligned}
-\int \nabla p \cdot u|u|^{\alpha-2} &= \int pu \cdot \nabla |u|^{\alpha-2} \\
&= \frac{\alpha-2}{2}\int p|u|^{\alpha-4}u \cdot \nabla |u|^2 \\
&\le c\int |p||u|^{\alpha-2}|\nabla u| \\
&\le c\int |p|^2|u|^{\alpha-2} + \frac{1}{2}\int |u|^{\alpha-2}|\nabla u|^2,
\end{aligned}$$

and rearranging yields (11.19).

We now appeal to the result of Exercise 11.2 to write

$$\begin{aligned}
\frac{1}{\alpha}\frac{d}{dt}\|u\|_{L^\alpha}^\alpha + \frac{1}{2c_\alpha}\|u\|_{L^{3\alpha}}^\alpha &\le c\int |p|^2|u|^{\alpha-2} \\
&\le c\|p\|_{L^\alpha}^2\|u\|_{L^\alpha}^{\alpha-2}.
\end{aligned}$$

We use the pressure estimate $\|p\|_{L^\alpha} \le c\|u\|_{L^{2\alpha}}^2$ and the Lebesgue interpolation

$$\|u\|_{L^{2\alpha}} \le \|u\|_{L^\alpha}^{1/4}\|u\|_{L^{3\alpha}}^{3/4}$$

to deduce that

$$c\|p\|_{L^\alpha}^2\|u\|_{L^\alpha}^{\alpha-2} \le c\|u\|_{L^{2\alpha}}^4\|u\|_{L^\alpha}^{\alpha-2} \le c\|u\|_{L^\alpha}^{\alpha-1}\|u\|_{L^{3\alpha}}^3,$$

and (11.20) follows.

11.5 With $\alpha = 3$ (11.20) becomes

$$\frac{1}{3}\frac{d}{dt}\|u\|_{L^3}^3 + \frac{1}{2c_3}\|u\|_{L^9}^3 \le c\|u\|_{L^3}^2\|u\|_{L^9}^3,$$

and therefore

$$\frac{1}{3}\frac{d}{dt}\|u\|_{L^3}^3 \le \|u\|_{L^9}^3\left(c\|u\|_{L^3}^2 - \frac{1}{2c_3}\right).$$

It follows that if $\|u_0\|_{L^3} < \varepsilon := (2cc_3)^{-1/2}$ then $\|u(t)\|_{L^3}$ is non-increasing.

11.6 Since $\dot{X} \le cX^\theta$ it follows that

$$\frac{\mathrm{d}X}{X^\theta} \le c\,\mathrm{d}t,$$

and so, integrating from t to τ for $t < \tau < T$ we obtain

$$\left[\frac{X^{-(\theta-1)}}{-(\theta-1)}\right]_{X(t)}^{X(\tau)} \le c(\tau - t)$$

which yields

$$\frac{X(t)^{-(\theta-1)}}{\theta-1} \le c(\tau-t) + \frac{X(\tau)^{-(\theta-1)}}{\theta-1}.$$

Letting $\tau \to T$ gives

$$X(t)^{-(\theta-1)} \le c(\theta-1)(T-t),$$

which on rearranging yields

$$X(t) \ge [c(\theta-1)(T-t)]^{-1/(\theta-1)} = \kappa(T-t)^{-1/(\theta-1)}.$$

11.7 Since $\alpha > 3$ we can use Young's inequality to estimate the term on the right-hand side of (11.20):

$$c\|u\|_{L^\alpha}^{\alpha-1}\|u\|_{L^{3\alpha}}^3 \le c\|u\|_{L^\alpha}^{\alpha(\alpha-1)/(\alpha-3)} + \frac{1}{2c_\alpha}\|u\|_{L^{3\alpha}}^\alpha.$$

Using this estimate we immediately obtain

$$\frac{1}{\alpha}\frac{\mathrm{d}}{\mathrm{d}t}\|u\|_{L^\alpha}^\alpha \le c\|u\|_{L^\alpha}^{\alpha(\alpha-1)/(\alpha-3)}$$

and (11.18) now follows from Exercise 11.6.

11.8 The rescaled solution $u_\lambda(x,t) = \lambda u(\lambda x, \lambda^2 t)$ will blow up at time T/λ^2, and if the constant in (11.22) is absolute we should have

$$\|u_\lambda(T/\lambda^2 - t)\|_{L^p} \ge ct^{-\gamma}$$

once more. We showed in Exercise 10.1 that

$$\|\lambda u(\lambda \cdot)\|_{L^p(\mathbb{R}^3)} = \lambda^{1-(3/p)}\|u\|_{L^p(\mathbb{R}^3)},$$

and therefore from (11.22) it follows that

$$\|u_\lambda(\lambda^2 T - t)\|_{L^p} = \|\lambda u(\lambda x, T - \lambda^{-2}t)\|_{L^p} \ge c\lambda^{1-(3/p)}(\lambda^{-2}t)^{-\gamma}$$
$$= c\lambda^{1-(3/p)+2\gamma}t^{-\gamma};$$

the factor of λ must vanish, otherwise we could decrease/increase c for appropriate rescalings. We therefore require that $\gamma = (p-3)/2p$, as claimed.

11.9 Since

$$\|u\|_{L^\alpha} \le \|u\|_{L^3}^{2/(\alpha-1)} \|u\|_{L^{3\alpha}}^{(\alpha-3)/(\alpha-1)}$$

it follows from (11.20) that

$$\frac{1}{\alpha}\frac{d}{dt}\|u\|_{L^\alpha}^\alpha + \frac{1}{2c_\alpha}\|u\|_{L^{3\alpha}}^\alpha \le c\|u\|_{L^\alpha}^{\alpha-1}\|u\|_{L^{3\alpha}}^3$$

$$\le c\|u\|_{L^3}^2\|u\|_{L^{3\alpha}}^\alpha,$$

i.e.

$$\frac{1}{\alpha}\frac{d}{dt}\|u\|_{L^\alpha}^\alpha \le \|u\|_{L^{3\alpha}}^\alpha\left\{c\|u\|_{L^3}^2 - \frac{1}{2c_\alpha}\right\}.$$

If $\|u\|_{L^3} < \varepsilon$ then the right-hand side is negative; it follows that if $\|u_0\|_{L^3} < \varepsilon$ then $\|u(t)\|_{L^3} < \varepsilon$ for all $t > 0$, and hence every L^α norm ($3 \le \alpha < \infty$) is non-increasing.

Chapter 12

12.1 Notice that the ith component of $(u \cdot \nabla)u$ is given by $u_j\partial_j u_i$. Similarly for $1/2(\nabla|u|^2)$ we have $u_j\partial_i u_j$. To complete the proof of the identity we need to show that $-u \times \omega$ is given $u_j\partial_j u_i - u_j\partial_i u_j$. We have

$$(u \times \omega)_i = \epsilon_{ijk}u_j\epsilon_{kmn}\partial_m u_n = [\delta_{im}\delta_{jn} - \delta_{in}\delta_{jm}]u_j\partial_m u_n = u_j\partial_j u_i - u_j\partial_i u_j,$$

as required.

12.2 Recall that $(A \times B)_i = \epsilon_{ijk}a_j b_k$ and $(\nabla \times C)_i = \epsilon_{imn}\partial_m c_n$. We start with the left-hand side of the equality; we have

$$
\begin{aligned}
(\nabla \times (A \times B))_i &= \epsilon_{imn}\partial_m(\epsilon_{njk}a_j b_k) = \epsilon_{imn}\epsilon_{njk}[(\partial_m a_j)b_k + a_j(\partial_m b_k)] \\
&= [\delta_{ij}\delta_{mk} - \delta_{ik}\delta_{jm}][(\partial_m a_j)b_k + a_j(\partial_m b_k)] \\
&= (\partial_k a_i)b_k - (\partial_j a_j)b_i + a_i(\partial_k b_k) - a_j(\partial_j b_i) \\
&= (B \cdot \nabla)A_i - B_i(\nabla \cdot A) + A_i(\nabla \cdot B) - (A \cdot \nabla)B_i,
\end{aligned}
$$

which completes the proof.

12.3 Using tensorial notation we find

$$\int f \cdot (\operatorname{curl} g) = \int f_i\epsilon_{ijk}(\partial_j g_k)$$

$$= -\epsilon_{ijk}\int (\partial_j f_i)g_k = \epsilon_{kji}\int (\partial_j f_i)g_k = \int (\operatorname{curl} f) \cdot g.$$

12.4 First let us assume that u is sufficiently smooth to carry out all the required computations. Since $\nabla \cdot u = 0$ the identity

$$\text{curl}\,(\text{curl}\,u) = \nabla(\nabla \cdot u) - \Delta u$$

yields (by multiplying by $\phi \in C_c^\infty(B)$ and integrating over B)

$$\int \text{curl}\,\text{curl}\,u\,\phi\,dx = \int -\Delta u\,\phi\,dx.$$

Integrating by parts on both sides yields

$$\int \omega \times \nabla\phi(x)\,dx = \int \nabla u(x) \cdot \nabla\phi(x)\,dx.$$

Notice that both the left-hand side and the right-hand side can be bounded in terms of $\|\nabla u\|_{L^r}$, and so we can extend the result to all u with $\nabla u \in L^r$.

Chapter 13

13.1 The result follows from successive changes of variables. We have

$$
\begin{aligned}
\|u_R\|_{L_t^{q'} L_x^q(Q_1^*(0,0))} &= \left(\int_{-1/2}^{1/2} \left(\int_{B_1} |Ru(Rx, R^2 t)|^q\,dx \right)^{q'/q} dt \right)^{1/q'} \\
&= \left(\int_{-1/2}^{1/2} \left(\int_{B_R} R^q |u(\bar{x}, R^2 t)|^q R^{-3}\,d\bar{x} \right)^{q'/q} dt \right)^{1/q'} \\
&= \left(\int_{-1/2}^{1/2} R^{q'-3q'/q} \left(\int_{B_R} |u(\bar{x}, R^2 t)|^q\,d\bar{x} \right)^{q'/q} dt \right)^{1/q'} \\
&= \left(\int_{-R^2/2}^{R^2/2} R^{q'-3q'/q} \left(\int_{B_R} |u(\bar{x}, \bar{t})|^q\,d\bar{x} \right)^{q'/q} R^{-2}\,d\bar{t} \right)^{1/q'} \\
&= R^{1-3/q-2/q'} \left(\int_{-R^2/2}^{R^2/2} \left(\int_{B_R} |u(\bar{x}, \bar{t})|^q\,d\bar{x} \right)^{q'/q} d\bar{t} \right)^{1/q'} \\
&= R^{1-3/q-2/q'} \|u\|_{L_t^{q'} L_x^q(Q_R^*(0,0))} = \|u\|_{L_t^{q'} L_x^q(Q_R^*(0,0))}.
\end{aligned}
$$

Chapter 14

14.1 Given two solutions u and v of the regularised equation with the same initial data, we consider $w = u - v$, which satisfies

$$\partial_t w - \Delta w + [(\psi_\varepsilon * w) \cdot \nabla]u + [(\psi_\varepsilon * v) \cdot \nabla]w + \nabla q = 0,$$
$$\nabla \cdot w = 0, \quad w(0) = 0.$$

Taking the inner product with w we obtain

$$\frac{1}{2}\frac{d}{dt}\|w\|^2 + \|\nabla w\|^2 = -\langle[(\psi_\varepsilon * w) \cdot \nabla]u, w\rangle,$$

since $\langle[(\psi_\varepsilon * v) \cdot \nabla]w, w\rangle = 0$ as $\psi_\varepsilon * v$ is divergence free. Therefore

$$\frac{1}{2}\frac{d}{dt}\|w\|^2 + \|\nabla w\|^2 \le \|\psi_\varepsilon * w\|_{L^\infty}\|\nabla u\|\|w\|$$
$$\le c\varepsilon^{-1/2}\|\nabla w\|\|\nabla u\|\|w\|$$
$$\le \frac{1}{2}\|\nabla w\|^2 + c\varepsilon^{-1}\|\nabla u\|^2\|w\|^2,$$

and so

$$\frac{d}{dt}\|w\|^2 \le c\varepsilon^{-1}\|\nabla u\|^2\|w\|^2.$$

Gronwall's inequality implies that

$$\|w(t)\|^2 \le \|w(0)\|^2 \exp\left(c\varepsilon^{-1}\int_0^t \|\nabla u(s)\|^2 \, ds\right),$$

and since $\|\nabla u\| \in L^2(0, T)$ and $w(0) = 0$ it follows that $w(t) = 0$ for every $t \in (0, T)$.

14.2 Take $\varphi \in \mathcal{D}_\sigma$ and write

$$\int_0^T \langle[(\psi_\varepsilon * u^\varepsilon) \cdot \nabla]u^\varepsilon - (u \cdot \nabla)u, \varphi\rangle \, dt$$
$$= \int_0^T \langle[(\psi_\varepsilon * u^\varepsilon) \cdot \nabla](u^\varepsilon - u), \varphi\rangle + \langle[(\psi_\varepsilon * (u^\varepsilon - u)) \cdot \nabla]u, \varphi\rangle$$
$$+ \langle[(\psi_\varepsilon * u - u) \cdot \nabla]u, \varphi\rangle$$
$$\le \int_0^T \|u^\varepsilon\|\|u^\varepsilon - u\|\|\nabla\varphi\|_{L^\infty} + \|u^\varepsilon - u\|\|\nabla u\|\|\varphi\|_{L^\infty}$$
$$+ \|(\psi_\varepsilon * u) - u\|\|\nabla u\|\|\varphi\|_{L^\infty},$$

since $\|\psi_\varepsilon * f\| \le \|f\|$. The right-hand side tends to zero, since $u^\varepsilon \to u$ in $L^2(0, T; L^2)$ and $\|\psi_\varepsilon * u - u\| \to 0$ in $L^2(0, T; L^2)$.

Chapter 15

15.1 If either a or b are zero then there is nothing to prove; so we can assume that $a, b > 0$. In this case, dividing by a^q and setting $x = b/a$ it suffices to show that

$$(1 + x)^q \le 2^{q-1}(1 + x^q). \tag{S.5}$$

To do this we find the maximum of

$$f(x) := \frac{(1+x)^q}{1+x^q},$$

which occurs when

$$f'(x) = \frac{q(1+x^q)(1+x)^{q-1} - qx^{q-1}(1+x)^q}{(1+x^q)^2} = 0,$$

i.e. when

$$q(1+x^q) - qx^{q-1}(1+x) = 0 \quad \Rightarrow \quad x^{q-1} = 1;$$

this implies that the maximum occurs when $x = 1$, and since $f(1) = 2^{q-1}$ this yields (S.5).

15.2 We use Hölder's inequality, noting that $1/q' = (q-1)/q$, where q' is the conjugate of q:

$$\int_U |\bar{p}|^q \, dx = \mu(U) \left| \frac{1}{\mu(U)} \int_U p \, dx \right|^q$$

$$\leq \mu(U)^{1-q} \left\{ \left(\int_U 1 \, dx \right)^{1/q'} \left(\int_U |p|^q \, dx \right)^{1/q} \right\}^q$$

$$= \mu(U)^{1-q+q/q'} \int_U |p|^q \, dx = \int_U |p|^q \, dx.$$

15.3 Since for any $\varepsilon < 1$

$$Q_\varepsilon^*(x, t) \subset B_{\sqrt{2}\varepsilon}(x, t)$$

it follows that $N(X, \sqrt{2}\varepsilon) \leq N_Q(X, \varepsilon)$ and arguing as in the proof of Lemma 8.6 it follows that $\dim_B(X) \leq \dim_P(X)$. A simple example showing that this inequality can be strict is provided by any set Y of the form $\{x\} \times T$, where $T \subset \mathbb{R}$ with $\dim_B(T) \neq 0$, since for such a set

$$N_Q(Y, \varepsilon^2) = N(Y, \varepsilon) = N_{\mathbb{R}}(T, \varepsilon),$$

where the last expression denotes the number of intervals of length 2ε required to cover T viewed as a subset of $\mathbb{R}$. Then

$$\dim_P(Y) = \limsup_{\varepsilon \to 0} \frac{\log N_Q(Y, \varepsilon^2)}{-2\log \varepsilon} = \limsup_{\varepsilon \to 0} \frac{N(Y, \varepsilon)}{-2\log \varepsilon} = \frac{1}{2}\dim_B(Y).$$

15.4 It is simple to show that $N_Q(X, \varepsilon) \leq M_Q(X, \varepsilon)$: if $z_j \in X$ are the centres of $M_Q(X, \varepsilon)$ disjoint cylinders, no ε-cylinder can contain more than one of the $\{z_j\}$ and so it will take at least $M_Q(X, \varepsilon)$ cylinders to cover X.

Now we show that $M_Q(X, \varepsilon/3) \leq N_Q(X, \varepsilon)$.

Given a maximal family $\{Q^*_{\varepsilon/3}(z_j)\}$, $j = 1, \ldots, M$, of disjoint cylinders with $z_j \in X$, if the cylinders $\{Q^*_{\varepsilon}(z_j)\}$ do not cover X then there is a point $z \in X$ such that $z \notin Q^*_{\varepsilon}(z_j)$ for each j; it then follows that $Q^*_{\varepsilon/3}(z)$ is disjoint from every $Q^*_{\varepsilon/3}(z_j)$, contradicting the maximality of M. Indeed, if $(x, t) \in Q^*_{\varepsilon/3}(z)$ and $(\xi, \sigma) \in Q^*_{\varepsilon/3}(z_j)$ then

$$|x - \xi| \geq |x - x_j| - |\xi - x_j| > \varepsilon - (\varepsilon/3) > \varepsilon/3$$

and

$$|t - \sigma| \geq |t - t_j| - |\sigma - t_j| > \varepsilon^2 - (\varepsilon/3)^2 > (\varepsilon/3)^2,$$

writing $z_j = (x_j, t_j)$.

The argument now follows that of Lemma 8.6 unchanged.

Chapter 16

16.1 We prove that there exists a constant C_4 such that for all $f \in W^{1,1}(B_1)$ with zero average,

$$\|f\|_{L^{3/2}(B_1)} \leq C_4 \|\nabla f\|_{L^1(B_1)}, \tag{S.6}$$

Suppose that (S.6) does not hold. Then there exist a sequence $f_n \in W^{1,1}(B_1)$ such that

$$\int_{B_1} f_n = 0 \qquad \text{and} \qquad \|f_n\|_{L^{3/2}} \geq n\|\nabla f_n\|_{L^1}.$$

Since $W^{1,1} \subset L^{3/2}$ there exists a constant C such that

$$\|f\|_{L^{3/2}} \leq C\,[\|\nabla f\|_{L^1} + \|f\|_{L^1}] \qquad \text{for all} \quad f \in W^{1,1}(B_1).$$

It follows that

$$n\|\nabla f_n\|_{L^1} \leq C\,[\|\nabla f_n\|_{L^1} + \|f_n\|_{L^1}],$$

which, provided than $n > C$, we can rearrange to give

$$\|\nabla f_n\| \leq \frac{C}{n - C}\|f_n\|_{L^1}.$$

If we consider the functions $g_n = f_n/\|f_n\|_{L^1}$ then

$$\int_{B_1} g_n = 0, \qquad \|g_n\|_{L^1} = 1, \qquad \text{and} \qquad \|\nabla g_n\|_{L^1} \leq \frac{C}{n - C}.$$

It follows from the fact that $W^{1,1}$ is compactly embedded in L^1 that there is a subsequence g_{n_j} such that $g_{n_j} \to g$ in L^1 and $\nabla g_{n_j} \rightharpoonup \nabla g$ in L^1; therefore $\|g\|_{L^1} = 1$ but $\int g = 0$ and $\nabla g = 0$, which is impossible, and so (S.6) holds.

Now to obtain the inequality on B_r define $f_r(x) := f(rx)$ and use the inequality on B_1:

$$
\begin{aligned}
\left(\int_{B_r} |f(x)|^{3/2}\, dx \right)^{2/3} &= \left(\int_{B_1} |f(rx)|^{3/2}\, r^3\, dx \right)^{2/3} \\
&= r^2 \|f_r\|_{B_1} \\
&\le C_4 r^2 \|\nabla f_r\|_{L^1(B_r)} \\
&= C_4 r^2 \int_{B_1} |\nabla f_r(x)|\, dx \\
&= C_4 r^3 \int_{B_1} |(\nabla f)(rx)|\, dx \\
&= C_4 \int_{B_r} |\nabla f(y)|\, dy.
\end{aligned}
$$

16.2 Set $E_k := E(\theta^k r)$. Then $E_{k+1} \le \alpha + \beta E_k$; if $E_k = \tilde{E}_k - \alpha/(1-\beta)$ then $\tilde{E}_{k+1} \le \beta \tilde{E}_k$ and it is clear that $\tilde{E}_k \le \beta^k \tilde{E}_0$, which implies that

$$
E_k \le \frac{\alpha}{1-\beta} + \beta^k \left(E_0 - \frac{\alpha}{1-\beta} \right).
$$

By choosing k sufficiently large it follows that $E_k \le 2\alpha/(1-\beta)$ as claimed.

16.3 If $\mathcal{P}^s(X) = 0$ then given $\varepsilon > 0$ there exists a δ_0 such that

$$
\mathcal{P}^s_\delta(X) = \inf \left\{ \sum_{i=1}^{\infty} r_i^s : X \subset \bigcup_i Q^*_{r_i} : r_i < \delta \right\} < \varepsilon/2 \qquad \text{for all} \qquad \delta < \delta_0.
$$

In particular, there exists a collection $Q^*_{r_i}$ such that $\sum_{i=1} r_i^s < \varepsilon$.

To show the converse, observe that if $\sum_i r_i^s < \varepsilon$ then $r_i < \varepsilon^{1/s}$ for every i, and so

$$
\mathcal{P}^s_{\varepsilon^{1/s}}(X) \le \varepsilon \qquad \Rightarrow \qquad \mathcal{P}^s_\varepsilon \le \varepsilon^s,
$$

from which it follows immediately that

$$
\mathcal{P}^s(X) = \lim_{\varepsilon \to 0} \mathcal{P}^s_\varepsilon(X) = 0.
$$

Chapter 17

17.1 Since

$$
\det M = \begin{vmatrix} m_{11} & m_{12} & m_{13} \\ m_{21} & m_{22} & m_{23} \\ m_{31} & m_{32} & m_{33} \end{vmatrix},
$$

the time derivative $\frac{\mathrm{d}}{\mathrm{d}t}(\det M)$ is the sum of three determinants,

$$= \begin{vmatrix} \dot{m}_{11} & \dot{m}_{12} & \dot{m}_{13} \\ m_{21} & m_{22} & m_{23} \\ m_{31} & m_{32} & m_{33} \end{vmatrix} + \begin{vmatrix} m_{11} & m_{12} & m_{13} \\ \dot{m}_{21} & \dot{m}_{22} & \dot{m}_{23} \\ m_{31} & m_{32} & m_{33} \end{vmatrix} + \begin{vmatrix} m_{11} & m_{12} & m_{13} \\ m_{21} & m_{2}2 & m_{23} \\ \dot{m}_{31} & \dot{m}_{32} & \dot{m}_{33} \end{vmatrix}.$$

The first of these is

$$\begin{vmatrix} a_{1j}m_{j1} & a_{1j}m_{j2} & a_{1j}m_{j3} \\ m_{21} & m_{22} & m_{23} \\ m_{31} & m_{32} & m_{33} \end{vmatrix} = a_{11} \begin{vmatrix} m_{11} & m_{12} & m_{13} \\ m_{21} & m_{22} & m_{23} \\ m_{31} & m_{32} & m_{33} \end{vmatrix},$$

since the terms with $j \neq 1$ yield two matrices in which one row is a multiple of the other. It follows that

$$\frac{\mathrm{d}}{\mathrm{d}t}(\det M) = (a_{11} + a_{22} + a_{33})\det M = (\mathrm{tr}A)(\det M).$$

17.2 Weak solutions of the 2D Navier–Stokes equations are smooth for $t > 0$, and so we can take the inner product with $-t\Delta u$ and write

$$t\frac{\mathrm{d}}{\mathrm{d}t}\|\nabla u\|^2 + t\|\Delta u\|^2 = \langle (u \cdot \nabla)u, t\Delta u \rangle = 0,$$

using the fact that $\langle (u \cdot \nabla)u, \Delta u \rangle = 0$. It follows that

$$\frac{\mathrm{d}}{\mathrm{d}t}(t\|\nabla u\|^2) + t\|\Delta u\|^2 = \|\nabla u\|^2,$$

and so integrating from 0 to T we obtain

$$T\|\nabla u(T)\|^2 + \int_0^T t\|\Delta u(t)\|^2 \, \mathrm{d}t = \int_0^T \|\nabla u(t)\|^2 \, \mathrm{d}t,$$

and the right-hand side is finite since $u \in L^2(0, T; H^1)$.

17.3 If $u \in H^{5/2}(\mathbb{T}^3)$ then $\partial_i u \in H^{3/2}(\mathbb{T}^3)$. Since $H^{3/2}(\mathbb{T}^3) \subset L^p(\mathbb{T}^3)$ for every $1 \le p < \infty$, it follows that for any such p we have $u \in W^{1,p}(\mathbb{T}^3)$. Morrey's Theorem now guarantees that $u \in C^{0,\gamma}(\mathbb{T}^3)$ for any $0 < \gamma < 1$ as claimed.

17.4 Using Hölder's inequality with exponents $(2/r, 2/(2-r))$ we can write

$$\int_0^T \|u(t)\|_X^r \, \mathrm{d}t = \int_0^T t^{-r/2}t^{r/2}\|u(t)\|_X^r \, \mathrm{d}r$$

$$\le \left(\int_0^T t^{-r/(2-r)} \, \mathrm{d}t \right)^{1-(r/2)} \left(\int_0^T t\|u(t)\|_X^2 \, \mathrm{d}t \right)^{r/2} < \infty$$

since $0 < r/(2-r) < 1$.

17.5 Fix s and consider

$$F(t) := t^\alpha - s^\alpha - (t - s)^\alpha.$$

Then $F(s) = 0$ and

$$\frac{\mathrm{d}F}{\mathrm{d}t}(t) = \alpha t^{\alpha-1} - \alpha(t - s)^{\alpha-1} \leq 0$$

since $0 < \alpha < 1$ and $t - s \leq t$.

17.6 Let

$$\dim_M(K) := n - \sup\{s : \ \mu(O(K, \varepsilon)) \leq C\varepsilon^s \text{ for some } C > 0\}.$$

The result of Lemma 17.11 shows that for any $d > \dim_B(K)$ there is a $C > 0$ such that $\mu(O(K, \varepsilon)) \leq C\varepsilon^{n-d}$, and so $\dim_M(K) \geq \dim_B(K)$. For the opposite inequality, if $d < \dim_B(K)$ then there is a sequence $\varepsilon_j \to 0$ such that there are at least ε_j^{-d} disjoint balls of radius ε_j with centres in K, from which it follows that

$$\mu(O(K, \varepsilon_j)) \geq \varepsilon_j^{-d} V_n \varepsilon_j^n$$

and so $\dim_M(K) \leq \dim_B(K)$.

17.7 Writing

$$\left| |k|^{3/2} - |j|^{3/2} \right| = \int_0^1 \frac{\mathrm{d}}{\mathrm{d}t} |j + t(k - j)|^{3/2} \, \mathrm{d}t$$

we can use the Mean Value Theorem to deduce that

$$\begin{aligned}
\left| |k|^{3/2} - |j|^{3/2} \right| &\leq \frac{3}{2} \left(\max_{t \in [0,1]} |j + t(k - j)|^{1/2} \right) |k - j| \\
&\leq \frac{3}{2} [|j| + |k - j|]^{1/2} |k - j|
\end{aligned}$$

as required.

17.8 Using the result of Exercise 1.7 we have

$$
\sum_{k \neq 0} \frac{1}{a|k|^2 + b|k|^4} \leq C \left(\frac{1}{a+b} + \int_1^\infty \frac{r^2}{ar^2 + br^4} \, dr \right)
$$

$$
= C \left(\frac{1}{a+b} + \int_1^\infty \frac{1}{a + br^2} \, dr \right)
$$

$$
\leq C \int_0^\infty \frac{1}{a + br^2} \, dr
$$

$$
= \frac{C}{\sqrt{ab}} \int_0^\infty \frac{dy}{1 + y^2}.
$$

References

Adams, R.A. & Fournier, J.J.F. (2003) *Sobolev spaces*. Academic Press, Kidlington, Oxford.

Agmon, S., Douglis, A., & Nirenberg, L. (1964) Estimates near the boundary for solutions of elliptic partial differential equations satisfying general boundary conditions. II. *Comm. Pure Appl. Math.* **17**, 35–92.

Aubin, J.P. (1963) Un théorème de compacité. *C. R. Acad. Sci. Paris* **256**, 5042–5044.

Bahouri, H., Chemin, J.-Y., & Danchin, R. (2011) *Fourier analysis and nonlinear partial differential equations.* Springer, Heidelberg.

Bardos, C., Lopes Filho, M., Niu, D., Nussenzveig Lopes, H., & Titi, E.S. (2013) Stability of viscous, and instability of non-viscous, 2D weak solutions of incompressible fluids under 3D perturbations. *SIAM J. Math. Anal.* **45**, 1871–1885.

Batchelor, G.K. (1999) *An introduction to fluid dynamics.* Cambridge University Press, Cambridge.

Beale, J.T., Kato, T., & Majda, A. (1984) Remarks on the breakdown of smooth solutions for the 3-D Euler equation. *Comm. Math. Phys.* **94**, 61–66.

Beirão da Veiga, H. (1985a) On the suitable weak solutions to the Navier–Stokes equations in the whole space. *J. Math. Pures Appl.* **64**, 77–86.

Beirão da Veiga, H. (1985b) On the construction of suitable weak solutions to the Navier–Stokes equations via a general approximation theorem. *J. Math. Pures Appl.* **64**, 321–334.

Beirão da Veiga, H. (1987) Existence and asymptotic behavior for strong solutions of the Navier–Stokes equations in the whole space. *Indiana Univ. Math. J.* **36**, 149–166.

Beirão da Veiga, H. (1995) A new regularity class for the Navier–Stokes equations in $\mathbb{R}^n$. *Chinese Ann. Math. Ser. B.* **4**, 407–412.

Beirão da Veiga, H. (2000) On the smoothness of a class of weak solutions to the Navier–Stokes equations. *J. Math. Fluid Mech.* **2**, 315–323.

Beirão da Veiga, H. & Secchi, P. (1987) L^p-stability for the strong solutions of the Navier–Stokes equations in the whole space. *Arch. Ration. Mech. Anal.* **98**, 65–69.

Benedek, A., Calderón, A.P., & Panzone, R. (1962) Convolution operators on Banach space valued functions. *Proc. Nat. Acad. Sci. U.S.A.* **48**, 356–365.

Bergh, J. & Löfström, J. (1976). *Interpolation spaces.* Springer-Verlag, Berlin.

Berselli, L.C. (2002) On a regularity criterion for the solutions to the 3D Navier–Stokes equations. *Differential Integral Equations* **15**, 1129–1137.

Berselli, L.C. & Galdi, G.P. (2002) Regularity criteria involving the pressure for the weak solutions to the Navier–Stokes equations. *Proc. Amer. Math. Soc.* **130**, 3585–3595.

Berselli, L.C. & Spirito, S. (2016a) On the construction of suitable weak solutions to the 3D Navier–Stokes equations in a bounded domain by an artificial compressibility method. *Commun. Contemp. Math.*, to appear.

Berselli, L.C. & Spirito, S. (2016b) Weak solutions to the Navier–Stokes equations constructed by semi-discretization are suitable. *Contemp. Math.*, to appear.

Biot, J.-B. & Savart, F. (1820) Note sure le magnétisme de la pile de Volta. *Annales Chim. Phys.* **15**, 222–223.

Biryuk, A., Craig, W., & Ibrahim, S. (2007) Construction of suitable weak solutions of the Navier–Stokes equations. In "Stochastic analysis and partial differential equations", *Contemp. Math.* **49**, 1–18.

Bogovskiĭ, M.E. (1986) Decomposition of $L^p(\Omega, \mathbb{R}^n)$ into the direct sum of subspaces of solenoidal and potential vector fields. *Soviet Math. Dokl.* **33**, 161–165.

Bourbaki, N. (2004) *Integration*. Springer-Verlag, Berlin.

Caffarelli, L., Kohn, R., & Nirenberg, L. (1982) Partial regularity of suitable weak solutions of the Navier–Stokes equations. *Comm. Pure. Appl. Math.* **35**, 771–931.

Calderón, C.P. (1990) Existence of weak solutions for the Navier–Stokes equations with initial data in L^p. *Trans. Amer. Math. Soc.* **318**, 179–200.

Cannone, M. (1995) *Ondelettes, paraproduits et Navier–Stokes*. Diderot Editeur, Paris.

Cannone, M. (2003) Harmonic analysis tools for solving the incompressible Navier–Stokes equations. In Friedlander, S. & Serre, D. (eds.) *Handbook of mathematical fluid dynamics, Vol. 3*. Elsevier, Kidlington.

Cao, C. & Titi, E.S. (2008) Regularity criteria for the three-dimensional Navier–Stokes Equations. *Indiana Univ. Math. J.* **57**, 2643–2661.

Castaing, C. (1967) Sur les multi-applications measurables. *Rev. France Inform. Rech. Oper.* **1**, 91–126.

Chemin, J.-Y. (1992) Remarques sur l'existence globale pour le système de Navier–Stokes incompressible. *SIAM J. Math. Anal.* **23**, 20–28.

Chemin, J.-Y. & Lerner, N. (1995) Flot de champs de vecteurs non lipschitziens et équations de Navier–Stokes. *J. Differential Equations* **121**, 314–328.

Chemin, J.-Y., Desjardins, B., Gallagher, I., & Grenier, E. (2006) *Mathematical geophysics*. Oxford University Press, Oxford.

Chen, C.-C., Strain, R.M., Yau, H.-T., & Tsai, T.-P. (2008) Lower bound on the blow-up rate of the axisymmetric Navier–Stokes equations. *Int. Math. Res. Not.* article ID rnn016.

Chernyshenko, S.I., Constantin, P., Robinson, J.C., & Titi, E.S. (2007) A posteriori regularity of the three-dimensional Navier–Stokes equations from numerical computations. *J. Math. Phys.* **48**, 065204.

Chorin, A.J. & Marsden, J.E. (1993) *A mathematical introduction to fluid mechanics*. Third edition. Springer-Verlag, New York.

Chung, S.-Y. (1999) Uniqueness in the Cauchy problem for the heat equation. *Proc. Edinburgh Math. Soc.* **42**, 455–468.

Constantin, P. (1986) Note on loss of regularity for solutions of the 3-D incompressible Euler and related equations. *Comm. Math. Phys.* **104**, 311–326.

Constantin, P. & Foias, C. (1988) *Navier–Stokes equations*. University of Chicago Press, Chicago, IL.

Dacorogna, B. (2004) *Introduction to the calculus of variations*. Imperial College Press, London.

Danchin, R. (2000) Global existence in critical spaces for compressible Navier–Stokes equations. *Invent. Math.* **41**, 579–614.

Dashti, M. & Robinson, J.C. (2008) An a posteriori condition on the numerical approximations of the Navier–Stokes equations for the existence of a strong solution. *SIAM J. Numer. Anal.* **46**, 3136–3150.

Dashti, M. & Robinson, J.C. (2009) A simple proof of uniqueness of the particle trajectories for solutions of the Navier–Stokes equations. *Nonlinearity* **22**, 735–746.

De Giorgi, E. (1957) Sulla differenziabilità e lanaliticità delle estremali degli integrali multipli regolari. *Mem. Accad. Sci. Torino. Cl. Sci. Fis. Mat. Nat.* **3**, 25–43.

De Lellis, C. & Székelyhidi Jr., L. (2010) On admissibility criteria for weak solutions of the Euler equations. *Arch. Ration. Mech. Anal.* **195**, 225–260.

DiPerna, R.J. & Lions, P.-L. (1989) Ordinary differential equations, transport theory and Sobolev spaces. *Invent. Math.* **98**, 511–547.

Doering, C.R. & Gibbon, J.D. (1995) *Applied analysis of the Navier–Stokes equations*. Cambridge University Press, Cambridge.

Duoandikoetxea, J. (2001) *Fourier analysis*. Graduate Studies in Mathematics **29**. American Mathematical Society, Providence, RI.

Escauriaza, L., Seregin, G., & Šverák, V. (2003) $L_{3,\infty}$-solutions of Navier–Stokes equations and backward uniqueness. *Russian Math. Surveys* **58**, 211–250.

Evans, L.C. (1998) *Partial differential equations*. American Mathematical Society, Providence, RI.

Evans, L.C. & Gariepy, R.F. (1992) *Measure theory and fine properties of functions*. CRC Press, Boca Raton, FL.

Fabes, E.B., Jones, B.F., & Rivière, N.M. (1972) The initial value problem for the Navier–Stokes equations with data in L^p. *Arch. Ration. Mech. Anal.* **45**, 222–240.

Fabes, E.B., Lewis, J.E., & Rivière, N.M. (1977) Singular integrals and hydrodynamic potentials. *Amer. J. Math.* **99**, 601–625.

Falconer, K.J. (1985) *The geometry of fractal sets*. Cambridge University Press, Cambridge.

Falconer, K.J. (1990) *Fractal geometry*. Wiley, Chichester.

Farwig, R. & Sohr, H. (1996) Helmholtz decomposition and Stokes resolvent system for aperture domains in L^q-spaces. *Analysis* **16**, 1–26.

Farwig, R., Kozono, H., & Sohr, H. (2005) An L^q approach to Stokes and Navier–Stokes equations in general domains. *Acta Math.* **195**, 21–53.

Fefferman, C.L. (1971) The multiplier problem for the ball. *Ann. of Math.* **94**, 330–336.

Fefferman, C.L. (2000) Existence and smoothness of the Navier–Stokes equation. In *The millennium prize problems*, 57–67. Clay Math. Inst., Cambridge, MA.

Feynman, R.P., Leighton, R.B., & Sands, M. (1970) *The Feynman lectures on physics*, Volume II. Addison Wesley Publishing Co., Reading, MA, London.

Foias, C., Guillopé, C., & Temam, R. (1981) New a priori estimates for Navier–Stokes equations in dimension 3. *Comm. Partial Differential Equations* **6**, 329–359.

Foias, C., Guillopé, C., & Temam, R. (1985) Lagrangian representation of a flow. *J. Differential Equations* **57**, 440–449.

Foias, C. & Temam, R. (1989) Gevrey class regularity for the solutions of the Navier–Stokes equations. *J. Funct. Anal.* **87**, 359–369.

Foias, C., Manley, O., Rosa, R., & Temam, R. (2001) *Navier–Stokes equations and turbulence.* Cambridge University Press, Cambridge.

Folland, G.B. (1999) *Real analysis. Modern techniques and their applications.* Second edition. John Wiley & Sons, Inc., New York, NY.

Friedlander, F.G. & Joshi, M. (1999) *Introduction to the theory of distributions.* Cambridge University Press, Cambridge.

Fujita, H. & Kato, T. (1964) On the Navier–Stokes initial value problem. I. *Arch. Ration. Mech. Anal.* **16**, 269–315.

Fujiwara, D. & Morimoto, H. (1977) An L_r-theorem of the Helmholtz decomposition of vector fields. *J. Fac. Sci. Univ. Tokyo Sect. IA Math.* **24**, 685–700.

Galdi, G.P. (2000) An introduction to the Navier–Stokes initial-boundary value problem. In Galdi, G.P., Heywood, J.G., & Rannacher, R. (eds.) *Fundamental directions in mathematical fluid dynamics*, 1–70. Birkhauser, Basel.

Galdi, G.P. (2011) *An introduction to the mathematical theory of Navier–Stokes equations. Steady state problems.* Second edition. Springer, New York, NY.

Galdi, G.P. & Maremonti, P. (1986) Monotonic decreasing and asymptotic behaviour of the kinetic energy for weak solutions of the Navier–Stokes equations in exterior domains. *Arch. Ration. Mech. Anal.* **94**, 253–266.

Gallagher, I. (1997) The tridimensional Navier–Stokes equations with almost bidimensional data: stability, uniqueness, and life span. *Int. Math. Res. Notices* **18**, 919–935.

Giga, Y. (1983) Time and spatial analyticity of solutions of the Navier–Stokes equations. *Comm. Partial Differential Equations* **8**, 929–948.

Giga, Y. (1986) Solutions for semilinear parabolic equations in L^p and regularity of weak solutions of the Navier–Stokes system. *J. Differential Equations* **62**, 186–212.

Gilbarg, D. & Trudinger, N.S. (1983) *Elliptic partial differential equations of second order.* Springer, Berlin.

Grafakos, L. (2008) *Classical Fourier analysis.* Graduate text in mathematics **249**. Springer, New York, NY.

Grafakos, L. (2009) *Modern Fourier analysis.* Graduate text in mathematics **250**. Springer, New York, NY.

Guermond, J.-L. (2006) Finite-element-based Faedo–Galerkin weak solutions to the Navier–Stokes equations in the three-dimensional torus are suitable. *J. Math. Pures Appl.* **85**, 451–464.

Guermond, J.-L. (2007) Faedo–Galerkin weak solutions of the Navier–Stokes equations with Dirichlet boundary conditions are suitable. *J. Math. Pures Appl.* **88**, 87–106.

Hale, J.K. (1980) *Ordinary differential equations.* Kreiger, Malabar, FL.

Hardy, G.H. (1932) A note on two inequalities. *J. London Math. Soc.* **11**, 167–170.

Hartman, P. (1973) *Ordinary differential equations.* Wiley, Baltimore.

Helmholtz, H. (1858) Über Integrale der hydrodynamischen Gleichungen, welcher der Wirbelbewegungen entsprechen. *J. Reine Angew. Math.* **55**, 25–55.

Heywood, J. (1976) On uniqueness questions in the theory of viscous flow. *Acta Math.* **136**, 61–102.

Heywood, J. (1988) Epochs of regularity for weak solutions of the Navier–Stokes equations in unbounded domains. *Tohoku Math. J.* **40**, 293–313.

Hopf, E. (1951) Über die Aufgangswertaufgave für die hydrodynamischen Grundliechungen. *Math. Nachr.* **4**, 213–231.

Hörmander, L. (1960) Estimates for translation invariant operators in L^p spaces. *Acta Math.* **104**, 93–139.

Iftimie, D., Karch, G., & Lacave C. (2014) Asymptotics of solutions to the Navier–Stokes system in exterior domains. *J. London Math. Soc.* **90**, 785–806.

James, R.C. (1964) Weakly Compact Sets. *Trans. Amer. Math. Soc.* **113**, 129–140.

Kahane, C. (1969) On the spatial analyticity of solutions of the Navier–Stokes equations. *Arch. Ration. Mech. Anal.* **33**, 386–405.

Kato, T. (1984) Strong L^p-solutions of the Navier–Stokes equations in $\mathbb{R}^m$ with applications to weak solutions. *Math. Zeit.* **187**, 471–480.

Kiselev, A.A. & Ladyzhenskaya, O.A. (1957) On the existence and uniqueness of the solution of the nonstationary problem for a viscous, incompressible fluid. *Izv. Akad. Nauk SSSR. Ser. Mat.* **21**, 655–680.

Koch, H. & Tataru, D. (2001) Well-posedness for the Navier–Stokes equations. *Adv. Math.* **157**, 22–35.

Kohn, R.V. (1982) Partial regularity and the Navier–Stokes equations. *Lecture Notes in Num. Appl. Anal.* **5**, 101–118. (Nonlinear PDE in Applied Science U.S.–Japan Seminar, Tokyo, 1982.)

Komatsu, G. (1979) Analyticity up to the boundary of solutions of nonlinear parabolic equations. *Comm. Pure Appl. Math.* **32**, 669–720.

Kozono, H. (1998) Uniqueness and regularity of weak solutions to the Navier–Stokes equations. In *Recent topics on mathematical theory of viscous incompressible fluid* (Tsukuba, 1996). *Lecture Notes Numer. Appl. Anal.*, **16**, 161–208, Kinokuniya, Tokyo.

Kozono, H. & Sohr, H. (1996) Remark on uniqueness of weak solutions to the Navier–Stokes equations. *Analysis* **16**, 255–271.

Kozono, H. & Taniuchi, T. (2000) Limiting case of the Sobolev inequality in BMO, with application to the Euler equations. *Comm. Math. Phys.* **214**, 191–200.

Kozono, H., Ogawa, T., & Taniuchi, T. (2002) The critical Sobolev inequalities in Besov spaces and regularity criterion to some semi-linear evolution equations. *Math. Z.* **242**, 251–278.

Krylov, N.V. (1996) *Lectures on elliptic and parabolic equations in Hölder spaces.* American Mathematical Society, Providence, RI.

Krylov, N.V. (2001) The heat equation in $L_q((0, T); L_p)$-space with weights. *SIAM J. Math. Anal.* **32**, 1117–1141.

Krylov, N.V. (2008) *Lectures on elliptic and parabolic equations in Sobolev spaces.* American Mathematical Society, Providence, RI.

Kukavica, I. (2008) Regularity for the Navier–Stokes equations with a solution in a Morrey space. *Indiana Univ. Math. J.* **57**, 2843–2860.

Kukavica, I. (2009a) Partial regularity results for solutions of the Navier–Stokes system. In Robinson, J.C. & Rodrigo, J.L. (eds.) *Partial differential equations and fluid mechanics*, 121–145. Cambridge University Press, Cambridge.

Kukavica, I. (2009b) The fractal dimension of the singular set for solutions of the Navier–Stokes system. *Nonlinearity* **22**, 2889–2900.

Kukavica, I. & Pei, Y. (2012) An estimate on the parabolic fractal dimension of the singular set for solutions of the Navier–Stokes system. *Nonlinearity* **25**, 2775–2783.

Ladyzhenskaya, O.A. (1959) Solution "in the large" of the nonstationary boundary value problem for the Navier–Stokes system in two space variables. *Comm. Pure Appl. Math.* **12**, 427–433.

Ladyzhenskaya, O.A. (1967) On uniqueness and smoothness of generalized solutions to the Navier–Stokes equations. *Zap. Nauchn. Sem. Leningrad. Otdel. Mat. Inst. Steklov.* (LOMI) **5**, 169–185; English transl. *Sem. Math. V.A. Steklov Math. Inst. Leningrad* **5** (1969), 60–66.

Ladyzhenskaya, O.A. (1969) *The mathematical theory of viscous incompressible flow.* Gordon and Breach, New York, NY.

Ladyzhenskaya O.A. (1970) Unique solvability in the large of three-dimensional Cauchy problem for the Navier–Stokes equations in the presence of axial symmetry. Seminar in Mathematics, V.A. Steklov Mathematical Institute, Leningrad, 7, *Boundary value problems of mathematical physics and related aspects of function theory*, Part 2, Edited by O.A. Ladyzhenskaya, 70–79.

Ladyzhenskaya, O.A. & Seregin, G.A. (1999) On partial regularity of suitable weak solutions to the three-dimensional Navier–Stokes equations. *J. Math. Fluid Mech.* **1**, 356–387.

Ladyzhenskaya, O.A., Solonnikov, V.A., & Uraltseva, N.N. (1967) *Linear and quasi-linear equations of parabolic type.* American Mathematical Society, Providence, RI.

Lemarié-Rieusset, P.G. (2002) *Recent developments in the Navier–Stokes problem.* Chapman & Hall/CRC, Boca Raton, FL.

Leray, J. (1934) Essai sur le mouvement d'un liquide visqueux emplissant l'espace. *Acta Math.* **63**, 193–248.

Lieberman, G.M. (1996) *Second order parabolic differential equations.* World Scientific Publishing Co., Inc., River Edge, NJ.

Lin, F. (1998) A new proof of the Caffarelli–Kohn–Nirenberg theorem. *Comm. Pure Appl. Math.* **51**, 241–257.

Lions, J.-L. (1969) *Quelques méthodes de résolution des problèmes aux limites non linéaires.* Dunod Gauthier-Villars, Paris.

Lions, J.-L. & Prodi, G. (1959) Un théoréme d'existence et unicité dans les équations de Navier–Stokes en dimension 2. *C. R. Acad. Sci. Paris* **248**, 3519–3521.

Lions, P.-L. (1994) *Mathematical topics in fluid mechanics, Volume 1: Incompressible models.* Oxford University Press, Oxford.

Lunardi, A. (2009) *Interpolation theory.* 2nd edition. Edizioni della Normale, Pisa.

Mahalov, A., Titi, E.S., & Leibovich, S. (1990) Invariant helical subspaces for the Navier–Stokes equations. *Arch. Ration. Mech. Anal.* **112**, 193–222.

Majda, A.J. & Bertozzi, A.L. (2002) *Vorticity and incompressible flow.* Cambridge University Press, Cambridge.

Marín-Rubio, P., Robinson, J.C., & Sadowski, W. (2013) Solutions of the 3D Navier–Stokes equations for initial data in $H^{1/2}$: robustness of regularity and numerical verification of regularity for bounded sets of initial data in H^1. *J. Math. Anal. Appl.* **400**, 76–85.

Masuda, K. (1967) On the analyticity and the unique continuation theorem for solutions of the Navier–Stokes equations. *Proc. Japan. Acad.* **43**, 827–832.

Masuda, K. (1984) Weak solutions of Navier–Stokes equations. *Tohoku Math. J. (2)* **36**, 623–646.

McCormick, D.S., Robinson, J.C., & Rodrigo, J.L. (2013) Generalised Gagliardo–Nirenberg inequalities using weak Lebesgue spaces and BMO. *Milan J. Math.* **81**, 265–289.

McCormick, D.S., Olson, E.J., Robinson, J.C., Rodrigo, J.L., Vidal-López, A., & Zhou, Y. (2016a) Lower bounds on blowing-up solutions of the 3D Navier–Stokes equations in $\dot{H}^{3/2}$, $\dot{H}^{5/2}$, and $\dot{B}_{2,1}^{5/2}$. arXiv:1503.04323. *SIAM J. Math. Anal.*, to appear.

McCormick, D.S., Fefferman, C.L., Robinson, J.C., & Rodrigo, J.L. (2016b) Local existence for the non-resistive MHD equations in nearly optimal Sobolev spaces. arXiv:1602.02588.

Mihlin, S.G. (1957) Fourier integrals and multiple singular integrals. *Vestnik Leningrad. Univ. Ser. Mat. Meh. Astr.* **12**, 143–155 (in Russian).

Miyakawa, T. & Sohr, H. (1988) On energy inequality, smoothness and large time behavior in L^2 for weak solutions of the Navier–Stokes equations in exterior domains. *Math. Z.* **199**, 455–478.

Montero, J.A. (2015) Lower bounds for possible blow–up solutions for the Navier–Stokes equations revisited. arXiv:1503.03063.

Mucha, P.B. (2008) Stability of 2D incompressible flows in $\mathbb{R}^3$. *J. Differential Equations* **245**, 2355–2367.

Muscalu, C. & Schlag, W. (2013) *Classical and multilinear harmonic analysis. Volume I.* Cambridge Studies in Advanced Mathematics. Cambridge University Press, Cambridge.

Nečas, J., Růžička, M., & Šverák, V. (1996) On Leray's self-similar solutions of the Navier–Stokes equations. *Acta Math.* **176**, 283–294.

Neustupa, J. (1999) Partial regularity of weak solutions to the Navier–Stokes equations in the class $L^\infty(0, T; L^3(\Omega)^3)$. *J. Math. Fluid Mech.* **1**, 309–325.

O'Leary, M. (2003) Conditions for the local boundedness of solutions of the Navier–Stokes system in three dimensions. *Comm. Partial Differential Equations* **28**, 617–636.

Pettis, B.J. (1938) On integration in vector spaces. *Trans. Amer. Math. Soc.* **44**, 277–304.

Planchon, F. (2003) An extension of the Beale–Kato–Majda criterion for the Euler equations. *Comm. Math. Phys.* **232**, 319–326.

Pooley, B.C. & Robinson, J.C. (2016) Well-posedness for the diffusive 3D Burgers equations with initial data in $H^{1/2}$. In Robinson, J.C., Rodrigo, J.L., Sadowski, W., & Vidal-López, A. (eds.) *Recent progress in the theory of the Euler and Navier–Stokes equations*, 137–153. Cambridge University Press, Cambridge.

Prodi, G. (1959) Un teorema di unicità per le equazioni di Navier–Stokes. *Ann. Mat. Pura Appl.* **48**, 173–182.

Raugel, G. & Sell, G.R. (1993) Navier–Stokes equations on thin 3D domains. I: Global attractors and global regularity of solutions. *J. Amer. Math. Soc.* **6**, 503–568.

Renardy, M. & Rogers, R.C. (2004) *An introduction to partial differential equations.* Second Edition. Springer-Verlag, New York, NY.

Robertson, A.P. (1974) On measurable selections. *Proc. Roy. Soc. Edinburgh Sect. A* **72**, 1–7.

Robinson, J.C. (2001) *Infinite-dimensional dynamical systems.* Cambridge University Press, Cambridge.

Robinson, J.C. (2006) Regularity and singularity in the three-dimensional Navier–Stokes equations. *Boletín de SEMA* **35**, 43–71.

Robinson, J.C. (2011) *Dimensions, embeddings, and attractors*. Cambridge University Press, Cambridge.

Robinson, J.C. & Sadowski, W. (2007) Decay of weak solutions and the singular set of the three-dimensional Navier–Stokes equations. *Nonlinearity* **20**, 1185–1191.

Robinson, J.C. & Sadowski, W. (2009) Almost everywhere uniqueness of Lagrangian trajectories for suitable weak solutions of the three-dimensional Navier–Stokes equations. *Nonlinearity* **22**, 2093–2099.

Robinson, J.C. & Sadowski, W. (2014) A local smoothness criterion for solutions of the 3D Navier–Stokes equations. *Rend. Semin. Mat. Univ. Padova* **131**, 159–178.

Robinson, J.C., Sadowski, W., & Sharples, N. (2013) On the regularity of Lagrangian trajectories corresponding to suitable weak solutions of the Navier–Stokes equations. *Procedia IUTAM* **7**, 161–166.

Robinson, J.C., Sadowski, W., & Silva, R.P. (2012) Lower bounds on blow up solutions of the three-dimensional Navier–Stokes equations in homogeneous Sobolev spaces. *J. Math. Phys.* **53**, 115618.

Roubíček, T. (2013) *Nonlinear partial differential equations with applications*. Second edition. Birkhäuser, Springer, Basel.

Rudin, W. (1991) *Functional Analysis*. McGraw-Hill, New York, NY.

Scheffer, V. (1976) Turbulence and Hausdorff dimension. In *Turbulence and Navier–Stokes equation, Orsay 1975*, Springer Lecture Notes in Mathematics **565**, 174–183. Springer, Berlin.

Scheffer, V. (1977) Hausdorff measure and the Navier–Stokes equations. *Comm. Math. Phys.* **55**, 97–112.

Scheffer, V. (1980) The Navier–Stokes equations on a bounded domain. *Comm. Math. Phys.* **73**, 1–42.

Scheffer, V. (1985) A solution to the Navier–Stokes inequality with an internal singularity. *Comm. Math. Phys.* **101**, 47–85.

Scheffer, V. (1987) Nearly one-dimensional singularities of solutions to the Navier–Stokes inequality. *Comm. Math. Phys.* **110**, 525–551.

Scheffer, V. (1993) An inviscid flow with compact support in space–time. *J. Geom. Anal.* **3**, 343–401.

Schonbek, M.E. (1985) L^2 decay for weak solutions to the Navier–Stokes equation. *Arch. Ration. Mech. Anal.* **88**, 209–222.

Seregin, G. (2014) *Lecture notes on regularity theory for the Navier–Stokes equations*. World Scientific, Hackensack, NJ.

Seregin, G. & Šverák, V. (2002) Navier–Stokes equations with lower bounds on pressure. *Arch. Ration. Mech. Anal.* **163**, 65–86.

Seregin, G. & Zajaczkowski, W. (2007) A sufficient condition of regularity for axially symmetric solutions to the Navier–Stokes equations. *SIAM J. Math. Anal.* **39**, 669–685.

Serrin, J. (1962) On the interior regularity of weak solutions of the Navier–Stokes equations. *Arch. Ration. Mech. Anal.* **9**, 187–195.

Serrin, J. (1963) The initial value problem for the Navier–Stokes equations. In *Nonlinear Problems* (Proc. Sympos., Madison, Wis.), 69–98. Univ. of Wisconsin Press, Madison, WI.

Shapiro, V.L. (1966) The uniqueness of solutions of the heat equation in an infinite strip. *Trans. Amer. Math. Soc.* **125**, 326–361.

Shatah, J. & Struwe, M. (1994) Well-posedness in the energy space for semilinear wave equations with critical growth. *Int. Math. Res. Notices* **1994**, 303–309.

Shnirelman, A. (1997) On the nonuniqueness of weak solution of the Euler equation. *Comm. Pure Appl. Math.* **50**, 1261–1286.

Shnirelman, A. (2000) Weak solutions with decreasing energy of incompressible Euler equations. *Comm. Math. Phys.* **210**, 541–603.

Simader, C.G. & Sohr, H. (1992) A new approach to the Helmholtz decomposition and the Neumann problem in L^q -spaces for bounded and exterior domains. In Galdi, G.P. (ed.) *Mathematical Problems Relating to the Navier–Stokes Equations*. Series on Advances in Mathematics for Applied Sciences, **11**, 1–35. World Scientific, Singapore.

Simader, C.G., Sohr, H., & Varnhorn, W. (2014) Necessary and sufficient conditions for the existence of Helmholtz decompositions in general domains. *Ann. Univ. Ferrara* **60**, 245–262.

Simon, J. (1987) Compact sets in the space $L^p(0, T; B)$. *Ann. Mat. Pura Appl.* **146**, 65–96.

Sohr, H. (1983) Zur Regularitätstheorie der instationären Gleichungen von Navier–Stokes. (German) [On the regularity theory of the nonstationary Navier–Stokes equations] *Math. Z.* **184**, 359–375.

Sohr, H. (2001) *The Navier–Stokes equations: an elementary functional analytic approach*. Modern Birkhäuser Classics. Birkhäuser/Springer, Basel.

Sohr, H. & von Wahl, W. (1984) On the singular set and the uniqueness of weak solutions of the Navier–Stokes equations. *Manuscripta Math.* **49**, 27–59.

Sohr, H. & von Wahl, W. (1986) On the regularity of the pressure of weak solutions of Navier–Stokes equations. *Arch. Math. (Basel)* **46**, 428–439.

Solonnikov, V.A. (1964) Estimates of the solutions of a nonstationary linearized system of Navier–Stokes equations. *Trudy Mat. Inst. Steklov.* **70**, in: Amer. Math. Soc. Translations, Series 2, Vol. **75**, 1–117.

Stein, E.M. (1970) *Singular integrals and differentiability properties of functions*. Princeton University Press, Princeton, NJ.

Stein, E.M. (1993) *Harmonic analysis: real-variable methods, orthogonality, and oscillatory integrals*. Princeton University Press, Princeton, NJ.

Struwe, M. (1988) On partial regularity results for the Navier–Stokes equations. *Comm. Pure. Appl. Math.* **41**, 437–458.

Takahashi, S. (1990) On interior regularity criteria for weak solutions of the Navier–Stokes equations. *Manuscripta Math.* **69**, 237–254.

Temam, R. (1977) *Navier–Stokes equations*. North Holland, Amsterdam. Reprinted by AMS Chelsea, 2001.

Temam, R. (1982) Behaviour at time $t = 0$ of the solutions of semi-linear evolution equations. *J. Differential Equations* **43**, 73–92.

Temam, R. (1983) *Navier–Stokes equations and nonlinear functional analysis*. SIAM, Philadelphia, PA.

Tychonoff, A.N. (1935) Uniqueness theorem for the heat equation. *Mat. Sb.* **42**, 199–216.

Vasseur, A. (2007) A new proof of partial regularity of solutions to Navier–Stokes equations. *Nonlinear Diff. Equ. Appl.* **14**, 753–785.

von Wahl, W. (1980) Regularitätsfragen für die instationären Navier–Stokesschen Gleichungen in höheren Dimensionen. *J. Math. Soc. Japan* **32**, 263–283.

von Wahl, W. (1982) The equation $u' + A(t)u = f$ in a Hilbert space and L^p-estimates for parabolic equations. *J. London Math. Soc.* **25**, 483–497.

von Wahl, W. (1985) *The equations of Navier–Stokes and abstract parabolic equations.* Friedr. Vieweg & Sons, Braunschweig.

Weinberger, H.F. (1999) An example of blowup produced by equal diffusions. *J. Differential Equations* **154**, 225–237.

Yosida, K. (1980) *Functional analysis.* Springer Classics in Mathematics, Springer, Berlin.

Zhou, Y. & Pokorný, M. (2010) On the regularity of the solutions of the Navier–Stokes equations via one velocity component. *Nonlinearity* **23**, 1097–1107.

Zuazua, E. (2002) Log-Lipschitz regularity and uniqueness of the flow for a field in $(W_{\mathrm{loc}}^{n/p+1,p}(\mathbb{R}^n))^n$. *C. R. Acad. Sci. Paris* **335**, 17–22.

Index